Engineering Reliability and Risk in Water Resources

NATO ASI Series

Advanced Science Institutes Series

A Series presenting the results of activities sponsored by the NATO Science Committee, which aims at the dissemination of advanced scientific and technological knowledge, with a view to strengthening links between scientific communities.

The Series is published by an international board of publishers in conjunction with the NATO Scientific Affairs Division

A	**Life Sciences**	Plenum Publishing Corporation
B	**Physics**	London and New York
C	**Mathematical and Physical Sciences**	D. Reidel Publishing Company Dordrecht, Boston, Lancaster and Tokyo
D	**Behavioural and Social Sciences**	Martinus Nijhoff Publishers Boston, Dordrecht and Lancaster
E	**Applied Sciences**	
F	**Computer and Systems Sciences**	Springer-Verlag Berlin, Heidelberg, New York
G	**Ecological Sciences**	London, Paris, Tokyo
H	**Cell Biology**	

Series E: Applied Sciences – No. 124

Engineering Reliability and Risk in Water Resources

edited by

Lucien Duckstein
Department of System & Industrial Engineering
University of Arizona
Tuscon, Arizona 85721
USA

Erich J. Plate
Institute for Hydrology and Water Resources
University of Karlsruhe
Kaiserstrasse 12
7500 Karlsruhe
FRG

1987 **Martinus Nijhoff Publishers**
Dordrecht / Boston / Lancaster
Published in cooperation with NATO Scientific Affairs Division

Proceedings of the NATO Advanced Study Institute on "Engineering Reliability and Risk in Water Resources", Tuscon, Arizona, USA, May 19-June 2, 1985

Library of Congress Cataloging in Publication Data

NATO Advanced Study Institute on "Engineering Reliability and Risk in Water Resources" (1985 : Tuscon, Ariz.)
Engineering reliability and risk in water resources.

(NATO advanced science institutes series. Series E, Applied sciences ; no. 124)
"Proceedings of the NATO Advanced Study Institute on 'Engineering Reliability and Risk in Water Resources', Tucson, Arizona, USA, May 19-June 2, 1985"--T.p. verso.
"Published in cooperation with NATO Scientific Affairs Division."
Includes index.
1. Water resources development--Statistical methods--Congresses. 2. Hydraulic engineering--Statistical methods--Congresses. 3. Reliability (Engineering)--Statistical methods--Congresses. 4. Decision-making--Statistical methods--Congresses. I. Duckstein, Lucien. II. Plate, Erich J. III. North Atlantic Treaty Organization. Scientific Affairs Division. IV. Title. V. Series: NATO advanced science institutes series. Series E, Applied sciences ; v. 124.
TC401.N285 1985 627 87-1722

ISBN 90-247-3492-4 (this volume)
ISBN 90-247-2689-1 (series)

Distributors for the United States and Canada: Kluwer Academic Publishers, P.O. Box 358, Accord-Station, Hingham, MA 02018-0358, USA

Distributors for the UK and Ireland: Kluwer Academic Publishers, MTP Press Ltd, Falcon House, Queen Square, Lancaster LA1 1RN, UK

Distributors for all other countries: Kluwer Academic Publishers Group, Distribution Center, P.O. Box 322, 3300 AH Dordrecht, The Netherlands

Printed in The Netherlands

CONTENTS

PREFACE

Hydraulic, hydrologic and water resources engineers have been concerned for a long time about failure phenomena. One of the major concerns is the definition of a failure event E, of its probability of occurrence P(E), and of the complementary notion of reliability. However, as the stochastic aspects of hydraulics and water resources engineering were developed, words such as "failure," "reliability," and "risk" took on different meanings for different specialists. For example, "risk" is defined in a Bayesian framework as the expected loss resulting from a precisely defined failure event, while according to the practice of stochastic hydraulics it is the probability of occurrence of a failure event. The need to standardize the various concepts and operational definitions generated numerous exciting discussions between the co-editors of this book during 1983-84 when L. Duckstein, under sponsorship of the Alexander von Humboldt Foundation (FRG), was working with E. Plate at the Institute of Hydrology and Water Resources of the University of Karlsruhe. After consulting with the Scientific Affairs Division of NATO, an organizing committee was formed. This comittee -- J. Bernier (France), M. Benedini (Italy), S. Sorooshian (U.S.A.), and co-directors L. Duckstein (U.S.A.) and E.J. Plate (F.R.G.) -- brought into being this NATO Advanced Study Institute (ASI).

Precisely stated, the purpose of this ASI was to present a tutorial overview of existing work in the broad area of reliability while also pointing out topics for further development. In order to provide a coherent set of lectures, substantial effort was undertaken to present a common mathematical framework for the various aspects of reliability and risk in water resources. This framework was to include the structural reliability aspects that refer to failure of structures such as dams, spillways, lakes, and locks, and the target-related reliability aspects that deal with water supply, both quantity and quality, for multiple purposes. The ASI program thus included sets of tutorial lectures on background, state-of-the-art, and pioneering aspects of reliability and risk. The present book is a reorganized and edited version of most of the lectures presented during the ASI.

It is remarkable that in November 1985, six months after the ASI, two meetings were held on subtopics of the ASI. The first was an Engineering Foundation-sponsored conference in Santa Barbara, California on "Risk-Based Decision Making in Water Resources." This conference dealt mainly with structural reliability. The second meeting was a workshop organized under the auspices of HYDROQUEBEC and the University of Montreal on "Risk Analysis in Hydrologic System Operations." As indicated by the title, only target-related reliability and risk were considered. Several participants in the NATO ASI also participated in at least one of these meetings.

It is hoped that other research endeavours will be developed from ideas presented at the ASI. The target readership of this book is a mix of university colleagues, practitioners from both the private and public sectors, and advanced graduate students working in various hydrology and water resources-related branches of engineering. This is also the background of the participants in the ASI.

The book is organized as follows: The Introduction begins with a paper in which Duckstein, Plate and Benedini develop a systems engineering framework for incident and failure-related criteria. This framework embeds hydraulic and hydrologic reliability, and can thus be applied to the study of both structure-related and target-related reliability (and risk). This introduction thus strives to provide elements of a common language for the remainder of the book, which consists of three main sections that focus on structural reliability, supply reliability and decision-making. Each of these sections is preceded by introductory material designed to put the papers into perspective.

We gratefully acknowledge our sponsors, namely, the NATO Scientific Affairs Division, the Salt River Project (Arizona), the City of Tucson, the Arizona Foundation (Tucson), and the University of Arizona Department of Hydrology and Water Resources and Department of Systems and Industrial Engineering. We also are grateful to all the ASI lecturers, including those whose papers could not be included in this book for a variety of reasons, the lively ASI participants, and many individuals who helped kindly before, during, and after the ASI. In particular, we wish to thank JoAnn Roof and Corla Thies, who handled most of the correspondence and logistics; colleagues Russ Ferrell and Moshe Shiedovich; graduate students Jené Hendrickson, Aregai Tecle, Juan Bernal, Anand Rakshit, Leo Salgado, and others; technical editor Sally Adams and administrative assistant Janolyn Berry, who handled the preparation and typing of the final manuscript; and last but not least, our families here in Tucson who hosted this event as their own.

I. INTRODUCTION

I. INTRODUCTION

WATER ENGINEERING RELIABILITY AND RISK: A SYSTEM FRAMEWORK

Lucien Duckstein
Department of Systems & Industrial Engineering, University of Arizona
Tucson, Arizona 85721, USA

Erich Plate
Institute for Hydrology and Water Resources, University of Karlsruhe
Kaiserstraße 12, 7500 Karlsruhe, FRG

Marcello Benedini
Institute for Water Resources Research, Via Reno 1
00198 Rome, Italy

ABSTRACT

A general discrete system representation is used to model the reliability and other incident- or failure-related characteristics of both hydrologic structures and supply schemes. The system model is composed of five sets: (1) input set (controllable and noncontrollable), (2) state, (3) state transition function, (4) output, (5) output function. Emphasis is placed on the latter two sets which consist of performance indices PI and combinations of PI or figures of merit. Examples of PI are type of incident, level of service, quality of service, and cost. Examples of figures of merit are engineering reliability and risk. The framework is applied to a dual-objective reservoir control problem (flood protection versus water supply) in which reservoir operation rules are compared from a multiple criteria viewpoint. Alternative approaches for calculating the criteria based on surplus incidents, deficit incidents or any one of the two types of incidents are discussed.

1. INTRODUCTION

The purposes of this paper are:

(1) to provide a system framework for embedding criteria related to incidents and failure of reservoirs, dams and other hydrologic systems, including supply schemes. By "incident" is meant any event, such as a low storage level, that falls outside of prespecified "normal" limits.

(2) to compare the performance of alternative operation schemes or rules from the viewpoint of multiple criteria which include those defined previously.

(3) to illustrate the concepts by means of a simple but realistic example based on bi-objective reservoir control.

It is hoped that the system model will provide a common basis for the papers in this book, as well as the embryo of a common language.

The system framework presented in this paper follows the model provided in Duckstein and Plate (1984). However, the emphasis in that paper was mostly on the evaluation of a design, with an example of culvert system design. In contrast, the present paper deals with the evaluation of the performance of an operation or management scheme. It should be noted, however, that the system framework itself remains essentially the same. Water resource operational problems have been examined by Simonovic and Mariño (1980, 1981, 1982), Burges and Lettenmaier (1982), Mariño and Mohammedi (1983), Simonovic and Orlob (1984) and Moy et al (1986). In order to relate the notion of incident and failure event in the physical (structure-related) sense to the parallel notion in a dynamic (target-related) sense, it may be useful to define a generalized "load" λ, a generalized "resistance" ρ, an incident mode" μ and the proposition:

Incident μ occurs if and only if $\lambda > \rho$ (1)

For example, in the case of water being supplied from reservoirs to satisfy users' requirements, the "load" λ would be the total requirement and the "resistance" ρ would be supply capacity. Other possible interpretations for λ and ρ are presented in Table 1, where the first two problems refer to hydrologic structures. This definition of incident is consistent with that of USCOLD (1975). The incident mode μ refers to the type of incident that may occur. For example, for a structure, an incident may correspond to failure of all or part of the structure; for a supply system, an incident may be a shortage or excess of water.

The specific example developed in the present paper evaluates and compares reservoir operation rules on the basis of a set of criteria or performance indices as defined in the next section.

2. SYSTEM FRAMEWORK

2.1. System Elements

Consider a discrete-time dynamic system Z as defined in Wymore (1976, 1982). For the sake of convenience, let this system be also discrete-state. An example of such a system may be found in Bogárdi and Bárdossy, Section IV of this book. Consider, here, a water supply scheme, represented by a single reservoir for the sake of simplicity. The system components are (T,S,X,Y,F,G) defined respectively as follows.

T : a time scale (j = 1, 2, ... , t, ... J); the horizon J may be infinite.

S : a discrete state vector describing the system at time j. The elements of this vector belong to set S, which may be infinite and may include storage volume, water quality indices, cumulative shortage and flood volumes as well as physical characteristics of a dam, such as resistance or fatigue. Generally speaking, let the state set S include

TABLE 1. Examples of generalized load λ and resistance ρ in hydrologic systems

Type of Problem	Load	Resistance	References
Bridge pier	Scouring	Pier pile depth	Laursen (1970) Davis & Dvoranchik (1971)
Flood levee Dam	-Flood stage -Flood duration -Flood exposure -Wind	Levee height Hydraulic and soil resistance to boiling sliding, erosion	Szidarovszky et al (1975) Duckstein & Bogardi (1981) Lee & Mays (1983) Plate (1984)
Water supply	Requirements or demand	Supply capacity	Klemes et al (1981) Shamir et al (1984)
Flood volume control	Flood volume	Reservoir flood storage	Loucks et al (1981)
Max. flood stage control	Incoming flood stage	Cresting capacity	Yazicigil, et al (1983)
Underground excavation	Piezometric pressure	Permeability of walls	Bogardi & Duckstein (1978)
Water quality (Streams, lakes)	Nutrients, sediments, pollutant loading	Cleaning capacity, low flow augm	Thomann (1972) Burges & Lettenmaier (1975)
Waste management	Hazards (chemical, radioactive)	Physical, individual, collective	Marsily, et al (1983)
Recreation	Number of visitor-days	Carrying capacity of facility	Knetsche (1974)

<u>performance indices</u> PI^k, k=1, ..., K, which measure how well the system performs. Let $\underline{PI}(j)$ be the performance index vector at time j. The state variable vector at time j is denoted $s(j) \in S$ and the state trajectory at time t is the sequence of vectors:

$$\underline{s}(t) = (s(o), s(1), \ldots, s(t)).$$

The performance trajectory is defined as the matrix:

$$\underline{\underline{P}}(t) = (\underline{PI}(1), \ldots, \underline{PI}(t)), \text{ with } \underline{PI}(j) \in s(j) \tag{2}$$

X = (U, W): input set or alphabet composed of a set of controllable elements U and a set of non-controllable elements W. Controllable or decision vector elements $u(j) \in U$ may include both those parameters which are fixed by the design, such as the height of a dam, and adjustable quantities, such as the opening of a spillway gate. Non-controllable elements $w(j) \in W$ include such quantities as the total net inflow into a reservoir or the aging of a structure.

A particular subset of U is the decision trajectory or control policy

$$\underline{u}(t) = (u(o), u(1), \ldots, u(t))$$

and a subset of W may be the random inflow sequence.

Y = an output set or alphabet comprising physical variables, such as water volume delivered and generalized loads λ. At every time j, the output usually includes an updated value of performance indices PI(j) which were defined as elements of the state set S, such as losses due to shortage or excess spills.

The output also includes <u>figures of merit</u> (FM), which are criterion functions defined over the components of the performance matrix $\underline{P}(J)$ (Equation (2)) where J is the horizon. For example, the mean value of $\underline{P}(J)$, the maximum or minimum component of $\underline{P}(J)$, may be FM's. Generally speaking, the outputs $y(j) \in Y$, j = 0,...,t, define the output trajectory

$$\underline{y}(t) = (y(1), \ldots, y(t)),$$

a subset of which may be the performance index trajectory $\underline{P}(t)$.

F : a state transition function which calculates the state at time j + 1 as a function of the previous states s(j) and the previous input x(j):

$$s(j + 1) = F(s(j), x(j)). \tag{3}$$

Note that by recurrence, s(j+1) depends on both state trajectory $\underline{s}(j)$ and input trajectory $\underline{x}(j)$; also function F in equation (3) includes the algorithm to calculate the performance index vector $\underline{PI}(j+1)$.

In the case of reservoir operation, one component of vector function F is frequently the mass balance equation:

$$v(j + 1) = v(j) - u(j) + w(j) \tag{4}$$

$v(j)$ = water volume in storage at time j

$u(j)$ = controlled outflow, including release

$w(j)$ = net random inflow;

Other components of F may include design parameters, sediment movement equations, trophic state transformation. An example of calculation of an economic performance index by means of equation (3) is as follows. Consider, as an element of $F(\cdot)$, the rule to calculate the benefit function $\emptyset(j)$ or the loss function $L(j)$ at time j. Note that both $\emptyset(j)$ and $L(j)$ depend, in general, on storage volume $v(j)$, release $r(j)$ and demand $D(j)$. A loss function under random demand $D(j)$ may be

$$L(j+1) = \begin{cases} r(j) - D(j) & \text{if } D(j) \le r(j) \le v(j) \\ (D(j) - r(j))^{\alpha} & \text{if } r(j) \le D(J), \text{ with } r(j) \le v(j),\ \alpha > 0 \end{cases}$$

This equation is a particular case of equation (3) with state variable $s(j) = v(j)$, and input $x(j) = (r(j), D(j))$. The noncontrollable input is the demand $w(j) = D(j)$.

G: an output function which provides the algorithm for calculating the present output $y(j)$ as a function of the present state $x(j)$:

$$y(j) = G(s(j)) \tag{5}$$

A simple example of output function is the transformation of a shortage incident into an economic loss.

TABLE 2: Behavior of a system under experiment $\underline{c}(t) = (\underline{x}(t), s(0), t)$.

TIME	INPUT	STATE	PERFORMANCE	OUTPUT
j∈J	x(j)∈X=(U,W)	s(j)∈S		y(j)∈Y
0	x(o)=(u(o),w(o))	s(o) (given)		y(1);
1	x(1)=(u(1),w(1))	s(1)	PI(1)∈s(1)	y(1)
.	.	.		.
.	.	.		.
.	.	.		.
t	x(t)=(u(t),w(t))	s(t)	PI(t)∈s(t)	y(t)

2.2 System Experiment

To operate, for example, a water supply system, the input vector or trajectory is manipulated to provide a desirable output vector or trajectory. The desirability of an output is measured by means of PI's and then FM's, calculated on the basis of a so-called system experiment $\underline{c}(t) = (\underline{x}(t), s(o), t)$, composed of an input trajectory (inflow) $\underline{x}(t)$, an initial state (storage) s(o), and a time t. The state transition function F(.) and the output function G(.) determine, respectively, the state trajectory $\underline{s}(t)$ and the output trajectory $\underline{y}(t)$, given any experiment $\underline{c}(t)$, as described in Table 2. Such a table may be used to describe the results of any digital simulation run, with decision trajectory $\underline{u}(t)$ and noncontrollable input trajectory w(t).

Although in principle a performance index PI, such as grade of service (defined below), may be calculated at every time j, one usually only evaluates a single performance index value PI(t) per system experiment and then one figure of merit value per ensemble of experiments. The figure of merit thus appears to be a "super-criterion."

2.3. Performance Indices

Ten incident-related PI's which will be used to assess the desirability of an input/output trajectory pair are now defined, namely:

PI^1 : grade of service

PI^2 : quality of service

PI^3 : speed of response

PI^4 : reliability

PI^5 : incident period

PI^6 : mission reliability

PI^7 : availability

PI^8 : repairability or resilience

PI^9 : vulnerability

PI^{10} : economic index vector (cost, loss, benefit)

The first three indices can be calculated solely from the system characteristics given so far and are respectively defined as:

PI^1: the grade of service is the relative frequency of providing a service, such as supplying water or reducing flood peak, when it is required, that is, when the demand arises or flood occurs. Shamir (this book) defines this PI as the "fraction of the demanded water which is supplied, computed over a specific period of time, such as a day, a month, a year".

PI^2: the quality of service or of response, measures for example the percentage of requirement satisfied. Shamir (this volume) speaks of a related but somewhat different quantity, namely, the "fraction of time the consumer is connected to a source and has some supply (even if less than demanded)". Alternatively, a negative quality of service PI^2 may be "the number of customers who have at least one failure event, greater than some value, once or more during the time period".

PI^3: <u>the speed of response</u> is the time elapsed between the occurrence of the demand for a service and the delivery of that service.

The other PI's can be calculated only after the incident mode μ, which includes failure modes, has been specified as in equation (1) and Table 1. The decision maker must make a subjective judgement to define the range of μ for which an incident is minor, major, or constitutes a failure.

Examples of incidents usually classified as minor incidents are:

- the inundation of a low-yield agricultural floodplain because of high reservoir release.

- a 10% agricultural water shortage occurring during a non-critical crop growth phase.

Examples of incidents which would generally be classified as failures are:

- a dam break causing heavy downstream damage

- an extreme drought causing 80% crop losses.

There appears to be a continuum between a small incident (a few per cent shortage) and large failure (total shortage). The main point is that incident or failure modes μ must be properly defined. Note that the definition may involve a Boolean combination of simple events, such as: $\{\mu{:}\in$ occurs if and only if $(PI^1 \leq 0.7$ and $PI^3 \geq 1)\}$.

PI^4: <u>The reliability performance index</u>, PI^4, under the condition that μ has been specified, is defined as an estimate of the relative frequency that the system is not in mode μ during the experiment $\underline{c}(t)$. To be precise, let $\delta(\mu,j)$ be an indicator function such that

$$\delta(\mu,j) = \begin{cases} 1 \text{ if } Z \text{ is in mode } \mu \text{ at time } j \\ 0 \text{ otherwise} \end{cases} \tag{6}$$

Then the reliability PI, with respect to mode μ, is by definition

$$PI^4(\underline{c}(t),\mu) = \frac{t+1 - \sum_{j=0}^{t} \delta(\mu,j)}{t+1} \tag{7}$$

This definition embeds both structural reliability (Szidarovszky et al., 1975, Plate, 1984; Section II of this book) and target-related reliability (Simonovic and Mariño, 1980, 1981, 1982; Burges and Lettenmaier, 1982; Mariño and Mohammedi; 1983; Moy et al, 1986; Section III of this book).

An operational definition of the remaining PI's is now given:

PI^5: <u>the incident period</u> is the mean interarrival time between entries into mode μ; it is also called the average recurrence time. If $d_5(\mu, n)$, $n \geq 1$, denotes the duration of the n-th interarrival time, then

$$PI^5(\underline{c}(t),\mu) = \frac{1}{N-1} \sum_{n=1}^{N-1} d_5(\mu,n) \tag{8}$$

where $N = N(\mu)$ is the number of incidents of mode μ during time horizon t. Shamir (this book) defines a similar PI as the "number of failure events per unit time", or else, as the "number of failures per unit time which are longer than some minimum".

PI^6: <u>The mission reliability</u> is an estimate of the probability that the system will not have an incident between the time that a non-null input requirement arrives and the corresponding service (water supply) is delivered. For a flood control problem, PI^6 may count the proportion of floods mitigated by the reservoir to the total number of incoming floods. Vogel (1986) gives real world examples of calculations of PI's.

PI^7: <u>The availability</u> is the probability that the system will not be in mode μ when the demand for a service occurs, for example, that the reservoir will not be above a critical control level when a flood arrives. PI^7 differs from PI^6 in that a large flood may still result in substantial spillway overflow and damage downstream. PI^6 and PI^7 are precise restatements of PI^1.

PI^8: <u>The repairability</u> is the average length of time that a system stays in mode μ during the experiment $\underline{c}(t)$:

$$PI^8(\underline{c}(t),\mu) = \frac{\sum_{j=0}^{t} \delta(\mu,j)}{t+1} = \frac{\sum_{n=1}^{N} d(\mu,n)}{N} \tag{9}$$

where $d_8(\mu,n)$ is the duration of the n-th mode μ incident (n=1,2,...,N). Index PI^8 is closely related to resilience which broadly speaking, measures the average recovery time from the occurrence of an incident mode μ given that one has occurred. For example, Hashimoto et al (1982) state that resilience measures "how quickly the system returns to a satisfactory state" and defines an index which is the limit of PI^8 as $N \rightarrow \infty$. A large resilience indicates a quick return to normal state or short times in incident mode; a small resilience indicates the opposite phenomenon. Further, Fiering (1982) proposes a set of alternative resilience indices; in the immediately preceding paper of the same journal issue, the concept itself is discussed and in the two papers that follow, alternative approaches for estimating resilience are presented.

PI^9: The vulnerability measures the average severity of an incident event. Shamir (this book) defines a similar PI as the "magnitude of the largest shortfall during the unit time". Here, mode μ must take on more than one value. Let $\underline{\mu}$ be a random variable such as shortage with known distribution function $F(\mu)$, then PI^9 may be the expected value:

$$PI^9\ (\underline{c}(t),\underline{\mu}) = E(\underline{\mu}) \tag{10}$$

or some other statistic.

PI^{10}: The economic PI is a vector whose components may include expected costs, losses and benefits, rates of return, and cash flows.

2.4 Figures of Merit

The PI's are usually calculated only at the end of each system experiment $\underline{c}(t)$ although, as stated before, they could be calculated at every time period if so desired. The FM's are then defined as a function of the PI's calculated over the ensemble of system experiments $[\underline{c}(t)]$.

If, for a given initial state s(o), $\underline{c}(t)$ is taken as the realization of a stochastic process with known distribution $F_{\underset{\sim}{c}}(c)$, then a set of possible FM may be defined by

$$FM^k\ (\mu) = \int PI^k\ (c,\mu)\ dF_{\underset{\sim}{c}}\ (c) \quad \text{for } k - 1, \ldots.K \tag{11}$$

An alternative definition may also be given in terms of a percentile or another statistic; furthermore, an FM may consist of a combination of PI's (for μ given) :

$$FM = FM\ (PI^1,\ PI^2, \ldots. PI^K) \tag{12}$$

Two particular FM's, implementation reliability FIR and engineering risk RI, are of special interest in the present study.

If a system is composed of a moderate number of independent elements, with known reliability, placed in a series-parallel configuration, then a so-called implementation reliability FIR may be calculated as in classical reliability theory. However, it should be realized that FIR represents only an upper or lower bound to the actual reliability because of the simplifying assumptions which have to be made in order to keep the calculations at a manageable level. Independence and lag one Markov dependence are examples of such assumptions.

Thus, if water is being supplied from a set of I independent sources, each with a reliability RE(i), i = 1, 2,...,I, then the implementation reliability of the sources placed in series and in parallel are, respectively:

$$FIR = \pi_i\ RE(i) \tag{13}$$

$$FIR = 1 - \pi_i\ (1 - RE(i)) \tag{14}$$

The reliability of series-parallel combinations can be calculated in a similar manner, but how realistic would such a model be? This point should be investigated on a case-by-case basis (Duckstein and Plate, 1984).

The other FM to be introduced at this point is engineering risk RI. The qualification "engineering" is needed to limit the scope of the word "risk" which could encompass concepts of purely social and psychological sciences that are quite important but outside of the scope of the present study. The comparison of two systems Z_1 and Z_2 is reduced to the comparison of two vectors of FM's.

$$\underline{FM}\ (Z_i) = (FM^1\ (Z_i),\ \ldots,\ FM^K\ (Z_i)),\quad i = 1, 2$$

where K is the number of FM's selected.

This comparison constitutes a multicriterion decision problem whose "cost-effectiveness" table (Goicoechea et al, 1982) or decision matrix (Hiessl et al, 1985) is an array of alternative systems versus FM's. Next, to each choice of system, associate a measure of the consequences of uncertainty in the system elements, such as uncertainty in the inflow sequence or in the requirements. Engineering risk is then defined as a joint probability distribution function over the set of all possible values of the FM's. As an example, the engineering risk of a supply scheme Z may be:

$$RI(Z,\ a(4),\ a(5),\ a(8)) =$$

$$Pr[\ FM^4(Z) < a(4);\ FM^5(Z) < a(5);\ FM^8\ (Z) > a(8)] \tag{15}$$

If we let reliability limit be a(4) = 0.95, mean recurrence time limit, a(5) = 30 months and proportion of down time threshold, a(8) = 0.10, then RI(Z) is the probability that the system Z will have a reliability below 0.95, a mean time between μ-events under 30 months and be in mode μ more than 10% of the time. Furthermore, the engineering risk RI(Z) is interpreted as the probability of being able to build a system with that specific combination of characteristics.

Note that system reliability RE = FM^4 is embedded into the definition of engineering risk, as shown for example in equation (15). Also, the particular case of that equation:

$$RI(Z,\ a(4)) = Pr\ (PI^4(Z) < a(4))$$

expresses that the simplest measure of risk, the probability of failure, is the complement of reliability.

3. EXAMPLE: BI-OBJECTIVE RESERVOIR CONTROL

3.1 Calculation of performance indices

In order to illustrate the above-presented concepts in a water resource management setting, consider the control of a single reservoir for the two objectives of flood control and water supply. Let h(j) be the water level in the reservoir at time j = 1,2,... and q(j) be the outflow, including release r(j) and spillway discharge as the case may be. The discretized time series of h(j) is sketched in Figure 1; under the minimum desirable level mh, there is a deficit incident; above the maximum level Mh, there is a surplus incident.

The maximum value of h is the height of the dam H, the initial state is s(o) = h(o). A similar sketch for q(j) is represented in Figure 2, but no upper limit of q is given; instead the maximum outflow for the n-th flood event (above Mq) is denoted Q(M,n) and the minimum outflow for the n-th drought event (below mq) is denoted Q(m,n).

The performance indices defined in the preceding sections may be calculated for high flows, low flows, or both. For this purpose, two indicator functions $\delta(\mu,j)$ need to be defined, using equation (6) with $\mu = M$, and then $\mu = m$.

Thus, the performance indices are respectively, for j = 0,1,2,...t

- Level of service PI^1:

High flow $PI^1(M,t) = t+1 - \Sigma\delta(M,j)$

Low flow $PI^1(m,t) = t+1 - \Sigma\delta(m,j) \qquad j = 1,\ldots,t$

Both $PI^1(t) = t+1 - \Sigma(\delta(M,j) + \delta(m,j))$

- Quality of service PI^2 and speed of response PI^3: these indices will not be utilized in the present study.

- Reliability PI^4:

High flow: $PI^4(M,t) = 1 - \dfrac{\Sigma\ \delta(M,j)}{t+1}$

Low flow: $PI^4(m,t) = 1 - \dfrac{\Sigma\ \delta(m,j)}{t+1}$

Both: $PI^4(\ t\ ) = 1 - \dfrac{\Sigma\ \delta(M,j) + \delta(m,j))}{t+1}$

- Period of incident PI^5:

High flow: $PI^5(M,t) = \dfrac{1}{N(M)-1} \sum_{n=1}^{N(M)-1} d_5(M-n)$

Low flow: $PI^5(m,t) = \dfrac{1}{N(m)-1} \sum_{n=1}^{N(m)-1} d_5(m,n)$

Both: $PI^5(m,t) = \dfrac{1}{N-1} \sum_{n=1}^{N-1} d_5(n)$

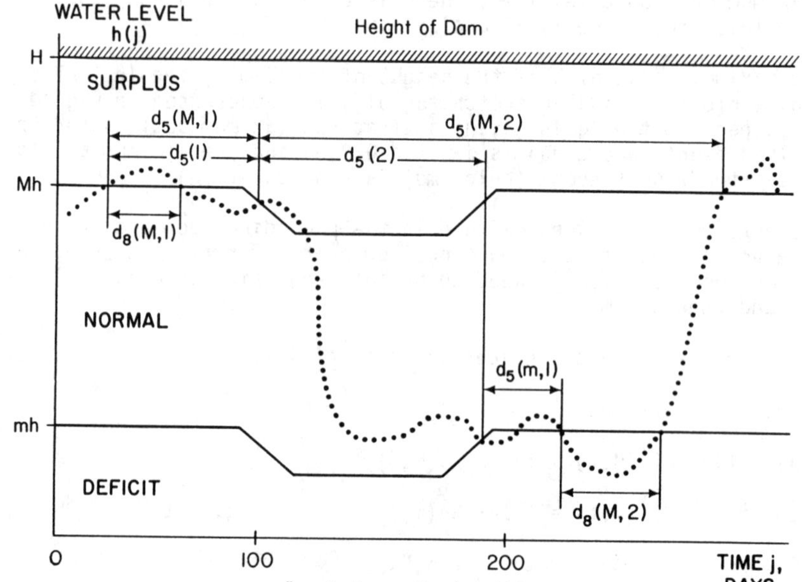

FIGURE 1: SKETCH OF WATER LEVEL (H) INCIDENTS

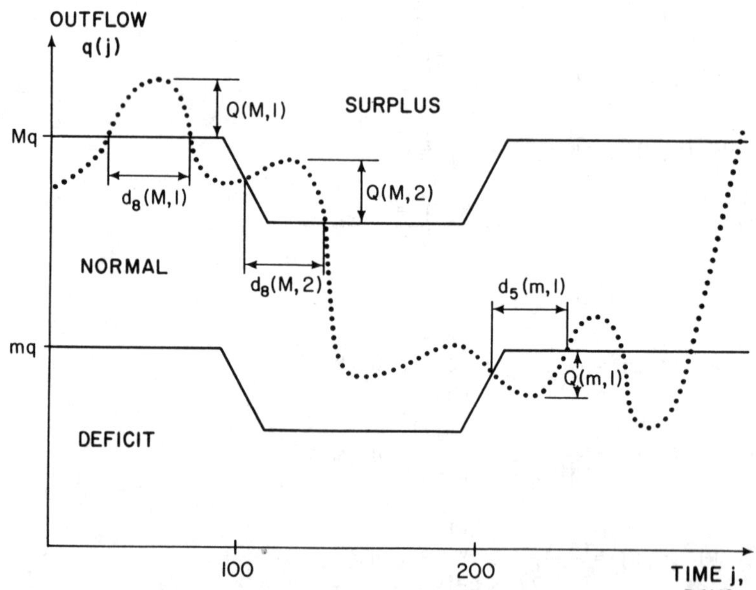

FIGURE 2: SKETCH OF OUTFLOW (Q) INCIDENTS

Incident number n	1	2	3	4	5	6	7	8	9	10	11	12	13	14
d_5 (n)	19	134	81	128	80	826	499	59	156	148	77	45	191	61
M-Incident nb n_1	1	2	3	4	5	6	-	7	8	9	10	11	12	13
$d_5(M,n_1)$	19	134	81	128	80	1325	-	59	156	148	77	45	191	61
$d_8(M,n_1)$	7	39	10	109	48	19	-	49	98	20	58	36	7	48
y (M,n_1)	12	49	2	2	3	6	-	163	6	5	22	2	5	1
m-Incident nb n_2							1							
d_5 (m,n_2)							3065							
d_8 (m,n_2)							120							
y (m,n_2)							113							

Incident number n	15	16	17	18	19	20	21	22	23	24	25	26	27	28
d_5 (n)	260	175	332	1062	91	127	321	813	542	41	279	234	162	
M-Incident nb n_1	14	15	16	17	-	-	-	-	-	18	19	20	21	22
$d_5(M,n_1)$	260	175	332	2955	-	-	-	-	-	41	279	234	162	
$d_8(M,n_1)$	114	48	73	112	-	-	-	-	-	17	87	123	116	48
y (M,n_1)	58	1	2	4						11	5	12	1	14
m-Incident nb n_2					2	3	4	5	6					
d_5 (m,n_2)					91	127	321	813						
d_8 (m,n_2)					75	38	132	109	53					
y (m,n_2)					86	18	76	51	133					

TABLE 3 : Surplus and Deficit Incidents for Operation Rule I where n(M) and N(m), are, respectively, the number of events of type M with interarrival times $d_5(M,n)$ and type m with interarrival times $d_5(m,n)$. The total number of incidents is N = N(M)+N(m), $d_5(n)$ in the interarrival time between any two consecutive incidents.

- Mission reliability PI^6 and availability PI^7: these two indices will not be used in the present study.

- Repairability or resiliency PI^8:

High flow: $$PI^8(M,t) = \frac{1}{N(M)} \sum_{n=1}^{N(M)} d_8(M,n)$$

Low flow: $$PI^8(m,t) = \frac{1}{N(m)} \sum_{n=1}^{N(m)} d_8(m,n)$$

Both: $$PI^8(t) = \frac{1}{N} \sum_{n=1}^{N} (d_8(M,n) + d_8(m,n))$$

Here $d_8(M,n)$ and $d_8(m,n)$ are, respectively, the duration of n-th incident M and n-th incident m. The other variables are as for PI^5.

- Vulnerability PI^9: For this index, the outflow as presented in Figure 2 is needed, because the maximum of h cannot be measured in case of spillway overflow. The degree of n-th incident M is measured by the maximum flow exceedance Q(M,n); the degree of n-th incident μ may be measured either by average minimum flow, by average maximum deficit, by average deficit or by absolute minimum outflow. The corresponding PI's are respectively:

High flow: $$PI^9(M,t) = \frac{1}{N(M)} \sum_{n=1}^{N(M)} Q(M,n)$$

Low flow:

(a) Average maximum deficit:

$$PI^9(m,t) = \frac{1}{N(m)} \sum_{n=1}^{N(m)} Q(m,n)$$

(b) Average deficit:

$$PI^9(m,t) = \frac{1}{\Sigma\delta(m,j)} \sum_{1}^{t} (Q(j)\ \delta(m,j)$$

(c) Minimum outflow:

$$PI^9 (m,t) = \min Q (m,n), n=1, \ldots N$$

$$= \min_j q (j), j=0, \ldots t$$

- Economic index vector PI^{10}: Separate counts may be kept for benefits, cost and losses. Figures of merit can then be defined as combinations or functions of the above PI's, as described previously (Equation (15)).

3.2. Comparison of operation rules

Consider, as a concrete case, the daily operation of a reservoir in the Black Forest in Germany, where both surplus and deficit incidents may occur. Specifically, a surplus occurs if the pool level h is above Mh = 18.40 meters and a deficit if h is below mh = 2.00 meters. Furthermore, surpluses or deficits lasting less than seven days are considered to be part of the same incident. Thus,

$$d_8(\cdot,n) \geq 7$$

and

$$d_5(\cdot,n) \geq 14 \text{ days}$$

Note that there is no seasonal variation of maximum level Mh and minimum level mh as is the case in Figure 1.

In the first phase of this example, a given operation rule labelled I is applied to the reservoir, starting with initial level h(o) = 18.40 m and using the available 21-year inflow sequence as input trajectory.

The results for operation rule I, shown in Table 3, lead to the following remarks:

(a) surplus incidents arrive consecutively; sequences of such incidents may be separated by very long deficit-prone time periods $d_5 (M,n_1)$

(b) a parallel statement may hold for deficit incidents.

On the basis of Table 3, the only statistical analyses that have some significance are as follows:

1. Interarrival time between any incidents $d_5(n)$, (28 Points):

$$\bar{d}_5 = 285.9 \quad , \quad s_5 = 304.0$$

An exponential distribution of $d_5(n)$ cannot be rejected on the basis of standard goodness-of-fit tests.

2. Interarrival time between surplus events $d_5(M,n_1)$

(a) for the whole sequence, (29 points):

$\bar{d}_5(M) = 330.8$; $s_5(M) = 659.7$

(b) during wet sequences only, (19 points):

$\bar{d}_5(M) = 140.4$; $s_5(M) = 88.9$

In either case, an exponential fit would be rejected, since the mean and standard deviation are far from being equal. It is thus seen that a wrong inference may be drawn (Poisson process) when the occurrences of all types of incidents are lumped

3. Duration of surplus events $d_8(M,n_1)$, (22 points):

$\bar{d}_8(M) = 58.45$; $s_8(M) = 39.14$

(Total duration: 1286 days)

4. Duration of deficit events $d_8\ (m,n_2)$, (6 points):

$\bar{d}_8(m) = 79.50$ (477 days)

5. Maximum surplus $y(M,n_1)$, (22 points):

$\bar{y}(M) = 16.64$; $s_y(M) = 35.99$

6. Maximum deficit $y(m,n2)$, (6 points):

$\bar{y}(m) = 78$

Some of the performance indices corresponding to 21 years (t = 7670 days) of operation can be calculated.

- Level of service PI^1 (days)

Surplus $PI^1(M,t) = 7670 - 1286 = 6384$

Deficit $PI^1(m,t) = 7670 - 477 = 7193$

Both $PI^1(\ t\ ) = 5907$

- Reliability PI^4

Surplus $PI^4(M,t) = \frac{6384}{7670} = 0.832$

Deficit $PI^4(m,t) = \frac{7193}{7670} = 0.940$

Both $PI^4 = \frac{5907}{7670} = 0.770$

- Period of incidents PI^5 (days)

Overall = 330.8

Surplus $PI^5(M,t) = 140.4$ during wet sequences

Deficit insufficient data

Both $PI^5(t) = 285.9$

Note here that the mission reliability of flood control may be taken as the probability of not having a surplus incident <u>during a wet period</u>.

- Repairability or resilience PI^8:

With the definitions given previously, PI^8 is calculated in days as:

Surplus $PI^8(M,t) = \overline{d}_8(M) = 58.45$

Deficit $PI^8(m,t) = \overline{d}_8(m) = 79.50$

Both $PI^8(\ t\ ) = \overline{d}_8 = 62.96$

However, another definition of resilience may be given in terms of the probability that the duration d_8 of an incident be over a given threshold. This threshold could be a percentile of the pdf of d_8.

- Vulnerability PI^9.

Here again, the straightforward values are in average number of days:

Surplus $PI^9(M,t) = \overline{y}(M) = 16.64$

Deficit $PI^9(m,t) = \overline{y}(m) = 78$

but another definition may be given in terms of the probability that the degree of an incident be above a certain threshold.

The next steps for developing the example are as follows:

- to simulate inflow sequences
- to calculate the various indices for other operation rules II, III,.... (including means and variances)
- to find an operation rule which constitutes a "good" compromise between the various conflicting performances using an appropriate multicriterion analysis technique (Gershon and Duckstein, 1984; Szidarovszky et al, 1986).

Specific examples of trade-offs between PI's are provided in Moy et al (1986).

4. CONCLUSIONS

(1) A mathematical system framework that embeds both structural reliability and target-related reliability has been defined.

(2) An incident is said to occur when a generalized load exceeds a a generalized resistance.

(3) The discrete-time discrete-state model is composed of five sets. The initial state and input trajectory uniquely determine the behavior of the system over time horizon t.

(4) The performance of the system is measured by means of a set of running criteria called performance indices PI, and a set of "super-criteria" called figures of merit FM. The FM's are combinations of PI's calculated over a set of observation sequences ("experiments"), such as simulation runs.

(5) Most performance indices can only be calculated after the incident modes have been defined.

(6) Reliability of a system component is a PI that can usually be calculated, but reliability of a complex system cannot in general be calculated from the reliability of the components.

(7) The bi-objective trade-off between water supply and flood control, for which incidents may be deficit or surplus, shows that wrong inferences may be made if the system is improperly specified.

ACKNOWLEDGMENTS

This research has been supported in part by funds from the Alexander von Humboldt Foundation, FRG, and the U.S. National Science Foundation under grant No. CEE 8110778.

REFERENCES

Burges, S.J., and D.P. Lettenmaier (1975) Probabilistic methods in stream quality management. Water Resources Bulletin, Vol. 11 pp. 115-130

Burges, S.J. and D.P. Lettenmaier (1982) Reliability measures for water supply reservoirs and the significance of long-term persistence in Decision-Making for Hydrosystems - Forecasting and Operation, T.E. Unny and E.A. McBean, Eds. WRP, Ft. Collins, Colo., pp. 299-323

Bogardi, I. and L. Duckstein (1978) Design under Stochastic Groundwater Fluctuation, J. Hydraul. Div. Amer. Soc. Civil Eng., (104), pp. 59-74

Davis, D.R. and W.M. Dvoranchik (1971) Evaluation of the worth of additional data, Water Resources Bulletin, Vol. 7, No. 4, pp. 700-707

Duckstein, L. and E.J. Plate (1984) A system framework for hydrologic reliability with application to the design of structures, Int. Assoc. for Hydr. Sciences, Publ. No. 47

Duckstein, L. and I. Bogardi (1981) Application of reliability theory to hydraulic engineering design, Proc. ASCE, HY7, July, pp. 799-815

Fiering, M.B. (1982) Alternative indices of resilience, Water Research, Vol. 18, No. 1, pp. 33-39, February

Gershon, M. and L. Duckstein (1984) A procedure for selection of a multiobjective technique with application to water and mineral resources, Applied Mathematics and Computation, April 14(3) pp. 245-271

Goicoechea, A., D. Hansen and L. Duckstein (1982) Multiple Objective Analysis with Engineering and Business Applications, J. Wiley, N.Y., p. 561

Hashimoto, T., J. Stedinger and D. Loucks (1982) Reliability, resiliency, and vulnerability criteria for water resources system performance evaluation, Water Resources Research, Vol. 18, No. 1, pp. 14-20

Hiessl, H., L. Duckstein, and E. Plate (1985) Multiobjective Q-analysis with concordance and discordance concepts, Applied Mathematics and Computation, Vol. 17 pp. 107-122, Vol. 17, pp. 107-122

Klemes, V., R. Srikanthan, and T.A. MacMahon (1981) Long-memory flow models in reservoir analysis: what is their practical value? Water Resources Research, Vol. 17, No. 3, pp. 737-751

Knetsche, J.L. (1974) Outdoor Recreation and Water Resources Planning, Water Resources Monograph Series, Vol. 3., American Geophysics Union Washington, D.C.

Laursen, M. (1970) Bridge design considering scour and risk, Transportation Engineering Journal, TE2, May pp. 149-164

Lee, Han-Lin and L.W. Mays (1983) Improved risk and reliability model for hydraulic structures, Water Resources Research, Vol. 19, No. 6, pp. 1415-1422 December

Loucks, D.P., J.R. Stedinger and D.A. Haith (1981) Water Resources Systems Planning and Analysis, Prentice-Hall, N.J., p. 559

Mariño, M.A. and B. Mohammedi (1983) Reservoir management: reliability programming approach, Water Resources Research, Vol. 19, No. 3, pp. 613-620, June

Marsily, G. de, E. Ledoux, and P. Masure (1983) Analyse des risques et geologie prospective en matiére d'enfouissement en profondeur des déchets de l'industrie nucléaire, Revue IRE Tijdschrift, Vol. 7, No. 2

Moy, W.S., J.L. Cohon and C.S. Revelle (1986) A programming model for analysis of the reliability, resilience, and vulnerability of a water supply reservoir, Water Resources Research, Vol. 22, No. 4 pp. 489-498, April

Plate, E.J. (1984) Reliability analysis of dam safety, in: W. Hall, C. Maxwell, L.R. Beard (eds) Frontiers of Hydrology, Fort Collins, Colo., pp. 287-303

Shamir, U., J. Bear and A. Gamliel (1984) Optimal annual operation of a coastal aquifer, Water Resources Research, Vol. 20, No. 4, April

Simonovic, S.P. and M.A. Mariño, Reliability programming in reservoir management, Water Resources Research
1. Single Multipurpose Reservoir, 1980, Vol. 16, pp. 844-848
2. Risk-Loss Functions, 1981, Vol. 17, pp. 822-826
3. System of Multipurpose Reservoirs, 1982, Vol. 18, pp. 735-743

Simonovic, S.P. and T. Orllob (1984) Risk-reliability programming for optimal water quality control, Water Resources Research, Vol. 20, No. 6, pp. 639-646, June

Szidarovszky, F., L. Duckstein and I. Bogardi (1975) Levee system reliability along a confluence reach, Journal of the Engineering Mechanics Division, ASCE, Vol. 101, No. EM5, Proc. Paper 11631, October, pp. 609-622

Szidarovszky, F., M. Gerson, and L. Duckstein (1986) Techniques for Multiobjective Decision-Making in Systems Management, Elsevier, Amsterdam, NL 506 p.

Thomann, R.V. (1972) Systems Analysis and Water Quality Management, McGraw Hill, New York

USCOLD (1975) Lessons from Dam Incidents, USA, Amer. Soc. Civil, Eng., New York

Vogel, R.M. (1986) Reliability indices for use in the design of water supply reservoir, this volume

Wymore, A.W. (1976) Systems Engineering Methodology for Interdisciplinary Teams, Wiley, Inc., New York, p. 362

Wymore, A.W. (1982) Design of service delivery systems. I. Working paper 82-10-08, Systems & Industrial Engineering Department, University of Arizona, Tucson, Arizona 85721, USA

Yazicigil, H., M.H. Houck and G.H. Toebes (1983) Daily operation of a multipurpose reservoir system, Water Resources Research, Vol. 19, No. 1, pp. 1-13, February

TABLE OF SYMBOLS

$a(\cdot)$	constant
$A(\cdot)$	action
$\underline{c}(t)$	system experiment of length t
$d(\cdot)$	duration
$D(j)$	water requirement
$E(\cdot)$	expected value
$f_w(w)$	probability density function of w
$F(\cdot)$	state transition function
$F_w(w)$	probability distribution functions of w
$FM(.)$	figure of merit
$g(j)$	physical output variable
$G(\cdot)$	output (or readout) function
$h(j)$	water level in reservoir
H	reservoir height
i	index, $i = 1,\ldots,I$
j	index for current time period, $j = 1,\ldots,J$
k	index, $k = 1,\ldots,K$
$L(\cdot)$	loss function
mh	target minimum of h
mq	target minimum of q
Mh	taret maximum of h
Mq	target maximum of q
n	index for number of events, $n = 1,2,\ldots,N(n)$
$N(\mu)$	number of events of type μ
Pr	probability
$q(j)$	net outflow
$Q(\mu,n)$	exceedance value of $q(j)$ in n-th event of type μ

r(j)	release
RE(.)	system reliability (figure of merit)
RI(•)	engineering risk
$\underline{s}$	state trajectory
s(j)	state (vector)
T	time horizon
u(j)	controllable input
U	controllable input alphabet
v(j)	water volume in storage
w(j)	non-controllable input
W	non-controllable input alphabet
x(j)	input
X	input set or alphabet
y(j)	output
Y	output set or alphabet
Z	system
α	constant
$\delta(\mu,i)$	indicator function for incident of type μ
ϵ	incident event
$\emptyset$	benefit function
λ	generalized load
μ	incident or failure mode
ρ	generalized resistance

II. RELIABILITY AND RISK IN STRUCTURES

II. RELIABILITY AND RISK IN STRUCTURES

Hydraulic design traditionally follows standard structural practice combined with a design load concept. The design load is either a single load or a load function that incorporates statistical aspects (for example, a design discharge based on an exceedance probability) as well as deterministic aspects (for example, a design freeboard). Although quantities such as the "n-year design flood" seem to imply a probability measure of the safety of the structure, they are not in fact related to the potential failure probability. Most hydraulic engineers are aware of this. For example, they do not imply that a spillway designed to discharge a 1000-year flood is safe for lesser floods or that it will fail for larger ones. Instead, they assume that the structure is practically safe under all loads if the design load is used and if all measures for increasing the safety are taken into account using "state of the art" methods.

There are a number of reasons why the design load concept should be reconsidered and design practice put on a sounder theoretical basis. An important reason is economics. Design by reliability is more cost effective for new structures, and reliability is also useful, as a criterion for the economics of refitting existing structures to meet increased safety requirements. A second reason is that structural engineers in many countries are working on standards which are no longer derived from empirical concepts, but which incorporate considerations of the probabilistic reliability for which both loads and resistance are random variables. If structural engineers can deal with loads whose randomness is fairly difficult to quantify, then hydraulic engineers, dealing with loads that are becoming more accurately quantified through the tremendous progress in hydrology, should strive to do the same. The most important reason, however, is that reliability theory offers a consistent and conceptually complete design framework. This fits well into our age of computers and sophisticated data gathering skills.

The assertion that reliability-based design is useful in hydraulic engineering must, however, still be demonstrated. There is no question that most of the structures designed according to conventional procedures have performed satisfactorily, while little experience exists for reliability-designed structures. We are presently at a stage where it is necessary to compare reliability-based designs with conventional designs, showing the relative advantages and disadvantages of each approach. In order to make these comparisons, it is necessary that engineers become familiar with the techniques of reliability analysis.

Accordingly, it is the purpose of Section II to present the concept of reliability analysis and its application to important structures. The section consists of two parts. In Part II.1, the basic principles of different types of design analysis, as expressed by four levels of design, are illustrated by means of simple examples. In Part 11.2, the safety of dams is considered as an important application of the principles and ideas developed in the first part.

Part II.1 begins with an introductory paper by Plate and Duckstein on the concepts of design by reliability. Fundamental to many design problems of hydraulic structures is the estimate of the probability of occurrence of rare events, as described by an extreme value probability distribution. Although the theory of extreme values was developed many years ago, it is only recently that more physically-based extreme value concepts have been explored. It is important that future developments in reliability-based design take into account these new developments. Reliability-based design concepts are illustrated in a paper by Correia et al, which is presented directly after the introductory material on risk analysis in hydraulic structures design.

The designers of structures under random loads distinguish four levels of design, which are illustrated in Part II.2 of this section. The common feature of the four design levels is that a decision variable, the design resistance r_D of a structure, is chosen in such a manner that the structure only fails if the load s exceeds a critical load s_D; the difference in the four levels lies in the specifications of r_D and s_D, and in the data required for the design. The higher the level, the more data are required. However, the complexity of analytical tools does not necessarily vary in the same direction.

Level 1 design is based on the concept of a safety factor. If r_D (for example, the minimum stress to which a structure can be subjected in its critical section) is specified by standards, then it is necessary to dimension the structure so that

$$s_D \leq \eta \cdot r_D$$

where η is the safety factor and s_D is the calculated stress due to the worst-case load. The concept is based on experience and risk averse decision-making and leads to a selection of η and r_D that yield an economically acceptable design. Sentürk presents a real-life case of level 1 design which has undergone major incidents, such as a near-failure of the spillway.

Level 2 design is based on a second moment analysis. If loads and resistances, as well as the parameters determining loads and resistances, are normally distributed, or if their distributions can be approximated by or transformed into normal density functions, then the probability of failure of the structure can also be determined from the normal distribution. It is fully specified by mean value and standard deviation. For this method, it is customary to define a random variable $z = r - s$, where r and s are, respectively, the random resistance and load. The coefficient of variation of the random variable z is then given by $C_z = \beta^{-1} = \sigma_z/\mu_z$, where μ_z and σ_z are the mean and standard deviation of z, respectively. Because $z = 0$ denotes the location of the failure surface, the "safety index β" determines the probability of failure. The parameter β is determined by performing a second moment analysis for the combination of the variables of a design problem. Some important concepts of this method of analysis with design examples are given in Yen's paper.

Level 3 design is concerned with the determination of the probability of failure without a prior specification of resistance and load probability density distributions. The most important case of level 3 design is based on the assumption that r and s are uncorrelated. This yields Freudenthal's design concept, from which a formula to calculate the probability of failure can be derived. A detailed explanation of this concept is given in Plate and Duckstein's paper. However, the case of uncorrelated r and s is not a necessary requirement for the applicability of a level 3 design. It is possible to determine the probability of failure by simulation, or to use empirically determined joint pdf's of s and r. Thus, level 3 design is more versatile, but also more complicated and requires a greater analytical or computational effort than lower level designs. A summary of important applications of level 2 and 3 reliability-based design to sea shore protection works is given in the paper by Vrijling, who also discusses and illustrates the problem of multiple causes of failure. This paper augments the concepts presented in the previous papers by introducing fault tree analysis.

A fourth level of design can be defined in which the economic, social, environmental and other consequences of a design are explicitly considered in the figure of merit (FM) used for the design. This concept is presented by Plate and Duckstein's paper in which the risk is introduced as an FM. Whereas the first three design levels use, as an FM, an externally imposed design criterion such as the safety factor or the probability of failure, the level four design does not specify a predetermined numerical value for the standard. Instead, the design value is found by optimizing a criterion function such as minimizing cost, minimizing the probability of fatalities, maximizing net benefit, or maximizing benefit cost ratio. Conflicting FM criteria or functions may be considered simultaneously in a multicriterion analysis, as exemplified in the last part of Section IV.

It is shown in von Thun's paper that the level 4 technique is well suited to find the most economical alternative for increasing the reliability of existing structures, or to find the design with the lowest probability of fatalities. The approach used by von Thun is one of a series of steps used in the assessment of the safety of dams in the USA. The concepts and ideas behind the different approaches are described in both Bowles and Stakhiv's papers. It becomes evident from the series of papers on dam safety that one of the key issues is the reliable estimation of the areas flooded downstream of a dam in the event of failure. This problem, which has been the subject of a large number of investigations, is illustrated by Ganoulis' approach to model the breaching of a dam as a stochastic process.

The series of papers of this section form, in our opinion, a comprehensive introduction to the theory of design of hydraulic structures by means of reliability or risk concepts, although theoretical details as well as multicriterion modeling had to be largely omitted. We hope that the list of references appended to most of the papers will serve as a guide to the reader who wishes more complete theoretical derivations.

RELIABILITY IN HYDRAULIC DESIGN

E.J. Plate
Institut für Hydrologie und Wasserwirtschaft (IHW) University of Karlsruhe, Kaiserstr. 12, D-7500 Karlsruhe 1, Federal Republic of Germany
and
L. Duckstein
The University of Arizona, Department of Systems and Industrial Engineering,Tucson, Arizona 85721

ABSTRACT

In hydraulic design, project reliability RE is defined as the probability that a project does not fail its purpose during its design life. It depends on the probability of failure P_F, which is the probability of the project to fail in any one year. In cases of high resilience (such as a reservoir not meeting its target release), P_F rather than RE is needed for design. In cases of low resilience (such as the dam of a reservoir failing through overtopping), the reliability is a more useful quantity because it incorporates the change of P_F (or of the hazard function, which is a related quantity) with time. The present paper is concerned with the reliability-based design of hydraulic structures.

In the first part of the paper, a general discussion of the concept of reliability-based design is given. It is shown that the probability of failure for a structure is not defined only through the hydrological input, but also depends on the parameters of the structure. The dependency can be expressed by means of generalized loads, and generalized resistances. In many cases of practical importance, loads and resistances are related by a joint probability distribution function (jpdf) and the failure probability can be obtained by elementary integrations. A discussion of P_F and of the factors contributing to it illustrates that P_F is an exact quantity for the ensemble of loads and resistances. For a sample of loads and/or resistances, or for other uncertainties, P_F follows a probability distribution.

The second part of the paper is concerned with the actual calculation of P_F. This calculation involves an extrapolation of probability distribution functions (pdfs) of loads and/or resistances to extreme values, accomplished by means of extreme value analysis. A short survey is given of the most important pdfs suitable for empirical extreme value analysis.

The results are illustrated by using an example of transformation of hydrologic data into hydraulic loads and calculating the reliability of the flood levee on a river which has been designed according to usual standards. It is shown that the actual reliability has very little to do with the exceedance probability of the design flood. A second example applies to the water quality of a river.

1. GENERAL DISCUSSION OF PROBABILISTIC DESIGN METHODS

1.1 Definition of Reliability

The general introduction in the first paper of the concepts of reliability analysis defined the reliability RE of an experiment as a performance index PI for the evaluation of the performance of a system. It is the purpose of this chapter to discuss reliability more directly by focusing on the design of structures. The reliability is considered to be the figure of merit for the design; it is the criterion which rates the performance of the system, as well as the value of the performance. We shall not consider more than one load, i.e. the generalized load vector λ reduces to $\lambda = s$ where s is a random variable expressing the load. Similarly, only one resistance is considered, so that the generalized resistance vector ρ reduces to $\rho = r$ where r is a random variable, as is shown schematically in Figure 1.

The definition of reliability as a performance index PI given in the introductory paper was

$$PI^4 = RE = 1 - \frac{n_f}{n} \qquad (1)$$

i.e. the reliability is expressed through the average number n_f of failures per number n of events that may cause failures. From Eq. 1, the reliability figure of merit (FM) is obtained as n goes to infinity i.e. for the ensemble of system experiments. Henceforth, RE is understood to be an FM

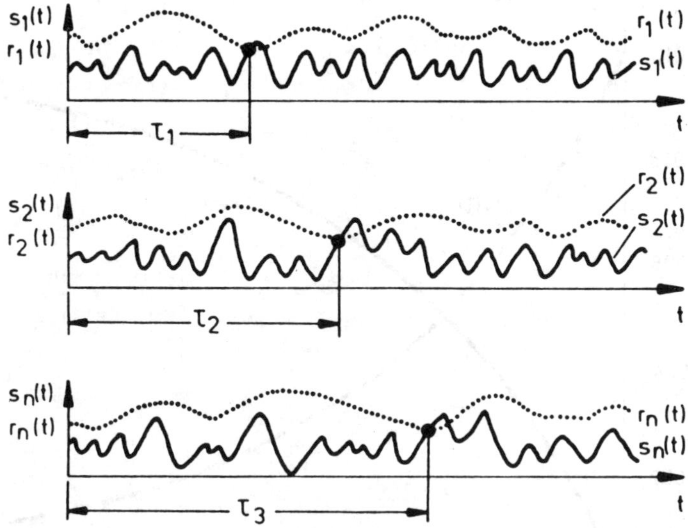

Figure 1: Definition of time to failure: experiment of m pairs of time functions.

except in section 1.3, as shall be discussed. However, for the continuous load function s(t) the reliability must be expressed differently as the probability of the first time that the structure fails, i.e. to reach the condition $s \geq r$ as function of time, or through the distribution function $F_\tau(\tau)$ where τ is the time from the beginning of service of the structure to first failure, as indicated in Figure 1. Such a function $F_\tau(\tau)$ is shown in Figure 2.

Based on Figure 2, the reliability of the structure at any time t:

$$RE(t) = 1 - F_\tau(t) \tag{2}$$

is defined through the probability density function $f_\tau(t)$. In reliability theory, it is useful to distinguish between $f_\tau(t)$, which is a function which varies with time, and $\beta(t)$, the hazard function or rate, also called the failure rate, which is defined as the conditional probability density for the failure event to occur in the time interval dt if it did not occur before t, thus:

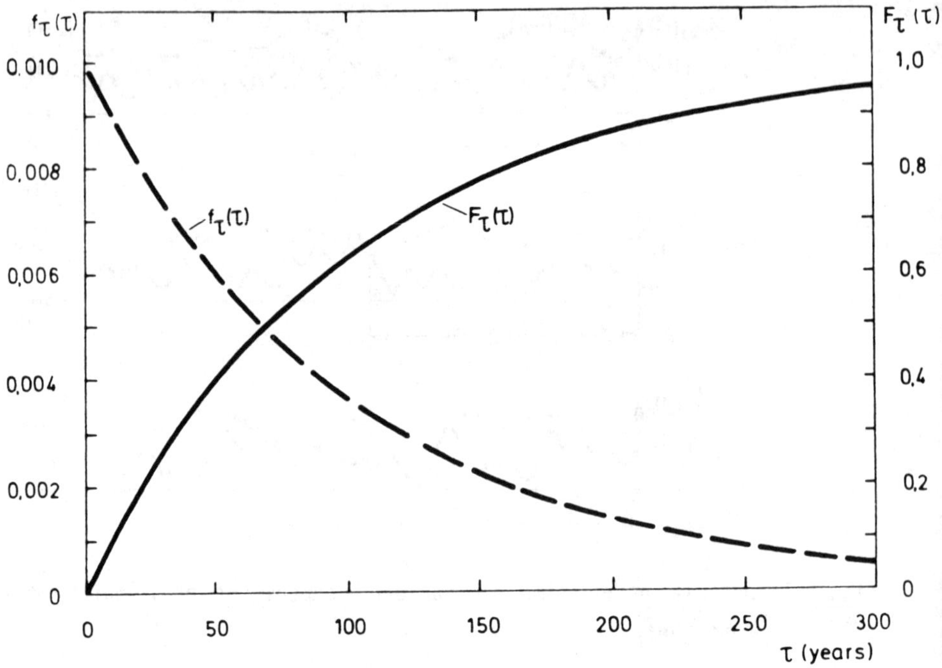

Figure 2: Probability density and probability function for time to first failure with constant failure rate β(t).

$$\beta(t) = \frac{f_\tau(t)}{1 - F_\tau(t)} \quad , \tag{3}$$

and since $f_\tau(t)$ is equal to $dF_\tau(t)/dt$ it follows that:

$$RE(T) = e^{-\int_0^T \beta(t)dt} \quad . \tag{4}$$

where T is a fixed time horizon.

For hydraulic structures it is not customary to use the failure rate β(t) in design but the so-called probability of failure P_F, which in our notation is a dimensionless quantity defined as:

$$P_F(T, \Delta T) = \int_T^{T+\Delta T} \beta(t)dt \quad . \tag{5}$$

The dependency on T here expresses the fact that, in general, P_F is nonstationary, as shall be discussed in section 1.2.1; the dependency on ΔT results from the time scale chosen for defining P_F. For example, if ΔT (as is customary in structural design) corresponds to one year, then $P_F = 0.01$ means that the failure event occurs once every $(1/P_F) \cdot \Delta T = 100$ years on the average, where $(1/P_F) \cdot \Delta T$ is called the recurrence interval.

One of the most serious shortcomings of traditional hydraulic design involving probabilistic components results from equating P_F with the exceedance probability $P_E(X_e)$ of a design load component $x = X_e$, for example a flood. This quantity is defined as:

$$P_E(X_e, \Delta T) = P\{x \geq X_e \text{ in } \Delta T\} = \int_{X_e}^{\infty} f_x(x, \Delta T)dx \qquad (6)$$

where $f_x(x, \Delta T)$ is the probability density of the random variable x for the interval ΔT.

In the general design case, P_E and P_F may be quite different because P_F is expressed as the probability of the event $s > r$ to occur in the interval ΔT, which depends on the joint statistical properties of the distributions of s and of r. For the general (one-dimensional) case, resistance and load follow a joint probability density function (jpdf), defined as $f_{sr}(s,r)$, and sketched in Figure 3. The design condition $P_E(s_e{=}X_e)$ is seen to be defined through the marginal distribution $f_s(s)$ only, whereas for P_F the entire jpdf is required. In terms of $f_{rs}(r,s)$, the failure probability P_F is defined as:

$$P_F(\Delta T) = P\ \{s \geq r, \Delta T\} \qquad (7)$$

$$P_F(\Delta T) = \int_0^{\infty} \{ \int_0^{s} f_{sr}(s,r)dr\}\ ds \qquad (8)$$

which is valid whether a relation exists between r and s or not.

As illustrated in Figure 3 the difference between P_E and P_F stems from different areas of the jpdf $f_{rs}(s,r)$ over which one integrates. P_E is the exceedance probability of the marginal density $f_s(s)$ corresponding to the

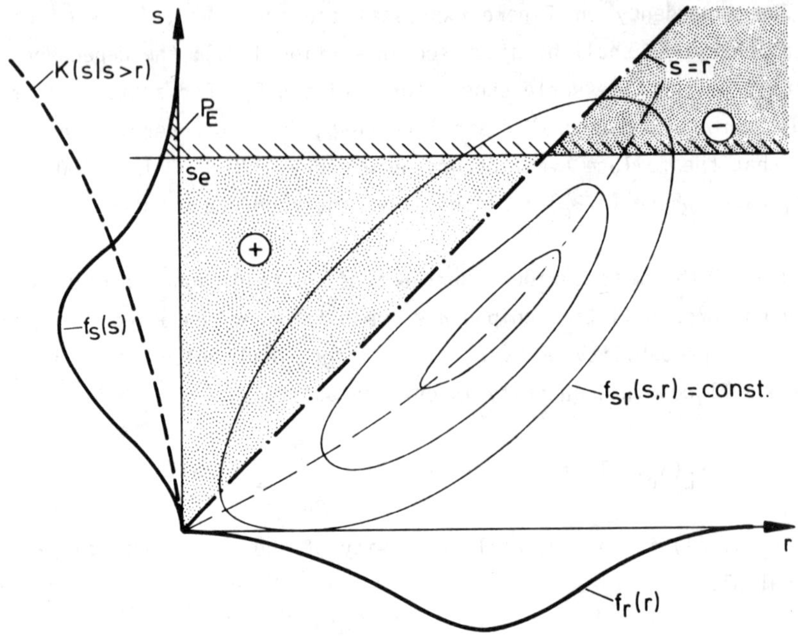

Figure 3: Definition of failure probabilities based on joint probability density function of load s and resistance r.

integral over the part of the (s,r) plane above $s = s_e$, whereas P_F is the integral over the s,r plane above s = r. Therefore, the relation between P_F and P_E is:

$$P_F = P_E + \Delta(+) - \Delta(-) \tag{9}$$

where $\Delta(+)$ is the integral of $f_{rs}()$ over the shaded area above the line through s_e in Figure 3, and $\Delta(-)$ is the integral of $f_{rs}()$ over the lower shaded area in Figure 3. Depending on the relative magnitude of $\Delta(+)$ and $\Delta(-)$, P_F may be either larger or smaller than P_E.

1.2 Design Based on Reliability

If reliability is taken as the design criterion, then the following design procedures is obtained. The structure must be designed so that the actual reliability RE based on the design life T_{Design} exceeds a reliability RE_{Design} which is given as part of the design information, or from standards, i.e.:

$$RE(T_{Design}) \geq RE_{Design} \quad . \tag{10}$$

In general, both s and r are random variables that can be modified by design paramters u_i. These design parameters are controllable by design decisions. The set of all parameters is combined into a decision vector U. For each choice of design parameters one obtains a particular value of P_F. Since the design condition Eq. 10 can be met for many sets of parameters U, it is useful to impose a second condition on U that optimizes some parameter. For example, one might find a set of parameters U* (consisting of u_i*) that leads to a minimum cost design. Furthermore, the set of parameters U* may be constrained by a condition $U \leq U_c$, i.e., $u_1 < u_{1c}$, $u_2 < u_{2c}$ ... $u_i < u_{ic}$ where u_{ic} are the constraints imposed by external or internal factors. With these conditions in mind, we may formulate the design problem in a slightly different way to read:

$$\begin{aligned} &\text{Find } \ U = U^* \ \text{ subject to } U^* \leq U_c \\ &\text{such that } RE(T_{Design}) \geq RE_{Design} \end{aligned} \tag{11}$$

Eq. 11 embodies the concept of design by reliability. Its improvement over previous design concepts based on P_E lies in the fact that it corresponds exactly to the failure which the design is supposed to avoid. Since the above discussion has shown that P_E bears little relation to P_F it is the opinion of the authors that design according to Eq. 11 is preferable.

1.2.1 Time variation of P_F. It should be noted that design by Eq. 10 or Eq. 11 is more general than using a design based on P_F only. The reliability is defined over the whole design horizon T, whereas $P_F(\Delta T)$ is only a temporal quantity which may vary during the life of the structure. This is illustrated in Figure 4 where the time dependence of the hazard function - and also of P_F, because of Eq. 5 - is shown. To illustrate, we consider the case of the dam of a reservoir which has been built to impound flood waters. For this case, the hazard function is not constant but varies with time. The time variation may have seasonal causes (floods do not occur during certain parts of the year) although this effect may be compensated by decreasing the flood storage volume) or the operation rule requires the reservoir to be empty at certain times (such as an irrigation reservoir at the end of the season). In such cases, β is a function of the season and long-term analyses must be concerned with that season or seasons in which β attains its maximum value.

The long-term variability of the hazard function is of great significance and requires careful consideration. Although there have not been enough dam failures on record to establish conclusively how the hazard function changes in the long run, it is reasonable to expect the pattern shown in Figure 4, which is the shape of the typical "bath-tub" curve well known in reliability engineering (see for example Dhillon and Singh, 1981, p. 28).

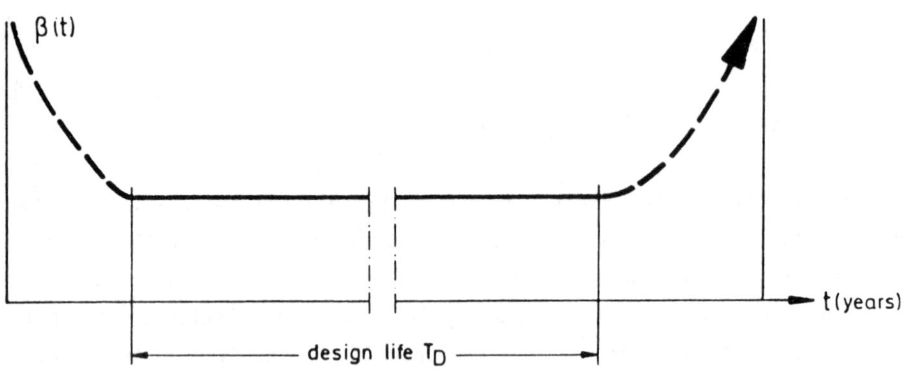

Figure 4: Standard "bath-tub" type failure rate (hazard function) $\beta(t)$.

During the first few years, the hazard curve is dominated by the type of failures that might occur after the dam has been put in operation. Examples are dam failures due to design mistakes or due to incomplete knowledge of underground conditions or of the type of rock on which arch dams abut. Also, problems associated with wetting of the banks (such as landslides), or with the weight of the water in storage (such as load-induced earthquakes, both of which might cause flood waves, hence damage or failure of the dam) are likely to occur in the initial period. It is extremely difficult to assume a shape for β in the initial period for an individual dam because the type of statistics published on dam failures (for example, Johnson and Illes, 1976) is not applicable to individual cases.

The hydrological hazard dominates the constant hazard region. The hazard caused by hydrological conditions remains constant unless changes occur in the climate or in the runoff characteristics. As long as these changes cannot be predicted, they only contribute to the uncertainty in the estimate of β.

Of some interest is the period of terminal hazard, which starts when neglect of dam maintenance sets in. This might occur in periods of prolonged social disturbance or when the purpose for which the dam was built no longer exists, so that the company or government which built the dam sees no reward in the maintenance of the dam. Barring these "human factors", the inevitable increase in hazard which leads to the ultimate destruction of the dam is due to the silting of the reservoir (if the sediment is not removed regularly). At the end of the constant hazard phase, the dead storage volume is completely filled and the active storage space decreases with time. Since no effective method of completely stopping the erosion process and of avoiding the reservoir silting is known sooner or later the active storage space is filled and a large flood leads to the destruction of the dam.

The bath-tub curve is not the only reasonable shape to be used in hydraulic structures. Consider for example the case of an off-shore structure which is subjected to the stationary force exerted by waves, so that the load probability distribution does not change with time. However, due to corrosion and occasional fatigue-caused cracks in the supporting elements

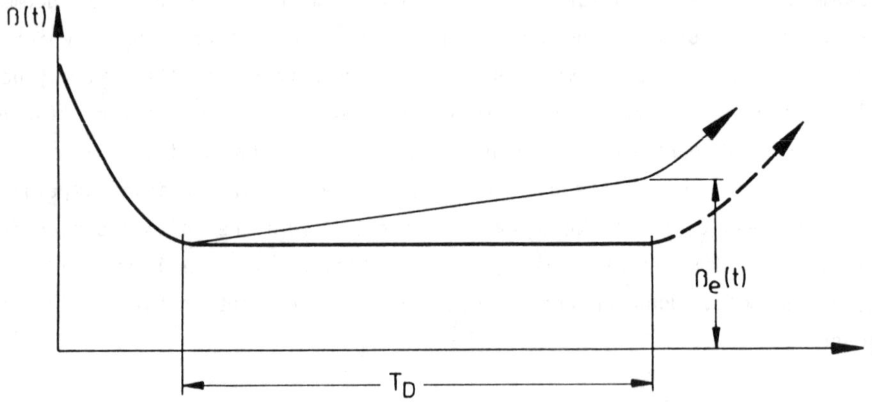

Figure 5: Development of the failure rate (hazard function) ß(t) due to slow degradation of performance.

of the structure such as the legs of the jacket, the resistance of the structure gradually decreases. The corresponding hazard function increases with time, as shown in Figure 5. Through the concept of reliability the overall probability of the structure to fail during its lifetime can be evaluated; or if the change of ß with time can be predicted, it is possible to design for a probability of failure $P_F(\Delta T)$ for the last time interval of the life-time T_D of the structure according to $\beta_e(t)$ in Figure 5.

1.2.2 Methodology for calculating the total reliability. We shall next turn to the problem of defining the steps which must be followed for calculating the total reliability of a hydraulic structure. We shall use reliability as defined by Eq. 3 and define for clarity the figure of merit "hydrologic reliability" RE_E

$$RE_E = e^{-\int_0^T \frac{P_E(t)dt}{\Delta T}} \tag{12}$$

where $\beta(t)$ has been replaced by $1/\Delta T \cdot P_E(t)$ (see Eq. 5). For the case P_E = const Eq. 12 becomes, with $T = n \cdot \Delta T$:

$$RE_E = (1 - P_E)^n \tag{13}$$

Analogously we define a "hydraulic reliability" RE_H with particular case P_H = const:

$$RE_H = (1 - P_H)^n \tag{14}$$

and a "structural reliability" RE_S in the particular case P_H = const:

$$RE_S = (1 - P_S)^n \tag{15}$$

In Eqs. 14 and 15, P_H is the probability of the annual extreme of loads s exceeding the design load s_H, and P_S is the probability of the annual extreme resistances r not exceeding the design resistance r_S. The relationship between these three quantities and the total reliability RE is illustrated in Figure 6, which includes both risk and uncertainty. In this concept, the reliability is a random variable because of the uncertainty in the causal processes. Therefore the quantities P_E, P_H, and P_S in the upper part of Figure 6 really are performance indices based on the set of observed experimental data. If in addition the jpdf of all the processes from the load data to the resistance quantities are known, then the risk is a quantity which has one value only for each set of design parameters U. Under this condition, the design stages as outlined in the top part of the scheme in Figure 6 yield RE as a figure of merit.

For an experimental situation yielding a set of time functions the general input vector X consists of the continuous functions $x_i(t)$, describing for example stage or rainfall. It is usually a time series of a continuous natural process which has been observed. From these inputs a hydrologic primary variable, for example a discharge Q, is calculated by means of a model. The model can be as simple as a measured stage-discharge relation, or as complicated as a simulation model of runoff in which the catchment model is subjected to time functions which have been generated artificially by means of input functions $x_i(t)$. The hydrologic reliability

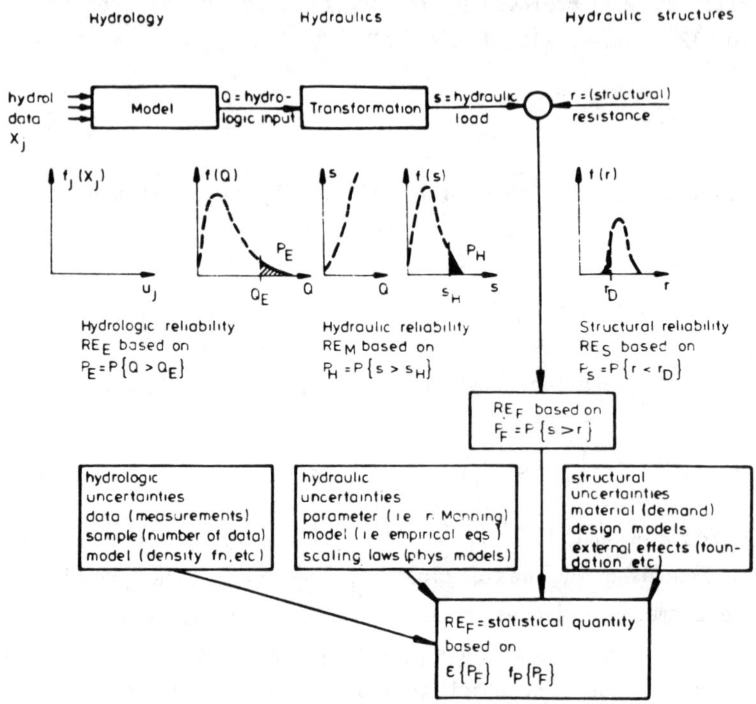

Figure 6: Schematic representation of the reliability concept and of the procedure of calculating reliability.

is then defined in the traditional manner by comparing the output variables from the hydrologic model, for example the peak discharges of floods Q, with a permissible discharge, for example a design discharge Q_E. In the traditional design procedure based on hydrologic reliability (or hydrologic risk, as $1\text{-}RE_E$ is sometimes called), RE_E is calculated from the exceedance probability P_E of the flood peaks by means of Eq. 13, in which n is now the number of years in the design life, and P_E is assumed to be constant, which would be true if all inputs x(t) as well as the catchment parameters of the model were stationary.

The hydrologic variable Q(t) is usually not the variable which is to be compared with the resistance in determining the probability of failure. It is transformed through hydraulic calculations into a design variable. A typical design variable to be determined through a hydraulic transformation is the stage in a canal which has to be dimensioned, such as:

the stage in an open channel (Duckstein and Bogardi, 1981)

the stage at an upstream point in a tidal river under the influence of a tidal surge at the mouth of a river (Plate and Ihringer, 1984)

the stage at a downstream point in an open channel when the stage at the channel inlet is given in terms of a flood wave (Kundzewicz and Plate, 1986

the stages in a system of sewers at points within the system when the statistics of the inputs due to rainfall are given from hydrological calculations.

The hydraulic transformation rule usually is determined either analytically, for example from mathematical models of the flow in open channels, or experimentally, or by means of a physical model of a river in a hydraulic laboratory. The result is a hydraulic load, i.e. a quantity that can be compared directly with the critical design variable. For the example cited, it is clear that the discharge is not the quantity that is comparable to the channel dimensions but it is the stage $s = h$ which must be compared with the design stage $s_H = h_{Design}$ where h_{Design} is that part of the levee height (for example) which is allotted to the canal discharge. The designer usually adds to h_{Design} a freeboard Δh whose first function is to provide a safety margin for other hydraulic processes, such as the effects of wind on the surface causing waves and wave run-up. Its second function is to allow for settling of the levee, or for the effect of parameter uncertainty. Parameter uncertainties may be due to changes in the roughness of the canal caused by irregular maintenance or weed growth as well as by erosion and sedimentation. If the freeboard were known exactly, then the hydraulic reliability would be defined by Eq. 14 where P_H is the probability that the design parameter h_D will be exceeded by an annual extreme load event.

In general, the hydraulic reliability is not equal to the hydrological reliability, because there usually is no equivalence between Q_E and s_H. To illustrate this point one must remember that Q_E usually is the peak flow at a gage, whereas s_H is the hydraulic design dimension at the construction site. A transformation of Q_E may therefore lead to a value of s which is lower than the value of s_H obtained from a discharge $Q < Q_E$, or in other words, to one value of Q_E there may correspond a probability distribution of s values.

Also, the hydraulic reliability is not equal to the engineering reliability, which is based on the probability of failure P_F, because it is not necessary that exceedance of s_H must lead to failure: it is possible that, under rare adverse conditions, failure occurs for a lower value of s than s_H. What this means is that r also obeys some probability distribution. It is therefore possible to define a structural reliability based on the probability P_S that the actual resistance is lower than the design resistance.

The resistance, as defined for the levee case, is indeed a true random variable because it is not possible to predict the exact properties of the levee at any one time. Its effective height varies unpredictably due to natural influences - such as settlement, animal actions, tree or other roots, or due to human factors - such as method of construction or maintenance problems. The situation is a little different, however, if we consider a material with well defined properties (such as steel). It is a fact that no two steels are exactly alike, and within narrow limits even steels produced from the same batch of ore are different. Consequently, when we construct many structures for the same design, the variability of the resistance leads to an expected value of $\varepsilon\{P_F\}$ which yields an average failure rate. However, for each individual structure, its probability of failure P_F is determined by the value of r for the steel used for that structure. That is, once we have chosen the steel, the variability of the actual value of P_F is reduced and becomes a specific property of the particular structure.

It is important to understand that there is a difference between the reliability based on a resistance, which is a random variable of unknown variability such as the levee height, and this type of variability. In the case of steel we may reduce the variability by determining, for our individual structure, the material resistances from tests on a piece of steel taken from the same batch as the steel used in the construction, whereas for the levee, we cannot determine in advance what the resistance may be.

Finally it should be noted that the uncertainty of the resistance can be controlled to some extent through the construction method, or by means of maintenance actions. Very careful dam construction using the principles of soil mechanics will reduce the variability of the performance of the levee under the effect of excessive stages, and a well maintained levee will prevent the rising of the hazard curve (see Figure 5). The design of the levee then poses the interesting problem of deciding to trade construction accuracy and maintenance cost against dam height: an inexpensively constructed dam of larger height might be a better solution to a levee problem than a low dam which is constructed with the most refined (and more expensive) technique and which is well maintained. Clearly both are possible ways of obtaining a structure of desired reliability. In this problem the design reliability is a constraint in an optimization problem, in which the decision on the levee parameters must lead to minimum construction and maintenance cost.

The combination of r and s finally yields the total reliability RE which is based on the probability of failure according to Eq. 8. We shall discuss the procedure for calculating P_F in section 2 of this paper.

1.2.3 The uncertainty of RE. The concept of calculating the probability of failure P_F outlined in 1.2.2 yields, in spite of the variability both of load and resistance, a single value of P_F for the set of experiments. If all the assumptions leading to the calculation of P_F were correct, then this value could be calculated. In actual cases, however this true value cannot be found because of uncertainties, and P_F, and thus also RE, is a random variable with pdf $f_p(P_F)$. This has been discussed in detail elsewhere, for example in Plate (1984b).

The uncertainty of P_F is mainly caused by two effects: the uncertainty in the data and the uncertainty of the models. The data uncertainty is due to measurement errors, or due to the fact that there is usually insufficient time to collect enough data to get stable estimates of parameters of input data pdfs. For example, for the design of a flood levee, we need long series of extreme values (for a discussion of these matters see Chow, 1964 or Yevjevich, 1972), or rainfall data which are converted into runoff data through conceptual models. The model uncertainty is caused by our not knowing, for example, the true extreme value distribution for our situation or by the fact that we approximate the essentially non-linear rainfall-runoff process by a linear model, such as the unit hydrograph.

The effect of uncertainty can be analyzed by different methods. The second moment analysis is extensively used (Tang and Yen, 1972), or Bayesian analysis (Davis et al., 1972) can also be applied. These methods serve to give ranges and perhaps pdfs for P_F, but the decision on what is acceptable cannot be made by any such method. Ultimately, the final decision of an acceptable design must come from engineering judgement of the individual, or from past experience of the profession as codified in standards and regulations. We shall not dwell on the uncertainty, except to point out that there is no principal difference between the different types of uncertainties. Hydraulic uncertainties exist because small effects of third variables, such as non-stationarity of the stage-discharge relationship or of time dependence of roughness are not identified. Data uncertainty is caused by measurement errors and sample size of the hydrologic input, whereas parameter uncertainties are due to inexactly known physical parameters, such as the roughness factor of a canal and others. Reliability analysis permits one to weigh the merit of trade-offs that can be made between physical knowledge and uncertainty, or between gathering more data or living with the data uncertainty. Clearly, deeper physical insight will decrease some of the scatter, but very often a decrease in uncertainty due to a better hydraulic (or hydrologic) model may lead to very little change in P_F. The same can be said for the value of additional data. Although uncertainty is strongly affected by the data base, it may be that the additional value of more data is not worth the effort. Of course, the opposite might also be true: a small improvement in the physical relations obtained perhaps through a model study in a hydraulic laboratory, or improvements by some field measurements of empirical parameters in hydraulic or hydrologic computational models, or improvements by simply gathering a few more years of discharge data may cause a large decrease in uncertainty and may lead to much better design parameters.

Because the analysis of uncertainty deals mostly with unknown probability distributions on all variables, one finds that there exists a certain arbitrariness in allocating the uncertainty to loads or resistances, or to hydrological or hydraulic models. This is exemplified in the freeboard for the flood levee, which includes allowances for many uncertainties. It is therefore suggested to proceed for the calculation of P_F as follows: all uncertainties which can be associated with non-hydraulic features (such as, for a levee, settlement, permeability, erodibility) are allocated to the resistance uncertainty, whereas all phenomena which influence hydraulic phenomena (such as, for a levee, roughness changes, waves and wind effects) are parts of the hydraulic uncertainty. In effect, this definition separates resistance and load according to the types of experts--again for a levee, the soil mechanics expert is responsible for the resistances and the hydraulic engineer for the loads.

1.3 Design Based on Loss Minimization

Related to the design according to a figure of merit RE is a design which is based on the selection of the set $U_L{}^*$ that minimizes a loss function L (ΔT). The function L(ΔT) quantifies the losses which are to be expected from selecting a set of parameters U. It is a figure of merit defined as the expected value of the consequences which occur if $s > r$. These consequences are usually costs, but other weighting factors are also possible if they can be quantified appropriately. A more general example of a loss function consists of the sum of the costs of the structures plus the costs which result from its operation and maintenance plus the expected cost should the structure fail. For such a definition, the consequences can be expressed by a continuous function K(s,r). Therefore, the consequence function is also defined in the (s,r) plane, and the risk is a figure of merit defined by means of an integral over the whole of the (s,r) plane:

$$L(\Delta T) = \int_0^\infty \int_0^\infty K(s,r) f_{sr}(s,r)\,ds\,dr \tag{16}$$

Frequently, the consequences for each load s take on one functional form $K_1(s)$ for $s < r$, and another form $K_2(s)$ for $s \geq r$, so that

$$L(\Delta T) = \int_0^\infty K_2(s)\{\int_0^s f_{sr}(s,r)dr\}ds + \int_0^\infty \{\int_0^r K_1(s) f_{sr}(s,r)ds\}dr \tag{17}$$

To complete the design according to a loss function it is necessary to formulate an optimization problem. In general, L(ΔT) varies with the choice of the design vector U, and the optimum choice of parameters U_L* is that set of parameters which minimizes L(ΔT). The design according to this quantity is:

$$\begin{aligned}&\text{Find } U = U_C^* \\ &\text{such that:} \\ &L^*(\Delta T) = \min_u \{L(\Delta T)\} = \min E\{K(u^*)\}\end{aligned} \tag{18}$$

The design based on this quantity is not always more desirable than the design according to Eq. 11, for two important reasons. The first lies in the fact that it mixes the concern for the safety of structures, and economic aspects. Safety and functionality are the prime responsibility of the designer, and are measured by his professional value system, which is part of the value system of the society in which he works. Economic aspects on the other hand belong to a different value system and are of concern to the designer only in the sense of the design principle expressed by Eq. 11. Whenever two different value systems must be combined, the loss-based design is less useful than the reliability-based design.

A second reason is that design by a loss function may mix responsibilities in a way which is not acceptable to the people involved in a decision process leading to a project or a structure. In the loss function one may have to include a number of factors which are outside the jurisdiction of the designer who can manipulate the design variables but cannot always decide what kind of costs resulting from failures should be included into the loss function. It is usually assumed that a loss function can be worked out in a collaboration between design engineer and the customer or "decision maker". But except perhaps for large-scale hydraulic projects, the "decision maker" does usually not want to be involved in the design process.

Therefore, we conclude that the reliability should be specified in consequence and as part of the value system of Society through rules and regulations. This has been the basis for all previous design rules based on a design load of a specified magnitude or of a specified recurrence interval. We advocate replacing these design criteria by a specified reliability, because of its exact correspondence with the probability of failure. By defining a permissible reliability RE_{Design}, the decision maker, or alternatively the writer of rules and standards, specifies the "state of the art" and gives legal protection to a design which, if based on some more or less variable concept of loss function K, might be very difficult to defend in court. Yet a comparison between a loss-based and reliability - based design may be useful, and trade-offs between the two criteria may be examined using a multicriterion analysis.

1.4 Design Based on the Concept of Risk

The special case $K_1 = 0$ inserted into Eq. 17 yields an often used definition for the risk associated with the design. In particular, the term risk is used with a binary function K, with value zero for $s < r$, and value K_{MAX} for $s \geq r$. A typical case is a sudden dam failure, or the breaking of a chain link: it does not matter by how much the load exceeds the critical value $s = r$, the damages which occur are the same. In such a case, $L(\Delta T)$ from Eq. 17 becomes the risk $RI(\Delta T)$, and:

$$RI(\Delta T) = L(\Delta T) = K \cdot P_F(\Delta T) \qquad (19)$$

which can be found by combining Eqs. 9 and 16.

The term risk defined in this way is much narrower in scope than the engineering risk RI defined in the introductory paper. According to the definitions used there, both RE and RI are performance indices PI which contribute to the figure of merit "engineering risk", and the definition of risk is:

$$RI = g(RE, L \ldots) \qquad (20)$$

where g( ) is a suitable function. Note that this definition incorporates both the concept $RI = 1 - RE$ and $RI = L$ as used here as special cases, although the general condition to be used on RI for design purposes is quite different for the two types of performance indices.

2. THE CALCULATION OF P_F

We now turn to the problem of actually calculating P_F. In this paper we shall consider only the case where no correlation exists between s and r, and where the marginal pdfs are known. This we call the Freudenthal approach. It corresponds to the problem of Level III design which is sometimes used in reliability design theory. For level II design based on extimates of the marginal pdfs using second moments, reference is made to Ang and Tang (1984).

2.1 The Approach of Freudenthal

The design based on reliability RE advocated in this paper depends very much on the ability to calculate P_F for the hydraulic design problems at hand, i.e. on the ability to perform the integration of Eq. 9. We have to face two prinicpally different problems. The first one is to specify the joint probability density for s and r, and the second one is to perform the integration required by Eq. 9.

Both problems are greatly simplified if it is possible to state that load and resistance are stochastically independent, so that

$$f_{sr}(r,s) = f_s(s)f_r(r) \tag{21}$$

and with this expression Eq. 8 becomes (Freudenthal, 1961):

$$P_F = \int_0^{\infty} F_r(s)f_s(s)ds \tag{22}$$

This equation permits a number of important conclusions. First we see that if the resistance is an exactly known constant value r_D, then $F_r(s)$ for $s < r_D$ is zero and for $s \geq r_D$ it is one, so that $f_r(r) = \delta(r - r_D)$ where $\delta(\ )$ is the Delta function. With this expression one obtains:

$$P_F = \int_{r_D}^{\infty} f_s(s)ds \tag{23}$$

i.e. P_F reduces to the hydraulic exceedance probability P_H.

A second observation is that the integration of Eq. 22 depends on those values of $F_r(s)$ which usually overlap with the extreme part of the asymptotic "tail" of $f_s(s)$. This again has two consequences: usually only that part of $f_s(s)$ is required which is not well defined by data, because an extreme value is a rare event and only few data points exist to determine empirically $f_s(s)$ for large values of s. A curve fit to the tail of $f_r(s)$ therefore is subject to large model uncertainty. On the other hand, the functional shape of $f_s(s)$ for small values of s, which is reasonably well defined, is really not very important. The engineer who works with reliability has to live with this problem. In particular, if the extreme value distribution is extrapolated beyond the (skimpy) data base to recurrence intervals far outside the range of human memory or experience, such as to 1000 or more years, one obtains estimates which are bound to be extremely unreliable. It is surely unreasonable to expect that a 100-year record - and only very few records of such length do exist - can be extrapolated 10-fold or even 100-fold without substantial errors.

Accuracy is, however, not the main point of design by reliability. In extrapolating the available records the engineer arrives at a design value. If this design value is obtained by a suitable method representing the state of the art, two very essential requirements will have been met:

a) the combined professional judgement of his peers considers this design value to be adequate

b) if a failure of the structure should ever occur because the design value is exceeded, a court of law can be convinced to blame the occurrence on forces beyond the engineer's control.

With these considerations in mind, the best way for obtaining an extreme value distribution is by fitting as many extreme value distributions to the data as are accessible, and by using that extreme value distribution which gives the best fit, and a conservative extrapolation.

2.2 Extreme Value Distributions

In considering extreme value distributions for calculating the reliability of hydraulic structures, it is useful to distinguish between purely empirical functions and those which are based on the theory of extreme values. The former are selected on the basis that predicted and observed distributions have shown good agreement. A recent survey of the applications of extreme value analysis may be found in Oliveira, 1983.

2.2.1 Empirical Distributions. Empirical extreme value distributions are almost always fitted to extreme values of fixed time intervals, for example to annual maxima or minima. We distinguish essentially two classes - those distributions which are based on the normal distribution

$$f(x) = \frac{1}{\sqrt{2\pi}\sigma_y} e^{-\frac{(y-y)^2}{2\sigma^2_y}} \cdot \frac{dy}{dx} \tag{24}$$

and those which are based on the gamma-distribution

$$f(x) = \frac{|\lambda|}{\Gamma(r)} |\lambda(y-y_0)|^{r-1} e^{-\lambda(y-y_0)} \cdot \frac{dy}{dx} \tag{25}$$

where $y = g(x)$ is a transformation of the observed or predicted variable x that is exceeded within each year with a probability $1 - F_x(x)$, and:

$$F_y(y) = F(g(x)) = \int_{-\infty}^{g(x)} f(x)dx \tag{26}$$

In Eqs. 24 and 25 y and σ_y are the mean value and standard deviation, respectively, of the transformed variable y, and λ, r, and y_0 are the three parameters of the three-parameter gamma distribution which depend on y and σ_y through the relations:

$$\lambda = \frac{\bar{y} - y_0}{\sigma^2_y}$$

$$r = \lambda(\bar{y} - y_0) \tag{27}$$

$$\gamma = \frac{m_{y3}}{\sigma^2_y} = \frac{4}{r}$$

where γ is the skewness coefficient defined through the third central moment m_{y3} of y, and through the variance $\sigma_y{}^2$.

Depending on the functional form of the transformation $y = g(x)$, one finds an infinite variety of functions in either class which might be suitable. The most elementary transformation is

$$y = x \quad \text{and} \quad \frac{dy}{dx} = 1 \tag{28}$$

in which case Eq. 24 becomes the normal distribution and Eq. 25 becomes the so-called Pearson III distribution. The normal distribution is quite unsuitable for extreme value analysis because it admits negative values of x.

A frequently used transformation is the logarithmic one:

$$y = \ln x \quad \text{and} \quad \frac{dy}{dx} = \frac{1}{x} \tag{29}$$

where ln() denotes the natural logarithm. It changes the normal distribution into the log normal (LN2) and the Pearson III (P3) into the log Pearson III (LP3) function, both of which are often used in extreme value analysis. The reason for this is that the original P3 function has the shape which is intuitively expected of an extreme value distribution, and that the three parameters λ, r, and y_0 allow a great flexibility for fitting experimental data. However, it is often not realized that in the course of the transformation to logarithms, the function LP3 changes its shape for almost all combinations of parameters. This becomes evident if the LP3 distribution is inspected closely, as was done by Bobée (1975); these results show that the shape of the LP3 function agrees with that of the P3 function only in the region $1 < r < 2$ for positive values of λ!

A transformation of the P3 function which allows keeping the three parameters while at the same time preserving more faithfully the shape of the function is given by:

$$y = x^s \; ; \quad y_0 = 0 \; ; \quad \frac{dy}{dx} = s x^{s-1} \tag{30}$$

which leads to the Kritsky-Menkel distribution:

$$f(x) = \frac{s \cdot \lambda^r}{\Gamma(r)} x^{sr-1} e^{-\lambda x^s} \tag{31}$$

for which tables are available (Kartvelishvili, 1969) and which is widely used in the USSR.

In contrast to LP3, the log normal distribution LN2 has, for all sets of parameters the required shape. However, because of its inflexibility with only two parameters y and σ_y, it should be extended to the three parameter distribution LN3 by means of the parameter x_o:

$$y = \ln(x - x_o) , \qquad \frac{dy}{dx} = \frac{1}{x - x_o} \tag{32}$$

The transformations 28, 29, 30 and 32 are only four out of an infinite number of possibilities, and many more transformations have been recommended; see for example Kartvelishvili (1969) or Yevjevich (1972) for more general discussions.

2.2.2 Extreme Value Theory for Annual Extrema. Classical extreme value theory starts from two important premises:

a) the data record can be cut into periods of equal length (for example, of one year), with the extreme values in each of these periods being uncorrelated,

b) the distribution of the extrema within each record period is known to be exponential in the transformed variable $y = g(x)$.

The first premise is generally accepted as valid when the average time between extrema is long, for example, one year, because it is unlikely that the extreme value of one year influences the extreme value of the next, unless of course they are part of the same meteorological event which takes place just around the time change from one period to the next. The second premise can often be verified experimentally to a good degree of approximation from flood records of many years duration (for example, Konecny and Nachtnebel, 1985).

If these two premises are met, then the probability of exceeding an event x in a large number of equal time intervals must obey the stability postulate of Frechet (Jenkinson, 1955) and a general solution is obtained (Jenkinson, 1955, 1977):

$$F(x) = 1 - \exp(-e^{-y}) \tag{33}$$

where F(x) is the probability function of the annual maximum value, and y is the reduced variate depending on x through the Jenkinson relation:

$$x = x_0 + \alpha\left(\frac{1 - e^{-ky}}{k}\right) \tag{34}$$

which include the Fisher - Tippett types of extreme value distribution as special cases:

$$\begin{array}{ll} \text{Type 1 (Gumbel) :} & k = 0 \\ \text{Type 2 (Frechet)} & k < 0 \\ \text{Type 3 (Weibull)} & k > 0 \end{array} \tag{35}$$

It is to be noted that the general function has three parameters to be fitted, so that it has the same flexibility as some of the empirical functions. Often one finds the Gumbel form with $\alpha = 1$ without shift parameter x_0, and then Eq. 34 reduces to the simple relation $y = x$ which does not have a disposable free parameter. As a consequence, the Gumbel distribution has little flexibility.

A difficulty in the use of the general extreme value formula (Eqs. 33 and 34) lies in the estimation procedure for the parameters α, x_0 and k which are not simply related to the moments of the distribution. Jenkinson (1977) gives methods, based on sextiles of the distribution, by means of which a solution can be found. For details, the original references should be consulted.

2.2.3 Extrema of continuous time functions. An entirely different approach based on the theory of continuous random functions yields distributions of the extrema which differ from the double exponential used by Gumbel. A theoretical solution can be found for the extrema of continuous random functions which depends on two properties:

a) on the distribution of the individual values of the function

b) on the degree of correlation between neighboring maxima.

Of considerable practical importance is the case of a time series with a normally distributed population of individual values, which are obtained for example by digitizing the recorded time function, and with a narrow band spectrum. For this type of time function, Longuet - Higgins (1952) has shown that the maxima are distributed according to a Rayleigh distribution:

$$f(x) = \frac{x}{\sigma_x^2} \exp\left(-\frac{x^2}{2\sigma_x^2}\right) \qquad x \geq 0 \tag{36}$$

with the cumulative distribution:

$$F(x) = \int_0^\infty f(x')dx' = 1 - \exp\left(-\frac{x^2}{2\sigma_x^2}\right) \tag{37}$$

The Rayleigh distribution gives the value of the maximum per wave. However the distribution of the maximum in a record of N waves is given by

$$F_N(x) = F^N(x) = 1 - e^{-N\, e^{-\frac{x^2}{2\sigma^2}}} \tag{37a}$$

which belongs again to normal extreme value theory and is empirically proven to be correct.

Physical processes which obey this distribution quite well are turbulence and water surface waves.

2.2.4 Intermittent processes. The analysis of continuous functions is based on records consisting of relatively short segments for which the assumption of stationarity applies. For such records, for example wind velocity or height of water waves, the fit by means of the Rayleigh distribution is usually quite good. However, a generalization or extrapolation to much longer records is in general not possible. A record with large waves or extreme gusts must be taken from a much longer record whose characteristics change drastically over time: periods without wind alternate with periods in which winds of moderate to storm force occur. An extreme case of such processes are floods, which usually are caused by rainfalls that are too short to be analysed as pieces of continuous stationary time functions, in hydrologic time series. The on-off rainfall phenomenon is a typical example of a true intermittent process, whereas turbulence and waves are only approximately intermittent. A consequence of intermittency is that a complete description of the occurrence of extreme values requires not only a probability distribution for the extrema of the stationary pieces of the record, but also a probability distribution for the stationary sections of the record.

In its simplest form this analysis proceeds by defining extrema as maxima x_{max} of those parts of a continuous function x which exceed a preset level x_s and by determining the probability distribution of the interval between consecutive events x_{max}. If the record is divided into sections of constant duration, for example of one day, and the probability of $x_{max} > x_s$ is the same for each day, then it follows that the time t during which the event $x_{max} > x_s$ occurs exactly n times follows a Poisson distribution:

$$f(t,n,x_s) = \frac{(\mu t)^n e^{-\mu t}}{n!} \quad (38)$$

where μ is the mean rate of occurrences of $x_{max} > x_s$.

More developed models use Poisson distributions to represent duration between stationary records, and this has been used for constructing a model of the sea surface valid for long records by Borgman (1963) and by Schueller and Choi (1977). It has also been used to analyse partial duration series in hydrology, for example, by Todorovic and his collaborators (Todorovic and Zelenhasic, 1970) in models in which it was combined with the exponential type extreme value distribution, and by Gupta et al. (1976) who on the same basis as Todorovic derived the joint probability distribution both for x_{max} and t. Konecny and Nachtnebel (1985) have investigated the interesting case of a seasonally variable extreme value distribution for a one-year period in which the exponential distribution, with λ represented by a Fourier series, is used as a basic function for which exceedance probabilities are calculated.

2.3 The Calculation of P_F

The last step in the calculation of P_F, neglecting uncertainty of the data, is the determination of the probability density for the resistance and the integration of Eq. 22. We shall illustrate this by means of two examples.

2.3.1 Example: P_F for a flood levee. Probability based designs of flood levees have been used in hydraulic engineering for many years. The levee height was determined by means of a design discharge Q_D which had a preassigned recurrence interval of T_n years, such as the T_n = 100 year flood. A definition sketch for a typical example is shown in Figure 7. The method of analysis by means of probabilistic concepts has proceeded as is shown in Figure 8a. The design discharge Q_D is determined, according to a predetermined exceedance probability P_E per year, by means of an extreme value analysis.

P_E corresponds to the recurrence interval $T_n = 1/P_E \; \Delta T$ in years. The corresponding value of Q_D is the design discharge which determines the dimensions of a flow channel. The discharge Q_D is transformed deterministically to a channel dimension, which is expressed by the design depth h_D. This transformation usually involves Manning's equation. The extreme value distribution of Q_D is thus transformed into an extreme value distribution $F_s(h)$ of the depth h. The levee height h is then determined to be

$$h_D < h \tag{39}$$

the difference $\Delta h = h - h_D$ being the empirical freeboard.

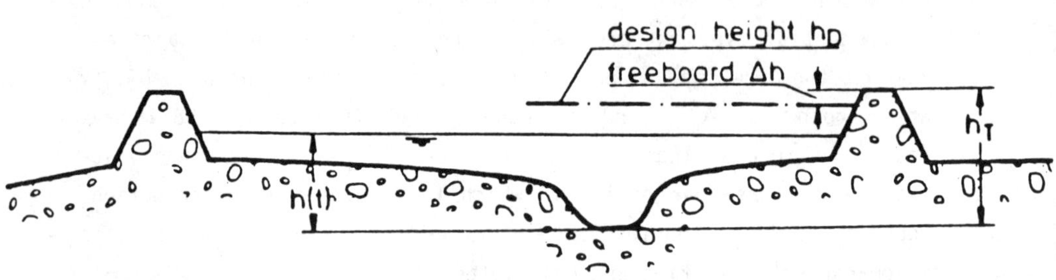

Figure. 7: Definition sketch for levee problem.

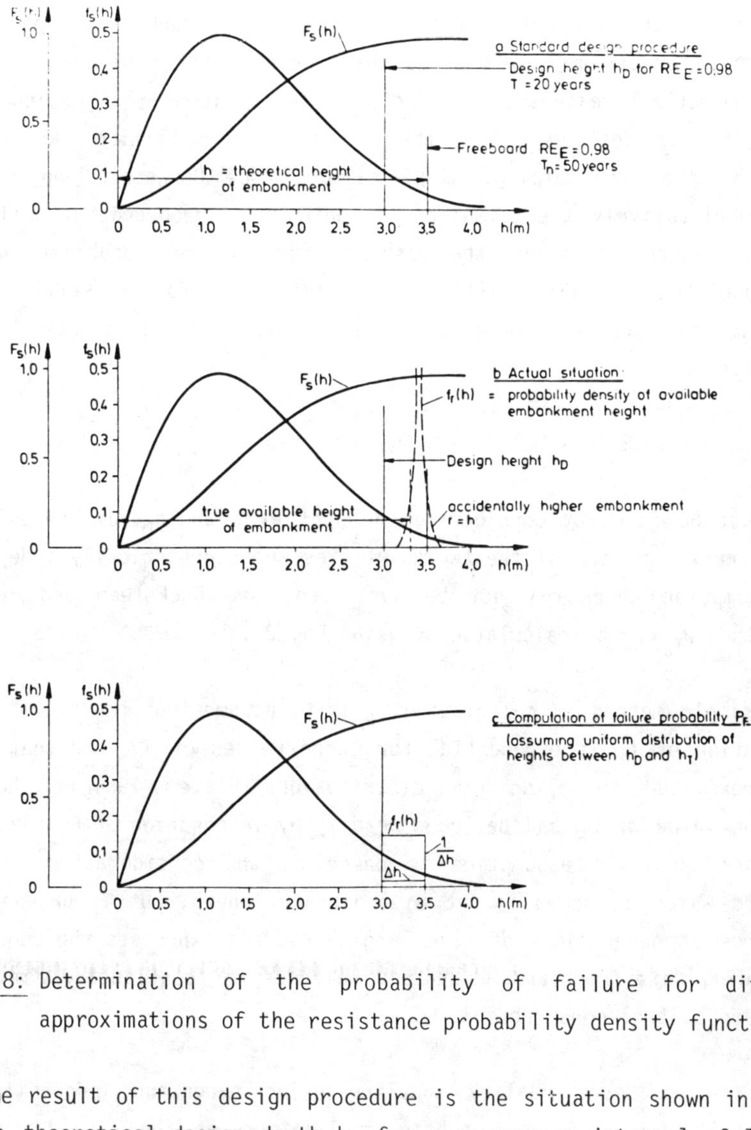

Figure 8: Determination of the probability of failure for different approximations of the resistance probability density function.

The result of this design procedure is the situation shown in Figure 8a. The theoretical design depth h_D for a recurrence interval of T_D = 20 years is 3.0 m. By adding a freeboard of 50 cm, and if this freeboard is not used for any other purpose, we add to the safety by going from a T_D = 20 years to T_D = 50 years.

Clearly, the freeboard will not be available under all conditions. Construction errors may even cause the freeboard to be higher. There are also a number of reasons why the dike could be lower than designed, for example due to settlement, or its height might be limited due to soil mechanical problems (piping, slope failure: see Duckstein and Bogardi, 1981). Qualitatively the situation including the freeboard variability is shown in Figure 8b, where the dashed curve is the probability density function of the available height h. The condition of failure still is $h<h_D$, but now this is a probabilistic event with an associated failure probability

$$P_F = P\{h < h_D\} \tag{40}$$

which must be found by considering the jpdf of h and h_D. If the pdfs of h and h_D are known, and if the two quantities are stochastically independent (an assumption which may not be justified, see Duckstein and Bogardi, 1981), then P_F can be calculated by using Eq. 22.

A simple method of calculating P_F is to assume that the "tail" of the distribution of f_s is exponential for large values of h_D, and that f_r can be approximated by a uniform distribution (Plate, 1984a). Then the resulting value of P_F can be found readily by integration of Eq. 22, based on Figure 8c. This assumption is based on the consideration that the freeboard which is added is not an arbitrary length, but is based on the experience of generations of river engineers, and expresses the conclusion that adding this freeboard adequately balances the conflicting demands of safety and cost of construction.

2.3.2 Example: River quality modeling using exceedance probability as design criterion. A more subtle application of the exceedance of a standard is found for the case of the concentration of a pollutant in a river. Let the pollutant input be a random emission M(t), to be expressed in mass flow per second, for example a continuous emission of pollutant from a chemical plant into a river. Let Q(t) be the discharge of the river, which is also a continuous random time function. It is assumed that M(t) and Q(t) will be uncorrelated. Let us furthermore assume that the fauna and flora of the river are not affected by the pollutant as long as the concentration C(t) of pollutant in the river remains below a critical level C_o. The supervising environmental protection agency decides that emission of the pollutant is permissible if the concentration C stays 99.9% of the time below a level $C_{oo} < C_o$. This leaves the proof of meeting this requirement to the polluter, who has to show that

$$P_F = P \{C(t) > C_{oo}\} \leq 0,001 \tag{41}$$

The condition $C(t) > C_{oo}$ can be formulated in terms of the random time functions M(t) and Q(t)

$$\frac{M(t)}{Q(t)} \geq C_{oo} \tag{42}$$

But the solution of this problem is not easy to obtain, as it involves the division of M(t) by Q(t) at all times. We can determine P_F in a conceptually simpler way by multiplying both sides of the inequality by Q(t), and thus:

$$M(t) \geq C_{oo} \; Q(t) \tag{43}$$

Therefore, the condition $C > C_{oo}$ is defined as a line, with slope C_{oo}, in the M-Q plane and intersects the joint probability density distribution f(M,Q), as is shown in Figure 9.

It is evident that the probability of failure P_F is equal to the volume formed by the jpdf above the crosshatched section of the M-Q plane in Figure 9, and in complete analogy to the case which was considered earlier we obtain:

$$P_F(C_{oo}) = \int_0^\infty F_Q\left(\frac{M}{C_{oo}}\right) f_M(M)\,dM \tag{44}$$

The application of these procedures to real data is left to the reader.

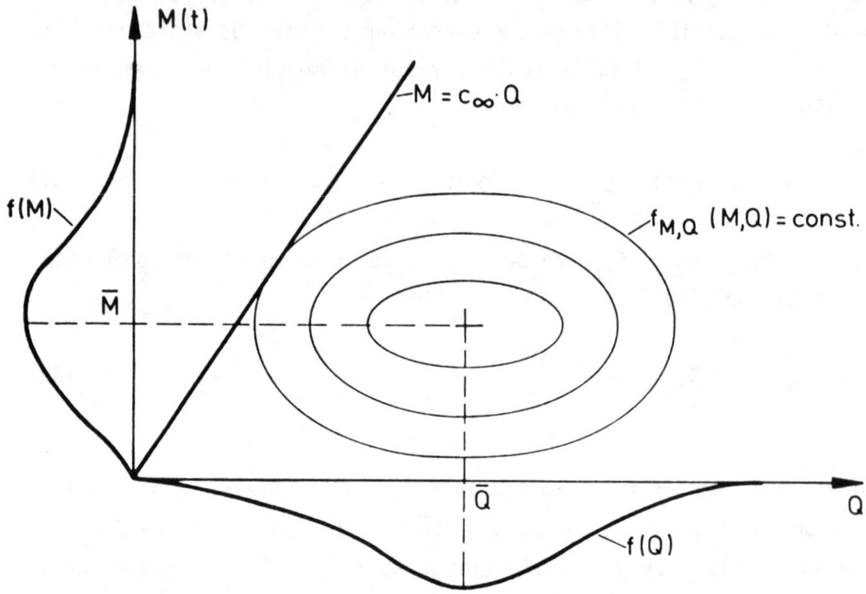

Figure 9: Failure probability for a pollution loaded river.

3. CONCLUSIONS

In this paper we have shown that the concept of reliability offers a systematic framework for the analysis of structural safety. Unlike the presently used hydrologic reliability (more usually called "hydrologic risk") it is directly related to the failure event. It also allows the separate treatment of reliability and uncertainty in a natural way. Because of these conceptual advantages, reliability is recommended as a design criterium. The paper concentrates on examples in which generalized loads and resistances are uncorrelated. Other papers in this book provide examples of correlated data.

REFERENCES

Ang, A.H.-S. and W.H. Tang (1984) Probability concepts in engineering planning and design, Vol. 2: Decision, Risk and Reliability, J. Wiley, New York.

Bobée, B. (1975:) The log Pearson Type III distribution and its application in hydrology, Water Resources Research, Vol. 11, 681-689.

Borgman, L. (1963) Risk criteria for offshore structures, ASCE, J. Waterways and Harbor Division, Vol. 89.

Chow, V.T. (ed), (1964) Handbook of Applied Hydrology, McGraw Hill, New York.

Davis, D.R., C.C. Kisiel, and L. Duckstein (1972) Bayesian decision theory applied to design in hydrology, Water Resources Research, Vol. 8, pp. 33-41.

Dhillon, B.S. and C. Singh (1981) Engineering Reliability: New Techniques and Applications, J. Wiley, New York.

Duckstein, L. and I. Bogardi, (1981) Application of reliability theory to hydraulic engineering design, Proc. ASCE, Vol. 107, J. Hydr. Div., pp. 799-815.

Freudenthal, A.M. (1961) Safety, reliability, and structural design, Proc. ASCE, Vol. 87, J. Struct. Div., pp. 1-16.

Gupta, V.K., L. Duckstein and R.W. Peebles (1976) On the joint distribution of the largest flood and its time of occurrence, Water Resources Research, Vol. 12, 295-304.

Jenkinson, A.F. (1955) The frequency distribution of the annual maximum (or minimum) values of meteorological elements, Quarterly Journal Roy. Met. Soc., Vol. 81, 158-171.

Jenkinson, A.F. (1977) The analysis of meteorological and other geophysical extremes, Meteorological Office, UK, Technical Note No. 58.

Johnson, F.A. and P. Illes (1976) A classification of dam failures, Water Power & Dam Construction, Dec. 1976

Kartvelishvili, N.A. (1969) Theory of stochastic processes in hydrology and river runoff regulation, (translated from Russian by Israel Program for Scientific Translations) Jerusalem.

Konecny, F. and H.P. Nachtnebel (1985) Extreme value processes and the evaluation of risk in flood analysis, Apl. Math. Modelling Vol. 9, 11-15.

Kundzewicz, Z. and E.J. Plate (1986) Selection of an optimum flood routing model by considering its effect on the extreme value of flood waves, (in press).

Longuet-Higgins, M.S. (1952) On the statistical distribution of the heights of sea waves, Journal of Marine Research, Vol. 11, 245-266.

Oliveira, J.T. (editor) (1983) Statistical extremes and applications, NATO ASI Series C; Vol. 131, D. Reidel Dordrecht.

Plate, E.J. (1984a) Reliability analysis of dam safety in W.H.C. Maxwell and L.R. Beard (editors) "Frontiers in Hydrology", Water Resources Publications. Littleton, Colo. USA, pp. 288-304.

Plate, E.J. (1984b) Reliability in hydraulic design, Proceedings, IAHR Symposium on Stochastic Hydraulics, Urbana, Ill.

Plate, E.J. and J. Ihringer (1984) Failure probability of flood levees on a tidal river, Proceedings, IAHR Symposium on Stochastic Hydraulics, Urbana, Ill.

Schueller, G.I. and H.S. Choi (1977) Offshore platform risk based on reliability function model, Proceedings, Offshore Technology Conference, Houston, 473-479.

Tang, W.H. and B.C. Yen (1972) Hydrologic and hydraulic design under uncertainties, International Symposium on Uncertainties in Hydrologic and Water Resources Systems, Proceedings Vol. 2, Tucson, Ariz. USA, pp. 868-882.

Todorovic, P. and E. Zelenhasic (1970) A stochastic model for flood analysis, Water Resources Research, Vol. 6, 1641-1658.

Yevjevich, V. (1972) Probability and Statistics in Hydrology, Water Resources Publication, Littleton, Colo. USA.

ENGINEERING RISK IN REGIONAL DROUGHT STUDIES

Francisco Nunes Correia
Maria Alzira Santos
Rui Raposo Rodrigues

Laboratório Nacional de Engenharia Civil, Av. do Brasil, 101,
1799 Lisboa, Portugal

ABSTRACT

Drought problems must be analyzed and solved at a regional level. Drought studies must focus not only on the hydrological characterization of the phenomena but also formulate the results in terms of engineering risk. In this paper concepts and methods to characterize droughts as regional processes are presented. The concepts of risk, reliability, resiliency and vulnerability are applied to characterize regional droughts and provide an engineering framework to deal with drought problems. A Portuguese river basin (rio Ave) is used to illustrate the concepts and techniques presented in the paper. A multivariate stochastic model is used to generate a synthetic series of monthly precipitation at six distinct regions defined in the river basin. Thirty-three years of monthly data are used to calibrate the parameters of this model. The generated series is used to perform the study of regional droughts from an engineering risk point of view.

1. INTRODUCTION

Droughts are frequently equated with shortages of water and considered as extreme events that can be studied in a way somewhat similar to floods. There are two misleading ideas in this common attitude. First, droughts are essentially long-lasting phenomena instead of short duration events, as is the case with floods. Second, droughts are essentially regional rather than local phenomena.

It is possible to conduct a study on the statistical distribution of annual minimum discharges based on extreme value theory. However, this approach has little to do with drought characterization.

The complexity of drought phenomena, and the "long-lasting" and "regional" nature of droughts may account for the different definitions and approaches in the literature and explain why no common terminology has been established so far.

Recently Santos (1983) and Santos et al (1984) presented a methodology for drought characterization thatgrasps the long-lasting and regional nature of the drought phenomenon and models this phenomenon as a stochastic process.

According to these authors there is a need to distinguish between point drought, if the drought is referred to a point in space (actually a very small region), and regional drought, if the drought is referred to a widespread region in which some portions may eventually be under drought conditions while some other portions may be not under these conditions. The authors propose a precise definition of regional drought and suggest some variables to characterize it, namely drought duration, deficit area, total areal deficit and drought intensity.

In these studies, precipitation is used as the drought determining factor since it is usually considered to be the single hydrometeorological variable having the most relevance to drought studies, it is of good quality and is very suitable for a regional approach, and abundant data are available.

The work that has been done so far is oriented more to a climatic characterization than to an engineering perspective in which "risk of failure," i.e. risk of drought, and "consequences of failure," i.e. drought consequences, become important concepts. Nevertheless, the formulation that was attempted by the authors, based on the stochastic characterization of regional droughts, can be easily reformulated in the framework of systems analysis. Concepts of risk and reliability, mission reliability, resilience and vulnerabilty, presented by Hashimoto et al (1982), Duckstein and Plate (1983) and Duckstein et al (1984), can be used to characterize droughts from an engineering risk standpoint.

The basic goal of this paper is to present a methodology to characterize droughts in the framework of an engineering risk analysis and to illustrate the potential of this methodology with a case study.

The purposes of the following sections are:

(i) to briefly review the concepts and definitions used for the stochastic characterization of regional droughts;

(ii) to formulate these concepts and definitions into an engineering risk framework, considering explicitly the concepts of risk, resilience and vulnerability as applied to droughts;

(iii) to illustrate the proposed methodology and the engineering risk formulation with the presentation of a detailed case study.

2. DROUGHT CONCEPTS AND DEFINITIONS

2.1 Literature Review

A number of indices designed to characterize historical droughts have been used from early times. A quite complete chronological list of drought indices can be found in WMO (1975). Santos (1981) presents a very detailed critical review of the literature on stochastic characterization of droughts.

The simpler drought indices make use only of the mean annual precipitation such as Lang's (1956) rain factor index and Martonne's

(1926) index of aridity. Others, in addition to the annual or monthly precipitation, make use of other climatic variables such as monthly temperature, annual evapotranspiration, annual potential evapotranspiration, and soil moisture condition.

Among this latter kind of index, the most used is the Palmer (1965) index which is a function of accumulated differences that are weighted between actual precipitation and precipitation requirements in terms of evapotranspiration.

Two indices of a different type, namely, the variability and annual precipitation deciles have also been frequently used in the drought literature.

The variability of annual precipitation, which is defined as the ratio of its standard deviation to its mean value, has been frequently taken as a good drought index. For instance, Chow (1964) shows that in the western United States, where frequent droughts occur, the coefficient of variation of annual rainfall exceeds 0.35, while in the eastern part of the country, where droughts are less frequent, this coefficient ranges from 0.15 to 02.5

The variability is also used by Sancho and Arias (1978), who state that in Mexico regions with high variability in precipitation are generally classified as arid or semi-arid.

However, considering the definition of variability, one cannot say, out of hand, that low values of variability are associated with humid regions and that high values of variability are associated with arid regions. Santos et al (1984) prove that in Portugal there is no direct relationship between high variability and dry regions.

Another drought index sometimes used as an indicator of drought occurrences is the k-th decile of annual precipitation. This is the value which is not exceeded by the lowest 10k percent of the annual precipitation observed in a sample of n years. The values between the (k - 1)th and k-th ranges constitute the k-th decile range.

Gibbs and Maher (1967) used annual values of precipitation, observed in a set of stations from 1885 to 1965, to characterize droughts in Australia. They classified the observed annual precipitation according to their sample frequency as very much below average (decile 1), much below average (decile 2), below average (decile 3), average (decile 4 to 7), above average (decile 8), much above average (decile 9) and very much above average (decile 10).

All the indices previously mentioned hardly lead to a quantitative analysis of droughts because they do not describe them in terms of duration, severity or intensity. Moreover, they do not address explicitly the regional nature of droughts. A more adequate methodology for drought characterization is required. In what follows, the concepts presented by Santos (1983) and Santos et al (1984) are reviewed.

2.2 Point Drought

The first concern is to define the phenomenon considered to determine the occurrence of a drought. This is commonly a climatic element and it is called drought determinant from here on. In this study, monthly precipitation is considered as the drought determinant and represented by $X(t)$. However, the methodology that follows can be applied to anything that is relevant to the drought problem at hand. Monthly or annual river discharge, evolution of soil moisture conditions in a cultivated area, and phreatic levels of an aquifer are good examples of phenomena that can be used as drought determinants.

Once the determinant is chosen, one can say that a point drought occurs whenever the value of the determinant is less than a function $f(t)$, which may or may not be a constant. This function is said to be the critical level or truncation level, and the nonnegative function

$$D(t) = \begin{cases} f(t) - X(t) & X(t) \leq f(t) \\ 0 & X(t) > f(t) \end{cases}$$

is called the deficit at time t. It measures the amount of water below a given critical level.

The critical level may be, for instance, the quantile of order p of the cumulative distribution function of the determinant at time t, $F_t(p)$. In this case the critical level is given by $f(5) = F_t^{-1}(p)$.

Point drought can be defined as the time interval during which the determinant is below the critical level.

The random variables used to characterize a drought are said to be characteristic variables of the drought or simply drought characteristics.

The most important characteristic variables associated with a point drought are the following:

- duration, L, corresponding to the length of the time interval $[t_o, t_e]$ in which the drought occurs, i.e.

$$L = t_e - t_o + 1$$

where t_o and t_e are, respectively, the onset and termination of the drought, i.e. the first and last time unit with $X(t) \leq f(t)$

- total deficit, D, that is, the total water deficit during the drought,

$$D = \sum_{t=t_o}^{t_e} [f(t) - X(t)]$$

- intensity, I, which is the ratio of the total deficit D to the drought duration L,

$$I = D/L$$

The intensity of a drought is quite an important variable since it measures the mean total deficit of a drought and its duration, and consequently allows the comparison of droughts with different total deficits and durations.

2.3 Regional Drought

From an engineering viewpoint, regional drought are more important than point droughts because they are more directly related to the problems of water shortages and their impacts. Nevertheless, their study has been relatively neglected in the literature when compared with the attention given to point droughts. This is why a very special effort has been made in this paper to characterize regional droughts.

In regional drought characterization, in addition to drought characteristics, some features of the affected region, namely its size and shape, must be considered.

To rigorously define regional drought, some necessary concepts have to be introduced and others have to be extended. This is the case with the determinant and the critical level, which must be taken as vectors. These vector coordinates are denoted $X_k(t)$ and $f_k(t)$, respectively, for $k = 1, \ldots, M$ with M the total number of stations in the region. $X_k(t)$ measures the precipitation at time t in station k, and $f_k(t)$ measures the critical level at the same time and same station.

The following definitions may be introduced:

- instanteous deficit area, $A_s(t)$, is the drought-affected area, within the region R, at time t. It is given by

$$A_s(t) = \sum_{k=1}^{M} a_k \; I_{K(t)} \; (k)$$

where $K(t) - \{k | X_k(t) \leq f(t)\}$ and $I_{K(t)}(k)$ is the indicator function; a_k is the effective area assumed for the station k. Generally, a_k is taken as the percentage of effective area to the total region area, and $A_s(t)$ takes values from 0 to 100, which gives the percentage of deficit area at time t.

- instantaneous areal deficit, D(t), is the weighted sum of deficits occurring in the region R. It is given by

$$D(t) = \sum_{k=1}^{M} a_k \, [f_k(t) - X_k(t)] \, I_{K(t)}(k)$$

Regional drought can be defined as the time interval during which the instantaneous deficit area equals or exceeds a given critical area, a_c.

The critical area a_c plays a role similar to the critical level f(t) in the definition of point droughts. It must be previously established and based on some physical or other meaningful characteristics of the region under study.

The following areal drought characteristics may now be introduced:

- duration, L, is the length of time interval $[t_o, t_e]$:

$$L = t_e - t_o + 1$$

where t_o is an instant such that the deficit area $A_s(t_o) \geq a_c$ but $A_s(t_o - 1) < a_c$ and t_e is the first instant after t_o such that $A(t_e) > a_c$ but $A_s(t_e + 1) < a_c$

- deficit area, A, is the mean deficit area during drought, that is, during the time interval $[t_o, t_e]$, i.e.

$$A = \frac{1}{L} \sum_{t=t_o}^{t_e} A_s(t)$$

Note that $A_s(t) \geq a_c$ for all $t \in [t_o, t_e]$.

- total areal deficit, D, is the sum of the instantaneous areal deficit D(t) along the drought duration

$$D = \sum_{t=t_o}^{t_e} D(t)$$

- intensity, I, is the ratio

$$I = D/L$$

For a complete understanding of the drought phenomena, these drought characteristics, which are random variables, need to be studied theoretically. For instance, one should know their probability distributions and some characteristic parameters in order to be able to find return periods of droughts or probabilities of occurrences. If this theoretical characterization is not feasible, one is limited to the estimation of the drought characteristics from the available data. These data are usually insufficient for a complete characterization of the statistical properties of drought characteristics, principally if large return periods need to be considered. However, if a stochastic model for the drought determinant can be established, a synthetic series can be generated, the statistical characterization of drought phenomena can be accomplished, and the characterization in terms of engineering risk can be performed. This is exactly what is done in the case study described in a later section.

3. RISK AND RELIABILITY CONCEPTS AS APPLIED TO DROUGHTS

3.1 Risk and Reliability

According to Hashimoto (1980) and Hashimoto et al (1982) the reliability of a system can be described as the probability that a system is in a satisfactory state. Risk can be defined as the complementary probability. For practical purposes it is necessary to establish a criterion to decide whether the system is in a satisfactory or unsatisfactory state. Denoting the set of all possible satisfactory states by S, the set of all unsatisfactory states by U, the state of the system at time interval t by X(t), the reliability of the system by α, and its risk by β, the following relations hold:

$$\alpha = \text{Prob}\,[X(t) \in S]$$

$$\beta = \text{Prob}\,[X(t) \in U]$$

$$\alpha + \beta = 1$$

This definition of risk is somewhat strict because it does not include any consideration of severity. This concept, however, is considered in the definition of resiliency and vulnerability.

In the previous section, a rigorous definition of regional drought was provided. This definition can be used to identify the unsatisfactory states of the system. The system is taken as being the region under study, and the occurrence of a drought is equated with an unsatisfactory condition.

Drought studies are often based on monthly data as is the case with the example presented in this paper. Under these circumstances, it is possible to assess risk and reliability in monthly terms. However, in many water supply and firm yield studies, a yearly analysis is preferred. Therefore, it is adequate to define dry year as a year in which at least one dry month occurs and to characterize drought risk and reliability in annual terms. In this way, it is possible to define concepts such as return period of a given drought.

As described in the previous section a regional drought is defined by two distinct variables, namely a truncation level, usually corresponding to a given non-exceedance probability, and a percentage of drought affected area. Given a numeric value for one of these variables, it is possible to study the relationship between the other variable and return period. In the case study presented later, the relationship between truncation level (in terms of non-exceedance probability) and return period is established for several percentages of drought affected area.

Duckstein et al (1984) introduce the concept of mission reliability as being the probability that the system will not fail between the time when a non-null requirement arrives and the corresponding service is delivered. This concept is easily incorporated into the framework of regional drought characterization.

In fact, it is very common that drought problems be relevant only during a given period of the year. This period will be called mission period.

Mission reliability can be considered as the probability that a drought does not occur during the mission period and, likewise, mission risk is the probability that a drought does occur. In this approach, a dry year can be defined as a year in which at least one dry month occurs during the mission period, and the study of annual mission risk can be easily done.

To study numerically the mission risk and reliability of a regional drought, it is sufficient to set the truncation level equal to zero for the period of the year in which drought is not relevant and different from zero for the mission period. These concepts are illustrated in the case study where agricultural needs are used to define the mission period.

3.2 Resilience

Resilience is a measure of the recovery time of the system. An immediate formulation of resilience in the context of regional drought characterization is to equate it with the duration of the drought and to study this duration statistically.

This definition, however, may not be very significant from a physical point of view. In fact, if a region is under a severe drought, it does not recover immediately when the drought condition ends. If a large total deficit occurred during the drought period, it is reasonable to consider that recovery is accomplished when a certain fraction of the total deficit is compensated by an excess of water that replenishes, at least partially, the exhausted resources. This excess of water has to be computed with respect to a given truncation level that can be called the recovery level. The recovery level may be eventually equal to the drought defining truncation level. However, it is more realistic that the recovery level is higher than the truncation level. An adequate value for the recovery level is the mean monthly precipitation.

The fraction of the drought total deficit that needs to be compensated may be called the recovery rate. If the recovery level coincides with the drought defining truncation level and the recovery rate is zero, the drought duration can be quated with resilience. If the recovery level is higher than the truncation level and/or the recovery rate is greater than zero, a recovery time must be considered and the study of resilience must be based on it. Naturally, the recovery time is equal to or greater than the drought duration and corresponds to a more precise definition of resilience. More formally, excess at time t can be defined as:

$$e(t) = \begin{cases} X(t) - r(t) & r(t) \leq t) \\ 0 & r(t) > X(t) \end{cases}$$

where r(t) represents the recovery level. The cumulative excess at time t' after drought termination, t_e, is given by

$$E(t) = \sum_{t=t_e}^{t'} e(t)$$

For a recovery rate α_r the <u>recovery time</u> t_r of a drought with total deficit D, can be defined as:

$$t_r = \text{Min} \left\{ (t-t_e): \frac{E(t)}{D} \geq \alpha_r \right\}$$

For any given values of r(t) and α_r, it is possible to study the statistical properties of t_r. This is fully illustrated in the case study.

3.3 Vulnerability

<u>Vulnerability</u> expresses the severity of failure in terms of its consequences. The concern is not how long the failure lasts, which is considered under resilience, but how costly it is.

In the context of regional drought studies it is possible to consider several formulations for the vulnerability analysis.

A first approach consists in considering the <u>total annual deficit</u> as a good measure of the severity of droughts. In this case it is possible to conduct a vulnerability study based on the statistical distribution of the total annual deficits. Proceeding in this way makes it possible to express a relationship between total annual deficits and return periods for a given drought definition (drought truncation level and drought affected area).

A second approach consists in focusing on <u>drought intensity</u> as the most relevant measure of drought consequences. To conduct a vulnerability study in annual terms, annual maxima of drought intensities can be used. It is possible, then, to establish a relationship between annual maximum drought intensity and return period.

A third and more attractive approach consists in establishing a <u>loss function</u>, L, to characterize the severity of a drought. The value of L may depend on any single drought characteristic or any combination of drought characteristics:

$$L = L(D, I, T, t_r)$$

In this expression L is the loss corresponding to a drought with a total deficit D, an intensity I, a duration T and a recovery time t_r.

If more than one drought occurs in one year, either the sum of losses or the largest loss can be selected to assess vulnerability in annual terms.

A simple but rather attractive example of a loss function is the following:

$$L = - 1/K \ln[1 - (D/D_{max})]$$

In this case only the total deficit D is considered. D_{max} represents a maximum value of total deficit corresponding to an infinite (catastrophic) loss and is a parameter determining how fast the loss L approaches its asymptotic value at D_{max}. The general shape of this particular univariate loss function is displayed on Figure 1.

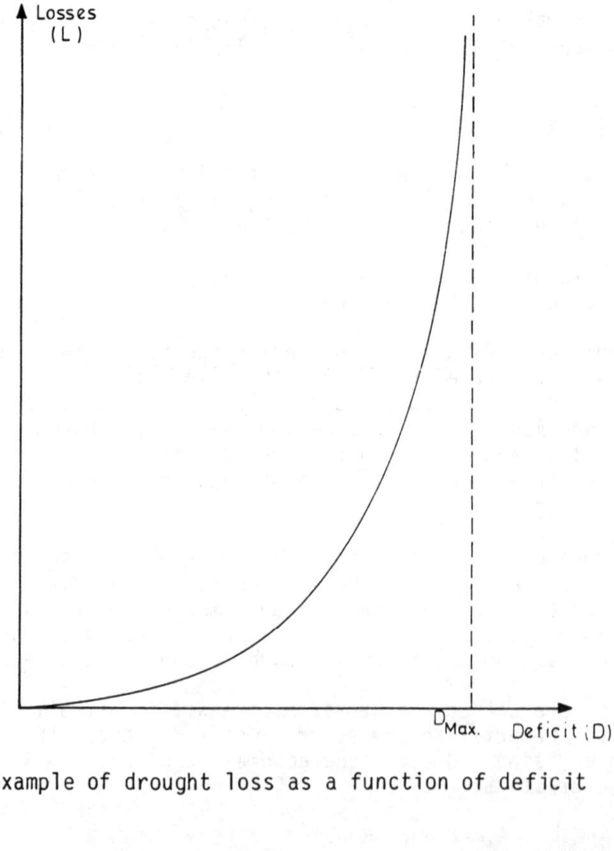

Figure 1. Example of drought loss as a function of deficit

In the case study presented in the following section, a loss function of this type is used to assess vulnerability in annual terms. Total annual deficits and annual maximum intensities are also used.

4. CASE STUDY: THE RIVER AVE BASIN

4.1 Location and Characteristics of the River Basin

The River Ave basin is located in the northwest of Portugal. The area of this basin is approximately 1,400 km^2 and it has a general northeast to southwest orientation, reaching the Atlantic Ocean close to latitude 41° 20' N. The average altitude is 280 m, with values ranging from zero to 1,261 m. Acccording to Thornthwaite (1945), the climate can be classified as wet in the downstream and flatter areas, and very wet in the upstream and mountainous regions.

The annual average precipitation ranges from 1,158 mm at Pedras Rubras (altitude 70 m) to 2,834 mm at Zebral (altitude 775 m). There is a large variability of the precipitation values within the year and from year to year. The lowest mean monthly precipitation corresponds to 14 mm for July at Pedras Rubras, and the highest corresponds to 431 mm for January at Zebral. The annual coefficients of variation range from 0.24 to 0.35, and the monthly coefficients of variation from 0.59 for May at Pedras Rubras to 1.51 for July at Zebral.

The characteristics of the relief and the influence of various climate factors may cause a portion of the watershed to be under drought conditions while the remaining area is not.

Human activity is very intense in the River Ave basin. Industry, agriculture, hydro-power generation and domestic supply rely heavily on surface water resources. These resources are sufficient under average conditions but become scarce during drought situations. Water storage in the watershed is not sufficient to provide an inter-annual regularization. Therefore, the effects of a drought are felt immediately in these circumstances.

Monthly precipitation is used to study regional droughts in the River Ave basin. The reason for selecting monthly precipitation as the drought determinant is that good quality data are readily available, it is very suitable for a regional approach, and it is the most important factor affecting surface water resources. Moreover, precipitation has a direct impact on agricultural activities, especially from October to January.

An extensive description of the River Ave basin can be found in Abreu (1985). The characterization of the monthly and annual precipitation of this river basin is presented in Rodrigues and Correia (1985).

Precipitation values in the watershed can be estimated from 15 rain gauges that have been operating continuously for the last 33 years.

The monthly values measured at some of these rain gauges are very highly correlated in space. Analysing the spatial structure of this correlation, and the first and second moments of monthly and annual values, six distinct regions were defined. The weighted monthly precipitation for every region was computed. There certainly is some degree of spatial correlation among the six regions. However, this correlation is much weaker than the correlation among rain gauges inside each region.

The area, the areal coefficient, the mean annual precipitation, and the number of rain gauges for every region are presented in Table 1.

Table 1. Characteristics of the six regions considered in the River Ave basin

Region No.	Area (km^2)	Areal Coefficient	Mean Annual Precipitation (mm)	Number of Raingauges
I	66.3	0.0478	1,158.2	1
II	563.0	0.4056	1,488.8	5
III	241.5	0.1740	1,818.0	3
IV	335.9	0.2420	1,612.9	2
V	160.9	0.1159	2,288.4	2
VI	20.4	0.0147	1,582.2	2

The location of the 15 rain gauges and the 6 regions that were considered are displayed on Figure 2.

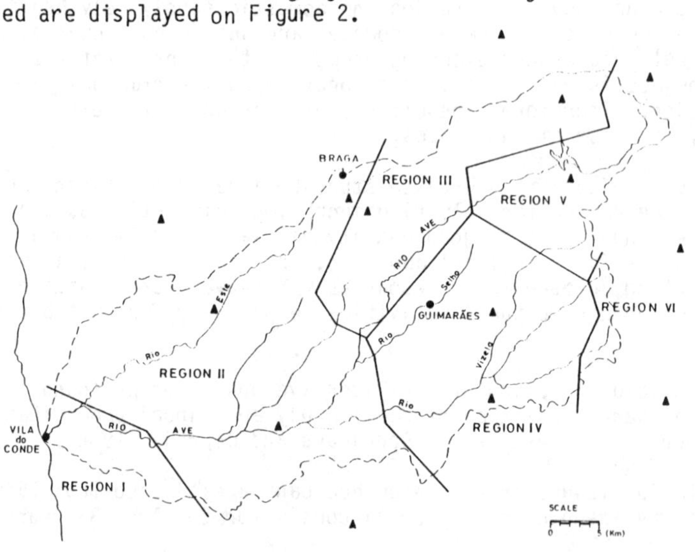

Figure 2. The six regions considered in the River Ave basin and the location of the 15 rain gauges.

4.2 Monte Carlo Generation of Multi-Site Monthly Precipitation

The 33 years of monthly data are clearly insufficient to draw conclusions on regional droughts beyond return periods of 20 or 30 years in the River Ave basin. However, these data are reasonably good enough to estimate the parameters of a multi-site stochastic model allowing for the Monte Carlo generation of a long series of monthly values, say 1,000 years.

It was found that the logarithm of the standardized monthly precipitation is normally distributed and serially independent. Let us call $X_i(m,k)$ the precipitation at region i during month m of year k ($i = 1, 2, \ldots, 6;\ m = 1, 2, \ldots 12;\ k = 1, 2, \ldots, 33$). The logarithm of the standardized values will be denoted as:

$$Y_i(t) = \ln \left[\frac{X_i\ (m,k) - x_i\ (m)}{s_i(m)} \right]$$

where $X_i(m)$ and $s_i(m)$, $m = 1, m\ 2, \ldots, 12$, are the mean and standard deviation for month m at region i, respectively. The dummy variable t corresponds to a monthly numbering:

$$t = 12\ (k - 1) + m$$

The vector $\underline{Y}(t)$, defined as:

$$\underline{Y}(t) = [Y_1(t), \ldots, Y_6(t)]^T$$

is multivariate normally distributed with mean μ and covariance Σ:

$$\underline{Y}\ (t) \cong N\ (\underline{\mu}, \underline{\Sigma})$$

Note that $\underline{\mu}$ and $\underline{\Sigma}$ are not time dependent since $Y_i(t)$ have been made stationary by transforming $X_i(m,k)$.

To build the stochastic model for data generation, we will use the method suggested by Santos (1981) based on the fact that, if $Z = [Z_1, Z_2, \ldots, Z_m]^T$ is a vector of m independent standard normal random variables, $\underline{d}$ an arbitrary m-vector and $\underline{C}$ an arbitrary (mxm) matrix, then the random vector:

$$\underline{Y}(t) - \underline{C}\ \underline{Z} + \underline{d}$$

has a multivariate normal distribution with mean $\underline{\mu}$ and covariance matrix $\underline{\Sigma}$ provided that:

$$\underline{\mu} = \underline{d}$$

$$\underline{\Sigma} = \underline{C}\ \underline{C}^T$$

Therefore to generate $\underline{Y}(t)$ we have only to find the matrix $\underline{C}$. To do so, we may apply the eigenvalue - eigenvector decomposition technique to $\underline{\Sigma}$, since every symmetric matrix has an eigenvalue - eigenvector decomposition of the form

$$\underline{\Sigma} = \underline{V}\ \underline{\Lambda}\ \underline{V}^T$$

where $\underline{\Lambda}$ is a (real) diagonal matrix whose diagonal elements are the eigenvalues of $\underline{\Sigma}$ and $\underline{V}$ is an orthogonal matrix whose column vectors are the eigenvectors of $\underline{\Sigma}$.

Let $\underline{D}$ be a diagonal matrix with diagonal elements equal to the square-roots of the diagonal elements of $\underline{\Lambda}$:

$$\underline{D} = \underline{\Lambda}^{1/2}$$

Then

$$\underline{\Sigma} = \underline{V}\ \underline{D}\ \underline{D}\ \underline{V}^T = (\underline{VD})\ (\underline{VD})^T = \underline{C}\ \underline{C}^T$$

Summarizing, we easily generate

$$\underline{Y}(t) = \underline{C}\ \underline{Z} + \underline{\mu}$$

taking $\underline{C} = \underline{V}\ \underline{D}$ with $\underline{V}$ and $\underline{D}$ defined above. After generating $\underline{Y}(t) = [Y_1(t), \ldots, Y_6(t)]$ for say 1,000 years, it is necessary to apply the inverse transformation to obtain monthly precipitation for every region:

$$X_i(m,k) = s_i(m)\ \mathrm{EXP}[Y_i(t)] + \overline{x}_i\ (m)$$

where i = 1, 2,...6, k = Int (t/12) + 1 and m = t - 12 (k - 1).

The values generated by this model can be used to characterize regional droughts employing the concepts and definitions presented in the previous sections.

4.3 Drought Characterization in Terms of Engineering Reliability and Risk

4.3.1. Risk and reliability. The generated series of 12,000 monthly precipitation values at every one of the six regions is used to assess the drought characteristics of the River Ave basin, using the engineering risk and reliability concepts presented in a previous section. A drought is defined by a critical affected area and a drought truncation level. The critical area is expressed in percentage of the total area. The truncation level is expressed as a non-exceedance percentile of the monthly preciptation distribution. A dry year is defined as a year with one or more dry months.

Three critical areas are considered, namely 50%, 75% and 100%. The relationship between drought truncation levels and return periods is studied and is presented in Table 2 and Figure 3.

It can be seen from this figure that for return periods larger than 5 years there is practically no difference in the corresponding drought truncation levels for the three critical areas. This means that if a severe drought occurs, it is likely to affect the entire watershed.

4.3.2. Mission reliability. Precipitation from October to January is very important for agriculture in the River Ave basin. If a dry year is defined as a year with one or more dry months in the October to January period, different results are obtained. These results are presented in Table 3 and Figure 4. It should be noted that this definition of a dry year is more exigent than the previous one. Therefore, a smaller number of droughts is observed in the generated series.

Table 2. Relationship between drought truncation level and return period for 50%, 75%, and 100% critical drought affected area

CRITICAL AREA	T=5	T=25	T=50
50%	0.0223	0.0037	0.0012
75%	0.0325	0.0058	0.0024
100%	0.0502	0.0140	0.0102

Table 3. Relationship between drought truncation level and return period for the mission period from October to January

CRITICAL AREA	T=5	T=25	T=50
50%	0.0523	0.0112	0.0061
75%	0.0703	0.0133	0.0100
100%	0.1178	0.0267	0.0157

4.3.3. Resilience. A very fast recovery can be observed if drought recovery time is equated with drought duration. The distribution function of the drought duration is presented in Table 4 and Figure 5. These results correspond to drought truncation levels of 0.20 and 0.01, and critical areas of 50% and 100%. Results for a recovery level of

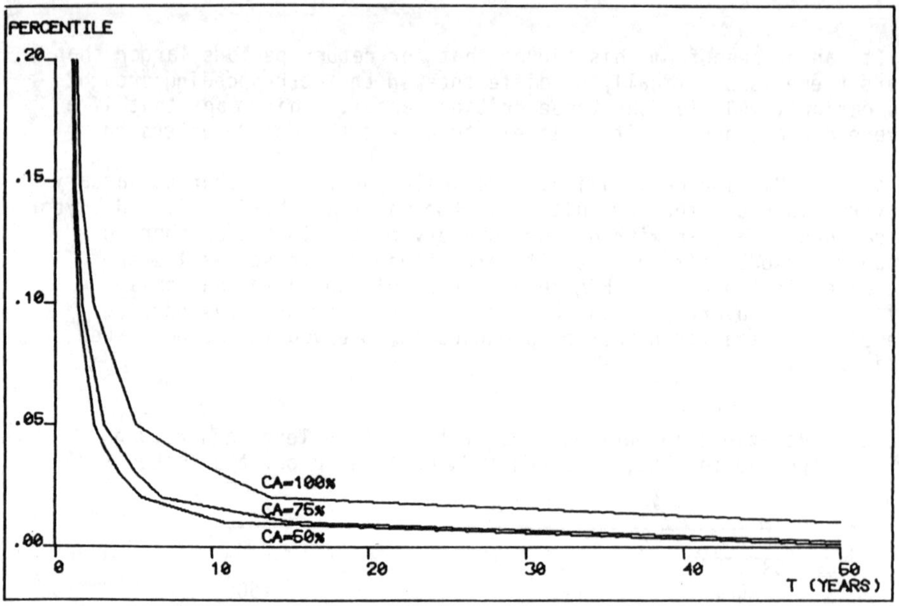

Figure 3. Drought risk for 50%, 75% and 100% critical drought affected areas.

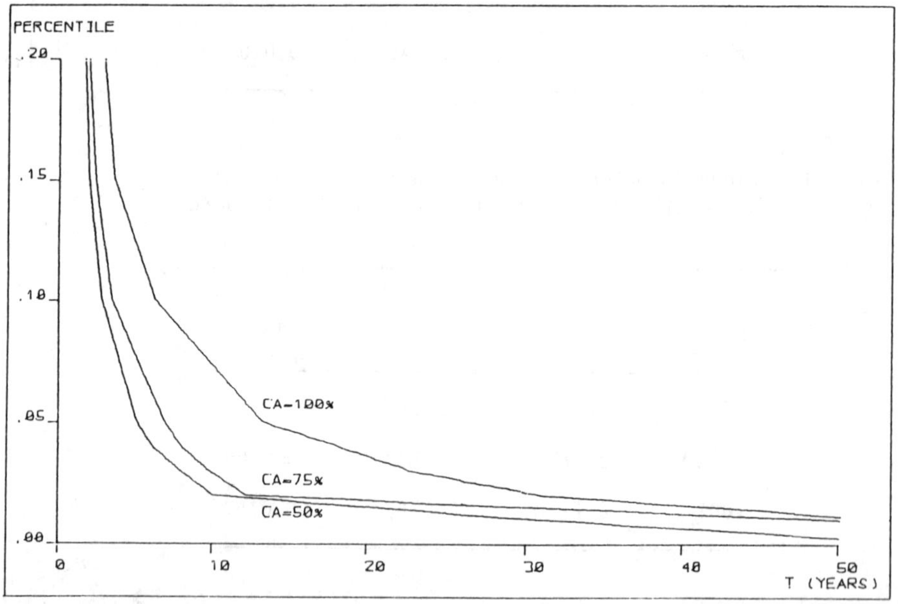

Figure 4. Mission drought risk for a mission period from October to January.

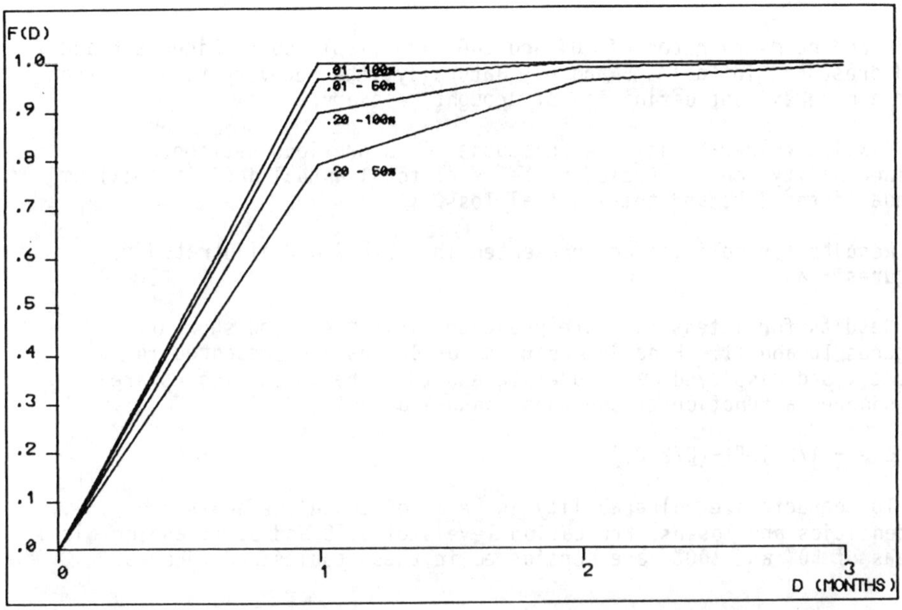

Figure 5. Drought resilience expressed in terms of drought duration.

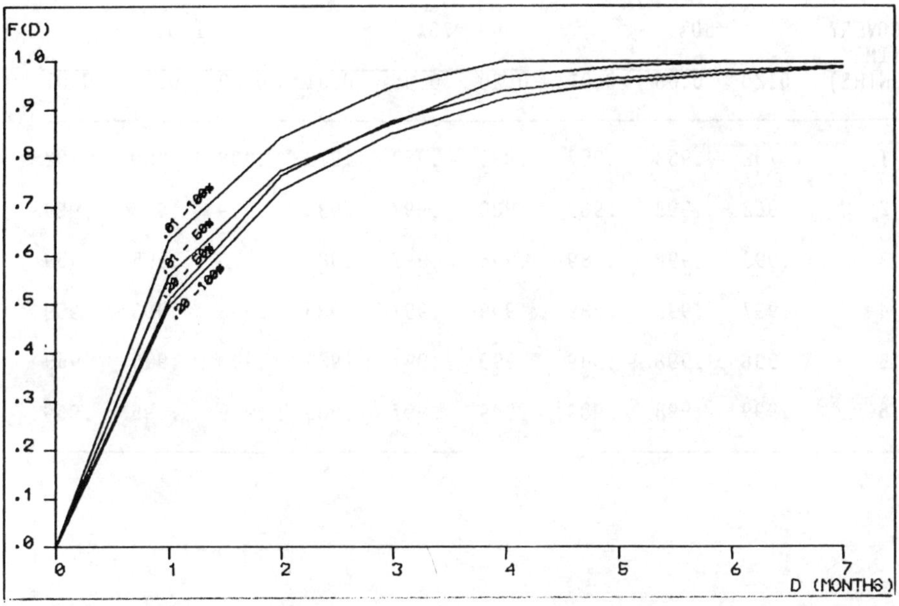

Figure 6. Drought resilience expressed in terms of drought recovery time for a recovery level of 0.50 and a 50% recovery rate.

0.50 and recovery rates of 50% and 100% are displayed on Figures 6 and 7 and presented in Tables 5 and 6. Naturally, the recovery time is larger for a more exigent definition of drought recovery.

4.3.4. Vulnerability. As discussed in a previous section, vulnerability can be studied in terms of total annual deficit, maximum annual intensity, and total annual loss.

Results for deficits are presented in Table 7 and illutrated on Figures 8 and 9.

Results for intensities are presented in Table 8 and shown on Figures 10 and 11. Finally, results for losses are presented in Table 9 and displayed on Figures 12 and 13. The annual losses are considered a function of the total annual deficit, D:

$$L = - 1/2 \ln[1-(D/260)]$$

To characterize vulnerability in terms of annual deficits, intensities and losses, truncation levels of 0.20 and 0.01, and critical areas of 50% and 100%, are considered in these tables and figures.

Table 4. Probability distribution function of drought duration

RECOVERY TIME (MONTHS)	50%			75%			100%		
	0.20	0.05	0.01	0.20	0.05	0.01	0.20	0.05	0.01
1	.792	.953	.957	.835	.962	.970	.898	.959	.792
2	.962	.998	.989	.980	.997	.935	.989	.995	.959
3	.990	.998	.989	.996	.997	.985	.998	.995	.959
4	.997	.998	.989	.999	.997	.985	.998	.995	.959
5	.998	.998	.989	.999	.997	.985	.998	.995	.959
6	.999	.998	.989	.999	.997	.985	.999	.995	.959

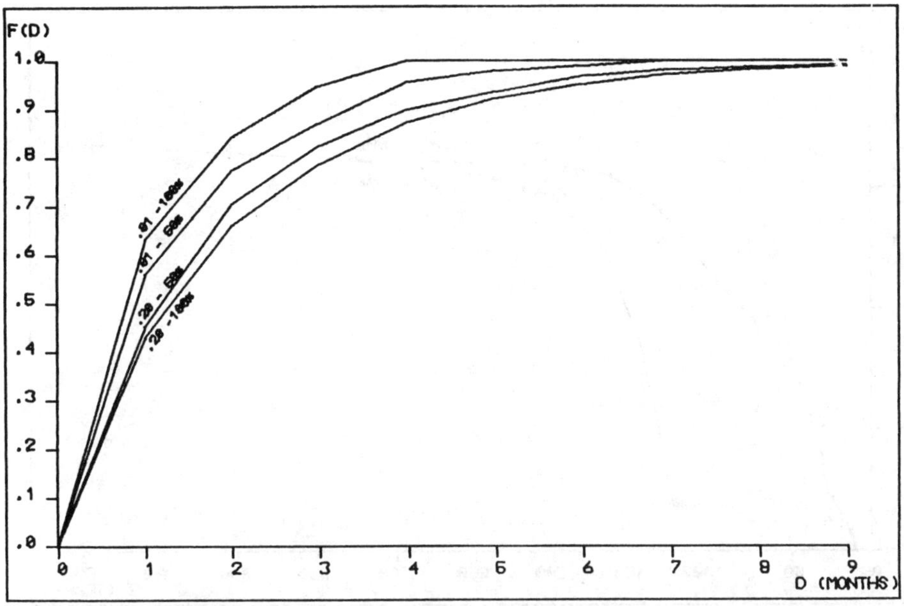

Figure 7. Drought resilience expressed in terms of 0.50 and a 100% recovery rate.

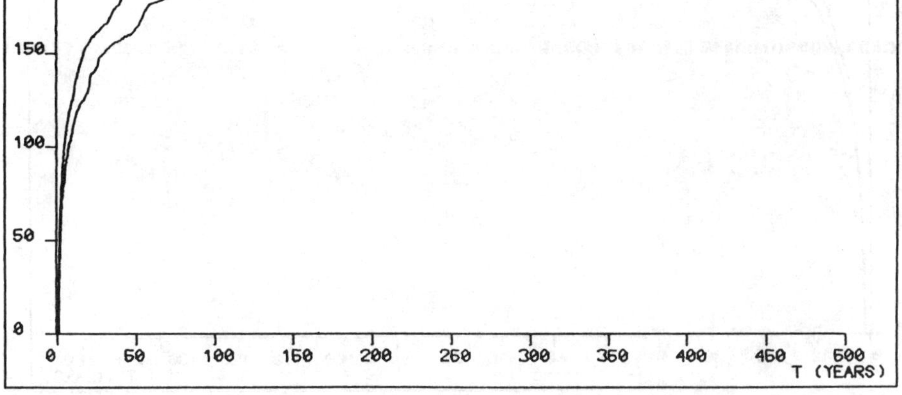

Figure 8. Drought vulnerability expressed in terms of total annual deficit for a 0.2 truncation level.

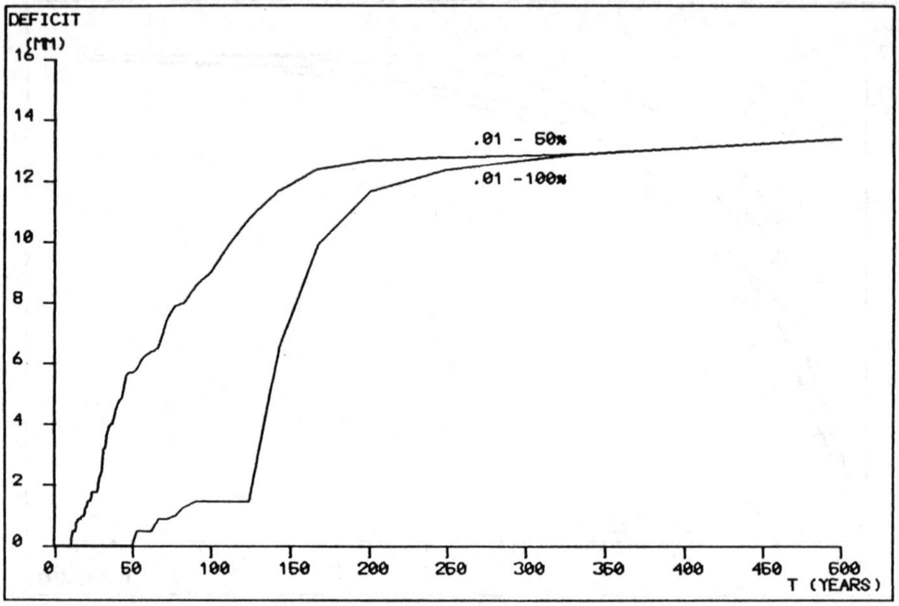

Figure 9. Drought vulnerability expressed in terms of total annual deficit for a 0.01 truncation level.

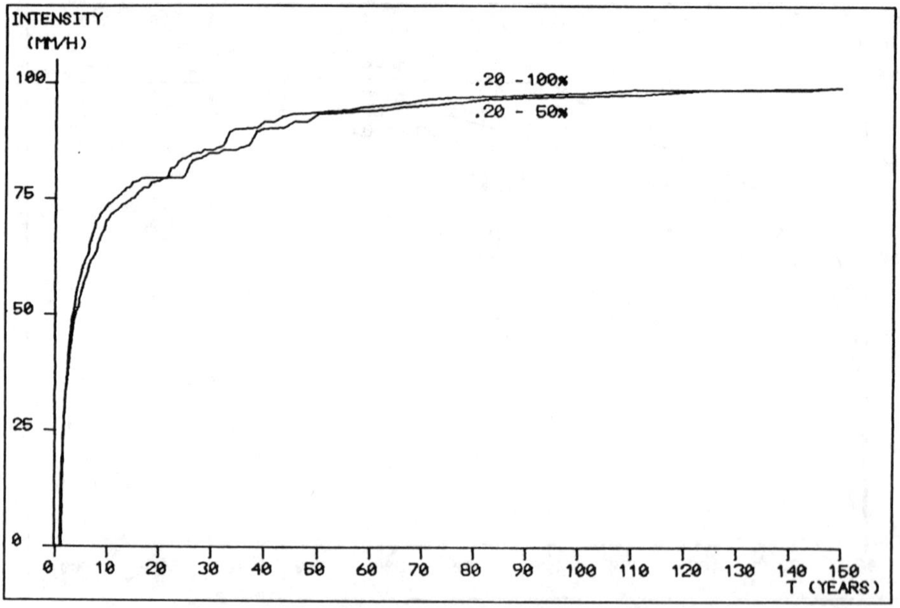

Figure 10. Drought vulnerability expressed in terms of maximum annual intensity for a 0.20 truncation level.

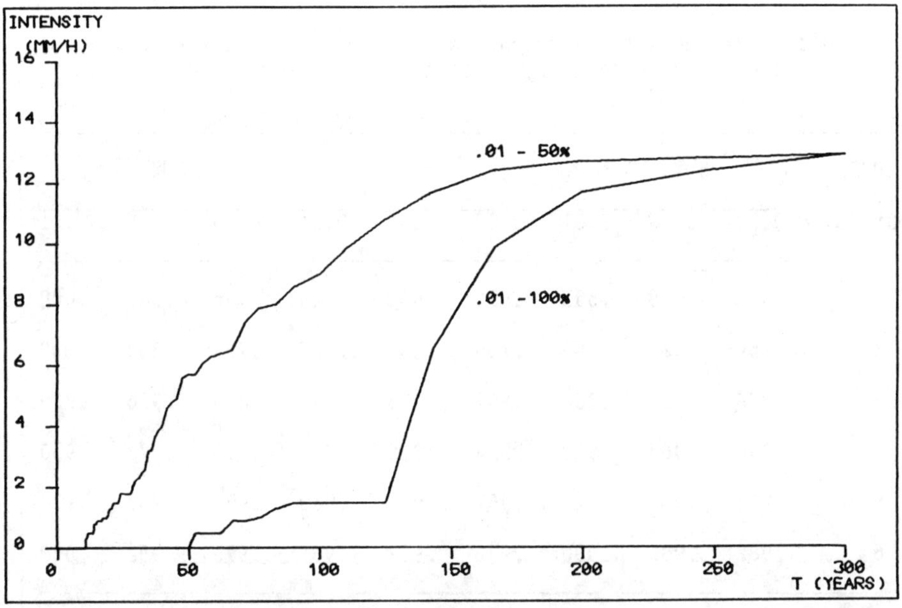

Figure 11. Drought vulnerability expressed in terms of maximum annual intensity for a 0.01 truncation level.

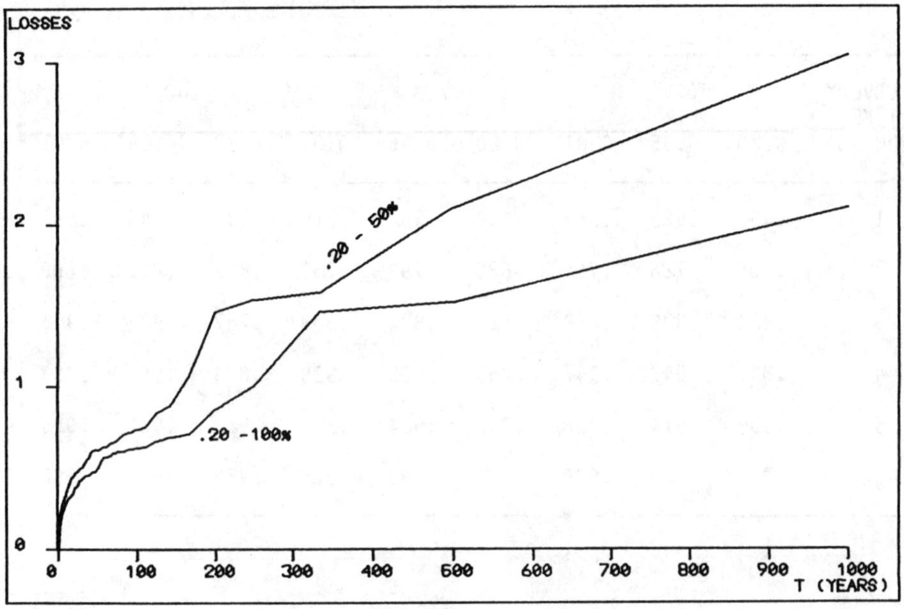

Figure 12. Drought vulnerability expressed in terms of total annual loss for a 0120 truncation level.

Table 5. Probability distribution function of drought recovery time for 0.50 recovery level and a 50% recovery rate

RECOVERY TIME (MONTHS)	50%			75%			100%		
	0.20	0.05	0.01	0.20	0.05	0.01	0.20	0.05	0.01
1	.508	.609	.553	.515	.588	.515	.496	.589	.600
2	.764	.823	.766	.755	.838	.757	.734	.854	.800
3	.876	.919	.862	.867	.923	.863	.850	.936	.900
4	.940	.963	.958	.934	.969	.954	.923	.973	.950
5	.966	.982	.979	.961	.987	.969	.957	.987	.950
6	.982	.986	.990	.978	.992	.984	.975	.992	.950

Table 6. Probability distribution function of drought recovery time for 0.50 recovery level and a 100% recovery rate

RECOVERY TIME (MONTHS)	50%			75%			100%		
	0.20	0.05	0.01	0.20	0.05	0.01	0.20	0.05	0.01
1	.452	.583	.553	.450	.562	.515	.430	.546	.600
2	.705	.786	.766	.689	.799	.757	.662	.801	.800
3	.824	.895	.862	.810	.898	.863	.787	.907	.905
4	.899	.947	.947	.889	.955	.939	.874	.958	.955
5	.935	.974	.968	.931	.984	.954	.923	.981	.955
6	.967	.980	.979	.964	.992	.969	.952	.990	.955

Table 7. Relationship between total annual deficit and return period

RETURN PERIOD (YEARS)	0.20			0.05			0.01		
	50%	75%	100%	50%	75%	100%	50%	75%	100%
5	100.9	97.4	82.5	11.9	11.3	0.0	0.0	0.0	0.0
25	160.3	154.6	141.8	32.5	31.6	27.5	1.8	1.2	0.0
50	183.7	182.4	162.6	37.8	36.5	35.9	5.7	3.2	0.0
100	200.6	194.4	185.5	41.8	40.9	39.3	9.0	7.9	1.5
500	256.1	256.1	247.6	72.5	72.5	53.0	13.4	13.4	13.4
1000	259.4	256.7	256.1	75.6	75.6	60.0	16.0	16.0	16.0

Table 8. Relationship between maximum annual intensity and return period

RETURN PERIOD (YEARS)	0.20			0.05			0.01		
	50%	75%	100%	50%	75%	100%	50%	75%	100%
5	54.7	57.3	58.5	11.3	10.8	0.0	0.0	0.0	0.0
25	80.0	83.2	83.7	26.1	26.1	25.3	1.8	1.2	0.0
50	93.3	93.7	93.7	31.6	31.6	31.6	5.7	3.2	0.0
100	97.2	97.9	97.9	37.9	37.9	37.9	9.0	7.9	1.5
500	100.1	100.1	100.1	45.2	45.2	45.2	13.4	13.4	13.4
1000	103.2	103.2	103.2	48.3	48.3	48.3	16.0	16.0	16.0

Table 9. Relationship between total annual loss and return period

RETURN PERIOD (YEARS)	0.20			0.05			0.01		
	50%	75%	100%	50%	75%	100%	50%	75%	100%
5	.246	.235	.191	.085	.080	.000	.000	.000	.000
25	.479	.451	.394	.279	.269	.225	.059	.038	.000
50	.613	.605	.491	.344	.327	.320	.217	.110	.000
100	.738	.689	.625	.399	.386	.364	.405	.334	.049
500	2.100	2.100	1.521	1.539	1.539	.598	.878	.878	.878
1000	3.036	2.183	2.100	2.624	2.624	.779	2.197	2.197	2.197

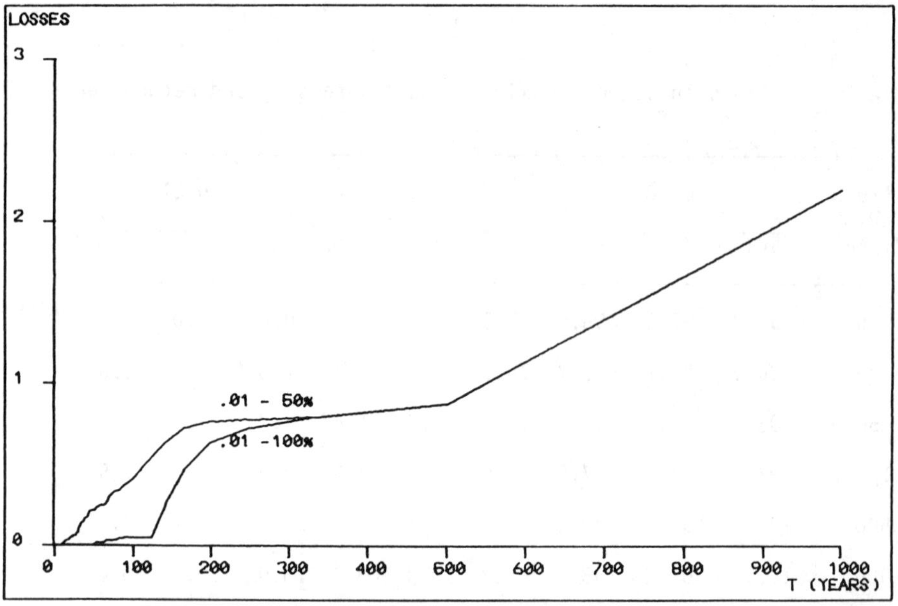

Figure 13. Drought vulnerability expressed in terms of total annual loss for a 0.01 truncation level.

5. CONCLUSIONS

A precise definition of regional drought is given. This definition is based on a drought determinant truncation level and a critical drought affected area. The concepts of drought deficit, drought intensity, drought duration, drought recovery time, and drought losses are discussed. These concepts are formulatd in the framework of an engineering risk analysis. In this framework, droughts are characterized in terms of risk and reliability, mission risk, resilience, and vulnerability.

The approach introduced in this paper is illustrated for the River Ave basin.

A series of 12,000 monthly values of precipitation at six regions is generated by a stochastic model. The parameters of the model are estimated for 33 years of monthly data.

Quantitative measures of drought risk, mission risk, drought resilience, and drought vulnerability are obtained for the River Ave basin and can be easily obtained for any region under study.

The proposed methodology proves to be very adequate for the characterization of regional droughts in terms of engineering risk. This methodology is very flexible, allowing for different definitions of drought and drought recovery. This feature makes it applicable to any region, whatever the climatic conditions may be.

ACKNOWLEDGEMENTS

This research was done under Project 62/11/8490, financed by PIDDAC (Central Administration Investment and Development Plan), and Project 62/12/7850, financed jointly by LNEC and NATO under the Science for Stability Program.

The Direcção Geral dos Recursos e Aproveitamentos Hidráulicos and the Instituto Nacional de Metorologia e Geofísica are acknowledged for providing data used in this study.

REFERENCES

Abreu, A. (1985 in preparation) Caracterização Geral da Bacia Hidrográfica do Rio Ave (General Characterization of the River Ave Watershed). Lisboa, Laboratório Nacional den Engenharia Civil,

Chow, V.T. (ed.) (1964) Handbook of Applied Hydrology. New York, McGraw Hill Book Co.

Duckstein, L. and E. Plate (1983) A System Framework for Hydrological Reliability with Application to the Design of Structures. Proceedings of the Hamburg Symposium, International Association of Hydrologic Sciences.

Duckstein, L., E. Plate and M. Benedini (1984) Incidents and Failures in Reservoirs and Dams: A Mathematical System Framework. Fall Meeting, American Geophysical Union.

Gibbs, W. and J. Maher (1967) Rainfall Deciles as Drought Indicators. Bulletin No. 48, Melbourne, Bureau of Meterology.

Hashimoto, T. (1980) Robustness, Reliability, Resilience and Vulnerability Criteria for Planning Water Resources Systems. Ph.D. Thesis, Ithaca, New York, Cornell University.

Hashimoto, T., J. Stedinger and D. Loucks (1982) Reliability, Resiliency and Vulnerability Criteria for Water Resource System Performance Evaluation. Water Resources Research, Vol. 18, No. 1, American Geophysical Union.

Lang, R. (1915) Versuch einer exakten Klassifikation der Boden in klimatischer und geologischer Hinsicht (a Proposed Exact Classification of Soils from the Climatic and Geological Viewpoints). Internat'l Mittlg. Fur Bodenkunde, Vol. 5.

Martonne, E. (1926) Une Nouvelle Fonction Climatologique: L'Indice d'Aridité (A New Climatological Function: The Aridity Index). Meteorologie, Vol. 2, Paris.

Palmer, W. (1965) Meteorological Drought. Paper No. 45, Washington D.C., U.S. Department of Commerce, Weather Bureau Research.

Rodrigues, R. and F. Correia (1985 in preparation) Caterização do Regime de Preciptações da Bacia Hidrográfica do Rio Ave (General Characterization of the Precipitation Regime of the River Ave Watershed). Lisboa, Laboratório Nacional de Engenharia Civil.

Sancho, C. and D. Arias (1978) Analisis de las Sequias en Mexico (Analysis of Droughts in Mexico). Ciudad de Mexico, Comissión del Plan Nacional Hidraulico.

Santos, M. (1981) On the Stochastic Characterization of Regional Droughts. Research Officer Thesis. Lisboa, Laboratório Nacional de Engenharia Civil.

Santos, M. (1983) Regional Droughts: A Stochastic Characterization. Journal of Hydrology No. 66. Amsterdam, Elsevier Science Publishers.

Santos, M., F. Correia, and L. da Cunha (1984) Drought Characterization and Impact Assessment, Case Study of Portugal, in Coping with Droughts, Edited by Yevjevich, V., L. da Cunha and E. Vlachos, Littleton, Water Resources Publications.

Thornthwaite, C. (1948) An Approach Toward a Rational Classification of Climate. Geographic Review, Vol. 38.

WMO (1975) Drought. Special Environment Report No. 5, Geneva, Secretariat of the World Meteorological Organization.

INCIDENTS AND FAILURES OF HYDRAULIC STRUCTURES SUBJECT TO INDEPENDENT FLOODS

Fuat Senturk
Hatboyu, Nilüfer Apt. 3/27 Suadiye
Istanbul, Turkey

ABSTRACT

Keban dam (Fig. 1) is built on the Firat river, the largest river in Turkey. The characteristics of the dam are listed in Table 1. The highest flood recorded at the dam site over a period of 45 years was 6,000 m^3/sec. Using standard hydrological procedures, the design flood has been determined to be 17,000 m^3/sec. As the dam is large and the volume of the reservoir is over 30 billion m^3, a decision tree has been prepared. As a result, the International Board responsible for the realization of the project increased the design flood to 19,000 m^3/sec, and then increased this figure again to 27,000 m^3/sec. The dimensions of the spillway were increased in proportion to enable safe evacuation in the event of a catastrophic flood. The spillway thus became a huge hydraulic structure equipped with six radial gates, each of which release 3,000 m^3/sec (the catastrophic flood of 27,000 m^3/sec was reduced to 17,000 m^3/sec as a consequence of flood routing). Not only was the spillway of Keban dam over-dimensioned, but the spillways of the large downstream dams were also designed accordingly.

The increase of the peak value of the design flood of the spillway of Keban dam seems to decrease the risk of failure of the rockfill by overtopping, but undoubtedly results in a cost increase of the spillways of the large downstream dams. Also while the probability of overtopping is minimized, the risks corresponding to two other phenomena are increased--namely, erosion of the riverbed downstream of the spillway and erosion due to cavitation on the surface of the spillway.

1. INTRODUCTION

The determination of the design flood is necessary for defining the dimensions of spillways. Probabilistic approaches are used for computing the peak value and the shape of the flood hydrograph. In some instances, these methods lead to over-dimensioned structures. Decision theory may be a better approach in such cases for estimating the final design flood. An increase of the design flood corresponds to an

Figure 1. Keban Dam

TABLE 1 Main characteristics of Keban dam at the dam site.

Hydrology	
Drainage area	64100 km^2
Drainage basin length	425 km
Average discharge	635 m^3/sec
Recorded maximum discharge	6800 m^3/sec
Recorded minimum discharge	145 m^3/sec
Flood peak discharge	27000 m^3/sec
Reservoir	
Maximum lake area	69 km^2
Maximum lake volume	30×10^9 m^3
Reservoir maximum elevation	845 m
Rockfill dam characteristics	
Type of dam: clay cored, compacted rockfill and concrete gravity	
Crest elevation	849-852 m
Maximum height above lowest foundation	211 m
Maximum height above river bed	167 m

Table 1. -- continued

Spillway and concrete dam characteristics	
Type of spillway	Gated spillway with radial gates
Crest elevation of the gates	828 m
Spillway discharge capacity	17,000 m^3/sec
Spillway gates	6 (16x24)

On the downstream course of the river four important dams were projected:

1- Karakaya dam (today under construction)	
Height of the dam	: 184 m
Reservoir capacity	: 9,59 billion m^3
2- Ataturk dam (today under construction)	
Height of the dam	: 184 m
Reservoir capacity	: 48,7 billion m^3
3- Birecik dam (today under consideration)	
Height of the dam	: 59,50 m
Reservoir capacity	: 1,22 billion m^3
4- Karkamis dam (today under consideration)	
Height of the dam	: 40,00 m

increased cost of the hydraulic structure and a decrease of the risk of failure of the main structure. A compromise value is thus to be chosen. A hydraulic structure is only 100 percent safe if its spillway can release an infinitely large flood. But its costs will also be infinitely large. This is true if the risk conditions are very well defined. Actually there are many reasons for the increased risk of failure when the spillway is over-dimensioned. Overtopping probability becomes negligible but downstream erosions risks increase accordingly. Engineers in charge of developing a decision analysis generally use data provided by decision makers or the owner of the hydraulic structure. In the case that these data are correct and complete, the analysis will also be complete. On the contrary, a decision based on incomplete information will be also incomplete and the result obtained may be deplorable.

In this paper an example of the latter case will be described, and the dilemma faced by the decision maker will be emphasized.

2. DEFINITION OF THE PROBLEMS

The design of the cofferdam of Keban dam constitutes the first problem. Flow measurement at the dam site covered a period of 27 years. The observed peak value was 6,800 m^3/sec (Table 1). The cofferdam itself is a 40.5 m high, earthfill dam, with reservoir capacity of one billion m^3. The choice of the design flood for the final determination of the dam height and the diameter of the derivation tunnels constitutes the subject of a decision making analysis. The first important city downstream is about 150 km below the dam. Therefore, the only important losses to be considered were those due to flooding of the foundation excavation. A decision tree was prepared. Two alternatives were investigated:

Alternative No. 1 $a_1 = Q_{50}$

Alternative No. 2 $a_2 = f(\theta_1,\theta_2)$

with $\theta_1 = Q_{6000}$ and $\theta_2 = Q_{25}$, respectively. The $u = u(a,\theta)$ values were computed accordingly. This decision tree was presented to the Board responsible for making the decision, and after time-consuming deliberations, the case $u(a_2,\theta_1)$ was chosen. The cofferdam was constructed, and the foundation pit was excavated after having rerouted the river. When a flood of 6,800 m^3/sec occurred, the cofferdam reservoir became completely filled, with no free board left, and overtopping was imminent. In order to save the cofferdam from collapsing, sand bags were placed on the crest of the cofferdam and a layer of clay was applied on the upstream face of this thin wall. This was an improvisation of the engineers in charge of the operation. As a consequence of this unfortunate event, the design of the spillway was completely changed.

The data presented to the Board related only to hydrology. Erosion that would take place downstream of the Keban spillway and possible sudden increases in leakage due to defects in the geological foundation were not mentioned. A new decision analysis using the available data was prepared, which enabled the Board to make the best judgment for the design flood of the spillway. The first design was based on a maximum inflow discharge of 17,000 m^3/sec. The Board then proposed higher θ values of 19,000, 25,000, and 27,000 m^3/sec for inflow discharges.

An increase in the design flood of Keban spillway means an increase of the design flood to the whole series of downstream dams. These facts were taken into consideration, and the decision tree was prepared accordingly. In the loss analysis, only the overflowing phenomenon was taken into consideration and very elaborate loss analyses were prepared.

The concern of this paper is not the preparation of the decision tree but the examination of the decision that was made based on this tree and the consequences of this decision.

3. THE SPILLWAY OF KEBAN DAM

The Board responsible for the construction of the dam and the spillway accepted Q_{max} = 27,000 m^3/sec as the inflow discharge. A flood routing was prepared, leading to Q = 17,000 m^3/sec for the design flood of the spillway. The characteristics of the hydraulic structure capable of releasing this flood are listed below:

Number of the gates	6
Discharge of each gate	3,000 m^3/sec
Number of bays	3
Discharge of each bay	6,000 m^3/sec
Dimensions of each gate	16 x 24 m
Crest elevation of the spillway	828 m
Minimum elevation of the bucket	726.39 m
Water level downstream of the spillway	691.60 m
Maximum difference of water level	153.40 m

The gates of the downstream hydraulic structures also became huge metallic structures. Thus the overtopping risk was eliminated. Keban dam has been in operation for 11 years.

4. CONSEQUENCES OF THE ADOPTED DECISION

Two consequences of the selection of design procedure and discharges are presented below. The first one is the near collapse of the spillway because of foundation scouring and surface erosion. The second one is the occurrence of an unexpectedly high infiltration.

4.1 Spillway Damage

The construction of Keban dam was completed in 1975. That year diluvial rains occurred in the eastern regions of Turkey and particularly in the watershed of Keban dam. It was assumed that the Keban reservoir would be filled in two years, but it was filled in only seven months; the spillway gates were then opened for the first time. They stayed open for about four months. The design discharge of the spillway was so great that the risk of overtopping was out of the question. But at the end of the four months, when the gates were closed, the disaster became apparent. The toe of the spillway was tremendously eroded, and the erosion reached the retaining wall of the flip-bucket (Fig. 2). Erosion of this type was detected on model studies and some protective action was taken by laying heavy rip rap to the toe of the spillway. A discharge of Q = 10,000 m^3/sec was the maximum discharge allowed for a short time, but in reality Q = 10,000 m^3/sec was released from the structure continuously for a period of four months.

Figure 2. Erosion downstream of Keban Spillway

Specifically, the rip rap layer was designed for a maximum discharge of 10,000 m^3/sec for a duration of only three days. A decision tree was not prepared nor was this particular risk taken into consideration in the main decision tree. The downstream rock, being a soft schist, was heavily eroded. Vibrations that occurred on the bucket were transferred directly to the foundations by means of pre-stressed cables. If these cables were eroded, this could have caused the collapse of the spillway and endangered the main structure.

Furthermore, large cavitation holes were apparent on the spillway face (Fig. 3). These holes, which covered a large area of the concrete pavement, were as deep as 1.20 m with width up to 2.4 m. Their repair has taken a lot of money and time. This loss also was not taken into consideration in the decision making analysis, although such a cavitation phenomena had been discussed in full detail. It is evident that it is difficult to include construction defects or incorrect design in the decision analysis. On the other hand, it appears that the beginning of cavitation can be related to flow on the spillway and the corresponding discharge. Thus, this kind of risk should also be taken into account and the resultant losses evaluated. The collapse of the spillway was imminent: the spilling stopped just in time, by chance. The decision tree with inclusion of the losses related to the erosion on the spillway may take another form and the design flood decreased to a more understandable value, which may avoid strong criticism after the fact.

Figure 3. Erosion of the face of Keban spillway

4.2 Increased Infiltration

The geology at the dam site is extremely distorted and cut by numerous faults. Although Keban marbles are the only visible outcrops at the dam site, on the right bank schists overlie these in the riverbed and underneath it. Furthermore, dolomites overlie and underlie the schists; these possess solution cavities which are cut in various directions by fissures. Karstification is intense; it is generally formed from the surface, except in cases of hydrothermal activity which occurs from depth to surface. As a result of tectonic activity, the faults that had formed at various depths brought the various geological units into contact with each other. Thus, the upstream impervious upper schists are in contact with Keban limestone along the main fault, and the block-containing schists are downshifted by approximately 100 m. Although at this location the upper schist levels appear to form a natural barrier against leakage, Keban limestone, which is exposed in the reservoir area, is found to be in contact with the fissured schists at the lower elevations of the main fault.

This geology caused an important incident. During the construction of the dam, existence of underground cavities were known. Grouting was undertaken and the karst in the foundation was tightened. Geological investigations were carefully conducted in order to build a dam of maximum height, because Keban is the key project of the river basin system. At the end of the first year, the discharge due to infiltration was acceptable. It was lower than 3 m^3/sec, but it increased suddenly and attained 25 m^3/sec. An investigation showed that a vortex had formed in the reservoir of Keban dam (Fig. 4).

Figure 4. Formation of vortex on the surface of Keban lake

Tracers (tritium) were used to trace this high infiltration and it was found that a 150 m deep cavity existed 350 m below the foundation. This was called "crabe cavity". This cavity was filled by specially chosen aggregate, and then grouted. The extra cost of this work turned out to equal one-third of the cost of the rockfill.

5. CONCLUSION

Many elaborate techniques of decision making exist. Particularly in hydraulic engineering, a Bayesian tree approach is often used. In order to reach the best results, the engineer responsible for the design must prepare a complete list of factors to be considered in the decision making analysis. All the facts involved must be carefully defined and given to the technicians who are responsible for preparing the decision tree. If this tree is not complete, risks that have not been considered may endanger the structure and the decision making Board may not arrive at a correct design.

REFERENCES

Aksoy, S., S. Ethembabaoglu (1979) Cavitation damage at the discharge channels of Keban Dam. ICOLD, New Delhi.

Sentürk, F. (1976) Comparison of the results obtained in nature and on the scale model of Keban Dam. Internal Report in Turkish, DSI, Ankara.

Turkish National Committee (1976) Investigation of leakages at Keban Dam. ICOLD, Mexico.

RELIABILITY OF HYDRAULIC STRUCTURES POSSESSING RANDOM LOADING AND RESISTANCE

Ben Chie YEN

Professor of Civil Engineering, University of Illinois at Urbana-Champaign, 208 N. Romine Street, Urbana, IL 61801, USA

ABSTRACT

The safety-threatening external loading from geophysical and nongeophysical sources and the capacity or resistance of a hydraulic structure both vary randomly in time and space. In the traditional analysis of the risk of a hydraulic structure, only the hydrologic factors are taken into consideration. Alternately, a safety factor could be arbitrarily assigned in order to provide some degree of protection against our ignorance. These approaches either provide us a partial view of the risk or are empirical and arbitrary. A true reliability evaluation cannot be accomplished unless <u>all</u> the factors are accounted for and properly analyzed and combined. At the other extreme, the method of direct integration for risk evaluation which would give exact, true reliability evaluation requires knowing the <u>exact</u> probability distribution of each of <u>all</u> the factors. This requirement is rarely satisfied in real situations. In this presentation a number of approximate techniques that could provide an estimate of the system reliability of a hydraulic structure are briefly reviewed. Emphasis is then placed on the mean-value first-order second-moment method and the advanced first-order method. These two methods, when applied conjunctively with a fault tree or event tree, provide a formal, albeit approximate, framework for considering quantitatively all the factors that influence total system reliability. It also gives a formal structure to combine quantitatively the risks evaluated separately by different hydrologic, hydraulic, geotechnical, structural and other specialists.

1. INTRODUCTION

The safety and reliability of a hydraulic structure to function properly depend on the capacity of the structure to resist external loadings. The external loading comes from hydrologic and other geophysical sources, as well as <u>non</u>-geophysical sources such as human intervention. Likewise, the resistance of the structure depends on various hydraulic, geotechnical, structural, and other technical and nontechnical factors. For a specified structure at a given location, both the loading and resistance vary with time, i.e., they are dynamic in both the mechanical and non-mechanical sense. Most of the factors that contribute to the loading and resistance follow random temporal variations rather than orderly patterns.

Consider, for example, the safety of a concrete dam. The resistance of the dam depends on the structural ability of the dam proper, the strength of the soil or rock foundation and abutment, and the proper functioning of the sluice and other regulating facilities. These strengths and characteristics are subject to natural time variability such as aging, fatigue and season and weather, in addition to human-imposed variability such as the schedule and procedure of maintenance and operation, and animal and vegetation-imposed variability. The loading consists of the hydrologic and hydraulic loads of floods of various return periods, the sediment in the reservoir, ice, debris and waves; the non-hydrologic geophysical loads of earthquake, wind, and landslide; and other human-imposed loads ranging from excess live loads to destruction and sabotage. Most, if not all, of these various factors that affect loading and resistance are random in nature. Despite the tradition of regarding dams to possess a time-invariant deterministic resistance, a dam may in fact fail under a small loading concurrent with a low resistance.

The objective of this presentation is to review briefly several traditional and recently developed techniques for quantitative estimation of the safety and reliability of hydraulic structures.

2. DEFINITIONS OF RISK AND RELIABILITY

Different definitions of safety, reliability, and risk appear in the literature. In this presentation, failure is defined as the situation in which the hydraulic structure is incapable of functioning to satisfy the desired objective. Thus, a failure event can be as serious as the collapse of a dam or an incident as non-critical as the capacity of a sewer or culvert being exceeded, causing some local flooding and property damage but without any behavior change or functional damage to the structure itself. Yen and Ang (1971) called the former type of failure a catastrophic type failure and the latter type a property damage type failure. Thus, risk is defined herewith as the probability of failure, P_f, that the loading L exceeds the resistance or capacity C:

$$\text{Risk} = P_f = P(L > C) \tag{1}$$

Hence, reliability is defined as

$$\text{Reliability} = P(L \leq C) = 1 - P_f \tag{2}$$

Let a performance variable Z be defined as

$$Z = C - L \tag{3a}$$

or

$$Z = (C/L) - 1 \tag{3b}$$

or

$$Z = \ln(C/L) \tag{3c}$$

then Equations 1 and 2 alternatively can be written as

$$\text{Risk} = P_f = P(Z < 0) \tag{4}$$

$$\text{Reliability} = 1 - P_f = P(Z \geq 0) \tag{5}$$

Both the loading L and resistance or capacity C can be a function of other component variables, i.e.,

$$L = g_L(X_i) \qquad i = 1,2,\ldots n \tag{6}$$

and

$$C = g_C(X_j) \qquad j = 1,2,\ldots m \tag{7}$$

After considering and combining the effect of the various influential factors, let the probability density function (pdf) of the loading be $f_L(\ell)$, and that of the resistance be $f_C(c)$, as shown schematically in Figure 1. Then, expressed in terms of the performance variable Z, the pdf of Z can be plotted as shown schematically in Figure 2. From Equations 3 and 4, the risk corresponds to the shaded area under the pdf curve and to left of the vertical axis $z = 0$; while the reliability is the corresponding (unshaded) area to the right of the axis.

Theoretically, the risk calculated according to Equation 4 is the same irrespective of which definition of Z (Equations 3a, 3b, and 3c) is used. However, if the exact pdf of Z is unknown and approximation is involved in the calculation, the estimated risk may vary according to the definition of Z. In engineering practice, there are situations where the pdfs' of L and C are difficult to find, whereas the pdf of Z can be estimated.

3. METHODS OF RISK CALCULATION

A number of methods have been proposed by various investigators for the calculation of the risk of failure of structures, including (but not limited to) the methods of: return period, safety factor, direct integration, Monte Carlo simulation, reliability index, mean-value first-order second-moment (MFOSM) analysis, and advanced first-order second-moment (AFOSM) analysis. A general comparison of these methods is given in Table 1. The safety factor method is not listed in the table because its characteristics depend on how the safety factor is defined. All these methods can provide some quantitative comparison between the relative reliability of different hydraulic structures. Except in very simple special cases, none of the methods can provide the actual total risk of a hydraulic structure, although when properly applied, the two first-order methods may give an acceptable estimate. A brief discussion is given below on the methods of return period, safety factor, direct integration, Monte Carlo, and reliability index. The two relatively new, more advanced and less familiar first-order methods are discussed in the following section. Those who are interested in the details of the methods should refer to the references cited in the discussion of the individual methods.

3.1 Method of Return Period

Among the different risk calculation methods, the return period method is the most familiar to hydrologists and hydraulic engineers. It considers the random nature of only the natural events, such as floods

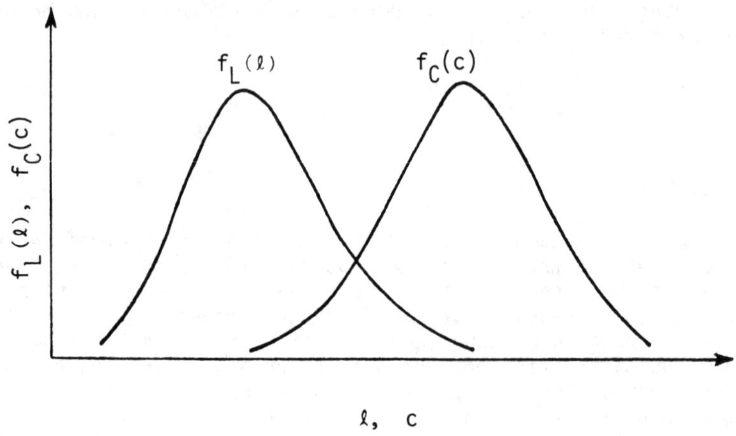

Fig. 1. Probability density functions for load and resistance.

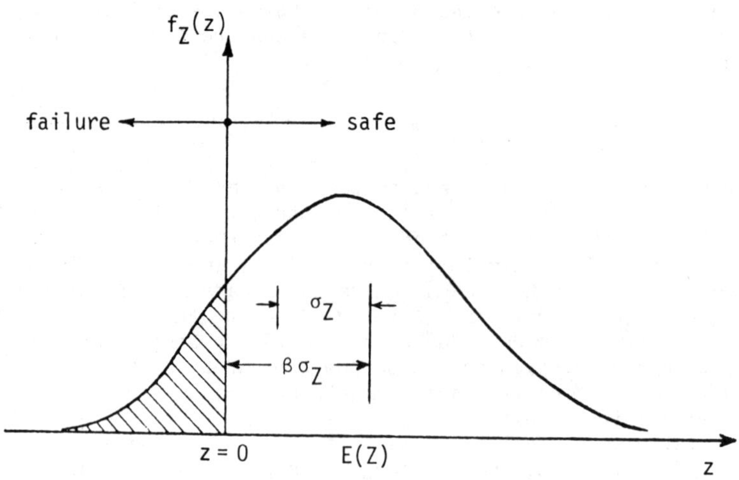

Fig. 2. Probability density function for performance variable Z.

Table 1. General comparison of risk calculation methods

Method	Return Period	Direct Integration	Monte Carlo	Reliability Index	MFOSM	AFOSM
Capability to account for different factors	very limited	limited	yes	yes	yes	yes
Information needed on probability distribution of factors	indirectly	extensive	moderate	first two statistical moments	only the combined distribution, for factors the first two statistical moments suffice	only the combined distribution, for factors the first two statistical moments suffice
Complexity in application	simple	complicated	moderately complicated	moderate	moderate	moderate
Amount of computations	simple	moderate to extensive	extensive	moderate to simple	moderate to simple	moderate
Capability to estimate total risk	no	difficult	extensive computations	no	yes	yes
Result adaptable for risk cost analysis	partial	yes	yes	no	yes	yes

or rainfall, and accounts for their statistical characteristics through a frequency analysis. The return period, T_r, is defined as the average length of time until the load equals or exceeds a specified magnitude of the resistance Q (Ang and Tang, 1975, p. 110). Thus, if T_r is in years, the probability that an event X equals or exceeds the design Q in a given year is

$$P(X \geq Q) = \frac{1}{T_r} \tag{8}$$

For hydrologic phenomena X is usually a continuous variable, hence $P(X = Q) = 0$. If the hydrologic risk is defined as the probability of occurrence of X greater than Q in any given year, then the non-failure annual probability is

$$P(X \leq Q) = 1 - \frac{1}{T_r} \tag{9}$$

Hence, the risk in an n-year period is

$$P(X > Q) = 1 - (1 - \frac{1}{T_r})^n \tag{10a}$$

or approximately,

$$P(X > Q) \approx 1 - e^{n/T_r} \qquad \text{for large } T_r \tag{10b}$$

$$P(X > Q) \approx n/T_r \qquad \text{for } T_r >> n \tag{10c}$$

In deriving Equations 8 through 10, two major assumptions are made (Yen, 1970):

(a) The occurrences of the random variable X are independent.
(b) The hydrologic system is time-invariant.

In other words, the occurrence of the events follow a Bernoulli sequence and the binomial distribution applies (Yen, 1970). However, in a natural system such as in a watershed, there is actually a seasonal variation, as well as geophysical long-term variations. The probability associated with the return period as expressed in Equations 8 through 10 nevertheless has been widely used as a measure of the risk for hydraulic structures (e.g., Borgman, 1963; Yen, 1970; Young et al., 1970; ASCE Task Committee, 1973; James and Lee, 1971). The hydrologic risk calculated according to Equation 10 for a specified return period does not require knowing the probability distribution of Q. The effect of the distribution of Q on the calculated risk using Equation 10 is indirect through the determination of Q for the specified T_r.

Although the risk calculation in Equation 10 is relatively direct and simple, it considers at most only part of the hydrologic risks. Other uncertainties associated with the loading and resistance are entirely ignored in this method. Hence, the total risk of a complex system cannot be evaluated by this method. In an attempt to account for other risks and uncertainties, sometimes a safety factor is adopted in the hydraulic analysis and design as a conjunctive supplement to the return period.

3.2 Method of Safety Factor

A safety factor has been used widely in various engineering fields as a protection against the engineer's ignorance, uncertainties and randomness of natural phenomena. By implication, a high safety factor implies a high reliability of the hydraulic structure, provided the values of the safety factor are measured in a consistent manner. The safety factor may be used alone or conjunctively with the return period to measure reliability. Yen (1979) has pointed out that at least six different types of safety factors have appeared in the literature, namely:

(a) pre-assigned safety factor
(b) central value safety factor
(c) estimated mean safety factor
(d) characteristic safety factor
(e) partial safety factor
(f) reduced safety factor

The arbitrary, pre-assigned safety factor is the type that is currently most often used in engineering design. On the other hand, in hydraulic projects the significant values of the parameters often are not the mean values but specified values. The characteristic factor is defined as

$$SF_c = R_o/L_o \tag{11}$$

in which R_o and L_o are the specified resistance and loading, respectively. Thus, SF_c is comparatively the most adaptable index for risk analysis of hydraulic structures.

However, the representativeness of the safety factor as a reliability measure depends on the extent that the factors contributing to the loading and resistance are accounted for. Yen (1979) described a first-order analysis that would provide a systematic, scientific, quantitative and comprehensive method for considering the various factors that influence system safety. Nevertheless, safety factors do not provide a direct measure of the total system risk; hence they are not adaptable for risk-cost analysis and cannot be incorporated directly into quantitative decision analysis.

3.3 Method of Direct Integration

In this method the risk is evaluated through a direct, analytical or numerical integration of the probability density function of the loading and resistance. In general, the risk can be expressed as

$$P_f = \int_0^\infty \int_0^\ell f_{C,L}(c,\ell)\, dc\, d\ell \tag{12}$$

in which $f_{C,L}(c,\ell)$ is the joint probability density function of C and L. For statistically independent C and L, Equation 12 can be simplified as

$$P_f = \int_0^\infty \int_0^\ell f_C(c)\, f_L(\ell)\, dc\, d\ell \tag{13}$$

or

$$P_f = \int_0^\infty f_L(\ell)\ F_C(\ell)\ d\ell \tag{14}$$

in which $F_C(\ell)$ is the cumulative distribution function of C. To facilitate the direct integration of Equations 12 or 14, the exact distribution functions of the loading, $f_L(\ell)$, and resistance, $f_C(c)$, or of the joint pdf, $f_{C,L}(c,\ell)$, should be analyzable and definable. If the distribution functions describe the loading and resistance correctly, then the risk evaluated by this method is exact.

Tang (1980) presented a procedure for incorporating the probability model uncertainty into risk evaluation. A direct integration was used to evaluate the hydrologic risk in his illustration. Wood (1977) assessed the overtopping and structural risks analytically with assumed probability density function of the flood and structural failure modes. No consideration was given to the uncertainties of hydrologic, hydraulic, and model error parameters. Tung and Mays (1980) evaluated the risks for a culvert or levee by estimating first the statistical parameters of loading and resistance from uncertainties of their parameters through a first order approximate formula, and then assuming distribution functions to the loading and resistance. Risk, as evaluated by using Equation 14, is found to be extremely sensitive to the distribution functions assumed. Therefore, an improper assumption of the distribution functions of the loading and resistance may negate the merit of accuracy of direct integration. Duckstein and Bogardi (1981) also studied the risk of a levee system due to various failure modes such as overtopping, boiling, slope sliding and wind wave erosion. The risk is estimated by direct integration of a joint density function of the load and resistance.

The biggest disadvantage of the direct integration method is the difficulty in the analytical derivation of the appropriate probability density functions of the loading and resistance, especially when the system is complex, such as is the case for a dam structure and its environment. The combination and numerical integration of a large number of influential factors to yield the total system risk may be a mathematical nightmare. Nevertheless, once $f_L(\ell)$ and $f_C(c)$ or $f_{C,L}(c,\ell)$ are defined, with the assistance of a computer, most of the integration can be performed numerically, although not necessarily economically in terms of computer time and cost. The direct integration method therefore is good only for simple systems or when highly accurate risk value is required.

3.4 Monte Carlo Simulation Method

The Monte Carlo method is a process using, in each simulation, a particular set of values of the random variable(s) artificially generated from the corresponding probability distribution(s). The simulation is generally performed by a computer. The expected value of risk can be estimated by examining the results of a large number of repetitive simulation runs and computed as the ratio of the number of failures to the number of simulations. For example, first, a set of sample values, X_i's, are generated according to the distribution function or statistical property of each X_i by means of a computer random number generator. Next, a corresponding value of the performance variable, Z in Equation 4, can be calculated. With sufficient repetitive sampling and calculation, a set of Z values are obtained. Finally, the risk of the failure event can be evaluated by the ratio of the number of negative Z to the total number of Z having been generated.

The Monte Carlo method has been used extensively in reliability problems of hydrologic and hydraulic projects and other engineering systems. For example, in a study on structural safety, Warner and Kabaila (1968) used the Monte Carlo method to obtain the distributions of the resistances, loads, and performance variables of a structure. In the design and operation of a multipurpose reservoir system, Askew et al. (1971) used the Monte Carlo technique to derive the optimum contract levels for the system, subject to the constraints imposed by stated maximum permissible risk to meet firm contract deliveries. Duckstein et al. (1981) estimated the probabilities of failure events for mine flooding by the Monte Carlo simulation method. Haan (1972) evaluated error probabilities of stochastic models as a function of the number of observations used in determining the parameters of the stochastic models of hydrologic problems with the Monte Carlo simulation method. Later, Matalas et al. (1975) and Wallis et al. (1977) applied this method to estimate the mean and standard deviation of skewness with several assumed distributions of flood sequences. Chow (1978) used the Monte Carlo method to generate sequences of hydrologic data for the study of the performance of a hydrologic system. Wen (1977) used Monte Carlo simulation results to verify his derivation of the statistics of combination of extreme loads.

The Monte Carlo method is a very flexible method that can be applied to a wide variety of problems. In fact, Monte Carlo simulation is perhaps the only solution technique for problems which cannot be solved analytically because of nonlinear behavior or complex system relationship. Despite its usefulness, the Monte Carlo simulation method has the following disadvantages:

(a) The risk estimated by using this method is not unique; it depends on the size of the samples and the number of runs, and is sensitive to the assumed distribution of the basic variables. It is never certain that the resultant statistical description indeed reflects the true moments of the joint probability distribution that is being simulated. The true risk is unknown and can only be approached by infinite samples or trials.

(b) The computer costs of Monte Carlo simulation increase substantially as the level of accuracy and the number of variables increase.

Hiller and Lieberman (1974) recommended that in general, if an equivalent analytical model is available, it should be used instead of Monte Carlo simulation.

3.5 Method of Reliability Index

As indicated previously in Equations 3 and 4, the failure of a system can be related to a performance variable Z which is a function of the resistance and loading variables, X_i,

$$Z = g(X_i) \qquad i = 1,2,\dots m \tag{15}$$

Hasofer and Lind (1974), among others, suggested that the reciprocal of the coefficient of variation of the performance variable Z may be considered as a convenient measure of the reliability of the structure, since additional information on the statistical characteristics of the component variables X_i are not necessary or may be unavailable. They called this quantity a reliability index β:

$$\beta = \frac{E(Z)}{\sigma_Z} \tag{16}$$

in which E(Z) and σ_Z are the expected value and standard deviation of Z, respectively. Thus, the reliability index may conveniently be considered as the distance from the origin (Z = 0) to the mean, E(Z), measured in standard deviation units. As such, β is a measure of the probability that Z will be less than zero, as shown in Figure 2. In other words, the reliability index behaves somewhat like a safety factor. As β increases, the risk, $P(Z < 0)$, decreases. But it does not give a direct estimate of the total system risk. There has been no known application of the reliability index to hydraulic structures.

4. METHODS OF FIRST-ORDER SECOND-MOMENT ANALYSIS

The first-order second-moment methods of risk analysis are recently developed powerful tools for estimating the total or component risks of a system. They can account for all the factors contributing to the system risk that one may conceive, whether their effects are large or small. The methods can consider factors from different disciplines, technical or nontechnical. Hence, they provide an avenue for dividing work among specialists in different disciplines and a framework for combining their results to give the total system risk, particularly when it is incorporated with fault tree or event tree analysis. These methods can simultaneously account for data of different quality from different disciplines or parameters, ranging from rudimentary assumptions to reliable statistical or theoretical values. They permit updating as new or improved data become available. For the individual factors, it is only necessary to know the estimated values of the mean and standard deviation of each parameter and not their respective probability distributions. Yet the required amount of calculation is moderate, usually far less than in either Monte Carlo simulation or, if it can be done numerically, direct integration.

The first-order analysis is an approximate analysis performed by truncating the second and higher order terms of the Taylor series expansion of the random variables. A second-moment analysis utilizes only the first two statistical moments, i.e., the expected value and variance (or coefficient of variation) of the random variables, evaluated at the point of Taylor's expansion (Ang and Tang, 1984). In actual engineering practice, the distributions of the constituent variables, X_i, i.e., $f_{X_1}(x_1)$, $f_{X_2}(x_2), \ldots, f_{X_{n+1}}(x_{n+1})$, $f_{X_{n+2}}(x_{n+2}), \ldots f_{X_m}(x_m)$, are usually not well defined; often, information on these variables is limited to means and variances. In such cases, approximations that are consistent with the type and quality of available information indeed are more sensible and the first-order second-moment method may be an appropriate approach.

4.1 Mean-Value First-Order Second-Moment (MFOSM) Method

In this method, the first-order Taylor's series for Z of Equation 15 is expanded about the mean values $\bar{x}_i$ of the variables X_i:

$$Z \approx g(\bar{x}_i) + \sum_{i=1}^{m} (X_i - \bar{x}_i) \frac{\partial g}{\partial X_i} \tag{17}$$

where the derivatives are evaluated at $\bar{x}_i = (\bar{x}_1 , \bar{x}_2 , \ldots \bar{x}_m)$. Taking the first and second moment of Z in Equation 17 and neglecting terms higher than the second order yields:

$$E(Z) \approx \bar{Z} = g(\bar{x}_i) \tag{18}$$

$$Var(Z) \approx \sum_{i=1}^{m} C_i^{\,2}\, Var(X_i) \tag{19}$$

where C_i's are the values of the partial derivatives $\partial g/\partial X_i$, evaluated at $\bar{x}_1, \bar{x}_2, \ldots \bar{x}_m$. The variables X_i's are assumed to be statistically independent. Accordingly,

$$\sigma_Z \approx [\sum_{i=1}^{m} (C_i\, \sigma_i)^2]^{1/2} \tag{20}$$

where σ_Z and σ_i are the standard deviations of Z and X_i, respectively.

The risk is evaluated from Equation 4. If Z is normally distributed, the risk is

$$P_f = 1 - \Phi\,(\frac{E(Z)}{\sigma_Z}) = 1 - \Phi(\beta) \tag{21}$$

in which $\Phi(\beta)$ is the cumulative standard normal distribution whose values can be found from standard tables (Ang and Tang, 1975; Benjamin and Cornell, 1970). In the MFOSM method the reliability index β given by Equation 16 is

$$\beta = \frac{g(\bar{x}_i)}{[\sum_{i=1}^{m} (C_i\, \sigma_i)^2]^{1/2}} \tag{22}$$

The risk evaluated from Equation 21 is approximate; it is exact when the basic variables, X_i's, are all normally distributed and when the functions g(.) can be expressed as a linear combination of the basic variables.

The MFOSM method as suggested in Equations 18, 20, and 22 has been used by engineers of different disciplines as a means of obtaining approximate probability of failure for complex engineering systems. Tang and Yen (1972) first introduced this method for hydraulic applications. Yen and Tang (1976) and Yen (1979) used this method to determine the relationship between the safety factor and the risk of a storm sewer failure by accounting for component uncertainties. Yen (1977) used this method to demonstrate how uncertainties due to pipe roughness could be systematically accounted for and how the risk associated with this design due to those uncertainties could be assessed. Later, Yen et al. (1980) applied this method to the hydraulic design of culverts with hydrologic and hydraulic uncertainties being considered. Prendergast (1979) adopted this method as the basis for evaluating the safety of a concrete gravity dam. The probabilities of sliding failure and overturning failure of the gravity dam were evaluated by considering various load effects which include reservoir stage, hydrostatic pressure, weight of the dam structure, uplift at the base of the dam and earthquake forces. The hydrologic and hydraulic aspects of dam safety, however, were not considered. Cheng (1982) applied the MFOSM method for a comparative study of overtopping risk of an earth dam. Tung and Mays (1980) used Equations 18 and 20 to calculate the mean and variance of the load and resistance from their component variables X_i's. The risk was evaluated by direct integration of Equation 14 with assumed probability distributions for the loading and resistance. It was shown that different assumed probability distributions of the loading and resistance could lead to significantly different values of risk.

The MFOSM method has been shown to be relatively simple, able to account for the various sources of uncertainties, adaptable for reevaluation with additional data and suitable for practical applications. However, this method has certain disadvantages and weaknesses, including the following (Cornell, 1972; Ditlevsen, 1973; Lind, 1977):

(a) In civil engineering projects, the events of failure often happen at extreme values, e.g., high floods or large earthquakes, rather than near the mean of the loads and resistance. Such variables are mostly associated with large variance and skewed probability distributions. Furthermore, civil engineering systems usually exhibit nonlinear behavior. The risk value estimated using the MFOSM method which linearizes the performance function g(.) and evaluates it at the mean values of variables X_i's may differ considerably from the actual risk.

(b) The risk value depends on how the performance function, g(.), is formulated. That is, different values of risk, p_f, (or reliability index, β) are obtained for different formulations of g(.). For example, as shown in Equations 3 and 4, defining Z as $C - L$, $(C/L) - 1$, or $\ln(C/L)$ could yield different values of p_f or β.

The aforementioned shortcomings could be overcome by considering the higher order terms in the Taylor series expansion of g(.) (Ang and Tang, 1975; Fiessler et al., 1979). However, this improvement requires additional statistical information, such as the skewness, which is usually inaccurate or unavailable in most situations. In addition, the calculation procedure with such improvement is cumbersome. One practical technique for improving the accuracy is the advanced first-order second-moment method proposed by Rackwitz (1976), which is presented in next subsection.

4.2 Advanced First-Order Second-Moment (AFOSM) Method

The essence of this method is linearizing the performance function $g(X_1, X_2, \ldots X_m)$ using the Taylor series expansion at a likely failure point $(\underline{x_1}^*, \underline{x_2}^*, \ldots \underline{x_m}^*)$ on the failure surface, rather than at the mean values as in the MFOSM method. The determination of the failure point, $\underline{x_i}^*$, however, is generally not a simple task; it is not known a priori, and therefore it needs to be determined iteratively. The algorithm for finding this point has been shown in a CIRIA (1977) report, has been improved by Cheng (1982) and is summarized as follows, with the assumption that the variables X_i's are uncorrelated.

The expected value and standard deviation of Z as approximated by first-order Taylor series expansion with respect to a point x_i^* can be written as

$$E(Z) \approx g(x_i^*) + \sum_{i=1}^{m} C_i(\bar{x}_i - x_i^*) \tag{23}$$

$$\sigma_Z \approx [\sum_{i=1}^{m} (C_i \ \sigma_i)^2]^{1/2} \tag{24}$$

in which $C_i = \partial g/\partial X_i$ are evaluated at the point $x_i^* = (x_1^*, x_2^*, \ldots x_m^*)$. The expression of σ_Z may be written in a linearized form as

$$\sigma_Z = \sum_{i=1}^{m} \alpha_i \ C_i \ \sigma_i \tag{25}$$

in which α_i's are sensitivity factors and are evaluated from

$$\alpha_i = \frac{C_i \ \sigma_i}{[\sum_{j=1}^{m} (C_j \ \sigma_j)^2]^{1/2}} \tag{26}$$

Substituting Equations 23 and 25 into Equation 16, the reliability index evaluated for the AFOSM method is

$$\beta = \frac{g(x_i^*) + \sum_{i=1}^{m} C_i(\bar{x}_i - x_i^*)}{\sum_{i=1}^{m} \alpha_i \ C_i \ \sigma_i} \tag{27}$$

If x_i^* is on the failure surface, then

$$g(x_i^*) = 0 \tag{28}$$

Substituting Equation 28 into Equation 27 and solving for the failure point x_i^*,

$$x_i^* = \bar{x}_i - \alpha_i \beta \sigma_i \qquad \text{for all } i \tag{29}$$

Determination of the failure point x_i^* from Equations 26, 27 and 28 requires iteration, or alternatively constrained nonlinear optimization as proposed by Cheng (1982). Equation 29 may be regarded as a series of equality constraints. The failure surface may be found by minimizing objective function $|g(x_i^*)|$ subject to the constraints of Equation 29 and the definitions of α_i and β as given by Equations 26 and 27, respectively. Cheng (1982) proposed to find the failure point and β by using the generalized reduced gradient (GRG) algorithm (Abadie and Carpentier, 1969).

The risk computed from Equation 21 using α given by Equation 27 at the failure point provides an improvement over that given by Equation 22 if the variables exhibit nonlinear behavior. The risk evaluated is exact, in the first order sense, only for linear failure surfaces (i.e., Z = 0) where all the basic variables are normally distributed. For convex surfaces where all the basic variables are normally distributed, the risk is bounded as

$$1 - \phi(\beta) \leq P_f \leq 1 - \chi_m^2(\beta) \tag{30}$$

in which $\chi_m^2(.)$ is a chi-squared distribution with m degrees of freedom. Rackwitz (1976) pointed out that if the distance function from the system mean to the failure surface shows distinct minima, then the lower bound will generally serve as a relatively good probability estimate within the accuracy of the first-order approximation.

However, for most real hydrologic systems, not all the basic variables are normally distributed. Rackwitz (1976) proposed a technique for transforming the non-normal variables into an equivalent normal distribution. The values of the cumulative distribution function, CDF, and the probability density function of the transformed equivalent normal distributions are the same as those of the original non-normal distributions at the failure point, i.e.,

$$F_{x_i}(x_i^*) = \Phi\left(\frac{x_i^* - \bar{x}_i^N}{\sigma_i^N}\right) \tag{31}$$

$$f_{x_i}(x_i^*) = f^N\left(\frac{x_i^* - \bar{x}_i^N}{\sigma_i^N}\right)/\sigma_i^N \tag{32}$$

in which $F_{x_i}(x_i^*)$ and $f_{x_i}(x_i^*)$ are the CDF and pdf of x_i at x_i^*, and $\Phi(.)$ and $f^N(.)$ are the CDF and pdf of the standard normal distribution, respectively. In order to do this, Rackwitz approximated the non-normal distribution function by a linear (i.e., first-order) Taylor series expansion. Thus, the mean, $\bar{x}_i^N$, and the standard deviation, σ_i^N, of the equivalent normal distributions become

$$\bar{x}_i^N = x_i^* - \Phi^{-1}(F_{x_i}(x_i^*))\ \sigma_i^N \tag{33}$$

$$\sigma_i^N = \frac{f^N[\Phi^{-1}(F_{x_i}(x_i^*))]}{f_{x_i}(x_i^*)} \tag{34}$$

Hence, the constraints for the nonlinear optimization problem become

$$\bar{x}_i^N - x_i^* - \alpha_1\ \beta\ \sigma_i^N = 0 \qquad \text{for all } i \tag{35}$$

where

$$\alpha_i = \frac{C_i\ \sigma_i^N}{[\sum_{j=1}^{m} (C_j \sigma_j^N)^2]^{\frac{1}{2}}} \tag{36}$$

Rackwitz and Fiessler (1977) pointed out that this normal transformation of non-normal variables is exact within the accuracy of the first-order analysis under consideration. Therefore, estimating risk as the lower bound in Equation 30 appears to be nearly exact within the accuracy of the first-order approximation for systems with convex failure surfaces where all the basic variables are either normally distributed or transformed normal.

4.3 Comparison of MFOSM and AFOSM Methods

Information on comparison between MFOSM and AFOSM can be found, for example, in Cheng (1982). One of Cheng's results is reproduced here as Figure 3. In his study the calculated risks of MFOSM and AFOSM are compared with the exact values obtained from direct integration and also with the risks obtained from Monte Carlo simulation. To make direct integration possible, the variables are assumed to be normally, uniformly, or log-normally distributed. Some of Cheng's significant findings are briefly given as follows.

(a) The computed values of the risk using the MFOSM method vary with the definition of the performance parameter Z used. For the conditions investigated, among the three definitions given in Equation 3, Z = C - L offers the best estimate, whereas Z = (C/L) - 1 is the worst. The differences become more pronounced when the risk becomes smaller. The risk calculated by the MFOSM method, irrespective of the definition of Z, is invariant with the distributions of the variables provided the values of the mean and standard deviation remain unchanged. The method gives acceptable estimates when the risk is high, e.g., $P_f > 0.1$. However, when P_f becomes small, only the results of Z = C - L may be acceptable.

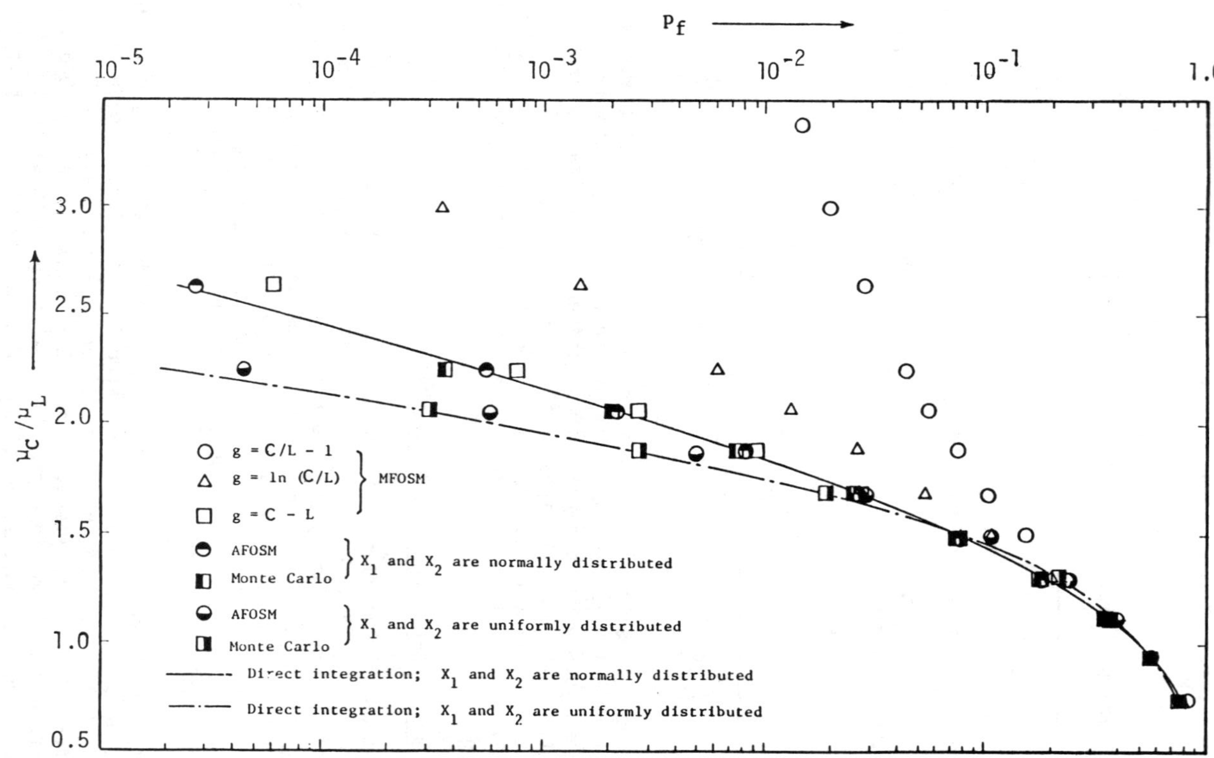

Fig. 3. Comparison of risk values calculated by different methods.

(b) The AFOSM method generally gives fairly accurate values of the risk irrespective of the definition of Z. However, the results do differ slightly for different distributions of the variables.

(c) The results of AFOSM with the failure point located either by using GRG or by iteration are practically the same when the risk is large. However, the iteration method diverges frequently when the risk is small. For large risks ($P_f > 0.1$), the computer time for GRG algorithm is longer than that of iteration, whereas for small risks ($P_f < 0.01$) the computer time for GRG is clearly shorter. The computer time required for AFOSM is consistently longer than that for MFOSM, but considerably shorter than that for Monte Carlo simulation.

(d) Considering the accuracy and consistency of the computed results, the sensitivity to the assumed probability distributions of basic variables and definition of the performance parameter, and the computer time and cost, Cheng recommended the use of AFOSM with GRG for dam safety analysis.

As mentioned previously, the MFOSM method has been applied by a number of researchers to various hydraulic and hydrologic problems. Cheng (1982) applied the AFOSM method to evaluate the overtopping risk of an existing earth dam in northern Illinois and compared the result with that of the MFOSM method. Melching and Yen (1984) recently applied AFOSM to a sewer design problem which was previously analyzed using MFOSM (Yen et al., 1976). In both cases the advantages of AFOSM were demonstrated.

REFERENCES

Abadie, J. and J. Carpentier (1969) Generalization of the Wolfe reduced gradient method to the case of nonlinear constraints, In: Optimization, ed. by R. Fletcher, Academic Press, London.

Ang, A. H.-S. and W. H. Tang (1975) Probability Concepts in Engineering Planning and Design, Vol. 1, Basic Principles, John Wiley & Sons, New York.

Ang, A. H.-S. and W. H. Tang (1984) Probability Concepts in Engineering Planning and Design, Vol. 2, Decision, Risk, and Reliability, John Wiley & Sons, New York.

ASCE Task Committee on the Reevaluation of the Adequacy of Spillways of Existing Dams (1973) Reevaluating spillway adequacy of existing dams, J. Hydraulics Div., ASCE, Vol. 99, No. HY2, pp. 337-372.

Askew, A. J., W. W.-G. Yeh and W. A. Hall (1971) Use of Monte Carlo techniques in the design and operation of a multipurpose reservoir system, Water Resour. Res., Vol. 7, No. 4, pp. 819-833.

Benjamin, J. R. and C. A. Cornell (1970) Probability, Statistics and Decision for Civil Engineering, McGraw-Hill, New York.

Borgman, L. E. (1963) Risk criteria (1963) J. Waterways and Harbors Div., ASCE, Vol. 89, No. WW3, pp. 1-35.

Cheng, S. T. (1982) Overtopping risk evaluation for an existing dam, Ph.D. thesis, Department of Civil Engineering, University of Illinois at Urbana-Champaign.

Chow, V. T. (1978) Evolution of stochastic hydrology, In: Applications of Kalman Filter to Hydrology, Hydraulics and Water Resources, ed. by C. L. Chiu, University of Pittsburgh, Pennsylvania, pp. 13-28.

CIRIA. (1977) Rationalisation of safety and serviceability factors in structure codes, Report 63, London.

Cornell, C. A. (1972) First-order analysis of model and parameter uncertainty, Proceedings, Intern. Symp. on Uncertainties in Hydrologic and Water Resources Systems, Tucson, Arizona, Vol. 2, pp. 805-825.

Ditlevsen, O. (1973) Structural reliability and the invariance problem, Solid Mechanics Div. Res. Rept. 22, University of Waterloo, Ontario.

Duckstein, L. and I. Bogardi (1981) Application of reliability theory to hydraulic engineering design, J. Hydraulics Div., ASCE, Vol. 107, No. HY7, pp. 799-815.

Duckstein, L., I. Bogardi and F. Szidarovsky (1981) Reliabiilty of underground flood control system, J. Hydraulics Div., ASCE, Vol. 107, No. HY7, pp. 817-825.

Fiessler, B., H.-J. Neumann and R. Rackwitz (1979) Quadratic limit states in structural reliability, J. Engineering Mechanics Div., ASCE, Vol. 105, No. EM4, pp. 661-676.

Haan, C. T. (1972) Adequacy of hydrologic records for parameter estimation, J. Hydraulics Div., ASCE, Vol. 98, No. HY8, pp. 1387-1393.

Hasofer, A. M. and N. C. Lind (1974) Exact and invariant second-moment code format, J. Engineering Mechanics Div., ASCE, Vol. 100, No. EM1, pp. 111-121.

Hillier, F. S. and G. J. Lieberman (1974), Operations Research, Holden-Day, Inc., San Francisco.

James, L. D. and R. R. Lee (1971) Economics of Water Resources Planning, McGraw-Hill, New York.

Lind, N. C. (1977) Formulation of probabilistic design, J. Engineering Mechanics Div., ASCE, Vol. 103, No. EM2, pp. 273-282.

Matalas, N. C., J. R. Slack and J. R. Wallis (1975) Regional skew in search of a parent, Water Resour. Res., Vol. 11 No. 6, pp. 815-826.

Melching, C. S. and B. C. Yen (1984) Slope influence on storm sewer risk, Proceedings, 4th Intern. Symp. on Stochastic Hydraulics, Urbana, Illinois, pp. 188-198.

Prendergast, J. D. (1979) Probabilistic concept for gravity dam analysis, Special Report M-265, U.S. Army Corps of Engineers, Construction Engineering Research Laboratory, Champaign, Illinois.

Rackwitz, R. (1976) Practical probabilistic approach to design, Bulletin 112, Comtie European du Beton, Paris, France.

Rackwitz, R. and B. Fiessler (1977) Non-normal vectors in structural reliability, SFB 96 Report 29, Technical University of Munich, pp. 1-22.

Tang, W. H. (1980) Bayesian frequency analysis, J. Hydraulics Div., ASCE, Vol. 106, No. HY7, pp. 1203-1218.

Tang, W. H. and B. C. Yen (1972) Hydrologic and hydraulic design under uncertainties, Proceedings, Intern. Symp. on Uncertainties in Hydrologic and Water Resources Systems, Tucson, Arizona, Vol. 2, pp. 866-882.

Tung, Y.-K. and L. W. Mays (1980) Optimal risk-based design of water resource engineering projects, Technical Report CRWR-171, Center for Research in Water Resources, University of Texas at Austin, Texas.

Wallis, J. R., N. C. Matalas and J. R. Slack (1977) Apparent regional skew, Water Resour. Res., Vol. 13, No. 1, pp. 159-182.

Warner, R. F. and Kabaila, A. P. (1968) Monte Carlo study of structural safety, J. Structural Div., ASCE, Vol. 94, No. ST12, pp. 2847-2859.

Wen, Y.-K. (1977) Statistical combination of extreme loads, J. Structural Div., ASCE, Vol. 103, No. ST5, pp. 1079-1093.

Wood, E. F. (1977) An analysis of flood levee reliability, Water Resour. Res., Vol. 13, No. 3, pp. 665-671.

Yen, B. C. (1970) Risk in hydrologic design of engineering projects, J. Hydraulics Div., ASCE, Vol. 96, No. HY4, pp. 959-966.

Yen, B. C. (1977) Uncertainties on roughness for pipe design, In: Stochastic Processes in Water Resources Engineering, (Proceedings 2nd IAHR Intern. Symp. on Stochastic Hydraulics, Lund, Sweden, 1976), Water Resources Publications, Littleton, Colorado, pp. 369-391.

Yen, B. C. (1979) Safety factors in hydrologic and hydraulic engineering design, In: Reliability in Water Resources Management, ed. by E. A. McBean, K. W. Hipel and T. E. Unny, Water Resources Publications, Littleton, Colorado, pp. 389-405.

Yen, B. C. and A. H.-S. Ang (1971) Risk analysis in design of hydraulic projects, In: Stochastic Hydraulics, (Proceedings 1st Intern. Symp. on Stochastic Hydraulics), University of Pittsburgh, Pennsylvania, pp. 694-709.

Yen, B. C. and W. H. Tang (1976) Risk-safety factor relation for storm sewer design, J. Environmental Engineering Div., ASCE, Vol. 102, No. EE2, pp. 509-516.

Yen, B. C., S.-T. Cheng and W. H. Tang (1980) Reliability of hydraulic design of culverts, Proceedings, IAHR APD Intern. Conf. on Water Resources Development, Taipei, Taiwan, Vol. 2, pp. 991-1001, Taipei, Taiwan.

Yen, B. C., Wenzel, H. G., Jr., Mays, L. W. and Tang, W. H. (1976) Advanced methodologies for design of storm sewer systems, Research Report 112, Water Resources Center, University of Illinois at Urbana-Champaign.

Young, G. K., R. S. Taylor and L. S. Costello (1970) Evaluation of the flood risk factor in the design of box culverts, vol. 1, theoretical development, Report No. FHWA-RD-74-11, U.S. Federal Highway Administration, Washington, DC.

PROBABLISTIC DESIGN OF WATER-RETAINING STRUCTURES

Drs. IR. J.K. Vrijling

Delft Hydraulic Laboratory, Ministry of Public Works, P.O. Box 177, 2600 MH Delft, The Netherlands

ABSTRACT

Holland has a rich tradition in the field of dikes. This knowledge, together with the history of dam disasters, shows that complete safety is unattainable. Realizing this, a method for assessing the probability of failure of a system of water-retaining structures has to be developed.

First, all possible failure mechanisms of the structures and all other possible causes (management error, human error, etc.) have to be determined.

Then, the relation between all possible failure mechanisms and the ultimate consequence of a flood or complete drawdown has to be analyzed. The fault tree is a very helpful tool for solving this problem.

Third, the probability of failure of the various mechanisms has to be determined by means of probabilistic calculations, and the probability of occurrence of the other causes has to be estimated on the basis of historical data.

Now, the probabilities of the base events (failure mechanisms and other causes) may be combined in the fault tree to derive the probability of failure of the water retaining system. And in the end, one must ask if this probability is acceptable from a socioeconomic point of view.

1. INTRODUCTION

In this paper, the developments in the field of the probabilistic design of water-retaining structures in Holland are outlined. Although the theoretical methods were already known, the practical application of probabilistic methods was stimulated by the design of the storm surge barrier Oosterschelde. Experience with these methods enabled the designers to unify the design of structures, mechanical equipment and management into one approach. This, in turn, led to developments in the field of dike design. At this stage, the methods are well known but the application is limited to difficult cases. Thus, the new dune design regulation is based on probabilistic reasoning. And a pump-storage scheme with 50 m high dams currently under design will be evaluated along probabilistic lines.

2. SYSTEM DESCRIPTION

A probabilistic analysis aims at identifying all possible causes of failure of the water-retaining system. Every cause may eventually lead to inundation of the hinterland or to the loss of precious water. The depth of the analysis depends on the description of the water retaining in detail. A complete analysis is only possible if a very detailed description (e.g. as construction files and on site measurements) is available. In the design stage the analysis is necessarily simplified.

In this paper we will confine ourselves to a very schematic description of the flood defence system of Holland. Holland is in principle lying at or below the low tide level of the sea. It is protected from sea floods by a system of defences that consists of dikes and dunes. The dikes are partly shielded from severe wave attack by shoals (see Fig. 1).

The system is continued along the river. Here the dike changes gradually from a sea dike into a river levee as the tidal movement dampens upstream. In the dike along the river, a sluice gives access to an old harbour. The sluice has to be closed by hand at water levels exceeding mean high water. The main harbours are situated outside the dikes at a high level, and are thus of no concern for this study.

The typical cross-section of a modern sea dike consists of a body of sand covered with mattresses and asphalt constructions in the zone attacked by waves and current. The crest and the inside slope are covered by a layer of clay with grass on top (see Fig. 2).

A dune is a naturally deposited mass of sand (see Fig. 3) that is in a state of dynamic equilibrium. During heavy weather, sand is lost to the sea in a reshaping process that enables the dune to withstand wave attack. During the summer the loss is regained by accretion. The waves transport sand to the beach and the wind takes it further inland.

A typical cross-section of a river levee is given in Figure 4. The dike stands mostly on a layer of alluvial clay and is also covered with clay. On the clay grows grass.

3. FAILURE MECHANISMS

Good engineering practice requires that attention should be given to all possible modes of failure of the construction under design. This is a common approach in the design of concrete or steel structures. In the design of water-retaining structures such as dams, dikes, and dunes, the approach is gaining ground, especially in combination with probabilistic reasoning. This is the result of the influence of the design of the storm surge barrier in the Oosterschelde.

A nonexhaustive overview of the failure mechanisms of dikes or dams is given below and illustrated in Figure 5.

- Overtopping is a well known mechanism, which leads to water entering the polder and to soaking of the dike.

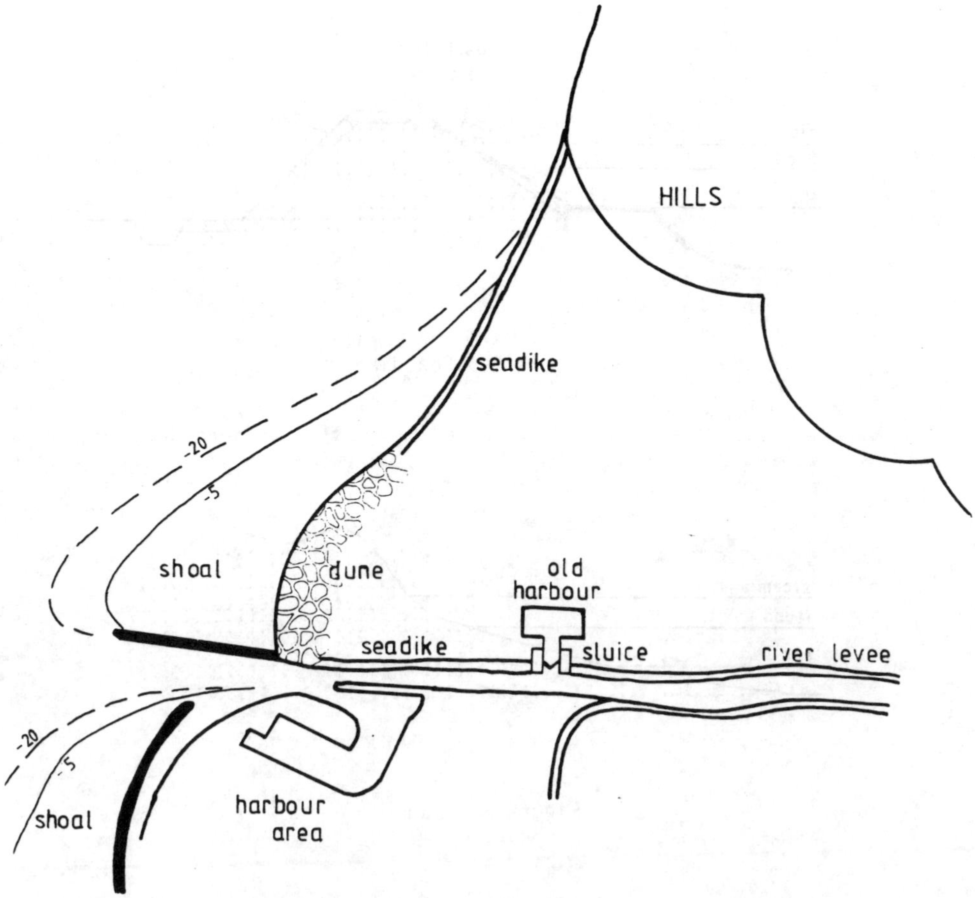

Figure 1. Schematic situation

- Wave overtopping is also a mechanism that gets a lot of attention in dike design. In this case, the amount of water entering the polder is negligible, so the dangerous consequences result from the soaking of the body of the dike and erosion of the inner slope.

- A slip circle at the inner slope may be caused among other things by a high phreatic plane in the dike. This will be the case when the duration of the high water level is long or permanent.

- Micro instability of the soil material at the inner slope may result from seepage and a high phreatic plane.

- Erosion of the outer slope may be caused by wave attack. The waves may be wind waves or displacement waves from ships.

- Erosion of the foreshore is caused by tidal or wave induced currents.

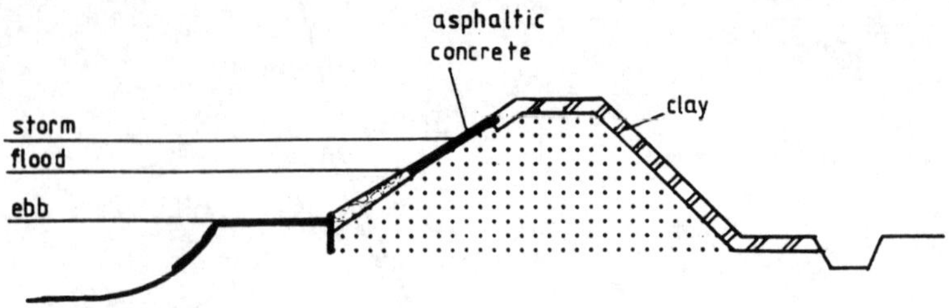

Figure 2. Sea dike.

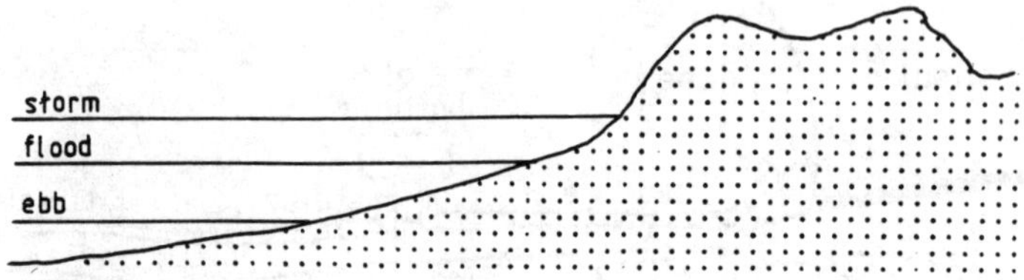

Figure 3. Dune.

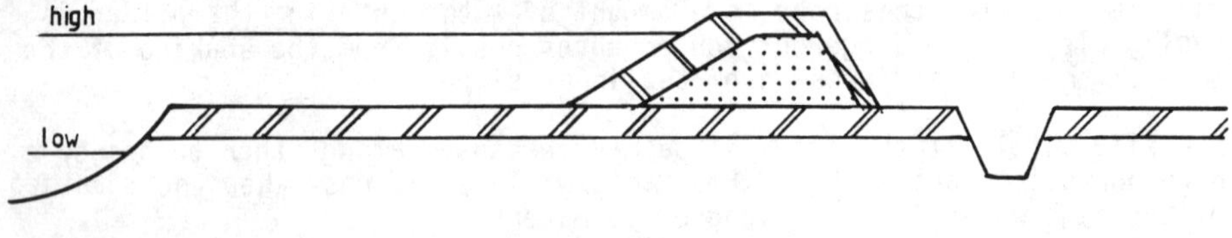

Figure 4. River levee.

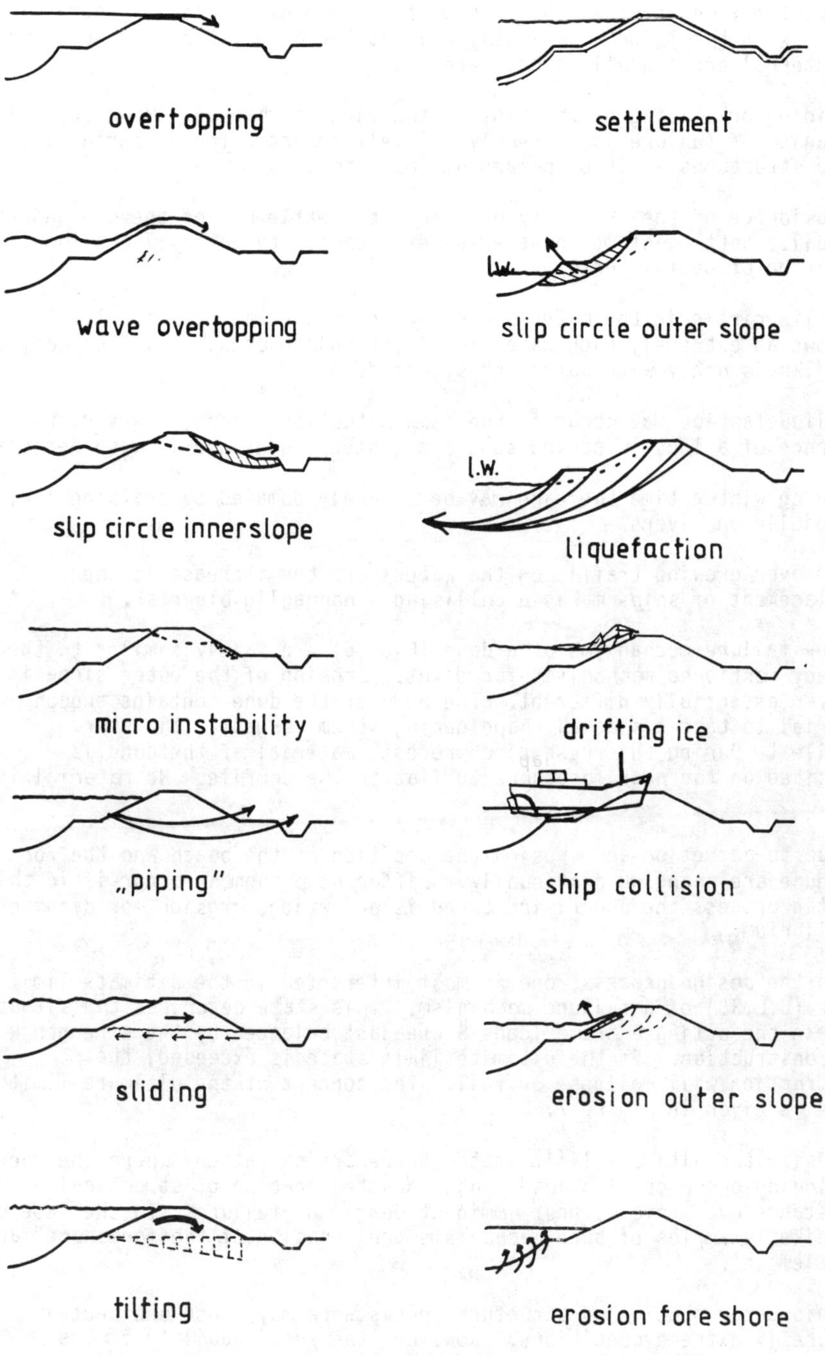

Figure 5. Overview of the failure mechanisms of a dike.

- Piping may occur, i.e. the gradual formation of a material entraining well. When the "pipe" eventually reaches the high water side, the process of internal erosion will accelerate.

- Sliding or tilting of the body of the dike may happen. However, this mechanism of failure is extremely unlikely to occur for an earth dam. For rigid structures it is of paramount importance.

- Subsidence of the crest may occur due to settlement of the dam and the subsoil. Settlement may however also be caused by internal erosion or by oxidation of peat layers.

- A slip circle in the outer slope may occur when a low water level follows an extremely high water level (or sudden drawdown). The body of the dike is heavy with water and slides down.

- A liquefaction may occur in the same situation. Here, however, the presence of a loosely packed sand and a steep foreshore is also necessary.

- During winter time the dike may be severely damaged by drifting ice, especially on rivers.

- The ever growing traffic on the waters and the increase in the displacement of ships makes a collision a nonnegligible risk.

The failure mechanisms of a dune (Fig. 6) are fairly similar to the already mentioned mechanisms for dikes. Erosion of the outer slope is however essentially different. The body of the dune contains enough material to take a special shape during storm surges: "the storm profile." During this reshaping process, material of the dune is deposited on the near foreshore to flatten the profile. No material is lost.

Due to accretion and erosion the position of the beach and the foot of the dune are changing continually. Different phenomena occur if in this erratic process the underlying trend is accretion, erosion, or dynamic equilibrium.

In the design process, one is most interested in the ultimate limit state (U.L.S.) of a failure mechanism. This state describes the situation wherein the acting extreme loads S are just balanced by the strength R of the construction. If the ultimate limit state is exceeded, the construction will collapse or fail. The concept of the ultimate limit state is given in Figure 7.

Beside the ultimate limit state, there are situations where the ever continuing presence of a load causes a deterioration of structural resistance over time without imminent danger of failure. In the case of the dike, examples of such mechanisms are "erosion of the foreshore" and "settlement".

This deterioration of structural resistance may cause unexpected failure in extreme conditions. However, the serviceability of the structure is often hampered before failure (excessive leakage due to

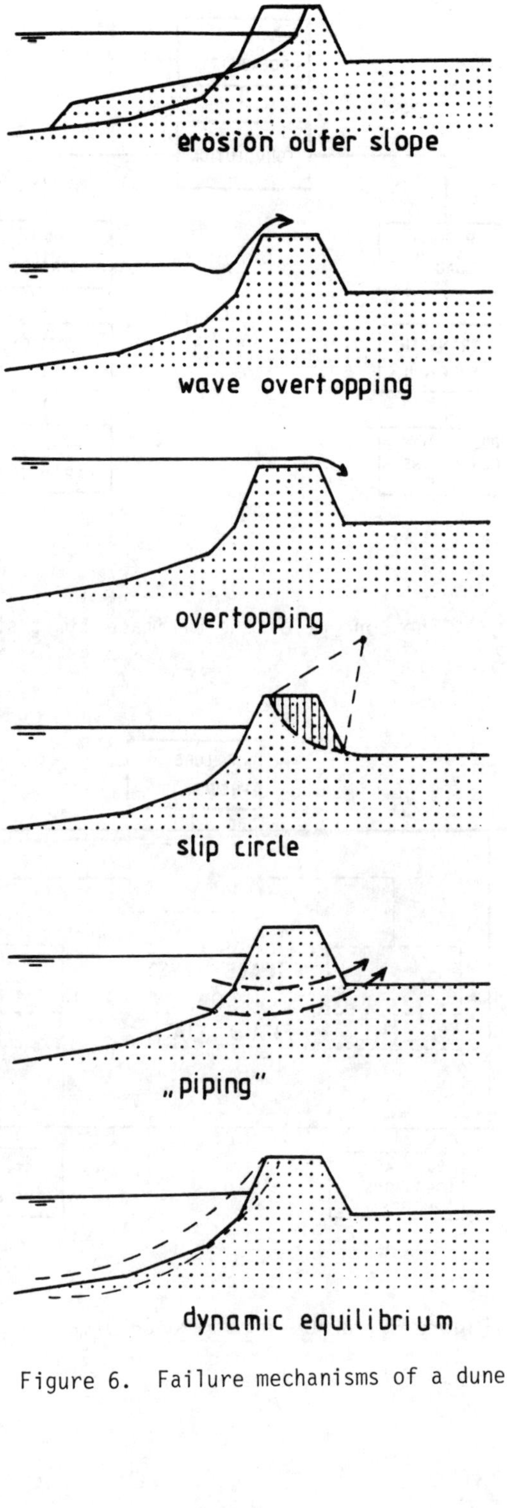

Figure 6. Failure mechanisms of a dune.

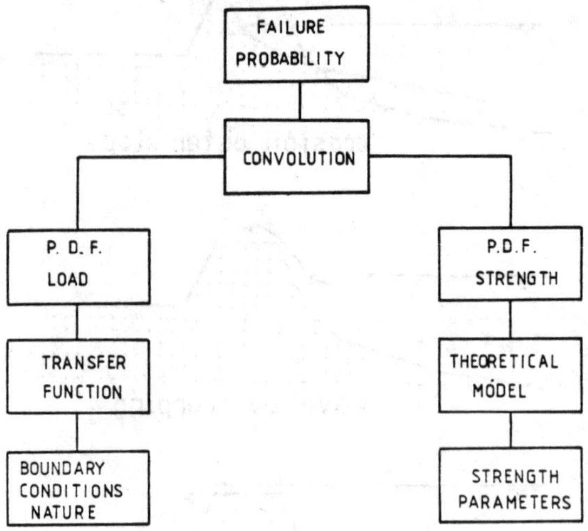

Figure 7. The concept of the ultimate limit state (U.L.S.).

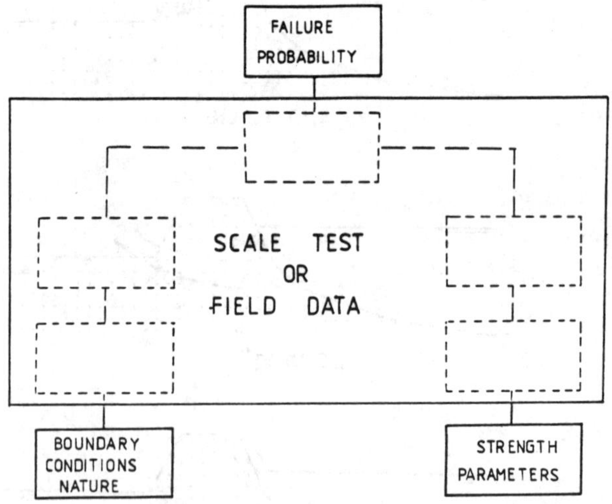

Figure 8. BLACK BOX APPROACH of a limit state.

piping). The serviceability limit state (S.L.S.) is treated in the same way as the ultimate limit state.

A point of great practical importance is that a serviceability limit state, i.e. a deterioration of structural resistance over time, can be improved in two ways:

1. by increasing the resistance to guarantee sufficient strength during the service life.

2. by controlling the deterioration of the resistance through inspection and maintenance procedures.

The second solution, although often economically feasible, introduces a certain risk because the structural safety now depends partly on the care of other people.

The application of limit state analysis presupposes in a certain way that the transfer functions necessary to transform boundary conditions into loads and the theoretical models defining the resistance (see Fig. 7) are mathematically known and manageable.

In the field of dike and dune design, this is not generally true. Especially for erosion and scour processes, neither the transfer function for transforming waves and current into forces nor the theoretical models for the stability of grains are exactly known. Here, the simulation of the limit state in a scale model may yield a solution (see Fig. 8).

A scale model of the structure is exposed to combinations of the natural boundary conditions. The amount of damage done to the model is correlated with the boundary conditions to develop a limit state equation. However model tests require utmost care in the interpretation of the results for a number of reasons. First, the scaling is done on the basis of the assumed physical laws of the limit state. If the assumptions are not correct, the results will be unreliable. Second, the scaling is only correct for one mechanism. That is, in breakwater tests, the armour units are far stronger than in prototype. So in reality the breakwater collapsed as a consequence of armour unit breakage. Third, limit state equations concentrate on equilibrium between load and resistance. In scale tests only damage is observable. Thus, loads have to exceed resistances by a margin in the test.

In some cases where the physical laws governing the phenomena are not exactly known, field data of boundary conditions, resistance parameters and damage are preferred as a basis for correlation. Difficulties may arise, however, in extrapolating the limit state equation to extreme loads (U.L.S) where field data are missing. For dune erosion, the field data were extrapolated by extensive scale model tests (Vellinga 1983).

4. RISK ANALYSIS OF THE SYSTEM BY MEANS OF THE FAULT TREE APPROACH

The goal of designing a flood defence system is to provide a certain safe protection against inundation for people and their property.

The modes of failure of dikes and dunes were listed above. Now the total probability of inundation of the polder (see Fig. 1) has to be assessed, taking into account the mechanisms described and the sluice.

Looking at the defense system, it is clear that it is a series system. If during a storm surge one of the elements fails, the sea dike, the dune, the sluice or the dikes along the river, then the polder will be inundated. For this simple case, the main fault tree is given in Figure 9. The link between failure of a dike section and the limit states of the failure mechanisms is analysed in a more detailed tree (see Fig. 10). The same fault tree is also suitable for describing the failure of a dune. The main difference is that erosion of the outer slope starts immediately under wave attack, as a protecting revetment is generally not present. Further, the actual position of the beach and the dune at the moment of the storm surge is uncertain due to the dynamic equilibrium of accretion and erosion under normal conditions. However, if the beach and the dune recede too far as a result of gradual erosion, the original profile will be re-established by beach nourishment (maintenance; see Fig. 11).

The influence of human action is evident in the case of the sluice. On the one hand, the sluice may fail due to technical failure mechanisms, such as loss of stability, collapse of the doors, or piping, or on the other hand due to human error the sluice may stay open during a storm (see Fig. 12). The analysis of the sluice is very schematic because no detailed description of the structure is available.

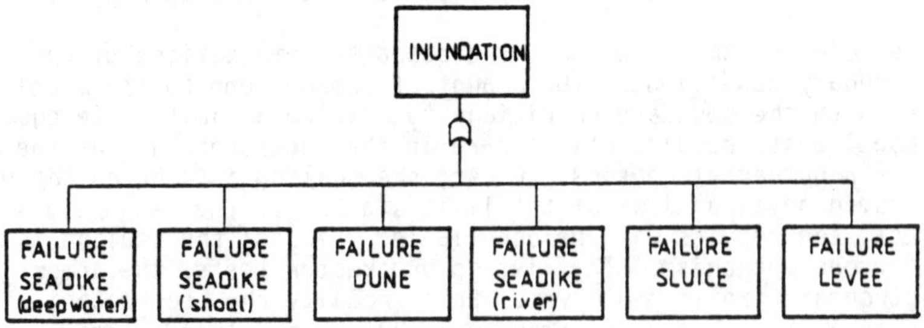

Figure 9. The main fault tree of the flood defence system

5. EVALUATION OF THE PROBABILITY OF FAILURE

To evaluate the probability of inundation, the probabilities of failure of all the mechanisms must be known. Every possible limit state contributes in principle to the total probability of a disaster.

The probability of failure of a mechanism may be found in two ways:

1. assessment of historical data

2. probabilistic calculations

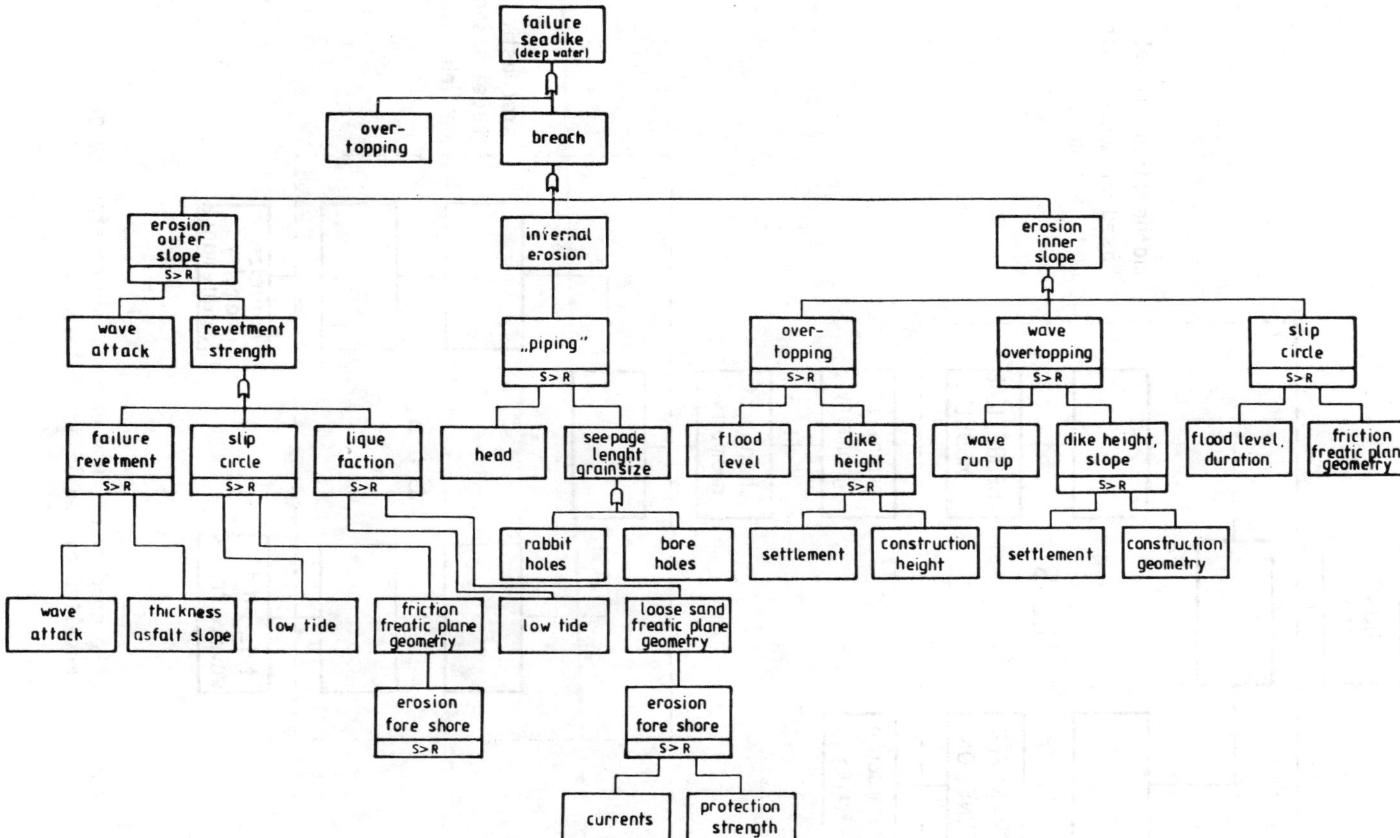

Figure 10. The fault tree of a dike section.

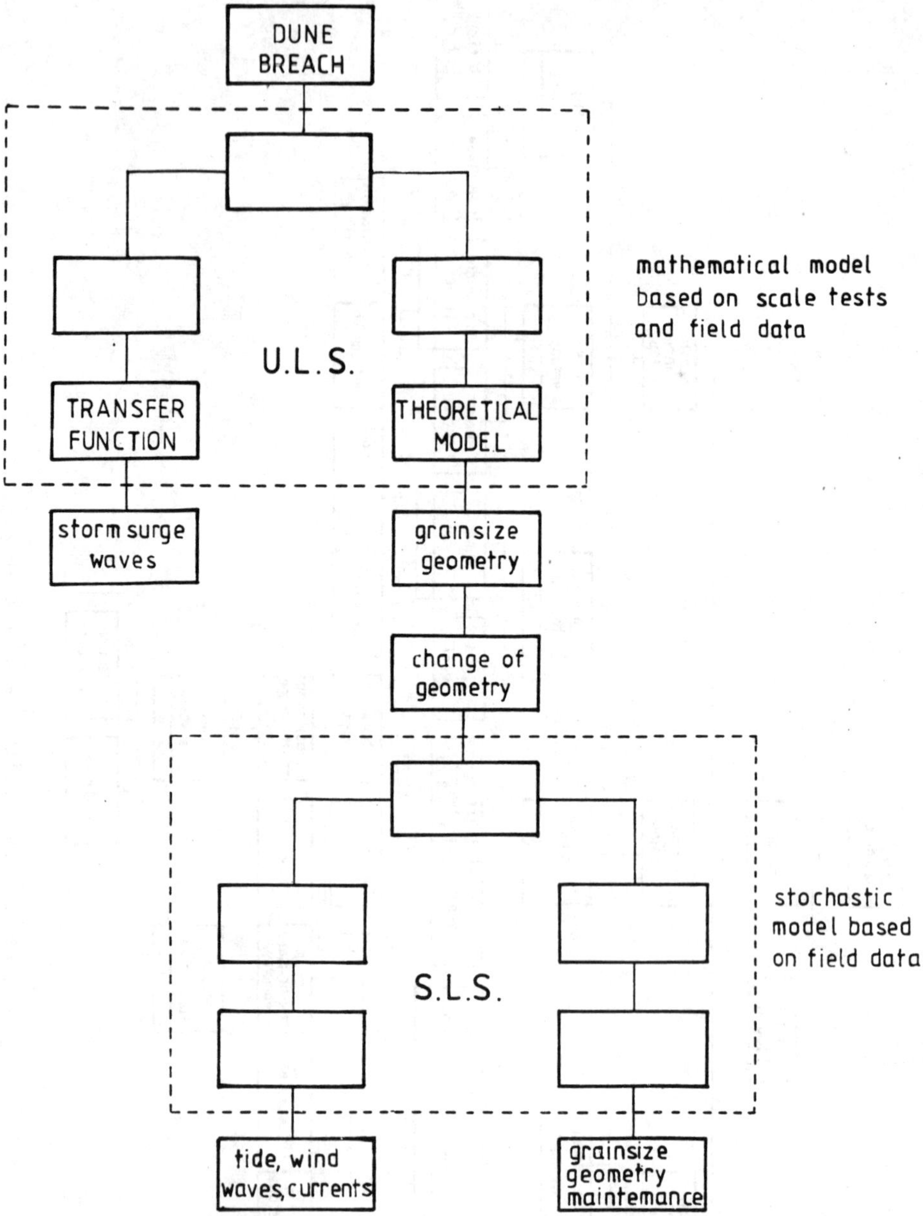

Figure 11. The process of dune erosion during storm surge modelled by two limit states.

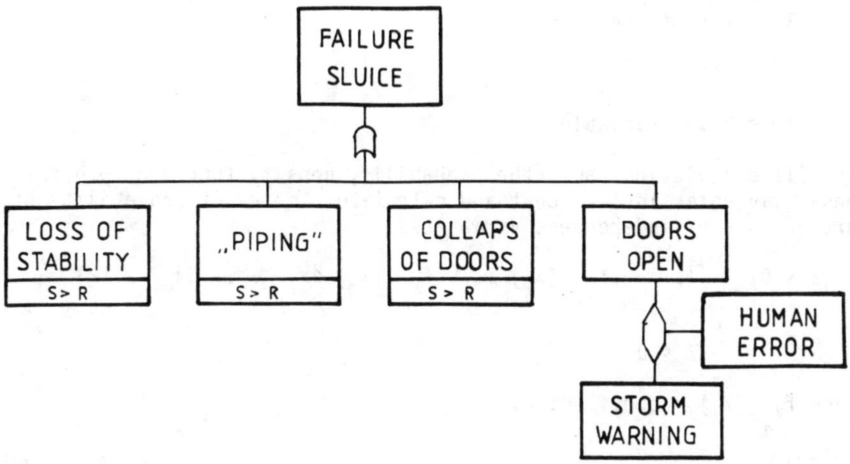

Figure 12. The fault tree of the failure of the sluice.

In the field of large dams, a lot of work has been done (Middlebrooks 1953) to derive average probabilities of failure from historical failure cases. Table I below gives some results. The average probability of failure is 10^{-4} per dam per year.

TABLE I. Causes of dam failure

cause	% of cases
overtopping	30%
internal erosion	38%
slipcircles	15%
other	17%

These data are not useful in the design of dikes because they do not reflect structural improvements or other circumstances. For electrical and mechanical components, historical failure frequencies are very important.

For structural design, the use of probabilistic calculations is preferred. There are three internationally agreed levels on which the limit state equations may be solved. The limit state equation is mostly written as

$$Z = R(X_1 \dots X_n) - S(X_{n+1} \dots X_m) = 0 \tag{1}$$

where R = resistance

S = load

X_i = basic variable

A level III calculation takes the probability density function (p.d.f.) of all basic variables into account and calculates the exact probability of failure in case of independent variables.

$$Pr\ (Z < 0) = \underset{Z<0}{\iint \ldots \int} f_{X_1}\ (x_1) \ \ldots . \ f_{X_m}\ (x_m)\ dx_1\ \ldots .\ dx_m \tag{2}$$

where $f_{X_i}\ (x_i)$ = p.d.f. of X_i

At level II the p.d.f.'s of the basic variables are if necessary approximated by a normal distribution in a more or less refined manner. Thereafter the expectation μ and standard deviation σ of Z are calculated by means of the equation

$$\mu(Z) = Z\ (\mu(X_1)\ \ldots .\ \mu(X_n)) \tag{3}$$

$$\sigma^2(Z) = \sum_{i=1}^{m} \left\{ \frac{\partial Z}{\partial X_i} \ . \ \sigma(X_i) \right\}^2 \tag{4}$$

and if Z is normally distributed

$$Pr\ (Z < 0) = 1 - \Phi\left\{\frac{\mu(Z)}{\sigma(Z)}\right\} = 1 - \Phi(\beta) \tag{5}$$

where Φ (..) = standard normal distribution

β = reliability index

Iterative computer programs are available that give good approximations of the exact failure probability for non-normal basic variables and non-linear Z-functions (advanced first order second moment approach, approximate full distribution approach). The normal design calculations, which use characteristic values of the basic variables and partial safety coefficients according to some format, are indicated with level I

$$\frac{R_{kar}}{\gamma R} > \gamma_S\ S_{kar} \tag{6}$$

where R_{kar}, S_{kar} = characteristic strength, load

The analysis of dikes is performed on levels II and III. The joint probability density function (j.p.d.f.) of wave spectra and storm surge levels is evaluated at level III. A physical model (See Fig. 13) is used to extrapolate the historical data set of storm surges to the mentioned j.p.d.f. (Vrijling and Bruinsma 1980).

The distribution of the storm surge level HW is of the Gumbel type

$$F_{HW}(h) = e^{-e^{-\frac{h-\alpha}{\beta}}} \qquad (7)$$

where $\alpha = 1.96$

$\beta = 0.33$

The conditional p.d.f. of significant wave heights is modelled by

$$f_{H_s|HW}(H_s) = N\,(\mu, 0{,}69^2) \qquad (8)$$

where $\mu = \sqrt{3.45\ HW - 7.67} + 4.50$

for $HW > 2.50$ m

The wave steepness is normally distributed.

$$f\left(\frac{H_s}{L}\right) = N\,(0.0375,\ 0.006^2) \qquad (9)$$

With this set of natural boundary conditions, the limit state of wave overtopping for the sea dike on deep water is calculated as follows.

$$Z = h_c - HW - Z_{2\%} - s - o - 1 \qquad (10)$$

where h_c = construction height

HW = storm surge level

$z_{2\%} = \frac{1.75\ H_s}{\sqrt{H_s/L}} \cdot \mathrm{tg}\ \alpha$; wave runup

H_s = significant wave height

s = settlement

o = oscillation

1 = sea level rise

The result of the level II calculation is summarized in Table II.

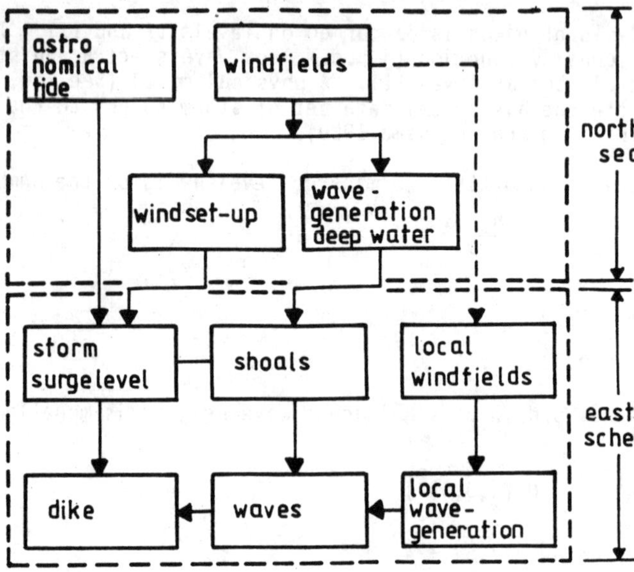

Figure 13. Model to predict wave spectra in conjunction with storm surge levels.

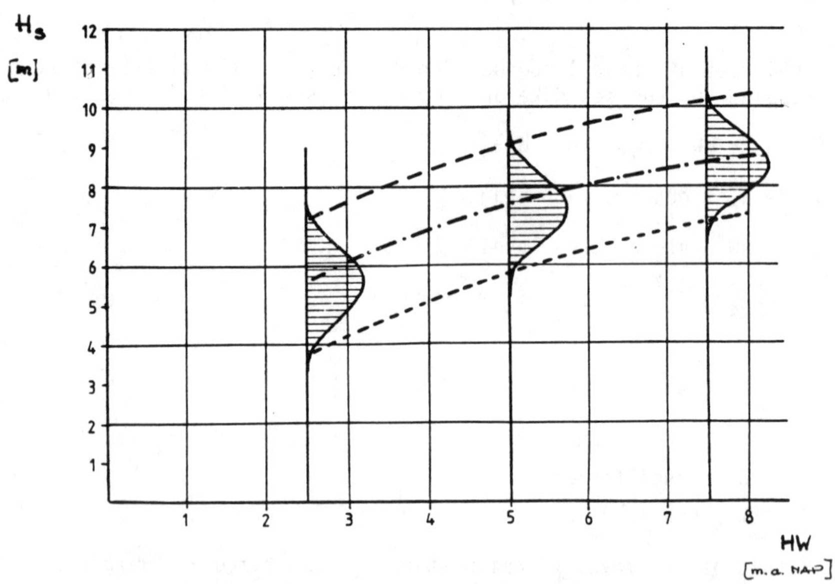

Figure 14. The conditional probability density function of wave energy on storm surge level.

Table II. The result of the level II calculations of the mechanism wave overtopping.

	μ	σ	X_i^*	%
h_c	15.8	0.10	15.80	0.00
HW	Gumbel		3.14	0.60
σH_s	0	0.69	0.753	0.20
H_s/L	0.0375	0.006	0.031	0.19
s	0.50	0.10	0.51	0.00
o	0.40	0.10	0.41	0.00
1	0.10	0.03	0.101	0.00

$\beta = 2.43 \quad p_f = 7.35\ 10^{-3}$

Dune erosion may be calculated along the same lines. The fact that the erosion is only known as a computer program is no problem.

$$Z = b - M * E\ (HW,\ H_s,\ o,\ 1,\ D_{50},\ \text{prof}) \tag{11}$$

where b = dune breadth (m)

E ( ) = erosion (computer program) (m)

prof = beach level (m^3)

D_{50} = grain size of the sand (μm)

M = model factor -

The level II result is given in Table III.

For a typical river levee, the probability of failure for the relevant limit states may also be calculated on level II (see Table IV).

The total probability of failure of the levee section lies between the following boundaries:

$$\max_{j=1}^{4} \left\{ P_{f_j} \right\} \leq P_{f_{sect.}} \leq \sum_{j=1}^{4} P_{f_j} \tag{12}$$

$$21.0\ 10^{-3} \leq P_{f_{sect.}} \leq 25.3\ 10^{-3}$$

Table III. The result of the level II calculation of the mechanism dune erosion.

	μ	σ	X_i^*	%
b	60.28	2.0	59.72	0.01
HW	Gumbel		4.51	0.86
σH_s	0	0.69	0.32	0.02
o	0.40	0.10	0.42	0.00
1	0.10	0.03	0.10	0.00
D_{50}	225	2	200	0.06
prof	0.0	60	-30	0.02
M	1.0	0.15	0.91	0.03

$\beta = 3.57 \qquad p_f = 1.78\ 10^{-4}$

Table IV. An overview of the probabilities of failure of the most important mechanisms of a river levee.

mechanism	p_{fj}	
overtopping	4.3	10^{-3}
slipcircle	0.04	10^{-3}
piping	21.0	10^{-3}
micro instability	0.007	10^{-3}

For practical purposes these boundaries are sufficiently narrow. If the correlations between the mechanisms caused by the storm surge level HW are taken into account, one finds a total probability of failure that equals the upper boundary.

The problem of the length of the total defence system is not so easily solved. Summing the probabilities of failure of the dike stretches, the dune, and the river levee leads to an unacceptably high probability of failure. Therefore, at this time each section is checked separately.

6. ACCEPTABLE RISK LEVELS FROM SOCIOECONOMIC POINT OF VIEW

The last question, but not the least important, is which probability of inundation is acceptable for society.

In Dutch dike design, the standard is a storm surge level with a return period of 10,000 years that must be fully withstood by the sea defense system.

To accommodate the probabilistic calculations as shown in this paper, an acceptable probability of a failure 10 times smaller than the design frequency is advised (i.e. 10^{-5}).

Studies are performed to find a more objective basis for the acceptable probabilities of inundation. In these studies two approaches are followed.

One approach translates into monetary units all damage caused by the inundation. Then the total cost of the investment in a saver dike and the present value of the risk is minimised.

The other approach considers only the number of people that will drown in case of inundation. This number may be looked at from two points of view.

The first is the point of view of the individual, who equates the probability of drowning with the normal risk of dying in an accident (10^{-4}). An acceptable probability of inundation from this viewpoint is:

$$P_{f_{acc}} \leq \beta^{*} \cdot \frac{10^{-4}}{P_{d|f}} \qquad (13)$$

where $0.1 < \beta^{*} < 10$ = policy factor

$P_{d|f}$ = probability of drowning given inundation

The second point of view is that of society. The assumption is that society finds a risk acceptable if the expected number of deaths is with some certainty below $\beta^{*} \cdot 100$ (for the Dutch situation; in general 10^{-5} population). Mathematically this is expressed as

$$E\,(N_d) + k \cdot \sigma(N_d) \leq \beta^{*} \cdot 100 \qquad (14)$$

where k = confidence limit (= 3)

For one large polder this leads to:

$$P_{f_{acc}} \leq \frac{\beta^{*2} \cdot 100^2}{k^2 \quad N_d^{\,2}} \tag{15}$$

If $\beta^* = 0.1$ and $k = 3$ the expression is

$$P_{f_{acc}} \leq \frac{11}{N_d^{\,2}}$$

Although this last criterion is more strict for central Holland than the already mentioned 10^{-5}, the criteria advised for environmental risks are generally far lower and of the order

$$P_{f_{acc}} \leq \frac{10^{-4}}{N_d^{\,2}}$$

This would lead however to somewhat unrealistic dike designs, which underscores the difficulties in this field. A lot more thought and discussion is needed before a clear view of the acceptable risk level is reached.

REFERENCES

Bakker, W.T. and Vrijling, J.K. (1980) Probabilistic design of sea defences, Coastal Engin. Conf., Sydney.

Middlebrooks, T.A. (1953) Earth dam practice in the United States. Transactions ASCE, Centennial Volume, pp. 697-722.

Vellinga, P. (1983) Predictive computational model for beach and dune erosion during storm surges, Coastal Structures Conf.

V.d. Graaff, J. (1983) Probabilistic design of dunes, Coastal Structures Conf.

Vrijling, J.K. and Bruinsma, J. (1980) Hydraulic Boundary Conditions, Symp. on Hydraulic Aspects of Coastal Structures, Delft.

USE OF RISK-BASED ANALYSIS IN MAKING DECISIONS ON DAM SAFETY

J. Lawrence Von Thun

Senior Technical Specialist, Bureau of Reclamation, Engineering and Research Center, P.O. Box 25007, Denver, Colorado 80225-0007

ABSTRACT

The problem of risk reduction due to dam failure of existing dams with a limited, reasonable expenditure of funds is modeled and solved through the use of quantitative risk-based decision analysis procedures. This approach aids in the identification of which dams are candidates for modification and given that modification is to occur the study can be used to aid in selecting the appropriate type and level of modification to be made. The elements of the procedure as required for practical application to the complex engineering, social, and environmental problem encountered in water resources reliability and risk investigations are illustrated.

1. INTRODUCTION

Interest in risk-based analysis for civil works was stimulated during the early years of the design and construction of nuclear power plants in the United States. Attempts to extend the methodology to water resources (specifically sizing of spillways) were harshly rejected by most engineers, largely because the procedures being suggested placed a specific value on human life. However, the general risk-based decision analysis framework for civil problems remained in incubation in colleges and universities throughout the 1970's. Periodically, a general paper on probability-based or risk-based decision analysis would be published extolling the merits of looking at the "whole problem" and evenhandedly examining alternative solutions. Such papers typically used hypothetical problems and only dealt with aspects of presenting the problem. Two major hurdles remained before the proposed application could become reality:

1. Development of the necessary data and probability judgments for real problems.

2. Acceptance of the procedure by agencies, organizations, or firms which typically had engineers and managers with widely diverse opinions on the usefulness and validity of the techniques (as well as concerns about the ethics and liability involved in using such techniques).

With the onset of the dam safety evaluation problem in the United States, new life was breathed into the use of the procedure. The need for a systematic procedure to: (1) discern which of more than 70,000 dams in the United States should be modified, (2) determine what "load level" should be selected for the modification, and (3) decide what the nature of any required modification should be was evident to dam owners, engineering firms, and risk analysts. Recognizing the potential merits of developing a procedure for practical application to real dam safety problems and the challenge in having this information "accepted" as input by decision makers, the Bureau of Reclamation has been, through a process of evolution, developing such a decision analysis framework. This framework has been and is being used in the process of reaching decisions on some dam safety modifications.

2. GENERAL APPROACH

The decision analysis process is approached in three phases, as follows:

- Phase I - Hazard Analysis
  - Results are used to evaluate the nature of the problem.
  - Information developed assists in establishing the appropriate modification "load" level if a modification is warranted. (The load level may refer to hydrologic load, reservoir level, seismic load, or all three.)
- Phase II - Risk Analysis of Alternatives
  - Presents a quantitative evaluation of the economic risk and life loss risk for existing conditions.
  - Compares quantitatively each alternative according to economic risk reduction, life loss risk reduction, and costs and benefits of alternatives (direct, environmental and any other factors).
- Phase III - Decision on Modification
  - Provides decision makers with an objective presentation of qualitative, quantitative, social, environmental, political, organizational, and other factors that relate to decisions on modification of each dam.
  - Through progressive experience with presentation of complete information in a common framework, decision makers are able to develop a relative, quantitative sense concerning the need, level, and type of modification that is appropriate for each site.

3. HAZARD ANALYSIS

After identification of the threshold levels of loading that may produce dam failure, the consequences of dam failure are estimated on the basis of inundation mapping. An analysis of the failure rate, failure type, flood wave travel time, and factors related to effective warning and evacuation is made to determine the potential for loss of life and property damage. In the general case, several types and levels of failure may have to be examined and the consequences plotted versus discharge level. For the very prevalent case of inadequate spillway capacity, the analysis determines the incremental effects between passage of a flood with and without dam or spillway failure. The steps in the hazard assessment for this type of problem are enumerated in Table 1.

Table 1. Outline of Steps in Hazard Assessment for Inadequate Spillway Capacity Problem

1. Perform flood routings to determine impact on dam as a function of flood level and establish outflow with no dam failure.
2. Identify failure modes and failure threshold for each.
3. Route floods with dam failure to obtain outflows.
4. Determine consequences as a function of flow.
5. Make hazard evaluation based on incremental consequences.

Separate relationships to illustrate the potential loss of life and the property damage versus flood discharge are required as the output of the hazard assessment. Sample relationships are shown on Figures 1a, 1b, 2a, and 2b.

The hypothetical case illustrated in Figure 1 shows a small incremental difference in population at risk and estimated loss of life for inflow flood levels above about 60 percent of the maximum discharge and significant incremental property damages for all flood levels with failure potential. Figure 2 illustrates a case where the incremental estimated loss of life is significant for all flood levels up to the PMF (probable maximum flood). For these hypothetical cases, the nature of the incremental hazard posed by potential dam failure can be discerned readily even with a rather cursory assessment of the damages and population exposure. More detailed assessments may be required for certain applications of these hazard assessments. For example, in some cases close attention must be given to details such as location of affected population, method of flood wave detection, warning dissemination, route and time for escape, and other factors. These data may be required to reach a decision as to whether or not a significant difference in potential loss of life exists between the case of no dam failure and dam failure.

In practice, the hazard assessment relationships take on quite variable forms that reflect the site-specific character of the dam

FIGURE 1a

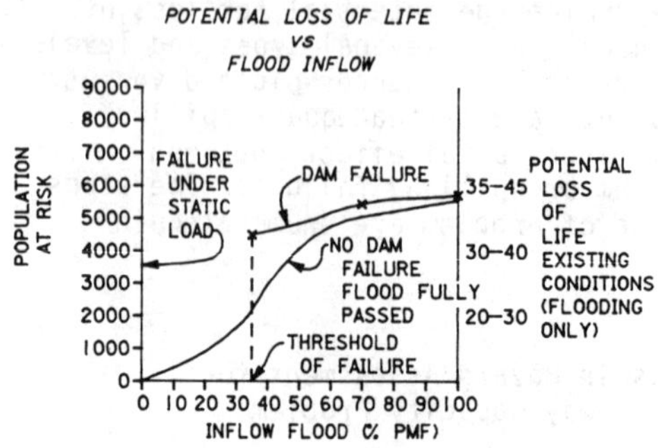

FIGURE 1b

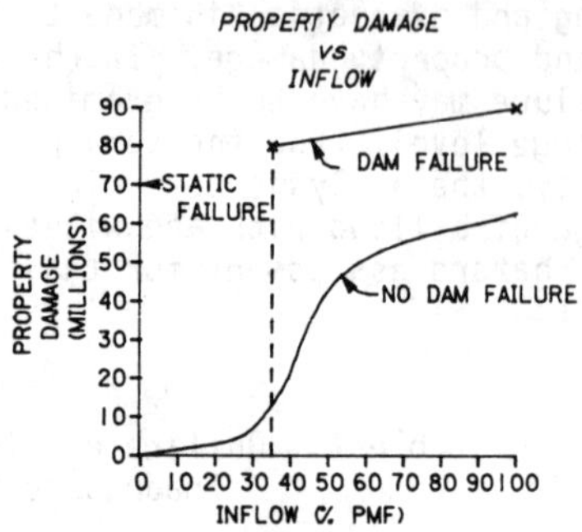

FIGURE 2a

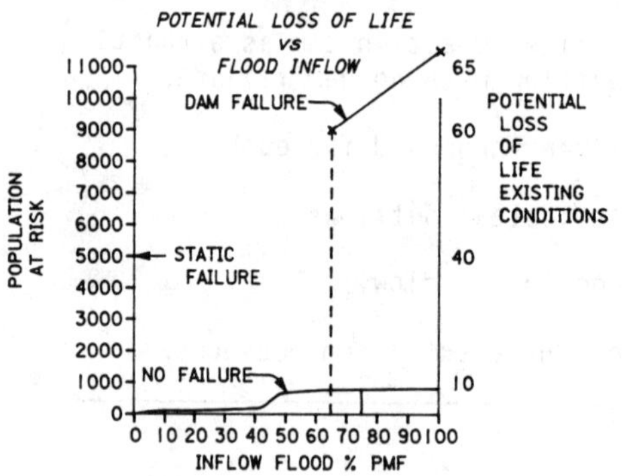

FIGURE 2b

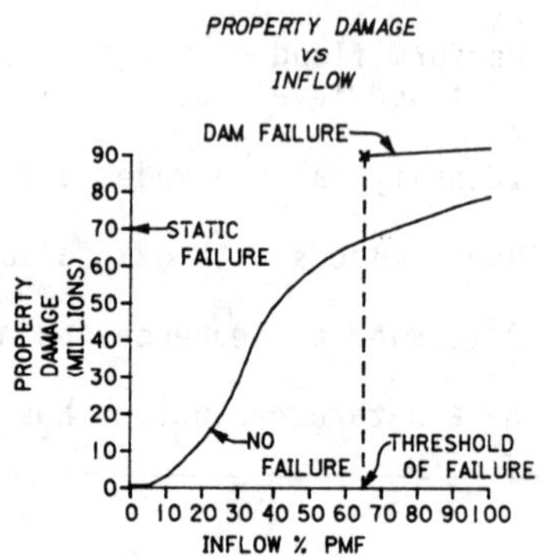

Note:

1. The incremental effects for dam failure due to static or earthquake loading are indicated at 0 discharge.

2. The relations between population at risk and estimated loss of life for the examples are purely a conjecture for this display and should not be considered an average or typical relationship.

safety problem. The hazard analysis presentation is considered by the author to be the single most important piece of information for the engineer, dam owners, regulators, public, and decision makers. Because it defines the nature of the problem in a visual way, it is easy to understand by everyone involved.

The minimum acceptable level of protection may also be identified at this stage of the process; however, recognition of the underlying principle used in developing the relationships must be maintained. Otherwise, alternatives that might alter the basic hazard assessment relationship may be inappropriately eliminated (e.g., relocation of a population segment at risk).

4. QUANTITATIVE RISK ANALYSIS

The quantitative estimate of the risks associated with a dam are computed according to the steps identified in Table 2. A commentary on these steps follows:

Table 2. Steps in Quantitative Risk Analysis Procedure

Perform for existing conditions and for each alternative modification or combination of modification.

1. Estimate likelihood of potentially damaging loads

   $P_{FLOOD1}, P_{FLOOD2}, P_{EQ1}, P_{EQ2}, P_{RES1}, \ldots$

2. Determine potential adverse response modes of the system

   $F_A, F_B, F_C, F_{PA}, F_{PB}, \ldots$

3. Estimate system response to loads

   $P_{F_{A/FLOOD1}}, P_{F_{PA/FLOOD1}}, P_{F_{C/EQ1}}$

4. Determine consequences of failure

   $LOSS/F_A, LOSS/F_B,$

5. Determine "Risk Cost" or "Expected Loss"

   $RISK\ COST = \Sigma(P_{LOAD} \times P_{F/LOAD} \times LOSS/F)$

Note: P = Probability, F = Failure Mode

Step 1. The loading conditions which usually need to be considered are static reservoir load, seismic load, and hydrologic (flood) load. However, for a specific dam, other loads may need to be considered and some loads may be ruled out as not having the potential for dam failure.

It is recognized that the frequency of hydrologic and seismic events having the potential to cause dam failure is difficult to estimate with confidence because of the lack of historic data on which to make projections of the return period of remote events. Nevertheless, such an estimate is part of any risk assessment whether it be subjective, qualitative, or quantitative.

Step 2. By examining the configuration, characteristics, and site-specific condition of a given dam-foundation-spillway system and considering the loading conditions to which it will be exposed, identify the potential modes of failure that would be the most likely result of adverse loading events.

The identification of the potential failure modes (full or partial) is the first step in trying to determine the response of the structure and also the first step in determining the consequences of failure. The detail to which the modes need to be identified is very much a site-specific determination. In some cases, the assessment of risk and consequences might be satisfactorily made by assuming a complete and instantaneous breach of the dam, while in other cases such an assumption may yield a completely unrealistic estimate of damages and resulting benefits of remedial measures.

Step 3. This step is one of the most difficult in the risk assessment process, but it is necessary to account for the entire range of loads to which a dam is exposed and the complexity in the determination (or knowledge) of how a dam responds to these loads. For example, engineers may be able to estimate that the threshold of damage for a given dam is a magnitude 6 earthquake at a source 10 km away (assume the source has a potential of up to magnitude 7.5). Similarly, the engineer's estimate that considerable damage (but not necessarily dam failure) would result from the magnitude 7.5 event. An estimate of response is required for all types of loads over the entire damage potential range of loads such that the total risk of failure may be realistically assessed. This step is one of the fundamental differences between a quantitative risk-based decision analysis and a "maximum event" analysis which is common practice.

Step 4. Once a mode of failure has been defined, the dam break (along with hydrologic event, if applicable) is routed through the flood-plain. Inundation maps (and flow rates) are used to determine the potential for loss of life, the level of property damage, and the environmental impact. (Note that this work is only an update of the work done in the hazard analysis.) Variables at the time of dam breach such as season of the year, time of day, reservoir elevation, and other local specific conditions that may be important should be taken into account at this stage of the analysis. Such items may be important in identifying appropriate remedial measures.

Step 5. The total risk cost is obtained by summing the product of the likelihood of the loading condition, the likelihood of a dam failure mode given the loading condition, and the cost of the damages resulting from that failure mode for each failure mode over the range of load levels for each loading condition, and for all applicable loading conditions, i.e.,

Partial Risk Cost = (likelihood of load level for specific load type) x (likelihood of specific failure mode given load level for specific load type) x (property damages produced by specific failure).

Total Risk Cost = sum of risk cost for all load levels of all load types and each relevant failure mode.

Because the numerical value of the parameters used in making the quantitative estimate (step 1 and 3) are (1) not accurately known as a single value and (2) the probability they may take on a certain value is also not known, sensitivity studies are used by the Bureau in making quantitative estimates of risk as opposed to the fully probabilistic approach often presented in the literature.

Table 3 illustrates the format for presenting the data developed in the quantitative analysis for a hypothetical dam under existing conditions. Table 4 illustrates the risks for the same dam with a modification that (1) prevents overtopping from floods and (2) provides structural improvements to prevent sliding. Each of these tables presents a risk of various loss-of-life conditions. The calculation to obtain this risk is described below and fully presented in Table 5.

For each load level and failure response type, a potential loss-of-life condition has been identified. A more serious potential loss-of-life condition is considered inclusive of lesser conditions. Since a probability has been estimated for each load level, for each failure response type given that load level, and for the probability of the adverse consequence coming to pass given failure, then the probability of occurrence of the potential loss-of-life condition is the product of the three probabilities. The sum of the products for all combinations producing a certain type of potential loss-of-life condition yields the probability of occurrence of that condition. For various modification alternatives, the response probability is revised but the computation format stays the same. For the computation of the potential loss-of-life condition under a warning system or plan, the probability of success of the system is identified. The potential for loss of life is considered to be reduced by that probability.

5. DECISION ANALYSIS PRESENTATION

In order for quality decisions on dam safety modifications to be made by decision makers (engineering management, dam owners, regulatory agencies, public bodies), an objective, comprehensive, and easily understood presentation of the relevant information is required.

This presentation needs to provide data or study results that help to answer the questions related to the following factors that influence the judgmental decisions on modification.

1. Loss of Life Potential

   - Does a significant incremental potential loss-of-life condition exist?

Table 3. Example Format for Display of Calculations on Risk Costs and Loss of Life Potential

Part I

Load range	Load range probability		Potential failure mode		Response probability		$ damages (millions) /life loss cond		Annual risk cost
	Step 1		Step 2		Step 3		Step 4		Step 5
40-60% PMF	(0.004)	x	Aux. spillway failure	x	(0.10)	x	(35) (A)	=	$14,000
60-80% PMF	(0.0008)	x	Aux. spillway failure	x	(0.30)	x	(35) (A)	=	8,400
60-80% PMF	(0.0008)	x	Dam overtopped	x	(0.05)	x	(80) (B)	=	3,200
60-80% PMF	(0.0008)	x	Service spill-way failure	x	(0.05)	x	(80) (B)	=	3,200
80-100% PMF	(0.0002)	x	Aux. spillway failure	x	(0.5)	x	(35) (A)	=	3,500
80-100% PMF	(0.0002)	x	Dam overtopped	x	(0.2)	x	(160) (C)	=	6,400
80-100% PMF	(0.0002)	x	Service spill-way	x	(0.1)	x	(160) (C)	=	3,200
Static (full reservoir)	(0.5)	x	Dam overtopped	x	(0.0005)	x	(80) (C)	=	20,000
Seismic M 5.5-6.5	(0.01)	x	Dam overtopped	x	(0.005)	x	(80) (C)	=	4,000
M 6.5-7.5	(0.001)	x	Dam overtopped	x	(0.1)	x	(80) (C)	=	8,000
M 7.5	(0.0001)	x	Dam overtopped	x	(0.3)	x	(80) (C)	=	2,400
					Total Annual Risk Cost			=	$76,300

Part II

Loss of life potential - exposure condition	Annual probability of the exposure condition occurring
(A) People in recreation pursuits within 10 miles (16 km) of the dam	0.00131
(B) Evacuation-related deaths (1 or 2 people)	0.00057
(C) Direct loss of life of up to 50 people for night failure (8 p.m. - 4 a.m.)	0.00016

Table 4. Modification Computation of Alternative Risk Costs and Loss of Life Potential

Part I

Load range	Load range probability		Potential failure mode		Response probability		$ damages (millions) - /life loss cond		Annual risk cost
Step 1			Step 2		Step 3		Step 4		Step 5
40-60% PMF	(0.004)	x	Aux. spillway failure	x	(0.10)	x	(35) (A)	=	$14,000
60-80% PMF	(0.0008)	x	Aux. spillway failure	x	(0.30)	x	(35) (A)	=	8,400
60-80% PMF	(0.0008)	x	Dam overtopped	x	(0.0)	x	(80) (B)	=	0
60-80% PMF	(0.0008)	x	Service spill-way failure	x	(0.05)	x	(80) (B)	=	3,200
80-100% PMF	(0.0002)	x	Aux. spillway failure	x	(0.5)	x	(35) (A)	=	3,500
80-100% PMF	(0.0002)	x	Dam overtopped	x	(0.0)	x	(160) (C)	=	0
80-100% PMF	(0.0002)	x	Service spill-way	x	(0.1)	x	(160) (C)	=	3,200
Static (full reservoir)	(0.5)	x	Dam overtopped	x	(0.0)	x	(80) (C)	=	0
Seismic M 5.5-6.5	(0.01)	x	Dam overtopped	x	(0.0)	x	(80) (C)	=	0
M 6.5-7.5	(0.001)	x	Dam overtopped	x	(0.0)	x	(80) (C)	=	0
M 7.5	(0.0001)	x	Dam overtopped	x	(0.1)	x	(80) (C)	=	800
					Total Annual Risk Cost			=	$32,300

Part II

Loss of life potential - exposure condition	Annual probability of the exposure condition occurring
(A) People in recreation pursuits within 10 miles (16 km) of the dam	0.00081
(B) Evacuation-related deaths (1 or 2 people)	0.00007
(C) Direct loss of life of up to 50 people for night failure (8 p.m. - 4 a.m.)	0.00001

Table 5. Calculation of Annual Probability of Loss-of-Life Exposure Condition Occurring

Modification Description	#	Load Prob.	Response Prob.	Potential Loss of Life Condition: 1 (Recreation Loss)	2 (Evacuation Loss)	3 (Direct Loss)
Existing dam	1	.004	.1	.0004	-	-
	2	.0008	.3	.00024		
	3	.0008	.05	.00004	.00004	-
	4	.0008	.05	.00004	.00004	-
	5	.0002	.5	.00010	-	-
	6	.0002	.2	.00004	.00004	.000013
	7	.0002	.1	.00002	.00002	.000007
	8	.5	.0005	.00025	.00025	.000083
	9	.01	.005	.00005	.00005	.000017
	10	.001	.1	.00010	.00010	.000033
	11	.0001	.3	.00003	.00003	.000010
				Σ .00131	.00057	.000163
Reduce over-topping + earthquake resist	3	.0008	(-.05)	-.00004	-.00004	-
	6	.0002	(-.2)	-.00004	-.00004	-.000013
	8	.5	(-.0005)	-.00025	-.00025	-.000083
	9	.01	(-.005)	-.00005	-.00005	-.000017
	10	.001	(-.1)	-.00010	-.00010	-.000033
	11	.0001	(-.2)	-.00002	-.00002	-.000007
				Σ .00081	.00007	.000010

- Does the modification proposed make a significant change?

- Can an improved warning system (detection, decision to warn, dissemination of warning, departure of population) reduce the risk of life loss?

- Are there load levels where dam failure causes no significant incremental effect?

2. Potential Property Damage Condition

- What is the level of the incremental damage due to dam failure?

- Does the modification make a significant change?

3. Probability of Dam Failure

- How does the estimated probability of failure compare to historical failure rates? How does the dam compare to the "average dam" of its type?

- What are the effects of the uncertainties in making this estimate?

- Is the predicted failure reasonably probable?

4. Magnitude of Impact

- If failure were to occur, would the region, area, country, dam owner, etc., be able to absorb the loss or recover?

5. Benefit-Cost Economic Analysis

- What is the risk cost reduction/modification cost ratio for each alternative for the range of values considered in the sensitivity studies?

6. Magnitude and Nature of Modification

- What percent of project cost is involved?

- Can funding be obtained?

- Is modification success assured?

7. Other Factors

- What are the site-specific considerations (environmental, technical, social, political)?

- What precedents will be established or broken?

- Does the proposed modification make sense?

In many cases, the decision analysis presentation phase and the decision itself will be fairly straightforward, while in other cases neither will be easy. However, in either case, the goal is to provide, through the decision analysis framework, an objective and comprehensive evaluation of the alternative courses of action that will allow the decision makers to have a full understanding of the problem and a recognition of the costs and benefits of the decision.

6. SUMMARY

Regardless of the analysis methods and decision process used, decisions on dam safety modification include consideration of hazard, risk, and judgment to various degrees by all involved in the effort.

The development of the dam safety modification question into the decision analysis framework described above provides a consistent, comprehensive, and systematic means for addressing the problem. Benefits of developing solutions to problems in this systematic way include a well-documented decision base, a good presentation base for public understanding, and an economic analysis to facilitate funding considerations.

A COMPARISON OF METHODS FOR RISK ASSESSMENT OF DAMS

David S. Bowles.

Utah Water Research Laboratory and Department of Civil and Environmental Engineering, Utah State University, Logan, Utah 84322-8200

ABSTRACT

Risk-based procedures for assessing appropriate safety levels for new and existing dams have been proposed for use in planning and design of dams and screening of unsafe dams. The risk assessment framework and its application to dams is presented. Approaches for estimating the various types of probabilities and consequences needed to perform a comprehensive dam risk assessment are described. Several methods for planning, screening, and design level risk assessment are summarized and compared. The lecture closes with a discussion of the advantages of, commonly stated objections to, and some unresolved issues related to, risk assessment of dams.

1. INTRODUCTION

1.1 Background

Interest in dam safety has grown in the past decade. According to the National Research Council (1985), reasons for this growth include: several disastrous dam failures or near failures in the United States and other countries; the classification of approximately 3,000 high-hazard dams as "unsafe" by the National Dam Inspection Program conducted by the U.S. Army Corps of Engineers; the high cost of improving these "unsafe" dams to meet current design standards; and the consideration, by many states, of more stringent regulation of privately-owned dams.

Simultaneously there has been a growing interest in the potential for using risk assessment procedures to provide a framework for addressing dam safety problems. Such procedures can provide a basis for evaluation of proposed safety improvements for new or existing structures. The purposes of this lecture are to present the risk assessment approach as it is applied to dams, to compare several alternative procedures for dam risk assessment, and to discuss some of the advantages, limitations, and other issues relating to this application of risk assessment.

In this lecture the term "risk" will be used to mean "the potential (probability) for the realization of unwanted consequences from impending events" (Rowe, 1977). According to this definition, the term risk has two dimensions associated with an undesirable event: the probability of its occurrence, and the magnitude of its consequences. By this definition economic damages resulting from a dam failure are considered to be a component of the consequences which may include potential life loss and environmental damage. The term "risk cost" is used to describe the expected value of economic damages on an annual basis.

1.2 Levels of Risk Assessment

Different levels of detail in the risk assessment procedures used for dam safety evaluations are appropriate at different stages in the life of a dam. As the data base for a dam grows and as the issues to be addressed change from general questions of site selection, to specific issues of the selection of design parameters, the degree of detail which can be justified in the risk assessment grows correspondingly (Bowles et al., 1978, Howell et al., 1980). For the purpose of this presentation, three levels of risk assessment applications to dams are distinguished. In order of increasing detail they are the planning level, the screening level and the design level.

At the planning level it is desirable to introduce an estimate of risk cost associated with dam failure into the benefit-cost analysis as a means of including societal risk into the process of deciding to build a dam (Pate-Cornell and Togaras, 1986). At this level the estimated probabilities and consequences of dam failure are only approximate and usually will rely heavily on historical information (U.S. Water Resources Council, 1979).

The screening problem is the identification and ranking of "unsafe" dams in order of priority for expenditure of limited funds to pay for remedial action. In this context the absolute values of probability and consequence estimates are less important than a consistent procedure for estimating them so that an accurate ranking will be achieved. At the screening level site-specific conditions would typically be evaluated using reconnaissance level investigations and only approximate engineering and economic analyses. An example of the screening level is the method developed for the Federal Emergency Management Agency (FEMA) by Stanford University (McCann et al., 1985).

A design level risk assessment involves detailed questions such as the selection of design standards and choices between design alternatives for the dam and its appurtenance structures. Carefully estimated probabilities and consequences must take into account site-specific conditions based on detailed site investigations and engineering and economic analyses. In addition the sensitivity of conclusions must be investigated with respect to uncertainties in the estimates of both probabilities and consequences. The only documented procedures at the design level are by the U.S. Bureau of Reclamation (USBR, 1986). Work which contributed to these procedures includes Howell et al. (1980).

1.3 Outline of Lecture

The lecture is divided into five sections. After this introductory section the overall framework for risk assessment and its application to dams is presented in Section 2. A description of approaches for probability and consequence estimation is provided in Section 3. Several methods of risk assessment which are currently in use or have been proposed are classified by risk assessment level and compared in Section 4. In Section 5 the advantages of, and commonly stated objections to, risk assessment of dams are discussed. The lecture is closed in Section 6 with a presentation of conclusions and some unresolved issues.

2. RISK ASSESSMENT FRAMEWORK AND ITS APPLICATION TO DAMS

2.1 General Framework

Risk assessment involves the identification, estimation and evaluation of risks associated with a natural or man-made system. The purpose of risk assessment is to evaluate whether the present margin of safety or reliability of the system is acceptable, or to select an alternative for controlling risk in terms of either the probability or consequences of system failure (see Fig. 1). Risk management comprises both the identification, estimation, and evaluation aspects of risk assessment and the implementation aspects of risk control (see Fig. 2). This lecture emphasizes risk assessment although some references are made to the larger problem of risk management.

Figure 3 illustrates the relationship between the steps of risk identification, estimation, evaluation and control for environmental risk management. The implementation of risk control measures, which could be structural or nonstructural in nature, would typically introduce new risk factors which could lead to the need for a second-order risk analysis. The first two steps of risk identification and estimation are usually performed by an analyst, such as an engineer or an economist. Risk evaluation and control are usually determined by a decision maker and typically involve political judgments as to risk acceptability (e.g., how safe is safe enough?). Examples of approaches used in each of these four steps in an environmental risk management problem are given in Figure 3.

2.2 Application to Dams

Dam engineering is not an exact science. The successful design and construction of dams requires the application of judgment by highly experienced engineers, geologists, hydrologists, and others. Traditionally the approach to dam design focuses deterministic analyses on extreme events, such as the probable maximum flood (PMF) or the maximum credible earthquake (MCE), and uses conservative estimates of such properties as concrete or soil strength. Safety factors are used to evaluate the ratio of resisting to overturning moments for such failure modes as slope instability. As a result, through the practice of the traditional approach, which is based on the accumulation of many decades of dam engineering experience, an impressive safety record has been achieved.

However, the traditional approach does not attempt to explicitly quantify all significant risk factors for a dam. Nor does it explicitly determine the degree of safety which can be justified for a particular structure considering the potential consequences of a sudden release of the contents of a reservoir following dam failure. The risk assessment approach provides the framework to make such a quantitative determination.

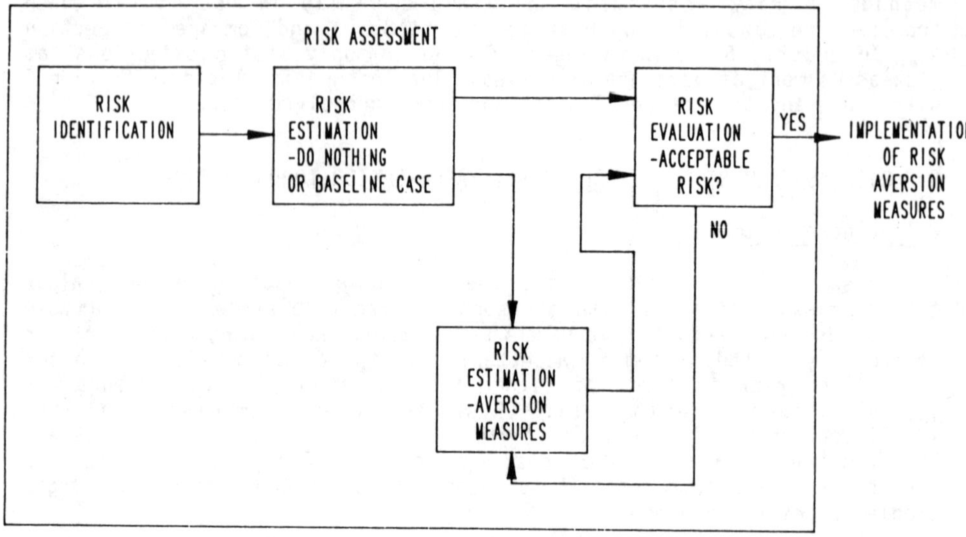

Figure 1. Risk assessment framework (after Bowles and James, 1986).

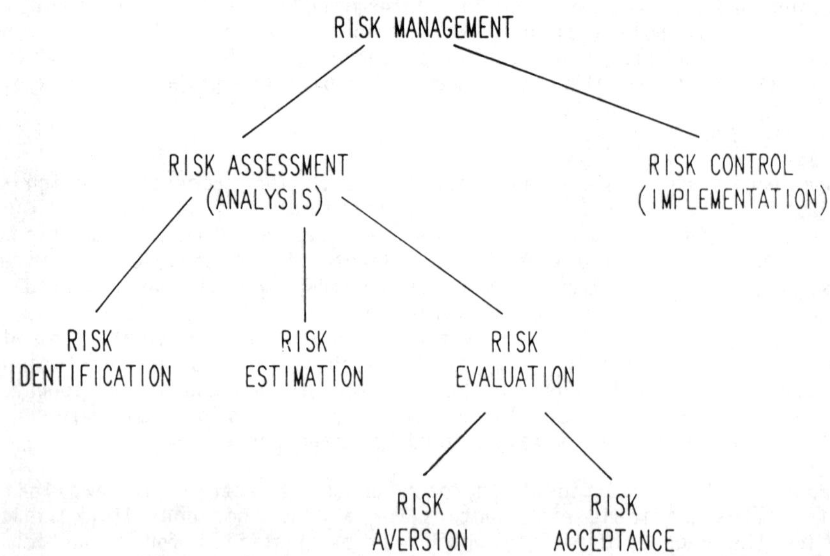

Figure 2. Components of risk management.

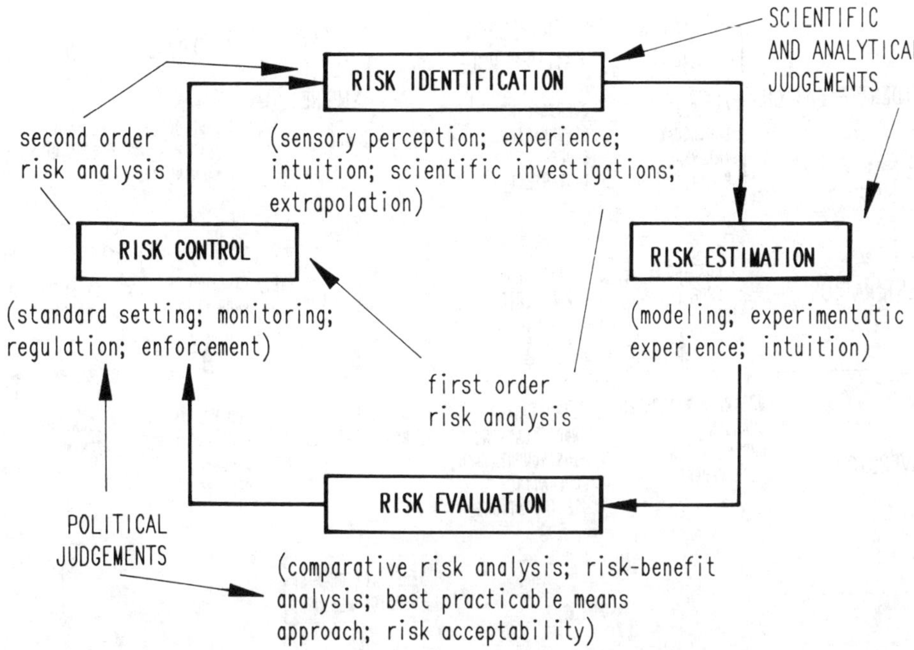

Figure 3. Environmental risk management (after O'Riordan, 1979).

The approaches to applying risk assessment to dams differ at the planning, screening and design levels. The approach appropriate at the most detailed, or design level, is presented below. This presentation follows the identification, estimation, aversion, and acceptance steps of the approach as shown in Figures 1 and 2.

2.2.1 Risk identification. Firstly a sequence of events is identified beginning with events that can initiate dam failure and ending with the consequences of the failure (see Fig. 4 and 5). Initiating events can be classified as external or internal. External events include earthquakes, floods, and upstream dam failure. Internal events include chemical changes in soil or concrete properties or latent construction defects. At low levels these events would not lead to dam failure. However, at high inflow rates a rapid rise in pool level could lead to overtopping, or a severe earthquake could result in structural deformation or liquefaction. These and other dam-foundation-spillway-reservoir system responses are failure modes which can lead to the outcome of the sudden release of the reservoir contents. The magnitude of the resulting property damage and life loss will depend on various exposure factors. These include flood routing to determine the path of the flood wave, the area of inundation, and the travel time; the time of day and season of the year; and the effectiveness of any warning systems and evacuation plans.

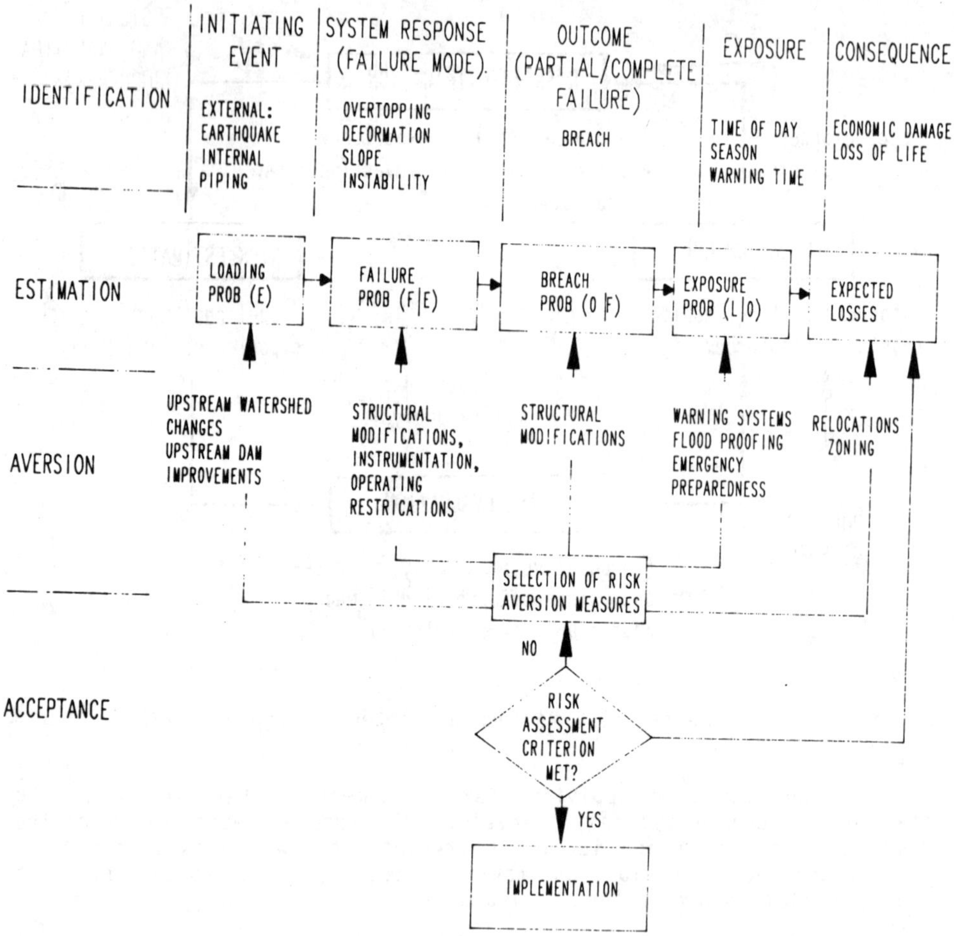

Figure 4. Risk-based method for assessing dam safety improvements (adapted from Bowles et al., 1984).

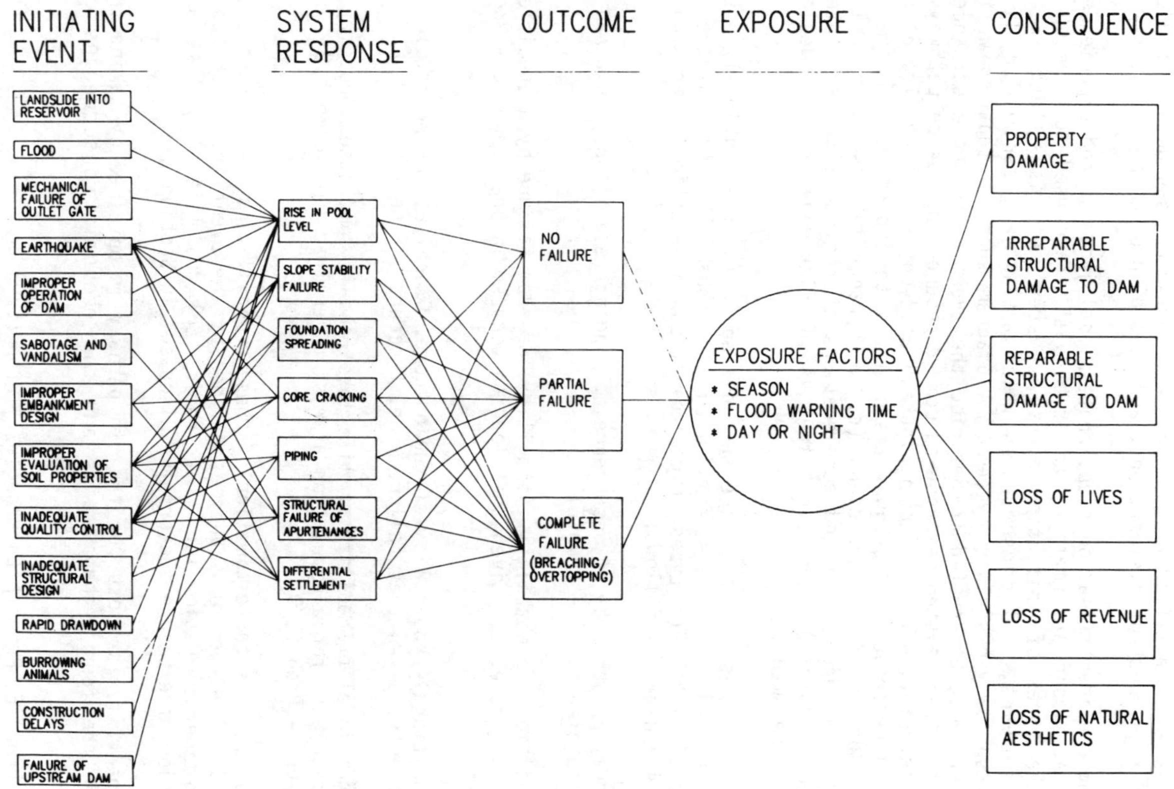

Figure 5. Event-system response-outcome-exposure-consequence diagram for an Earth Dam (after Howell et al., 1980).

Consequences are classified as life loss and economic loss which includes property damage, cost of dislocations, and loss of project benefits.

During the identification step, professional judgment and experience, review of available information, and site visits are used to develop a list of the types of initiating events, system responses, outcomes, exposure factors, and consequences which apply to a particular dam-foundation-spillway-reservoir system. A diagram such as that shown in Figure 5 is then constructed to describe the event-consequence sequences or initiating event-system response-outcome-exposure-consequence pathways. Using the information assembled in Figure 5 an event tree (see Fig. 6) is developed to describe each pathway associated with a particular range of magnitudes of an initiating event (e.g., a range of reservoir inflow magnitudes or a range of ground accelerations at the dam site associated with seismic activity). The event tree is the risk model for the design level risk assessment.

2.2.2 Risk estimation. The second step in the (design level) risk assessment procedure is the estimation of the probability and consequence components of risk. The types of probabilities to be estimated are shown on Figure 4 and are as follows:

- Annual probability of occurrence of loading (e.g. flood) in a range of magnitudes, E - Prob(E)
- Conditional probability (fragility) of dam failure by a specified failure mode, F, given that loading occurs in the range, E - Prob (F/E)
- Conditional probability of the outcome, O, release of reservoir contents, given that failure mode, F occurs - Prob (O/F)
- Conditional probability of life loss (and in some cases property damage), L, for a population at risk given that the outcome O occurs - Prob (L/O)

Alternative methods for estimation of these probabilities are discussed in Section 3.1.1. All event sequences defined by the event tree risk model are considered. Estimation of economic (L_E) and life (L_L) loss and consideration of exposure factors are discussed in Section 3.2.

The partial risk cost for the ith pathway is obtained by taking the product of the four probabilities and the economic consequences as follows:

$$c_i = \text{Prob (E) Prob (F/E) Prob (O/F) Prob (L/O) } L_E \qquad (1)$$

The total risk cost is obtained by summing the partial risk costs over all N mutually exclusive pathways as follows:

$$C = \sum_{i=1}^{N} c_i \qquad (2)$$

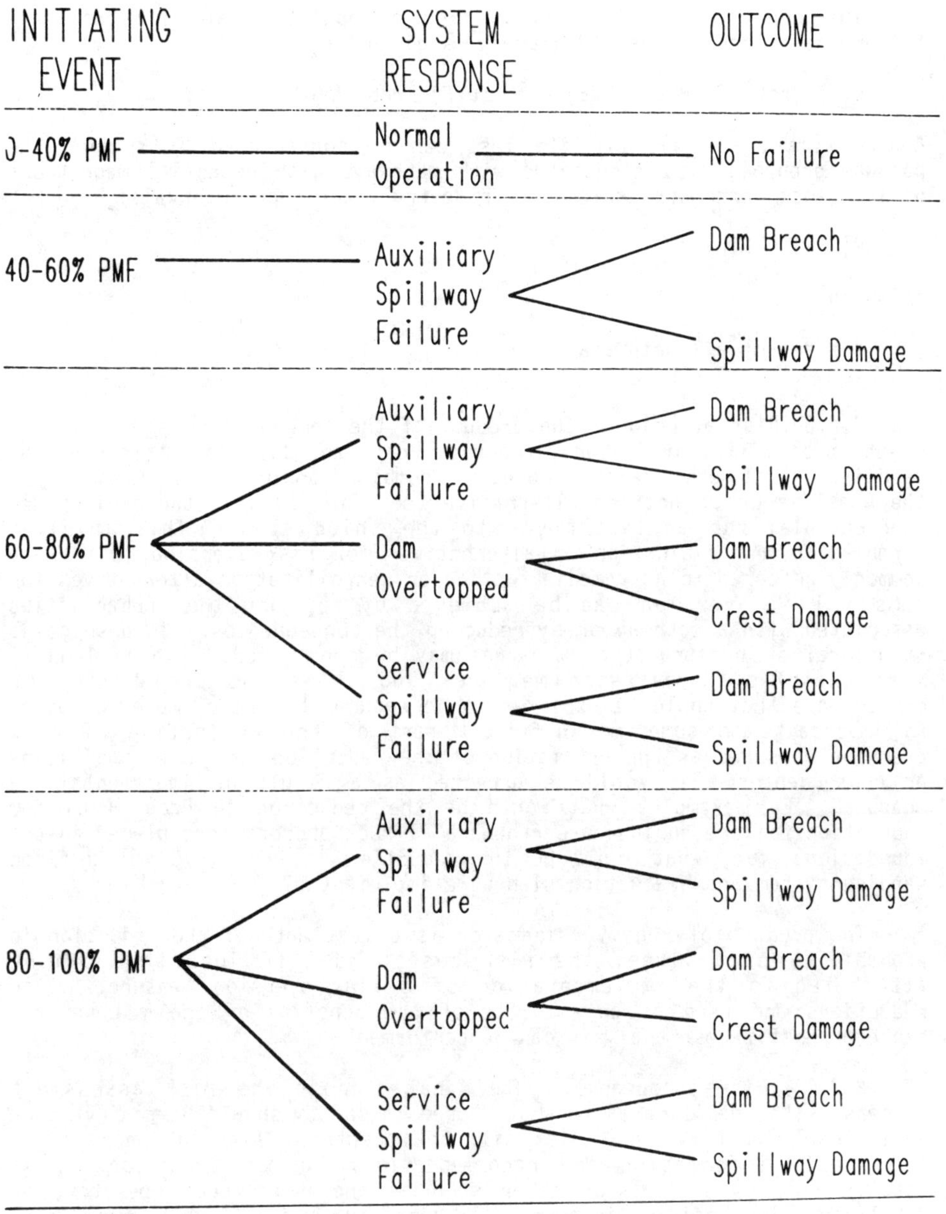

Figure 6. Event tree for hypothetical dam considering hydrologic loading only.

The probability of life loss for a population at risk due to a failure described by the ith pathway is given by:

$$p_i = \text{Prob (E) Prob (F/E) Prob (O/F) Prob (L/O)} \quad (3)$$

A histogram, $P(L_{Lj})$, of life loss can be constructed by summing the pathway probabilities associated with pathways with potential magnitudes of life loss consequences falling in several magnitude intervals:

$$P(L_{Lj}) = \Sigma p_i$$

in which

s = set of all pathways for which $L_{Lj} < L_L < L_{Lj+1}$

2.2.3 Risk aversion. The product of the second step is an estimate of the probability of failure, the risk cost, and life loss that would be associated with each failure mode, or combination of failure modes, for the baseline or do nothing alternative (see Fig. 1). If these risks are unacceptable, the analyst moves to the third step. This involves formulation and evaluation of alternatives for risk aversion, which are commonly referred to as remedial action or rehabilitation alternatives for dams. Risk aversion can be achieved by reducing the probabilities associated with a pathway or by reducing the consequences. In both cases structural or nonstructural measures may be considered. Figure 4 lists some examples of aversion measures and shows the probability or consequence that would be expected to be reduced by their implementation. An important and sometimes difficult part of the evaluation of risk aversion measures is the estimation of the reduction in the probabilities or consequences that would be expected as a result of implementing a measure. For example, what would be the reduction in Prob (F/E) for foundation failure during an earthquake if rock anchors were placed in the foundation? Or, what would be the reduction in Prob (L/O) if a flood warning system and evacuation plan were implemented?

The product of the aversion step is an estimate of the reduction in probability of failure, the risk cost, and life loss that can be attributed to the implementation of a risk aversion measure. Such reductions are used as an estimate of the benefits of the measure and hence a benefit-cost analysis can be performed.

2.2.4 Risk acceptance. The final step in the risk assessment process is the decision as to what degree of safety should be achieved, or equivalently what residual risk will be accepted. Although the analyst can supply information and recommendations for this decision, it is usually made by a decision maker such as the dam owner, operator, or regulator. The decision is especially sensitive and difficult where lives are at risk and where large investments will be required to improve safety with little or no effect on the project benefits, except of course to their expected longevity considering the reduced likelihood of dam failure.

3. ESTIMATION PROCEDURES

3.1 Probability Estimation

3.1.1 General approaches. There are three general approaches to probability estimation: historical/empirical, judgmental, and analytical. The historical/empirical approach utilizes historical frequencies of events as probability estimates. The larger the available sample size used the better the estimates are expected to be. Thus a long record of flows or seismicity can be expected to yield better (i.e., less uncertain) estimates of extreme floods or earthquakes than a short record. Similarly a large data base of dam failures, categorized according to such factors as the type, size, age, and location of the dam, can be expected to provide better estimates of the probability of failure due to a specified failure mode than would a small data base. However, the available data base for dam failures is usually small for a particular category of dam, especially large dams, and in any case will provide only an estimate of probability of failure for "average" conditions rather than for the specific conditions of the dam which is under evaluation.

In order to incorporate additional information about the study dam obtained from field work, laboratory testing, engineering analyses and expert judgment, the historical/empirical probability estimates are treated as initial estimates which can be updated judgmentally using the additional information. The Bayesian approach to estimation provides a formal method for updating historical probability estimates using "judgmental" information, as follows:

$$\text{Prob}''(F/X) = \frac{\text{Prob}'(F)L(X/F)}{\text{Prob}'(F)L(X/F) + \text{Prob}'(\bar{F})L(X/\bar{F})} \qquad (4)$$

in which

$\text{Prob}''(F/X)$ = updated estimate of probability of failure given additional information X about the dam

$\text{Prob}'(F)$ = prior estimate of probability of failure

$\text{Prob}'(\bar{F})$ = prior estimate of probability of no failure

$L(X/F)$ = likelihood of observing information X given that the dam were to fail

$L(X/\bar{F})$ = likelihood of observing information X given that the dam were not to fail

McCann et al. (1985) provide values for the likelihood functions, L, for several dam failure modes although it is not clear what is the basis for the suggested values.

The third method of probability estimation is the analytical method in which models are used to transform probability distributions of loads and strength parameters into an estimate of the probability of failure. For example, the probabilistic slope stability method of Sharp et al.

(1981) can be used to estimate the probability of slope failure due to the spatial variability in shear strength in a zoned embankment. Yen (1986) presents four methods for obtaining analytical probability estimates: direct integration, mean-value first-order second-moment method (MFOSM), advanced first-order second-moment method (AFOSM) and Monte Carlo method. Unfortunately the analytical approach usually requires additional expensive field work in order to estimate distributions for strength parameters. Also, site specific information is often lacking and realistic models necessary for estimating failure probabilities for such phenomena as piping are not available. At this time the analytical approach is generally considered to be an approach which needs additional research and development before it is ready for use in practical dam risk assessments. Therefore, the historical/empirical approach with judgmental updating where appropriate is most commonly used.

The four types of probabilities to be estimated in a dam risk assessment were listed in Section 2.2.2 as Prob (E), Prob (F/E), Prob (O/F) and Prob (L/O). Estimation of the first three probability types is briefly discussed below and estimation of the fourth type is discussed in Section 3.2.

3.1.2 Probability of initiating events, Prob (E). Most commonly, three types of loading conditions are considered for a dam risk assessment: static reservoir load, hydrologic (flood) load and seismic load. Other loading conditions may be identified for a particular structure but only these three will be considered here.

Static loading can lead to various failure modes such as piping or embankment instability. The probability of reservoir loading (i.e., reservoir stage) being in a range that could lead to such failure modes can be evaluated from information on the operating policies for the dam.

Hydrologic loading is evaluated in terms of peak inflow design rates. Where measured flow data are available, a flood frequency analysis can be performed to estimate the probability of occurrence of more frequent events. Depending on the length of record it might be reasonable to extend such a frequency curve to events with return periods of 100 or 200 years (see for example Bulletin No. 17B, U.S. Department of Interior, 1982). However, the probable maximum flood (PMF) is associated with much rarer events and its return period cannot be determined with confidence. Since a probability must be estimated for flow ranges up through the PMF in order to perform a quantitative risk assessment it is customary to make an arbitrary probability assignment for the estimated PMF. Typically this would be in the range of 10^{-4} to 10^{-6} per year. USBR (1986) suggests an approach whereby the 5% and 95% confidence limits on the PMF are assigned probabilities of 10^{-6} and 10^{-4}, respectively. The frequency curve is extended to the plotted point corresponding to the PMF, and thus a probability of peak reservoir inflow being in any flow range up to the PMF can be obtained. Typically in a risk assessment the dividing points between flow ranges are on critical flows associated with thresholds for failure modes of the existing dam or its rehabilitation alternatives, or with changes in the downstream stage-damage relationship.

The estimation of probabilities of seismic load ranges has certain similarities to the case of estimating the probabilities of hydrologic load ranges. Frequency analysis is performed on available historical seismic data and then extensions are made to the maximum credible earthquake (MCE). All known faults (and random locations if appropriate) which could affect the dam site after attenuation is taken into account must be considered and the probability of accelerations at the dam being in various ranges are estimated. As with the case of hydrologic loading, acceleration ranges are based on critical values of acceleration associated with thresholds for failure modes of the existing dam or its rehabilitation alternatives.

3.1.3 Conditional system response probability or fragility, Prob (F/E). For each loading type the event tree constructed during the risk identification step (see Fig. 6) specifies failure modes or system responses which could occur during initiating events with magnitudes in each loading range. The relationship of Prob (F/E) and (E) is referred to as the fragility. In some cases such as overtopping of an embankment dam this relationship may approximate a step function (see Fig. 7a). In other cases, such as a seismically induced failure mode in a region of low seismicity, the fragility may be a smooth curve which does not reach a probability of 1.0 at the maximum acceleration at the site.

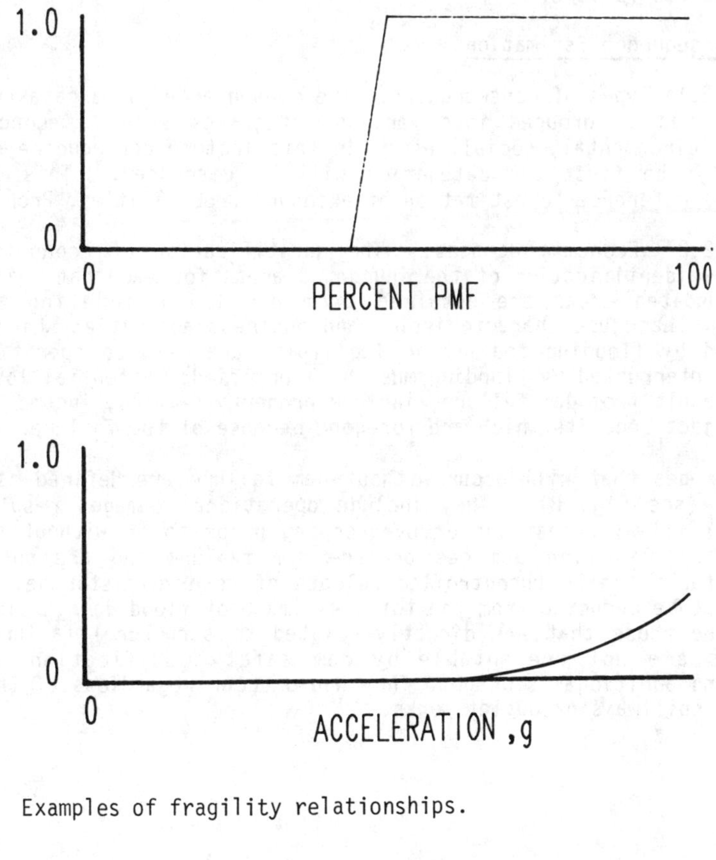

Figure 7. Examples of fragility relationships.

Fragility relationships are based on engineering analysis, experience, and judgment. Research in the form of scale model studies, numerical modeling, and evaluation of historical dam failures is needed in order to improve our current estimates of these relationships. Uncertainty in estimates of the threshold at which Prob (F/E) significantly departs from zero, or the steepness of the fragility curve, leads to wider confidence limits on estimated probabilities of dam failure and risk costs.

3.1.4 Outcome probabilities, Prob (O/F). The outcome of a failure mode may be a partial or complete failure of the structure. A complete failure implies a breach with release of the reservoir contents and a partial failure implies damage to the structure but no catastrophic release of the reservoir contents.

For a seismically induced failure mode the type of outcome would typically be dependent on the level of the reservoir at the time of the failure. Thus an evaluation of the reservoir operating policy can lead to an estimate of Prob (O/F). For a hydrologically or statically induced failure mode the outcome probability may be set equal to 1.0 since the probability of reservoir level is usually accounted for in Prob(E) or it can be used to account for other factors such as the duration of an overtopping event which could determine the difference between a partial and complete failure.

3.2 Consequence Estimation

3.2.1 Types of consequences. The consequences of a catastrophic dam failure can be grouped into various categories such as economic, life loss, environmental, social, etc. In this lecture consequence estimation for only the first two categories will be discussed. This discussion includes reference to estimation of exposure probabilities, Prob (L/O).

3.2.2 Economic damages. The quantification of economic damages requires identification of the inundated areas for each dam failure mode. The inundated areas are obtained from dam break modeling and flood routing. Land use characteristics and business activities which would be affected by flooding and public facilities and service operations which may be interrupted by flooding must be identified. Potential losses which could result from dam failure, include property damages, incomes foregone, and project benefits which are foregone because of the failure.

Damages that will occur without dam failure are defined as baseline damages (see Fig. 8). They include operational damages resulting from normal spillway operations or overtopping prior to or without structural failure. Baseline damages precede dam failure and are not properly attributable to the uncontrolled release of reservoir storage. As such, they must be deducted from the total estimate of flood damages in order to determine those that are directly related to structural failure. These damages are not preventable by dam safety modification, except by providing additional storage. They would occur regardless of the changes made to spillways or outlet works.

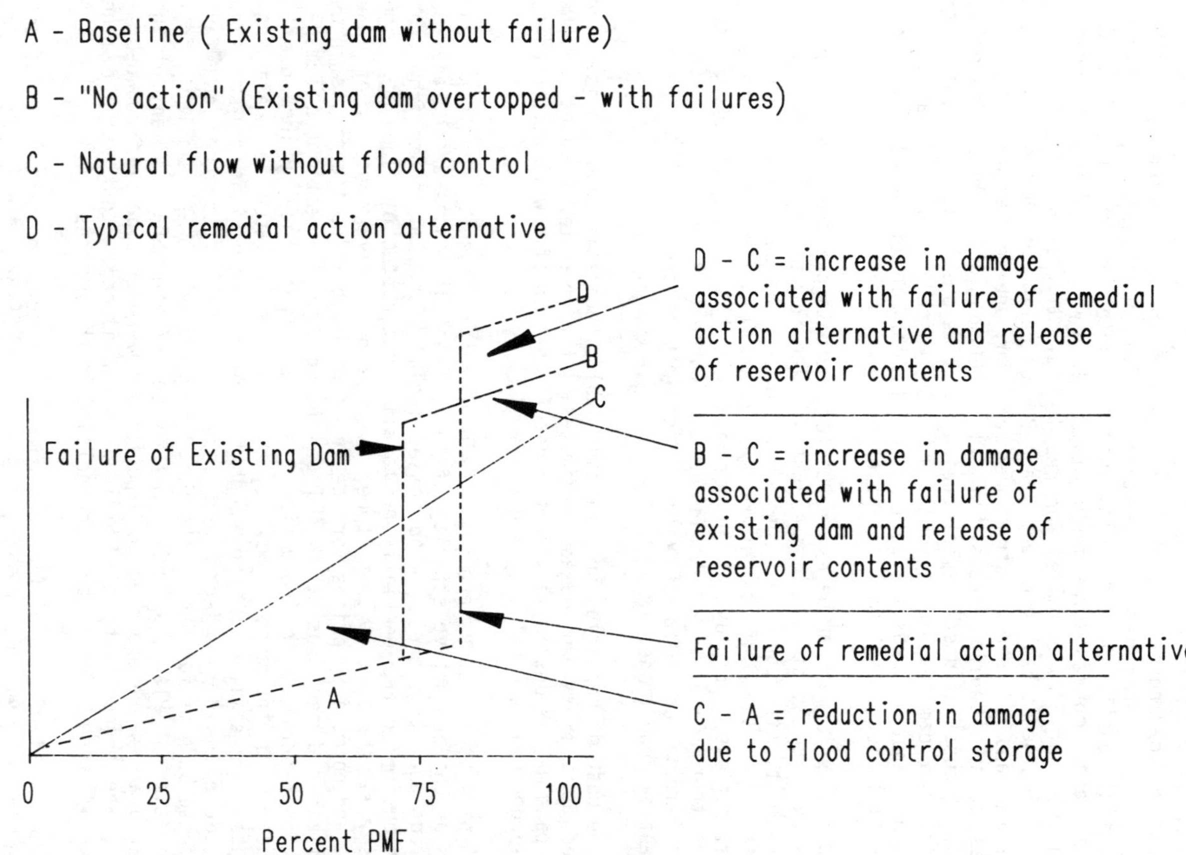

Figure 8. Example of determination of differential damages (adapted from USBR, 1986).

Damage estimates for a given failure mode must be made for alternative loading conditions (e.g., flow range as a percentage of the PMF). These damages are estimated for agricultural, instream, residential, public and industrial losses incident to dam failure. Stage-damage ratio curves have been developed by the Federal Insurance Administration (FIA) and used for residential structures. The U.S. Army Corps of Engineers has developed stage-damage ratios which can be used for commercial and industrial structures and for public facilities. The losses to residential areas from flood inundation include damages to dwellings, costs of displacement during the time dwellings are repaired or rebuilt, and damages to personal property which can be estimated from local insurance records. Dwelling values can be obtained for broad residential districts from local records.

Business losses can be estimated for two categories of loss: capital losses and employment and income losses. The former involves the losses of equipment, structures and contents. Values are taken from industrial indices for various standard industrial code (SIC) industry classifications. Employment losses involve the jobs and associated incomes that are lost due to flooding. Current population reports and census reports are also of value in estimating commercial as well as residential conditions.

The cost of repairing or replacing public facilities and services are assumed as a proxy for damage estimates. For example, bridges in the flood path may be destroyed with the result that traffic will be forced to take longer routes until they are replaced.

Agricultural losses are divided into capital losses, crop losses and losses of the productive capacity of land. Capital losses are damages to equipment and structures. Crop losses include foregone income from crop operations which are partially or totally interrupted by flooding. The losses are estimated by crop type, average productive capacity of land bases, and normalized prices for the crops involved. Input costs, such as fertilizer and seed costs, should be included in the estimated damages since their services in production are partially or totally lost due to destruction of the crop. Most of the analyses related to these loss estimates are similar to the economic benefit evaluations of standard water development projects and loss estimation procedures and farm enterprise budgets for use in the U.S. Such procedures are available from the Division of Planning Technical Services of the USBR. If any of the inundated areas are expected to experience severe erosion or flood debris damage, the productivity of the land is diminished. Lost value of the land should then be estimated and included in the damages.

The amount of warning time and emergency evacuation procedures affect the amount of property damage. Given sufficient time, people can generally evacuate with some amount of property, including equipment, personal cars and other high value but mobile items. Livestock may also be moved to higher ground or away from areas where they may be isolated by flood. These actions can reduce damages considerably. Therefore, an assessment of any currently installed or proposed emergency warning system and subsequent warning times and the emergency evacuation procedures needs to be completed in order to apply reduction factors (exposure probabilities) to the damage estimates, particularly capital and personal property estimates. Figure 9 shows graphically a probabilistic analysis of a monitoring system and warning system.

In addition to property damages, loss of reservoir benefits must be estimated and included in the damage calculations. These losses are included for the period of time that reservoir storage is lost or reduced. If the structure is to be repaired, then these losses are included for a partial rather than project life basis, but the repair costs are included in the estimates of losses.

3.2.3 Potential life loss. The importance of considering the potential impact of a dam failure on downstream populations cannot be over emphasized. This issue is discussed in Section 5.2. In order to estimate potential life loss the population at risk must be identified from inundation maps. Often there will be more than one population at risk (e.g., recreational use below dam, ranchers below dam, downstream town) where populations are distinguished by location (and warning time) and exposure characteristics (e.g., sparse such as farmers, dense as in a town, seasonal as for campers, residential or business with different day and night time characteristics).

Each population at risk must be evaluated separately with respect to the effectiveness of a warning system, the length of warning time considering the travel time for the flood wave, and the effectiveness of evacuation plans. An evaluation of historical dam failures by USBR (1986) indicates a very significant reduction in exposure probability for smaller communities with adequate warning time over similar communities without adequate warning time. USBR (1986) found an exposure probability of 0.0002 for situations in which the warning time exceeded 1.5 hours. Estimates of exposure probabilities for large metropolitan areas, or for other special situations, can be improved by using evacuation modeling (Jaske, 1985). Differences in exposure probabilities associated with different failure modes must be considered. For example, if failure occurred as the result of an earthquake, warning time would likely be less than a failure resulting from hydrologic causes for which inflow forecasts might provide forewarning of exceptionally large inflows.

3.3 Sensitivity Analyses

Traditionally, estimates of loads on material strengths in engineering design are made on a conservative basis. In a risk based approach additional estimates are made to describe a range of estimates of each load or probability or consequence. With this "internal" estimate approach the effects of uncertainty in the estimation of the inputs to a dam safety decision can be explored and documented through sensitivity analysis. Where the recommended decision is sensitive to uncertain factors additional efforts to reduce the degree of uncertainty can be considered. Alternatively, if the level of uncertainty cannot be economically reduced the conservative design approaches can be used to achieve acceptable safety standards. Since uncertainty exists in all the inputs (e.g., probability and consequences) to a risk assessment the use of sensitivity analyses is an important part of a comprehensive dam risk assessment.

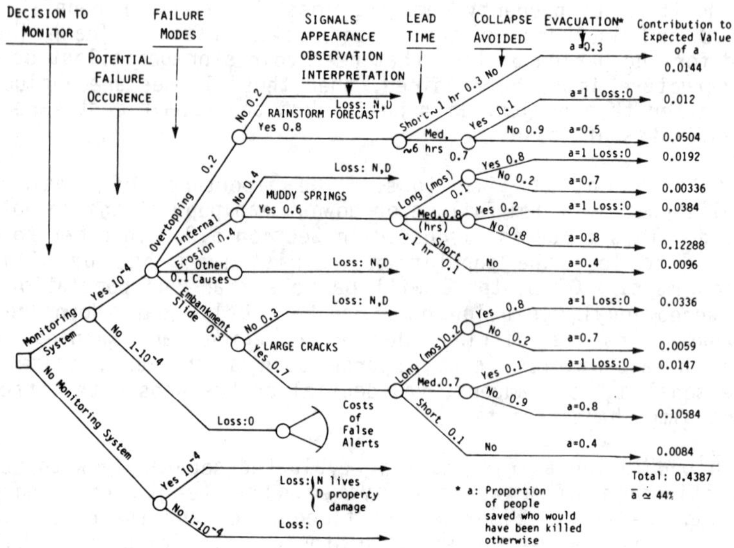

Figure 9. Probabilistic analysis of a monitoring and warning system (after Pate-Cornell and Togaras, 1986).

4. COMPARISON OF METHODS

4.1 Selection of Methods for Comparison

Methods compared are grouped into three categories based level of risk assessment. The traditional or subjective engineering design approach which does not explicitly quantify risk or assess its acceptability is excluded from this comparison. Neither does the comparison include generic evaluation approaches such as surrogate worth trade-off (Haimes and Hall, 1984), ELECTRE (Gershon and Duckstein, 1983), or multi-attribute utility analysis (Keeney and Raiffa, 1976). A recent evaluation for the U.S. Army Corps of Engineers (University of Southern California, 1985) includes the traditional engineering design and generic evaluation approaches with an apparent emphasis considering hydrologic loading.

4.2 Planning Level Methods

At the planning level an estimate of risk cost associated with dam failure can be introduced into the benefit-cost analysis which is used to provide the economic justification for a decision to build a dam. Baecher et al. (1979, 1980) describe a method for estimating the risk cost at the planning stage based on an average (or default) historical/empirical failure rate of 10^{-4} per year for large dams. In a recent paper Pate-Cornell and Togaras (1986) presented three case studies of planning level risk assessments and extended the procedure to the case of sequential dam failure. The level of detail used in these planning level methods is consistent with the level of effort which is expended on other planning studies and the availability of site specific information. Therefore, this level would not be suitable for comparing the risks associated with specific design alternatives for a proposed structure or for remedial action alternatives at an existing structure.

4.3 Screening Level Methods

The screening problem is the identification and ranking of "unsafe" dams in order of priority for expenditure of limited funds to pay for remedial action. By its very nature the screening problem implies that several, and perhaps many, dams must be assessed. Therefore, the level of detail required for analysis must be limited, and consistent procedures will be important to achieve an accurate ranking. Screening level methods are divided into index or qualitative methods and quantitative methods.

4.3.1 Index (qualitative) methods. Both the U.S. Army Corps of Engineers and the U.S. Bureau of Reclamation have developed index methods for ranking existing dams in terms of a subjective measure of "risk." Hagen's (1982) method, developed for the Corps of Engineers, defines a "relative risk index," R as the sum of an "overtopping failure score," O and a "structural failure score," S. The overtopping failure score is defined as follows:

$$O = O(1) \times O(2) \times O(3) \tag{5}$$

in which

O(1) = additional number of homes endangered by failure

O(2) = project flood capability as percentage of current design flood standard (PMF)

O(3) = project capability to resist failure by overtopping

and the structural failure score is defined as:

$$S = S(1) \times S(2) \times S(3) \tag{6}$$

in which

S(1) = number of homes endangered by failure at normal maximum probability

S(2) = evidence of structural distress

S(3) = potential seismic activity

The value assigned to each of the six factors is determined subjectively based on site inspection, review of design and construction records, and other available information. It varies between 1 and 5 with 1 being the most favorable. Thus the maximum or worst case index value would be 250 and a high score would lead to assignment of a high priority for remedial action to reduce the risk of dam failure.

The Bureau of Reclamation index method was developed as part of their Safety Evaluation of Existing Dams (SEED) (USBR, 1980) program. The evaluation process includes a site inspection and a review of available information on dam design, construction, and operation. A detailed evaluation report is prepared and a numerical measure of the dam's condition and damage potential, referred to as a site rating (SR) is assigned. The SR is obtained by summing the scores assigned to four elements relating to the conditions of the dam (age, general condition, seepage problems, and structural behavior measurements) and four elements describing the damage potential (capacity, hydraulic height, hazard potential/hydrologic adequacy, and seismic zone). Each element is scored on a scale of 0 to 9 with 0 being the most favorable value.

During the evaluation a list of recommendations for upgrading the dam is made and their significance is represented by a weighting system (USBR, 1980). The sum of the weights for all recommendations is added to the SR to give a "SEED" value which is used to rank the dam with respect to other dams.

4.3.2 Quantitative methods. Quantitative procedures for screening level risk assessment of dams have been developed by Stanford University for the Federal Emergency Management Agency (McCann et al., 1985) and by MIT (Vanmarcke and Bohnenblust, 1982). The framework for both procedures is similar to that described in Section 2.2. Each is summarized below.

The Stanford procedure is divided into three parts. The first part involves identification of the expected losses from dam failure and utilizes the information which is usually required for a Phase I inspection of the National Dam Safety Inspection Program. It involves collection of data, estimation of the probability of failure of the dam due to various initiating events, dam breach modeling and flood routing, and estimation of direct economic losses.

In the second part of the procedure the expected loss due to dam failure is estimated from the failure probability and economic loss estimates in the first part. Under the third part the cost and degree of safety improvement for each rehabilitation alternative is considered. This information is presented on a cost-effectiveness basis in order to rank a portfolio of unsafe dams for the expenditure of limited remedial action funds. Since ranking is the objective of the Stanford procedure it emphasizes standardized methods of analysis which can be expected to lead to reasonably consistent and reproducible results. Typically these methods are simplifications of what would normally be done at the design stage.

The MIT Screening Procedure is similar to the Stanford approach. It is illustrated by a case study in which 16 Vermont dams are ranked for remedial action using information available in inspection reports from the National Dam Inspection Program. The updating of historical frequency/empirical probability estimates of dam failure using information contained in the inspection reports using a Bayesian procedure is illustrated.

4.4 Design Level Methods

A framework for design level methods of risk assessment for dams was proposed by Bowles et al. (1978) and was detailed by Howell et al. (1980). A related approach was developed by the U.S. Bureau of Reclamation (USBR) and has most recently been described by the USBR (1986). The design level approach has been in use by the Engineering and Research Center of the USBR for several years and by the writer in conjunction with risk-based assessments of remedial action alternatives at several dams.

The USBR approach is divided into two major phases: a hazard assessment and a risk cost analysis of rehabilitation alternatives for existing dam or design alternatives for new dam. In the first phase the consequences of dam failure are evaluated in order to establish the minimum acceptable level of protection for an existing structure. For most dams this minimum protection level will be determined largely by the consideration of potential life loss assuming that an effective warning system and emergency action plan has been adopted. For a new structure the maximum loading conditions of the PMF for the inflow design flood and the MCE for seismic loading would normally be used.

Once the minimum acceptable level of (protection) loading before failure has been selected, the risk cost analysis is performed to determine whether or not higher levels of protection can be economically justified. The risk assessment follows the basic steps described in

Sections 2.2 and 3 of this paper and is illustrated by a hypothetical example in USBR (1986).

Although the overall framework for the USBR approach resembles that of the Stanford or MIT screening approaches the methods of analysis and probability and consequence estimation are more detailed and correspond to those normally used at the design stage of dams. A crucial factor in the successful performance of design level risk assessments is the incorporation of professional judgment in the identification of rehabilitation alternatives, identification and evaluation of failure modes and outcomes, the subjective estimation and updating of probabilities, the performance of meaningful sensitivity studies, and the interpretation of risk assessment results. Therefore, it is essential that qualified and experienced professionals, who would normally be responsible for dam design, should also be responsible for design level dam risk assessments.

5. ADVANTAGES AND COMMON OBJECTIONS

5.1 Advantages

Proponents of quantitative risk assessment of dams argue that it provides a comprehensive framework for the evaluation and presentation of dam design or remedial action alternatives. It requires the consideration of the interaction between initiating events, failure modes, possible outcomes, exposure conditions and consequences over the full range of magnitudes of each. Therefore, it aids in the identification of which factors (e.g., loading types, failure modes, and exposure factors), or ranges of factors, contribute most to the total risk associated with dam failures, and which options are most promising or cost effective for achieving risk reduction.

Presentation of the results of a quantitative risk assessment provides a concise and systematic display of the economic and noneconomic consequences of each alternative which is considered. The open presentation of all alternatives requires a high degree of objectivity in both the presentation and the supporting analyses. The effects of uncertainty in both the analyses and the exercise of professional judgment can be displayed in a manner which allows for understanding of the degree of technical confidence in the risk assessment. In fact the rational framework of risk assessment leads to clear documentation of the information necessary for a good appreciation by lay people of the significance of complex technical issues. It, therefore, facilitates constructive debate among opposing parties in the decision making process. The information obtained from a risk assessment is an input to the decision making process, and not the decision itself.

The decision maker can readily compare risk of failure at one dam with that at another, or with the risk associated with other types of public works projects. Also he can compare estimated risks with acceptable risk criteria and can consider such quantitative measures as estimates of risk reduction, residual risk, benefit cost ratios, cost effectiveness or cost to save a life. Such measures can provide a useful input to decision making on dam safety which is often a very emotional and politically sensitive issue. As the information base available to the analyst grows, a risk assessment can readily be updated to include revised estimates of probabilities and consequences, changes in uncertainty bounds, new insights into the likely performance of the structure, or new alternatives to be evaluated.

5.2 Common Objections

Frequently those unfamiliar with, or unconvinced about, the merits of the risk assessment approach to the evaluation of dam safety will focus attention on the limitations of the approach. Also, on occasion their objections are based on misconceptions about the approach.

An example of the latter is the statement that "risk assessment diminishes the role of engineering judgment." This statement has been made by some highly experienced engineers who fear that risk assessment is an attempt to replace professional experience and judgment with automatic decision rules which imply a level of knowledge of dam structure performance which they know is currently beyond our reach. The truth is that there can be no meaningful risk assessment of a dam without the involvement of highly competent and experienced dam designers, geological engineers, hydrologists, and others. A risk assessment can only be as good as the input provided by these professionals. The risk assessment approach requires the quantification of the judgments of these professionals and this can only be effectively achieved through involving them in the entire risk assessment process. The risk analyst can help to manage the collection and synthesis of information provided by dam engineering professionals, but he can never substitute for the role of experienced professionals. Under the risk assessment approach the role of such professionals is expanded rather than diminished, since a much wider range of loadings and other conditions is considered than would traditionally be the case under the traditional "worst case" approach to design.

In a similar vein, it is sometimes pointed out that "deterministic approaches to analysis and design are time-tested and that no competent professional can be expected to replace them with new probabilistic methods until these have been similarly proven." It is unlikely that anyone would disagree with this statement. However, it should be pointed out that an important basis for the inputs to a comprehensive risk assessment is the deterministic analyses of slope stability, dynamic structural stability, dam break, flood routing, seepage, etc. Where new analytical probabilistic methods are available they should be run in parallel with the deterministic approaches in order to establish a basis for their evaluation for potential future use.

Other commonly expressed objections to the risk assessment approach focus on its present limitations. These include the fact that probability and statistics are not broadly understood and are viewed with suspicion by many. This objection is not a fundamental limitation of the approach, but rather a handicap to its rapid acceptance which must be addressed by those who pioneer its use. In the long term it presents a challenge to educators to better prepare the public to understand and deal rationally with risk which impacts virtually every aspect of our lives. Another objection is that engineers, geologists and other professionals are generally inexperienced at making subjective probability estimates. This is certainly a valid concern and one that the pioneers in this field are having to address. Sensitivity analysis should be used to assess the relative importance of errors in subjectively updated or estimated probability estimates.

Others express concern that the use of risk assessment implies acknowledgment that there is "some chance of a dam failure occurring." However, this fact is unchanged whether or not a risk assessment is performed. It would seem preferable to systematically identify and examine all "reasonably probable" failure mechanisms and to invest wisely in safety improvement measures rather than deceive ourselves that the chance of dam failure is mysteriously reduced if we do not acknowledge that such a possibility exists.

Currently, certain causes of dam failure, such as inadequate grouting of an abutment, or piping of an embankment cannot be adequately assessed with analytical tools and therefore the basis for their inclusion in a risk assessment is highly subjective. However, this is not an inadequacy of risk assessment per se, but rather of the current state-of-the-art in dam engineering.

Another topic which is highly sensitive is the incorporation of life loss potential into a risk assessment. Dam failure frequently results in the loss of human life. No one would dispute the undesirability of this anymore than they would dispute the undesirability of people losing their lives from any other accidental cause. However, risk is unavoidable. It is only the degree of risk which we can control, and then we are subject to limitations of how much safety we can afford to purchase. In the case of dam safety the lives threatened are not necessarily those of the direct beneficiaries of the presence of the dam and reservoir, and therefore they are usually considered to be exposed to the risk of dam failure involuntarily.

At one extreme of this sensitive topic are those who would advocate placing an economic value on life and thereby integrating life loss into an overall economic assessment of the consequences of dam failure. At the other extreme are those who would require that the risk of life loss from dam failure should be reduced to zero, which of course, is an impossible requirement, although it can be approached as the investment of funds to save a life become "infinitely" large. Obviously as individuals and as a society we cannot afford such levels of safety. However, the simple inclusion of the value of life into an economic evaluation appears to be equally unattractive, and even immoral. On the other hand, to ignore the issue implies a zero value of life which is obviously an unacceptable alternative also.

Risk assessment does not require its users to take either extreme position, or any other position on this issue. But as with the idea of accepting some, albeit very small probability of dam failure, it does enable us to openly address the issue in such a way that high levels of human safety should be achieved with cost effectiveness. If the potential life loss from a dam failure is significant, then a hazard evaluation would suggest the use of probable maximum flood and maximum credible earthquake design criteria. In other cases the most significant reductions in the probability of life loss often can be made through the nonstructural approach of a monitoring system, emergency warning system, and evacuation plan. Such systems are typically quite inexpensive to design, install, and operate relative to the cost of structural measures. If these systems can be used to reduce the probability of life loss to very small levels, then the main basis for a decision on selection between structural alternatives for rehabilitation can be economic.

6. CONCLUSIONS

None of the limitations to the risk assessment approach which have been identified to date appear to warrant dismissal of the approach. Further research is needed in order to develop improved approaches for dealing with them, but meanwhile application of risk assessment to dams can provide valuable insights into the choices available to the decision maker. Such research can be effectively conducted through collaboration between universities and experienced practicing dam engineers in the public and private sector. Joint case studies on actual dams can be an effective means for developing improvement in the risk assessment approach.

The need for providing engineers and other professionals with better backgrounds in probability and statistics has already been identified. For the next generation of professionals this need can be addressed through the undergraduate and graduate curricula, but for the present generation of professionals, workshops and short courses will be needed to familiarize these individuals with the utility of risk-based approaches.

Related to the topic of education is the challenge of communicating the results and conclusions of a risk assessment to the lay public and to decision makers. In this regard, carefully prepared graphics and tabular presentations combined with the use of analogies to more familiar business and insurance situations can be very effective. Even when provided with quantitative risk estimates, individuals are highly influenced by their perceptions of the risk even if these are at variance with the estimates obtained from a formal risk assessment performed by highly experienced professionals. An excellent example of this is in the nuclear field. Just what the role of perceived risk should be in the field of dam safety decision making needs to be defined.

Risk assessment should be recognized as a part of the overall cycle of risk management referred to in Section 2.1 of this lecture. The cost effectiveness of nonstructural measures for improving and "managing" dam safety should be remembered (see Fig. 4). Such measures include operator training for detection of problems with a dam, frequent safety inspections by qualified professionals, steps to limit growth of downstream development in the potential inundation area, and implementation and updating of an emergency action plan.

Successful application of risk assessment and risk management approaches to evaluating dam safety requires a combination of the skills of experienced dam design professionals and risk analysts. At this time pioneering work has been completed or is currently underway on the application of design level risk assessment approach to evaluating remedial action alternatives on several dams. It is a field with both technical and communications challenges but when properly used it can be a valuable tool, particularly for those responsible for deciding remedial action at existing dams. Risk assessment can be expected to lead to improved dam safety and more effective use of limited funds available for rehabilitating existing dams.

ACKNOWLEDGMENTS

The ideas presented in this paper have developed over approximately ten years as the result of a series of research projects and consulting assignments in which the writer has been involved. The writer wishes to acknowledge the important role that the following of his colleagues have had in the development of these ideas: Dr. Loren R. Anderson, Dr. Ronald V. Canfield, and Dr. Terry F. Glover.

REFERENCES

Bowles, D. S., L. R. Anderson and R. V. Canfield (1978) A systems approach to risk analysis for an earth dam, Proceedings of the International Symposium on Risk and Reliability in Water Resources, Waterloo, Ontario, Canada, 13 pp.

Bowles, D. S. and H. Y. Ko, Eds. (1984) Probabilistic characterization of soil properties, American Society of Civil Engineers, 182 pp.

Gershon, M. and L. Duckstein (1983) Multiobjective approaches to river basin planning, J. Water Resour. Plan. and Mgt., Vol. 109, No. 1, pp. 13-28.

Hagen, V. K. (1982) Re-evaluation of design floods and dam safety, Transactions of 14th International Congress on Large Dams.

Haimes, Y. Y. and W. A. Hall (1974) Multiobjectives in water resources systems analysis: the surrogate worth trade off method, Water Resour. Res., Vol. 10, No. 4, pp. 615-624.

Howell, J. C., L. R. Anderson, D. S. Bowles and R. V. Canfield (1980) A framework for risk analysis of earth dams, Report submitted to Water and Power Resources Service (U.S. Bureau of Reclamation), Engineering and Research Center, Denver, Colorado, 87 pp.

Jaske, R. T. (1985) FEMA's computerized aids for accident assessment, Preprint No. IAEA-SM-280/85, International Symposium on Emergency Planning and Preparedness for Nuclear Facilities, Rome, Italy.

Keeney, R. L. and H. Raiffa (1976) Decisions with Multiple Objectives, John Wiley, New York.

O'Riordan, T. (1979) Environmental impact analysis and risk assessment in a management perspective, In: Emergency Risk Management, Edited by G. T. Goodman and W. D. Rowe, Academic Press, New York, p. 23.

Pate-Cornell, M. E. and G. Togaras (1986) Risk costs for new dams: economic analysis and effects of monitoring, Water Resour. Res., Vol. 22, No. 1, pp. 5-14.

Rowe, W. D. (1977) An Anatomy of Risk, Wiley, New York.

Sharp, K., L. R. Anderson, D. S. Bowles and R. V. Canfield (1981) A model for assessing slope reliability, Invited paper in Frost Action and Risk Assessment in Soil Mechanics, Transportation Research Record 809, pp. 70-78.

U.S. Bureau of Reclamation (1980) Safety Evaluation of Existing Dams (SEED), Government Printing Office, Washington, D.C.

U.S. Bureau of Reclamation (1986) Guidelines to decision analysis, ACER Technical Memorandum No. 7, Denver, Colorado.

U.S. Department of Interior (1982) Guidelines for determining flood flow frequency, Bulletin No. 17B, Hydrology Subcommittee, Interagency Advisory Committee on Water Data.

University of Southern California (1985) Evaluation of risk assessment methodologies for prioritizing remedial dam safety projects, Final Research Report submitted to National Science Foundation.

Vanmarcke, E. H. and H. Bohnenblust (1982) Risk-based decision analysis in dam safety, Report No. R82-11, Massachusetts Institute of Technology.

Yen, B. C. (1986) Reliability of hydraulic structures possessing random leading and resistance, In: Engineering Reliability and Risk in Water Resources, Edited by L. Duckstein and E. Plate, Martinus Nijhoff Publishers BV, Dordrecht, The Netherlands.

RISK ANALYSIS CONSIDERATIONS FOR DAM SAFETY

David A. Moser* and Eugene Z. Stakhiv*

U.S. Army Engineer Institute for Water Resources
Casey Building, Fort Belvoir, VA 22060

ABSTRACT

Risk analysis has become a fashionable enterprise, displacing a broad and diverse segment of traditional deterministic and standards-based approaches to evaluation and decision making for public health and safety. The general view is that risk analysis provides a formal basis to rationally evaluate much of the uncertainty inherent in difficult social choices about investments for improving health and safety that has heretofore been implicitly factored into conservative design standards. In water resources engineering there is as much a lack of consensus about what risk analysis means and how it is to be applied as is found in other disciplines. A number of perceived and practical difficulties of applying conventional risk analysis methods are investigated in the context of evaluating remedial measures for improving the hydrologic safety of dams.

1. INTRODUCTION

The purpose of this inquiry is to review some of the perspectives underlying risk analysis in a public decision making setting and to illustrate the difficulties of applying conventional techniques of risk analysis to a dam safety problem as an example. There seems to be a growing divergence in the burgeoning practice of risk analysis. This divergence appears to be between the theoretical approaches of academic researchers and the pragmatic orientation of practitioners, who are assigned the responsibility for designs based on such analyses and who are held accountable for their decisions and actions by the public. The dichotomy stems from the pursuit of traditional and more refined, but narrowly focused, analytical techniques that allow decision makers to become increasingly precise about very restricted and constrained formulations of complex social problems. Also, decisions concerning risk related actions are routinely made by Federal and state regulatory agencies that are responsible for the health and safety of citizens. Such decisions are no longer simply administered through reliance on professionally defined standards and criteria.

Nevertheless, risk analysis is generally conducted in a relatively conservative manner. Legal liability concerns, and the fact that the public expects a higher degree of safety in public domain technology, account for

---

*The views expressed in this paper are solely those of the authors and do not necessarily reflect the policy of the U.S. Army Corps of Engineers.

much of this conservatism. Academicians admonish the practitioners to incorporate available, but highly esoteric techniques, into the evaluation of public projects that impose substantial health and safety risks. This admonition is normally sidestepped by practicing engineers, who develop ad hoc methods tailored for specific tasks. As a consequence, there does not appear to be much coherence among the various existing risk evaluation methods, even for similar problems and technologies.

Three recent inquiries demonstrated the confusion in approaches to risk analysis. Although numerous standard, acceptable techniques are available for application to a diversity of public planning and regulatory problems, there is not yet a commonly shared philosophy for risk analysis. Nor is there a convergence of opinion on how specific risk analysis techniques are to fit within broader benefit-cost or risk-cost evaluation frameworks. Practical, intangible political considerations often constrain the use of valid methods. Much of what is termed risk analysis is merely a sensitivity test of the uncertainty surrounding the deterministic computation of an engineering design standard or a health standard, for example. In this presentation, "risk" carries the commonly accepted connotation of the probability of occurrence of a hazard multiplied by the consequences of the hazardous situation.

Kunreuther and Linnerooth (1983) have shown how risk analysis is often used to examine very narrowly defined technical components of a problem, often ignoring the larger social and economic issues at stake. They studied the application of risk analysis to four similar problems of siting liquefied natural gas facilities in four different countries. They found that the risk analysis parts of the studies merely addressed the variability of consequences of certain assumed design standards, but did not incorporate the results into a broader, systematic evaluation framework. Another common problem encountered was the consideration of risk analysis on a highly constrained problem, which did not allow for the evaluation of a full range of alternatives. The questions derived from the study addressed the general issue of whether or not risk (however it is defined and measured) is to be an objective that is traded off against other objectives to achieve an optimal risk-cost design. Or else, is risk analysis to be used in its more confined form of a sensitivity analysis about the uncertainty surrounding the choice of traditional design standards?

Another study, on dam safety criteria, conducted by the National Research Council (NRC; 1985) revealed that "... current dam design practices reflect great diversity in the standards used for classifying dams relative to the hazards and in the criteria for evaluation of safety from extreme floods and earthquakes" (pg. 2). The report also noted three general approaches used to determine the magnitude of floods, which is the basis for the evaluation of hydrologic dam safety and the risk of failure. These are the deterministic, probabilistic and risk analysis approaches. The NRC committee further recommends that components of a reservoir safety evaluation "... should strike a balance among such considerations as project benefits, construction costs, social costs, and public safety ..." (pg. 3).

The third study conducted jointly by the National Science Foundation (NSF; 1985) and the Corps of Engineers, was linked to the NRC study. A group of experts were assembled to look at the various methods currently in use or potentially available for dam risk analysis. The conclusion of the group was that "... although substantial research is being done to develop theoretical risk analysis methodology, there is currently no comprehensive method which could routinely provide a formal risk-based analysis for dam safety evaluation." Bowles and James (1984) arrived at a similar conclusion in their overview of risk-based design and evaluation methods for dam safety.

A major problem of the NSF-Corps group of experts was that they were asked to compare essentially incommensurable risk analysis methods, some of which were selected to represent more generic approaches. The difficulty encountered by the group stems from the same problem defined by Kunreuther's analysis, i.e. there are numerous legitimate, scientifically supportable ways of conducting the technical components of traditional engineering reliability analysis or of the statistical, probabilistic or deterministic formulations of hydrologic inputs which are the loading factors of a typical dam safety problem. There are not, however, a set of agreed upon decision rules and evaluation approaches that can combine engineering reliability analysis with economic efficiency and social well-being criteria. It is the combination of these considerations that serve to define "acceptable social risk."

More often than not, the choices of a particular engineering reliability approach are complicated by, if not often in conflict with, the broader-based economic evaluation framework. The melding of economics, reliability and social risk often comes about in the use of an "expected value" figure to represent the risk. Expected value essentially ignores the uncertainty of many loading factors and implicitly weights the outcomes in a manner which may not be relevant to the evaluation of a low probability, high consequence event, such as dam failure. For such events a "min-max" or "worst case" approach seems to be more appropriate than an "expected value" viewpoint.

Risk analysis, then, is not to be construed merely as traditional engineering reliability analysis, which tends to focus on the performance of a system in terms of target outputs. Rather, the full range of social and economic benefits and costs need to be considered in the formulations of alternative courses of action, including "no action." In the case of dam safety, for example, breaching a dam may be considered if that is the economically efficient and safer course of action. The different dam safety risk analysis methods reviewed by the NSF-Corps expert group are arranged in Figure 1 to show the variability in methods considered. The figure shows only two dimensions of a much larger set of criteria used in evaluating the methods and models. The hydrologic and hydraulic inputs to risk analysis (loading factors) are arrayed against the particular approach in which the objectives and economic decision rules are incorporated as part of the evaluation strategy. A number of approaches that were considered combine Bayesian analysis and more standard engineering reliability analysis, especially the FEMA/Stanford method (McCann et al, 1984).

The conventional understanding of engineering reliability analysis is exemplified by Yen (1979) and Duckstein and Plate (1984). Yen describes the six common approaches of engineering design for reliability through the use of safety factors to account for both hydrologic uncertainty and the structural design uncertainty. The "characteristic safety factor" approach perhaps best describes the design approach that federal water resources

FIGURE 1. ALTERNATIVE APPROACHES TO DAM SAFETY/RISK ANALYSIS

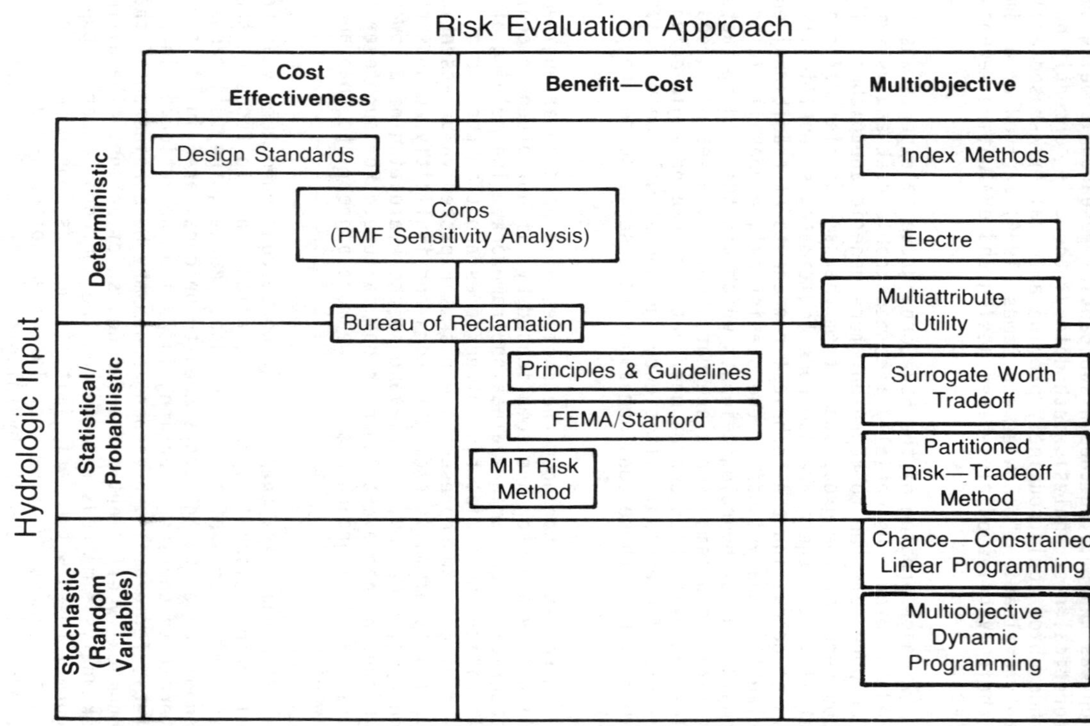

agencies apply to dam safety and spillway design. A prespecified flood event of a given magnitude (and an implied n-year return period) is chosen as the basic design criterion. Yen, however, proposes that the selection of the safety factor or design criterion itself can be determined rationally using reliability analysis. In Yen's model, reliability analysis of the safety factor is considered through a first-order, second-moment method. This considers the mean (expected) value of a variable and the coefficients of variation of the factors under consideration.

Duckstein and Plate (1984) advance Yen's engineering reliability approach by moving from explicit consideration of the risk and reliability implied in engineering design safety factors to a more general consideration of discretely identified system performance indicators. Thus, engineering reliability analysis is seen to emphasize the desired outputs of a project, measured in terms of predetermined targets. A new, "supra" performance indicator is introduced, the "figure of merit" (FM), which integrates various selected combinations of performance indicators over a given time horizon. Engineering risk, then, may be defined in terms of such a set of FM's. In this sense, engineering reliability is imbedded into engineering risk, which avoids the problems of oversimplification.

These engineering reliability methods, of course, rely on the precise knowledge of the probability density functions of the uncertain variables and parameters associated with the loadings on, and resistance of a structure. For low probability events (e.g. less than 10^{-3} per year) the outcomes have been shown to be very sensitive to the distribution functions, and the sensitivity increased with the coefficient of variation (Yen, 1979). A direct confirmation of this sensitivity to assumed distributions for a "probable maximum flood" (PMF) was conducted by Stedinger and Grygier (1985) with respect to risk-cost analysis of spillway design for dam safety. They demonstrated that a risk-cost analysis of spillway alternatives was greatly dependent on the assumed distribution of the extreme flood (PMF) event (Gumbel,log-normal, log-Gumbel, etc.). There is a reluctance among practicing engineers to move away from traditional, deterministically based designs based on safety factors, especially when such rather large uncertainties of loss-of-life and economic consequences of dam failure are added to a full risk-cost analysis. After all, these design safety factors have undergone a long evolutionary history based on failure and adaptation. These factors are not arbitrarily determined by engineers and are not entirely subjective.

Engineering reliability analysis presents a rational analytical approach to the formal consideration of uncertainty within a traditional perspective of engineering design criteria and prespecified project outputs. A large component of a typical project-related risk analysis relies on subjective judgements and tradeoffs among incommensurable and intangible objectives. Thus, the contention that risk is merely the complement of reliability does not fully account for all the decision variables that a public water resources agency, for example, must evaluate.

Reliability analysis may very well comprise a more rational approach towards providing a more complete spectrum of information needed in a risk analysis. Many socio-economic issues which are subsumed in pre-assigned target outputs as part of reliability analysis, are actually the crux of risk analysis. The heart of risk analysis appears to be the determination of "acceptable" social risk. This is done at the programmatic level of a public water agency as well as at the individual project level. One of the key issues of the debate within the engineering profession is whether "acceptable risk" is to be predetermined as part of standard engineering practice or as a consequence of economic decision rules, or even on the basis of "zero fatalities." Several risk evaluation-decision frameworks encompass the more specific project oriented reliability analysis which may provide a more refined basis for the necessary tradeoffs.

2. RISK EVALUATION FRAMEWORKS

Upon reviewing the extensive literature on risk analysis and commensurate evaluation frameworks, a large number of variations of several fundamental approaches become apparent. Risk analysis appears to be embedded in one of three generic evaluation philosophies:

(a) regulatory type decision process based on fixed standards and criteria (e.g., cost-effectiveness analysis; least cost design for a given standard; comparative/relative risk assessments).

(b) normative decision making with explicit or implicit decision rules (e.g., benefit-cost analysis; maximization of net benefits promoted by the U.S. Water Resources Council's "Principles and Guidelines" (1983) and applied by all Federal water agencies).

(c) eclectic, relativistic decision making based on multiple objectives, criteria or decision attributes (e.g., multiobjective optimization) where the relative importance of various objectives is explicitly derived through examining preferences of decision makers.

The important point to note about the application of economic efficiency (maximizing net benefits) as the prime decision criterion of U.S. federal water resource agencies is that the "acceptable" level of risk is defined implicitly by the alternative which maximizes net economic benefits. If the problem is correctly formulated, however, with all the proper constraints specified, there ought to be a convergence between all three approaches. In reality, however, cost effectiveness analysis is used most often where the largely intangible benefits of an action are difficult to specify, but the risk-costs are known fairly well. This is especially true in human health and safety issues related, for example, to safety provisions for automobiles, the work place, household appliances, and drugs. It may also be true for low probability-high consequence events such as dam failure and nuclear power plant failure.

Each of the three broad, philosophically distinct, evaluation approaches have been used in the context of dam safety analysis by the federal water agencies. In fact, many agencies employ some features of each of the approaches for different components of a problem. Consider, for example, a dam rehabilitation situation necessitated by changes in hydrologic information or increased downstream population, numerous alternatives can be considered and typically include: (1) raising the dam, (2) widening the spillway, (3) altering the operating rules, (4) reallocating storage, (5) breaching the dam, (6) improving warning and evacuation plans, (7) doing nothing, (8) building a new upstream dam, (9) various combinations of the above.

Each alternative has associated with it a number of incommensurable and intangible benefits and consequences. These include differences in the population at risk and expected fatalities, as well as longer term regional economic consequences if failure occurs. Cost-effectiveness analysis may be used early in a screening process to eliminate obviously inferior alternatives. Multiobjective analysis may be useful in exposing the tradeoffs and relaxation of constraints that need to be made when comparing one alternative with another. Net benefits analysis may be used to optimize the scales of an alternative which has been selected on the basis of a multiobjective tradeoff.

3. SOME RISK AND UNCERTAINTY CONSIDERATIONS IN DAM SAFETY

Historical experience provides little data for assigning probabilities and values to many of the important variables in the dam safety problem identified so far. All evidence points to the size and frequency of a PMF event as one of the key factors that influence the evaluation of alternative measures to improve dam safety and to reduce the risk of failure from a hydrologic event. The concept of a PMF event as the dominant factor determining "safety" may not be a particularly useful one in risk analysis, however. The PMF peak flow is defined as the hypothetical flood that is considered to be the most severe, but still reasonably possible, event. This definition implies that it is possible for peak discharges (Q) to exceed a PMF, but that these higher discharges are "unreasonable." In other words, $P[Q > PMF] > 0$, but this exceedance probability is ignored. There is considerable uncertainty, however, as to the peak discharge and volume of a PMF event. Prakash (1983) argues that depending on the method of computation, the value of PMF peak can vary up to 200 percent. Even when the PMF is treated as a fixed maximum, it is typically assigned an exceedance frequency of between 10^{-3} and 10^{-6}. In the determination of the likelihood of dam failure from a PMF event, this assumption ignores the possibility of peak flows in excess of the estimated PMF by cutting off the tail of the flow distribution. Typically, however, the simplifying assumption by the Corps and other federal and state water agencies is that the probability of dam failure is at least equal to the exceedance frequency of the PMF as the critical loading event.

Some analysts seem to believe that the uncertainty about the PMF discharge can be incorporated into a risk analysis by conducting a sensitivity analysis. This is usually done by varying the frequency of a PMF alone without including the uncertainty about the size of the PMF. Again, this ignores the fact that the probability of dam failure depends upon both the probability of the inflow event and the size of the event (peak flow and total volume).

Newton (1983), on the other hand, proposes that the probability of the PMF be based on the probability of each of the components of the PMF computation. The probable maximum precipitation (PMP) itself is a fairly rare event. Newton assigned the PMP a probability of 10^{-6} per year based on a Gumbel extreme value plot of 35 years of record of maximum annual 72-hour precipitation depths. Since the PMF is based on the PMP occurring after an antecedent storm, the probability of the PMF is computed as the probability of the storm couplet occurring within three days of each other. Overall, Newton estimated that a more realistic value for the probability of occurrence of the PMF was on the order of 10^{-10} per year. Nevertheless, as more recent data on maximum precipitation are collected, there have been an increasing number of rainfall events which have approached within 70-90 percent of the estimated PMP figure (National Research Council, 1985). This has resulted in the U.S. National Weather Service making a series of upward revisions to the PMP and has acted as the catalyst for reexamining the issue of the safety of existing dams in terms of hydrologic adequacy.

The second source of uncertainty identified above deals with the physical resistance of a dam to withstand the stress of an inflow event and the nature of the dam failure if the event exceeds the resistance. A dam's ability to withstand a given inflow influences the conditional probability of failure. This, in turn, affects the marginal probability of failure. The ability of an earthfill dam to survive overtopping is not well documented. In many models dealing with dam safety, a critical depth of overtopping is assumed, such as "the dam fails if overtopped by 1 foot or more." Therefore, the conditional probability of failure--given an inflow that overtops the dam by 1 foot or more--is one (1). In general, there is considerable uncertainty about the level of overtopping that a dam can withstand without failure. In addition, the length of time that a dam is overtopped is also likely to influence the likelihood of failure.

The breaching characteristics of a dam failure are important in determining the speed and extent of breach formation and the volume of sediment load in the flood wave. Recent analysis of earthfill dam failures indicates that the time required for the breach to reach its maximum development varies from 0.25 to 6 hours (MacDonald and Langridge-Monopolis, 1984). The dam resistance and breaching characteristics will influence the downstream direct flood damages, the number of people inundated and the length of warning time to those at risk. One of the most important components of any investigation of dam safety and measures to improve safety is likely to be better estimates of the time between overtopping and incipient failure, given the characteristics of the dam. This will provide information on the ability to develop effective, reliable warning systems.

The full extent of the economic impact of a dam failure is not certain. The historical experience from 100 to 200-year floods may not be transferable to the estimation of damages from a dam failure flood or even a non-failure (large spillway) flood well beyond the historical record. The physical damages from such floods will be much more extensive due to the higher velocities and sediment loads. In addition, many areas will be inundated that have never experienced flooding. Extensive damage to the public and private infrastructure, such as transportation, communication, and public utility facilities is likely to be unique to dam failure floods. These damages may cause long lasting disruptions to the economy of the affected region and, in some instances such as navigation, may affect the national economy as well. Current flood damage estimation methods, furthermore, focus on direct physical structural damage, ignoring to a large extent the regional economic impacts of failure.

Little research has been done in this area, although the research on the economic impacts of earthquakes may be of value (Ellson et al., 1984).

In addition to the direct, indirect, and induced economic losses of downstream areas, the failure of a dam also causes the loss of future services of the dam. These services could include navigation flow storage, flood damage control, hydropower, water supply, and recreation. The loss of these services would result in economic losses to industries dependent on, or related to, these outputs. Recreation, for example, is highly site specific and clearly dependent on the reservoir and maintenance of downstream flows.

The potential for loss of life and human health impairment are of prime concern in dam safety. In terms of fatalities, the problem is two-fold: determining the "population at risk," or number of people inundated by both "failure" and "non-failure" floods, and relating that population to the likely number of fatalities. The population at risk is typically considered as the permanent and transient population potentially inundated by the flood. The number of people in this group who are actually threatened by the flood, however, may depend on the month of the year or even the time of day. Some individuals may work in the threatened area but do not live there. In other instances, the inundated area may have high seasonal use but little or no permanent population.

Obviously, the number of fatalities, if failure occurs, can not be known with certainty. This uncertainty must be incorporated into risk analysis by evaluating the possible number of fatalities and the probability distribution of fatalities. An additional significant problem is incorporating the non-fatal physical and emotional damages to those who survive the dam failure flood. Typically, these effects are ignored or are assumed to be a linear transformation of the number of the fatalities.

Warning and evacuation systems may be an effective means for reducing the injuries and loss of life from a dam failure. Analysis of fatalities from previous dam failures suggests that providing a warning time of 1.5 hours or more can significantly reduce the risk of fatalities from failure (Bureau of Reclamation, 1984). The effectiveness of the warning system depends on the credibility of forecasting systems and communications network as well as on the response of those downstream in terms of their willingness and ability to evacuate to safe locations. Experience from hurricanes indicates that even advance warnings do not induce all those at risk to evacuate to safety. The National Oceanic and Atmospheric Administration (NOAA) is currently using a system model for evaluating flood forecasting response schemes developed by Krzysztofowicz and Davis (1983). In addition, the transportation system used by those leaving an area may not be adequate for a rapid evacuation of a heavily populated urban area. Therefore, more advanced warnings may be required to allow all those who wish to leave to be able to do so. Issuing warnings sufficiently in advance of the actual dam failure to allow evacuation, however, increases the likelihood of false alarms. Therefore, the evaluation of the effectiveness of warnings should consider the economic and other impacts of false alarms (e.g. see Ellson et al., 1984).

Improvements in public safety are desirable goals that are largely undisputed. Improving the safety of a dam, or anything else that poses a potential safety or health risk, is often very costly. In addition, safety is never an absolute: there is always some residual risk. Therefore, the essential question of risk-based decision analysis in dam safety is "how safe should a dam be?" As noted previously, the answer to this question has commonly been attempted through the application of several different analysis/decision frameworks. However, there is considerable promise that risk-cost/risk-benefit approaches promoted by the American Society of Civil Engineers (1973), Baecher, et al., (1980), the Bureau of Reclamation (1984), Langseth and Perkins (1983) and the National Research Council (1985) all seem to be converging on a common evaluation perspective, at least for dam safety.

4. COMPARATIVE RISK ANALYSIS FOR DAM SAFETY

Comparative risk analysis is an often practiced approach in the health hazard field. Analogous risk situations are identified and the implied safety/health standard of prior decisions is used to support the selection of the acceptable level of risk (safety) for the current problem. An example for the problem of potentially unsafe dams is to upgrade every dam to meet the current nuclear regulatory standard for nuclear power plants, an annual probability of failure of about 10^{-6}. Both dams and nuclear power plants share the characteristics of imposing involuntary risk on the population. If society, through public regulatory agencies and congressional oversight, judges that nuclear power plant health and safety risks should be no greater than 10^{-6}, then dams ought to be equally failure proof. The use of this approach assumes that a "socially acceptable" level of risk has been identified in prior decisions about comparable technological hazards. Of course, in order to fulfill even this simple decision rule, one needs to know the probability of dam failure and the effectiveness of the "fix" in probabilistic terms.

Comparative risk analysis is relatively simple conceptually, since it avoids several potentially troublesome problems that include the value of human life and health and the monetization of economic losses. In economic terms, comparative risk analysis assumes that, up to the prior decision risk standard, the social benefits from improved safety equal or exceed the social costs. It also assumes that preventing one fatality by increasing safety is equivalent to preventing a thousand fatalities: the number of people at risk is irrelevant to the safety decision. Finally, the expenses of meeting the risk standard are irrelevant to the decision. Aside from the avoidance of certain decisions, the main problem with this approach is that it is difficult to apply to a complex technology like a nuclear power plant. The reliability of each component must be known, as well as the conditional probabilities of failure of the entire system.

Applying the comparative risk approach to the problem of dam safety still requires determining the technically feasible alternatives to achieving the "social standard" of safety. Assuming that more than one alternative is available, and that each alternative has economic advantages and disadvantages, the comparative risk approach essentially reduces to cost-effectiveness analysis. The investment decision is the one that achieves the safety standard for the least cost.

A variation of the comparative risk approach is the agency "portfolio problem." For example, the U.S. Army Corps of Engineers has an inventory of about 500 dams. Each has a different annual probability of failure, ranging from perhaps 10^{-4} to 10^{-9} from a combination of hydrologic, seismic and structural risks. Taken as a group, however, the probability of any failure of the 500 dams over the next 50 years could be fairly high. A choice of a socially acceptable risk of failure for the group over a 50-year planning horizon can be used to infer the minimum safety standard for each dam. For instance, assume that the socially acceptable risk of failure for the 500 dams is defined such that the probability of one or more failures over the next 50 years is not to be greater than 10^{-3}. This implies that the risk of failure of any dam in any year must be less than 4.0×10^{-8} (Table 1). Thus, each dam should be designed or upgraded to ensure that its annual probability of failure is no greater than 4.0×10^{-8}. Again, if more than one alternative to achieve the prescribed degree of safety is available, this variation of comparative risk analysis becomes a cost effectiveness problem. Table 1 provides an example of how policy analysts may use comparative group risk information to identify a broader measure of safety. The problem, of course, is that the probabilistic measures of individual dam reliability are themselves highly uncertain, and it is difficult to translate abstract measures of reliability into discrete remedial actions.

5. COST EFFECTIVENESS ANALYSIS FOR DAM SAFETY

Cost effectiveness is another commonly used approach in risk-based decision analysis. In its simplest form, cost effectiveness would lead a decision maker to choose the least expensive alternative or combination of alternatives to achieve the social safety standard. A more advanced cost effectiveness analysis would forego a safety standard and instead examine the tradeoffs between the safety expenditures and some measure of the degree of safety. The measure of the amount of safety could be the probability of failure or the number of expected fatalities. Cost effectiveness emphasizes the expenditures on the investment to improve safety and typically ignores other attendant costs.

Therefore, the appropriate level of investment in safety may be the minimum expense necessary to achieve a desired level of safety from an agency's budget standpoint. This chosen alternative, however, may not be economically efficient when all opportunity costs are considered. Cost effectiveness can be used where the objective minimum level of safety is not prespecified. In this case, more or less ad hoc decision rules are applied to approximate the appropriate level of safety. These rules could include that the expenditure for safety can be increased as long as the incremental cost per life saved is less than X (implied or explicit value of human life). An alternative rule could be to choose the level of expenditure where the marginal expenditure per life saved begins to accelerate. Thus, the shape of the expenditure function and the safety function determine the appropriate level of spending on safety improvements.

6. RISK-BENEFIT ANALYSIS FOR DAM SAFETY: AN EXAMPLE

The discussion of cost effectiveness indicates that decisions on investments for safety must eventually consider the tradeoffs between the cost and level of safety. The most complete consideration of all the economic costs as well as risks in a traditional water resource decision framework is risk-benefit or risk-risk/cost analysis. The benefit and cost components of these decision approaches are based on the economic principles

TABLE 1. PORTFOLIO COMPARATIVE RISK

PROBABILITY OF DAM FAILURE PER EACH DAM YEAR	PROBABILITY OF NO DAM FAILURES IN 25,000 DAM YEARS	PROBABILITY OF ONE OR MORE DAM FAILURES IN IN 25,000 DAM YEARS	ACCEPTABLE PROBABILITY OF ONE OR MORE FAILURES IN 25,000 DAM YEARS	REQUIRED ANNUAL PROBABILITY OF DAM FAILURE: EACH DAM	PROBABILITY THAT A DAM WILL FAIL ONCE IN 50 YEARS
P_F	$(1-P_F)^{n \cdot t})$	$(1-(1-P_F)^{n \cdot t})$	P_{FA}	P_{FAi} $(1-(1-P_{FA})^{1/n \cdot t})$	$(P_{FAi}(1-P_{FAi}^{49})50$
10^{-2}	~0	~1	10^{-2}	4.0×10^{-7}	2.0×10^{-5}
10^{-3}	1.4×10^{-11}	~1	10^{-3}	4.0×10^{-8}	2.0×10^{-6}
10^{-4}	8.2×10^{-2}	9.2×10^{-1}	10^{-4}	4.0×10^{-9}	2.0×10^{-7}
10^{-5}	7.8×10^{-1}	2.2×10^{-1}	10^{-5}	4.0×10^{-10}	2.0×10^{-8}
10^{-6}	9.7×10^{-1}	2.5×10^{-2}	10^{-6}	4.0×10^{-11}	2.0×10^{-9}
10^{-7}	9.9×10^{-1}	2.5×10^{-3}	10^{-7}	4.0×10^{-12}	2.0×10^{-10}
10^{-8}	~1	~0	10^{-8}	4.0×10^{-13}	2.0×10^{-11}

n = NUMBER OF DAMS IN PORTFOLIO, (n = 500)

t = NUMBER OF YEARS IN PLANNING HORIZON, (t = 50)

embodied in benefit-cost analysis. Thus, the application of the "with" and "without" improvements and "willingness-to-pay" principles provide a more complete description of the economic gains and losses from increments in safety. Risk-benefit and risk-risk/cost analyses go beyond traditional benefit-cost analysis by explicitly considering risk (or safety) as an objective equal to the maximization of net economic benefits or minimization of net economic costs. Typically, risk is measured in terms of expected adverse health or safety effects. In addition, equity considerations can be included such as when:

1. The direct beneficiaries of the outputs of an investment do not bear the risks of investment failure, or

2. The safety improving investment reduces the risk for some but increases it for others.

Uncertainty about the economic costs and benefits or a risk reducing investment must be incorporated in the analysis. Although the costs and benefits can be identified, the monetary value of the gains and losses, and whether or not they will be realized, is uncertain. The standard approach is to convert both the costs and benefits into an expected value term based on the probability distributions derived in the risk assessment.

One of the difficulties of applying a formalized risk analysis to dam safety is the determination of a credible probability distribution for rare large floods. Faced with this uncertainty, the standard approach noted in the discussion on risk assessment is to apply sensitivity analysis by varying the probability distribution. One method is to interpolate the frequency distribution between a 100-year event and a PMF event where the probability of the PMF is varied in the sensitivity analysis (National Research Council, 1985). Assume that flood events follow an exponential distribution, so that the probability of an event with a peak flow less than or equal to Q_I is:

$$F[Q_I] = 1 - \exp[-r\,(Q_I + q_o)] \qquad (1)$$

where r and qo are parameters to be determined. To employ the interpolation method, the values of r and qo must be determined such that:

$$1 - 1/T = F[Q_T] = 1 - \exp[-r\,(Q_T + q_o)]$$

and

$$1 - 1/100 = F[Q_{100}] = 1 - \exp[-r\,(Q_{100} + q_o)]$$

The solution of these simultaneous equations yields values for r and q_o:

$$r = (\ln 100 - \ln T) / (Q_{100} - Q_T)$$

$$q_o = (Q_{100} \ln T - Q_T \ln 100) / (\ln 100 - \ln T)$$

where:

$1/T$ = exceedance probability for the PMF event = $P[QI > QT]$,

Q_T = peak flow rate for the PMF event,

Q_{100} = peak flow rate for the 100 year event,

1/100 = exceedance probability for the 100 year event = $P[QI > Q100]$, and

T = the return interval for the PMF.

The above relationships determine the exceedance probability for any peak flow event between Q_{100} and Q_T given Q_{100}, Q_T and T:

$$1 - F[Q_I] = \exp\{1/(Q_T - Q_{100}) \\ [Q_{100} \ln T - Q_T \ln 100 + (\ln 100 - \ln T)\; Q_I]\} \qquad (2)$$

In our example problem, which is based on real reservoir data, assume that:

$$Q_T = 432{,}000 \text{ cfs}$$

$$Q_{100} = 43{,}200 \text{ cfs}$$

and allowing T to take on the values of 10^{-4} and 10^{-6} year the exceedance probabilities of flood events as proportions of the PMF can be determined and are shown in Table 2. The probability density function for Equation (1) using the derived values of r and q_o is:

$$f[Q_I] = dF/dQ_I = [(\ln 100 - \ln T) / (Q_{100} - Q_T)] \\ \exp\{[1 / (Q_T - Q_{100})] \\ [Q_{100} \ln T - Q_T \ln 100 + (\ln 100 - \ln T)Q_I]\} \qquad (3)$$

Downstream flood losses can be represented as a function of the damageable capital and the level of inundation at each point downstream from the dam. The level of inundation at any point downstream is determined by the downstream distance from the dam, the peak outflow from the dam and the hydraulic characteristics of the river valley. Finally, the peak outflow from the dam is a function of the inflow event, represented by the inflow hydrograph, and the reservoir and dam characteristics. Thus, assuming that the dam passes the flood safely, the downstream damage function can be represented as:

$$D_i = D_i\{K_i,\ h_i[Q_o(Q_I,\underline{R},\underline{Y}),\ \underline{M}]\} \qquad (4)$$

where:

D_i = downstream flood losses along a reach i miles from dam,

K_i = damageable capital along a reach i miles from dam,

$h_i[\ ]$ = level of inundation by flood water along a reach i miles from dam,

$Q_o(\ )$ = peak rate of reservoir outflow,

TABLE 2. EXCEEDANCE PROBABILITIES OF FLOOD EVENTS

PROPORTION OF PMF	PEAK INFLOW (cfs)	EXCEEDANCE PROBABILITIES OF FLOOD EVENTS GIVEN THAT THE RETURN PERIOD OF PMF = T	
		$T=10^4$	$T=10^6$
0.01	4320	0.85000000	0.85000000
0.05	21600	0.09000000	0.09000000
0.10	43200	0.01000000	0.01000000
0.20	86400	0.00599484	0.00359381
0.30	129600	0.00359381	0.00129155
0.40	172800	0.00215443	0.00046416
0.50	216000	0.00129155	0.00016681
0.60	259200	0.00077426	0.00005995
0.70	302400	0.00046416	0.00002154
0.80	345600	0.00027826	0.00000774
0.90	388800	0.00016681	0.00000278
1.00	432000	0.00010000	0.00000100

Q_I = peak rate of reservoir inflow,

$\underline{R}$ = vector of reservoir characteristics, e.g. dam crest length stage-storage relationship, etc.,

$\underline{Y}$ = vector of dam safety improving choice variables, e.g. spillway width, dam height, spillway crest height, etc., and

$\underline{M}$ = vector of river hydraulic characteristics.

Total downstream flood losses resulting from a reservoir inflow event are the sum of the losses along all the downstream reaches and are represented for convenience as a function only of the peak inflow rate and the vector of dam safety choice variables, $D(Q_I, \underline{Y})$.

Given the characteristics of the reservoir and dam, $\underline{R}$ and $\underline{Y}$ in Equation (4), there is a critical inflow event, represented by its peak inflow rate, that overtops the dam. This overtopping may lead to the formation of a breach and dam failure. If failure occurs, downstream flood losses are likely to exceed those that occur if the dam passes the same flood safely. Total downstream flood losses if the dam fails are:

$$D = D(Q_I, \underline{Y}) + F(Q_I, \underline{Y}) \tag{5}$$

where F is the increment in downstream flood losses due to failure of the dam. Representing the critical or threshold flood that overtops the dam as Qc and assuming the overtopping results in dam failure, then failure and nonfailure downstream flood losses are:

$$D = D(Q_I, \underline{Y}) \qquad \text{if } 0 < Q_I < Q_c$$

and

$$D = D(Q_I, \underline{Y}) + F(Q_I, \underline{Y}) \qquad \text{if } Q_c < Q_I < Q_T$$

where Q_T is the peak flow rate of the 1 PMF flood event as before. The threshold flood that overtops the dam depends on the dam safety hydrologic characteristics, $\underline{Y}$, at the time of the flood event.

7. IMPLICATIONS OF RISK-BENEFIT EVALUATION FOR DAM SAFETY

The impact of different alternative dam safety modifications on "failure" and "non-failure" downstream flood losses is significantly different. Consider a dam with a 500 foot spillway that can currently pass a 0.65 PMF event safely (threshold flood) but will be overtopped and threatened with failure for any larger events. Two commonly used alternatives to improve the hydrologic safety of the dam are to widen the spillway or to raise the dam crest elevation. For flood events larger than the current threshold flood, widening the spillway has little impact on the downstream losses due to dam failure since downstream peak failure flows will be similar to those with failure under the existing, inadequate spillway conditions. The wider spillway will, however, increase the size of the flood event the dam can safely pass so that the probability of failure is reduced. Raising the dam crest also reduces the likelihood of failure but would also result in larger peak flows if failure occurs compared to the existing dam height.

For flood events less than the current threshold flood of 0.65 PMF, the impact on downstream damages of the two common dam safety modifications is just the opposite from that described above. Raising the dam crest elevation will not alter downstream flood losses compared to the current conditions (original spillway). Widening the spillway, however, increases the peak rate and frequency of outflows and, consequently, increases downstream economic losses compared to the current spillway width. Figure 2(a) shows the representative "failure" and "non-failure" downstream flood losses from various flooding events for the existing spillway and for increments in spillway width up to one large enough to pass a 1 PMF event. Figure 2(b) also shows the representative "failure" and "non-failure" downstream flood losses from various flooding events, but the comparison is between the current dam height with those of increments of raising the dam height up to an elevation which would pass a 1 PMF event with the existing spillway.

An examination of Figure 2(a) reveals that spillway widening would increase flood damages in the portion of the flood probability distribution that is more likely to occur. Langseth and Perkins (1983) show this implicit tradeoff should be considered in spillway design. Conversely, Figure 2(b) shows that raising the dam height affects flood damages only in the very low probability tail of the flood event distribution. Although "failure" damages increase with increased dam height, "non-failure" damages decline. Both of these effects occur in the low probability tail of the probability distribution as shown by the crosshatched area in Figure 2(b).

The implications for risk-cost tradeoffs between management options is clear: decision makers must consider an incommensurable substitution of more frequently occurring economic damages (widening the spillway) for a more remote possibility of an even greater magnitude catastrophic dam failure (raising the dam), along with the possible increase in upstream flooding costs. The use of expected value contributes to tradeoff difficulties by implicitly weighting parts of the frequency spectrum and potentially skewing choices among solutions. In addition, the potential loss of life is different under both circumstances, compounding the incommensurable tradeoff problem. Presumably, either alternative (widening the spillway or raising the dam) has the same anticipated reliability, measured in terms of the ability to pass a 1 PMF flood. However, raising the dam may increase the "failure" flood stage substantially, placing a higher proportion of the downstream population at risk. The expected fatalities of dam failure under the alternative of raising the dam is then larger than the expected fatalities of failure under the alternative of widening the spillway.

A complete risk assessment evaluation of the benefits and costs of the dam safety modifications requires the evaluation of the costs of the modifications as well as the expected foregone project benefits if failure occurs. The determination of expected foregone project benefits, if failure occurs, accounts for the fact that the benefits of the project accrue over time and the year of failure is uncertain. Assume that the project benefits include flood control, plus additional benefits from hydropower and

FIGURE 2(a). FAILURE AND NON-FAILURE FLOOD DAMAGES AT INDICATED SPILLWAY WIDTH ADDITION

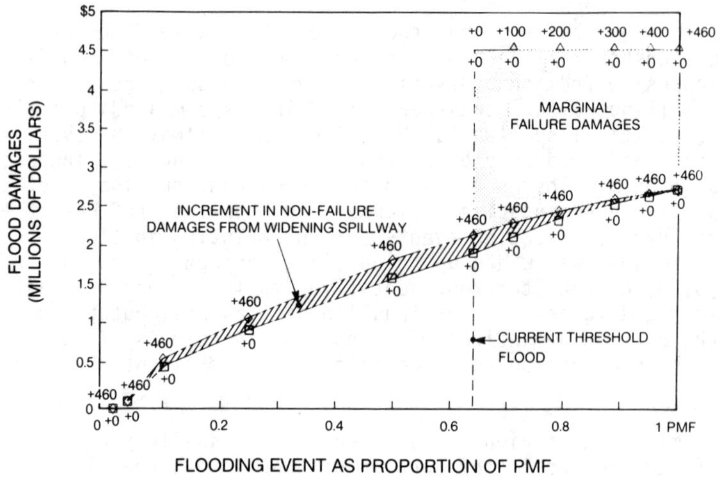

FIGURE 2(b). FAILURE AND NON-FAILURE FLOOD DAMAGES AT INDICATED DAM HEIGHT ADDITION

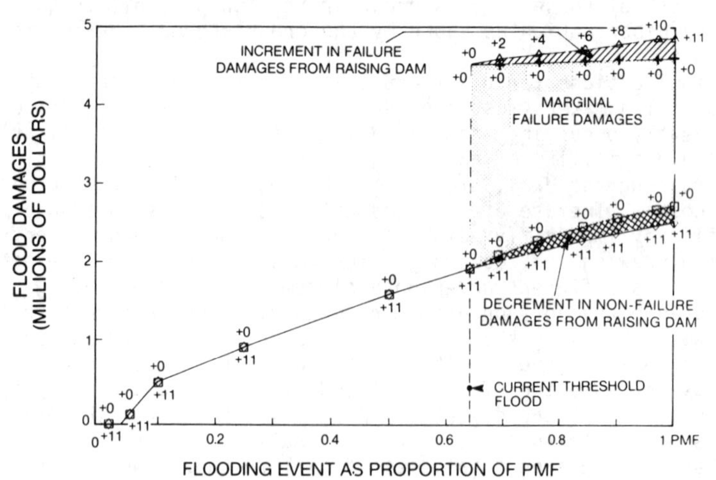

recreation benefits. Without any dam safety modifications annual flood control benefits are the difference between the expected flood damages without the dam less the expected residual flood damages with the existing spillway-dam height combination. In our terminology residual damages are those which occur when a flood is of sufficient magnitude to pass through the spillway and exceeds the capacity of the downstream flood channel. In evaluating the alternative modifications, the project benefits should be adjusted for changes in residual flood damages due to the safety modification. Assuming that the non-flood control project benefits are constant each year, the present value of these benefits in each possible year of failure for each scale of safety modification can be calculated using the standard annuity formula:

$$PV_t = A \{[1 - (1 + r)^{-(n - t)}] / r\}, \quad t = 0,\ldots,n \tag{6}$$

where:

PV_t = present value in year t if dam fails in year t,

A = annual project benefits,

r = interest rate,

n = useful life of the dam, and

t = year of failure.

The determination of the expected value of the lost project benefits in each year requires the probability of dam failure in each year. The annual probability of failure is assumed, here, to be the exceedance probability of the threshold flood for each type and scale of safety modification. (The probability of dam failure is much more complex and consists of point probabilities of many independent events.) This probability is determined from Equation (2) once the threshold flood is determined for each safety modification. Following standard present value concepts suggested by Baecher et al. (1980), the expected present value of foregone project benefits can be calculated for each type and scale of safety modification as:

$$PV_{E[B]} = \sum_{t=0}^{n} [PV_t / (1+r)^t]\, P[F] = P[F] \sum_{t=0}^{n} [PV_t / (1+r)^t] \tag{7}$$

where:

$PV_{E[B]}$ = expected present value of lost project benefits for the safety modification, and

P[F] = annual probability of dam failure with the safety modification, and PV_t,r,t defined as above.

There are two common ways of evaluating the relative and absolute worth of alternative dam safety/rehabilitation solutions. The U.S. Water Resource Council's (1983) "Principles and Guidelines" for project evaluation call for a net economic benefits maximization decision rule. Consideration of risk-net benefits tradeoffs would fulfill this requirement. Many health and safety related policy and program decisions are made on risk/cost principles, since the benefits are intangible or difficult to quantify. In the case of dam safety, costs are known with greater certainty than benefits projected out over a 50-year horizon. Ideally, however, risk/cost analysis is comparable to risk-net benefits analysis. The risk/cost of the dam safety problem is the sum of the expected present value of:

(1) the residual downstream flood losses,

(2) the increment in "failure" compared with "non-failure" downstream flood losses,

(3) the lost project benefits, and

(4) the modification construction costs and increment in operations and maintenance for each type and scale of safety modification.

The expected present value of net benefits, of course, is the difference in expected benefits minus expected costs. It is important here to note a subtle distinction between risk/costs and net benefits. Net benefits for the various scales of modifications (wider spillway or higher dam) measure the expected net present value of reducing the risk of failure. Risk/costs measure the expected total costs corresponding to the level of the risk of failure for each scale of modification.

Skeptics of the net benefits approach argue that this will show that the expected value of the costs of safety improvements will invariably exceed the benefits: the more certain investment expenditures will dominate the expected economic benefits from improved safety. Risk-benefit and risk/cost analysis, however, are not single objective approaches. They emphasize the tradeoffs between economic gains and losses and risk: there is no requirement for the safety investment to be justified on economic efficiency grounds alone. A more serious argument against the use of expected values is that the use of a single parameter description of the benefits and costs of safety, as well as the level of safety, is inappropriate in the analysis of extremely low probability, high consequence events. Alternative methods of incorporating uncertainty within the benefit (cost) risk framework, such as the "partitioned multiobjective risk method" (Asbeck and Haimes, 1984) can be used. Although these methods may not necessarily provide a "better" decision, they are designed to explicate the nature of the decisions which need to be made in a more uniform manner.

Figure 3(a) shows the expected values of net benefits and risk/cost assuming that the annual probability of a PMF is 10^{-4}, while Figure 3(b) assumes a PMF probability of 10^{-6}. These values represent alternative end points of the probability distribution. Comparison of these figures indicates the dominance of the flood frequency distribution in the determination of net benefits (risk/cost). In Figure 3(a), the spillway capacity that maximizes expected net benefits (minimizes the risk/cost) will pass a 0.71 PMF flood without overtopping the dam. The increment in dam height that maximizes net benefits results in a dam safety modification that prevents overtopping from 0.89 PMF or smaller event. If the probability of a PMF is 10^{-6}, however, Figure 3(b) shows that the "economically justified"

FIGURE 3(a). PRESENT VALUE OF RISK-COST AND NET BENEFITS PROBABILITY OF PMF = 10^{-4}

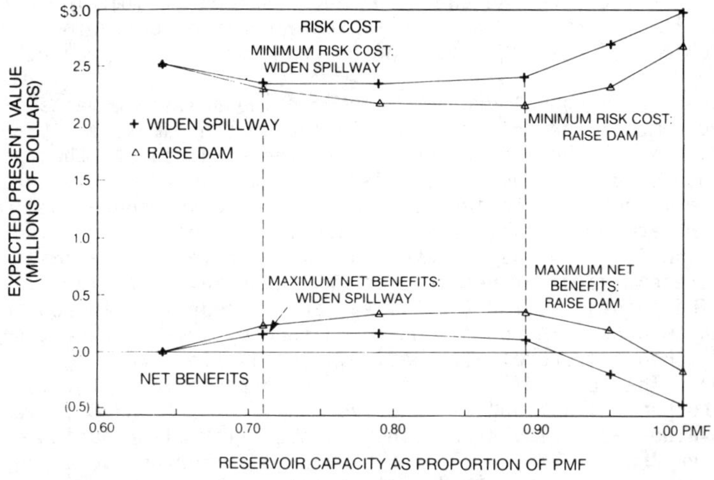

FIGURE 3(b). PRESENT VALUE OF RISK-COST AND NET BENEFITS PROBABILITY OF PMF = 10^{-6}

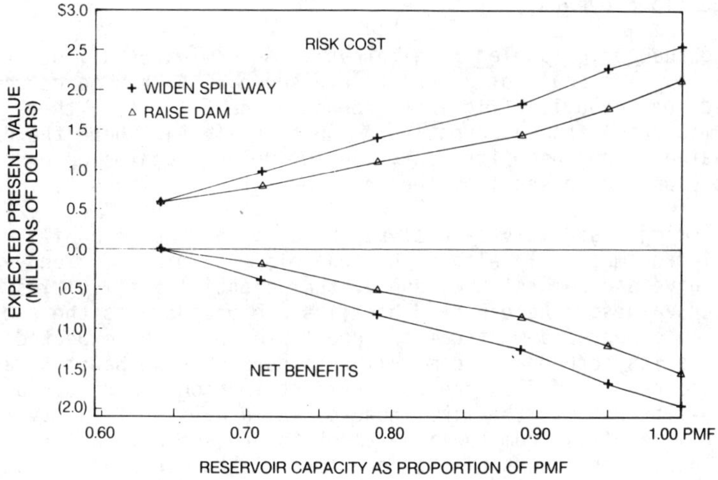

spillway as well as dam height would be those of the existing dam. Stedinger and Grygier (1985) have shown that the choice of a particular frequency distribution representing the spectrum of hydrologic events can have similar effects on the choice of the risk/cost minimizing alternative. This sensitivity of the economic efficiency decision criterion indicates the likely problems of relying solely on an expected value approach.

Figure 4 shows an alternative, but more generalized, means of displaying some of the information relevant to the dam safety problem. It shows the tradeoffs between economic efficiency and safety that are likely to be encountered in a dam safety risk evaluation. In this figure, several alternative investments to improve dam safety and/or reduce the impacts of dam failure have previously been evaluated. Each point along the solid line curves, I and II, show the scale of the alternative that maximizes the expected present value of net benefits while achieving a given level of safety. The points labeled A_1, A_2, A_3, and A_4 represent increasing scales of spillway widths. The points on the dashed B curve represent different scales of a non-structural alternative, such as flood warning and evacuation. The level of safety is the expressed number of lives saved compared to the "no action" alternative. Along the A and B curves, the expected values are calculated using a probability distribution of flood events where $P[Q > 1 \text{ PMF}] = 10^{-4}$. The A' and B' curves use a probability distribution where $P[Q > 1 \text{ PMF}] = 10^{-6}$. Note that the evaluation of benefits and costs requires the application of the "with project" and "without project" principle. Thus, the "no action" alternative neither saves lives nor generates economic benefits or costs. Therefore, the origin represents the "no action" alternative. Note, also, that although net benefits are zero in the "no action" alternative, there are residual expected costs and expected fatalities due to the possibility of dam failure from large flood events.

Along the curve labeled I, alternative A_2 dominates all other alternatives. The scale of alternative A shown at A_2 can be economically "justified" on economic efficiency grounds alone. In fact, there are still net economic gains from alternative A_2 up to scale A_4, where the expected present value of net benefits equals zero. Only at scales beyond A_4 are there net economic losses from the investment to save lives.

Performing a sensitivity analysis by changing the probability distribution dramatically alters the feasible economic efficiency-risk tradeoff curve and demonstrates one of the potential difficulties of relying on expected values. The expected benefits are measured as the avoided increment in expected downstream failure losses plus the expected future project benefits foregone if dam failure occurs. These benefits are highly sensitive to the flood flow probability distribution. A reduction in the PMF probability reduces the likelihood of failure, significantly reducing the expected benefits from the investment to prevent dam failure. In addition, one of the costs of the investment is the increase in the expected value of the "no-failure" downstream losses compared to the existing condition. Although this expected value also declines as the probability distribution is shifted, the reduction is less than proportional to the decline in expected benefits. This stems from the fact that widening an uncontrolled spillway increases reservoir outflows for higher probability events as well as the low probability event that threaten the safety of the

FIGURE 4. NET BENEFITS-RISK TRADEOFF

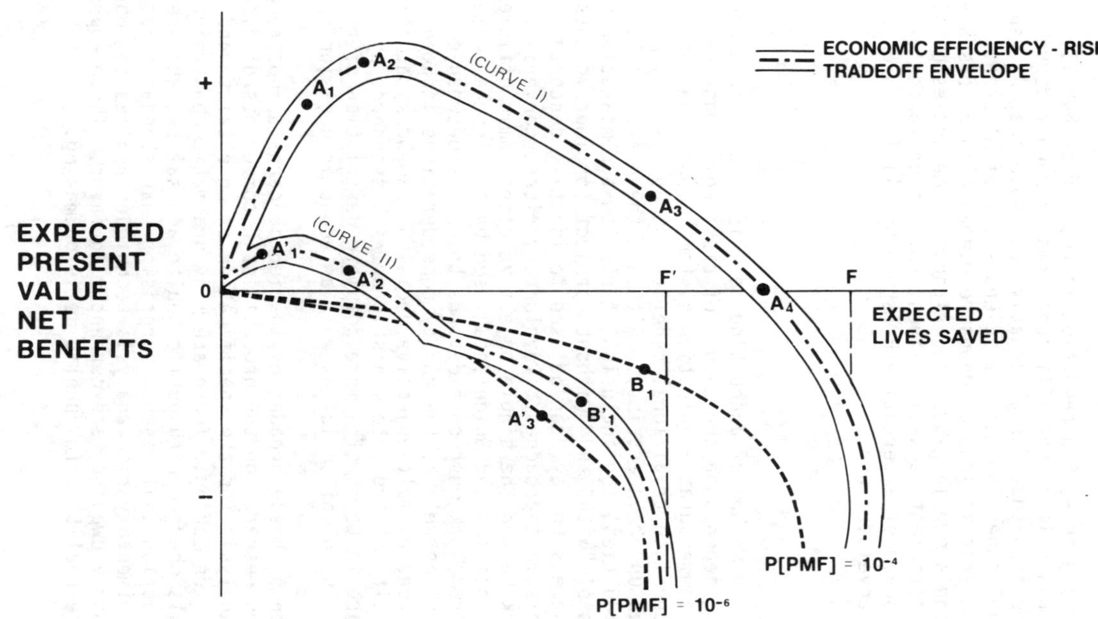

A_1, A_2, A_3 = INCREASING SCALES OF AN ALTERNATIVE WIDENING SPILLWAY
∀B = INCREASING SCALES OF A NON-STRUCTURAL ALTERNATIVE
F = EXPECTED FATALITIES FROM DAM FAILURE PROBABILITY OF PMF = 10^{-4}
F' = EXPECTED FATALITIES FROM DAM FAILURE PROBABILITY OF PMF = 10^{-6}

dam. Thus, downstream "non-failure" losses increase compared to the "no action" alternative.

Returning to Figure 4, the reduction in the PMF probability to 10^{-6} (solid curve II) has shifted the efficiency-risk tradeoff curve for alternative A more than that for alternative B. Note that A_2 (shown as A_2' on curve II) is no longer the most efficient scale of Alternative A. This stems from the increase in induced downstream non-failure losses as discussed above and shown in Figure 4. Although some scales of alternative A are still most efficient at saving lives, this is no longer globally true. Thus, the scale of alternative B (shown by B_1') is now on the economic efficiency-risk envelope curve.

8. CONCLUSION

The appropriate method of accounting for risk and uncertainty in water resources planning depends on the nature of risk, which must be further transformed to a commensurate evaluation system. Individual-voluntary risk can be incorporated into traditional benefit-cost analysis by modeling individual choices under uncertainty. Although there are practical measurement difficulties, the inclusion of risk on a comparative or incremental basis of "with" and "without" project (or improvements) individual willingness to pay, can be used to identify conceptual errors in standard deterministic approaches. Involuntary risks, conversely, require a decision framework beyond the single objective of economic efficiency, which underlies the "Principles and Guidelines" used by the federal water agencies. This framework must consider the tradeoffs between economic efficiency and risk as well as the equity considerations induced by risk shifting. This broader, multiobjective decision framework sacrifices the analytical simplicity of the deterministic or risk standards approach. The gains, however, are in terms of a more complete description of the gains and losses and the inclusion of social as well as economic decision criteria.

Consideration of these broader evaluation concerns do not reduce the need for continued research on the probabilistic components of risk analysis and sensitivity analysis for the specification of uncertainty and its influence on decision making. There are many pragmatic constraints imposed on public officials charged with public health and safety, who are legally liable for their actions. It is not surprising that standards based approaches and rudimentary risk analysis techniques applied to "worst case scenarios" (minimax viewpoint) are currently in favor. These approaches simply reflect the realities for public decision making.

REFERENCES

American Society of Civil Engineers, 1973. Reevaluating Spillway Adequacy of Existing Dams. Report of ASCE Task Committee. J. Hydraulics Div., HY2, 337-372.

Asbeck, E.L. and Y.Y. Haimes, 1984. The Partitioned Multiobjective Risk Method (PMRM). In: Large Scale Systems, 6, 13-38.

Baecher, G.B., M.E. Pate, and R. DeNeuvfille, 1980. Risk of Dam Failure in Benefit-Cost Analysis. Water Resources Research, 16(3), 449-456.

Bowles, D.S. and L.D. James, 1984. Risk-Based Design and Evaluation of Dam Safety: An Overview. Presented at the 1984 Fall Meeting of the American Geophysical Union, San Francisco. December, 1984.

Bureau of Reclamation, 1984. Guidelines to Decision Analysis, Chapter 14. In: "Design Standards, No. 13 for Embankment Dams". U.S. Dept. of Interior, Bureau of Reclamation.

Duckstein, L. and E. J. Plate, 1984. A System Framework for Hydrologic Reliability with Application to the Design of Structures. Special Hamburg Issue, Bulletin of the Int. Assoc. of Hydrol. Sciences, IAHS Publ. 147.

Ellson, R.W., J.W. Milliman and R. B. Roberts, 1984. Measuring the Regional Economic Effects of Earthquakes and Earthquake Predictions, Journal of Regional Science, Vol. 24 (4) 559-579.

Krzysztofowicz, R. and D.R. Davis, 1983. A Methodology for Evaluation of Flood Forecast-Response Systems. Part I: Analyses and Concepts. Water Resources Research, Vol. 19(6), 1423-1429.

Kunreuther, H.C. and J. Linnerooth, 1983. Risk Analysis and Decision Processes. Springer-Verlag, New York.

Langseth, D.E. and F. E. Perkins, 1983. The Influence of Dam Failure Probabilities on Spillway Analysis. In: Proceedings on a Conference on Frontiers in Hydraulic Engineering, H.T. Shen (Ed.) ASCE, 459-464.

MacDonald, T.C. and J. Langridge-Monopolis 1984. Breaching Characteristics of Dam Failures, Journal of Hydraulic Engineering, 110 (5) 567-586.

McCann, M. M. Franzini, J.B., Kavazanjian, E. and H. C. Shah, 1984. Preliminary Safety Evaluation of Existing Dams, Vol. I. Prepared for: Federal Emergency Management Agency, By: Dept. Civil Engineering, Stanford Univ.

National Research Council, 1985. Safety of Dams: Flood and Earthquake Criteria. By Committee on Safety Criteria for Dams, National Academy of Sciences. National Academy Press.

National Science Foundation, 1985. Incorporating Risk Analysis into Dam Safety Evaluation. Prepared by Division of Policy Research and Analysis for U.S. Army Corps of Engineers Institute for Water Resources. PRA Issue Paper 84-71.

Newton, D. W., 1983. Realistic Assessment of Maximum Flood Potential. J. Hydraulic Engineering. 109(6), 905-918.

Prakash, A., 1983. Deterministic and Probabilistic Perspectives of the PMF. In: Proceedings on the Conference on Frontiers in Hydraulic Engineering, H.T. Shen (Ed.) ASCE. 535-540.

Stedinger, J. and J. Grygier, 1985. Risk-Cost Analysis and Spillway Design. In: H. C. Torno (Ed.). Computer Applications in Water Resources. Proceedings, Specialty Conference of the ASCE, pp. 1208-1217.

U.S. Water Resources Council, 1983. Economic and Environmental Principles and Guidelines for Water and Related Land Resources Implementation Studies. Superintendent of Documents, U.S. Government Printing Office.

Yen, B.C., 1979. Safety Factors in Hydrologic and Hydraulic Engineering Design. In: Reliability in Water Resources Management, E.A. McBean, K.W. Hipel and T.E. Unny (Eds.) Water Resources Publ., Littleton, Co. pp. 389-407.

CONSEQUENCES OF THE FAILURE OF A WATER STORAGE SYSTEM

Jacques Ganoulis

School of Technology
Aristotle University of Thessaloniki
54006 Thessaloniki
Greece

ABSTRACT

Evaluation and prediction of socio-economic consequences of an accidental failure of a water storage system must be based on hydrologic impact studies of the area downstream. Such studies are necessary for developing emergency plans and, based on these, for issuing flood warnings sufficiently early that the population downstream can be evacuated and damage to industrial and agricultural properties can be reduced.

In view of the inaccuracy involved in determining the flood hydrograph for the moment of failure of the water storage system, the problem is formulated here in stochastic terms. This allows us to study the effects of varying the size of the breach and the possible shapes of the hydrograph entering the downstream site. Considering the characteristics of the discharge hydrograph at the dam site as random variables, inundation maps are produced for consecutive periods of time after the breach of the water storage structure. Finite difference equations expressing the mass and momentum balance are used to compute the probability and the statistical moments of the flood wave propagation. Linear systems analysis and Monte Carlo simulation are applied assuming either a uniform distribution or a lower and upper triangular probability distribution of the initial breach of the dam. An illustrative example showing the principal features of the proposed methodology is presented for the case of a two-dimensional hypothetical plain.

1. INTRODUCTION

The term "failure" as used in this paper refers to the structural failure of a water retaining system when, for various causes, the resistance of the structure fails to withstand the applied loads. In hydraulic design, various "failure" modes can be defined for every particular hydraulic system (Duckstein and Plate, 1983; Plate, 1984). In a general sense, failure is the inability of the system to meet the *a priori* specified purpose, such as discharging a certain quantity of water or maintaining a fixed water level in the reservoir.

In case of accidental failure of a water-retaining structure, flood waves are generated, propagate downstream and can cause severe damage, both from the economic and social point of view. Water storage reservoirs are usually located, for geologic and topographic reasons, in the upper reach of valleys and plains. Intensive urban, agricultural and industrial development often occur in the downstream areas. However, such sites are highly vulnerable to flooding in case of the partial or total accidental failure of the water storage system.

Dams or flood levees are, of course, designed to withstand the most unusual conditions and loads caused by extreme natural events, such as abnormal precipitation, snowmelt, earthquakes or landslides. However, past experience in various countries where dam failure accidents have occurred indicate the usefulness of studies dealing with the hydrologic, economic and social impacts resulting from these rare events. In the case of important water storage projects, legislation in many European countries requires analysis of the consequences of an accidental breach of the water-retaining structure. A prior estimation and prediction of the severity of flooding in the event of such an accident are necessary for the organization of emergency plans.

Estimation of economic consequences in the form of flood damage functions, opportunity losses, and the organization of emergency plans are based on prior hydrologic impact studies for the downstream area. These studies involve the prediction of various parameters of dam failure floods, such as the extent of inundated area versus time, the time evolution of the flood wave front, or the stage and discharge hydrographs at various characteristic locations. The time scale and the hydraulic properties of the floods are a function of the reservoir characteristics (volume, water level) and the plain topography (one- or two-dimensional, surface slope and roughness). In the case of an initially substantial volume of water in the reservoir and steep riverbeds downstream, the wave front will be abrupt, propagating with high wave velocities as is typical for dynamic waves. In this case acceleration and non-linear interactions are important and must be taken into account in the mathematical simulation. When the initial water level in the reservoir is not very high and the floodplain is nearly flat, i.e. it can be considered as two-dimensional, then the flood wave amplitude decreases relatively quickly. Linearity of the hydraulic system can be assumed so that the description of the flood propagation process is greatly simplified, both in the deterministic and stochastic approaches.

Over the past decades, the deterministic approach has been used extensively to predict the dynamic flooding which results from an accidental dam break. In fact, the mathematical computation using shallow-water equations is an old hydraulic problem. An overview of previous works on this subject is given in the books by Abbott (1979) and Cunge et al (1981). Analytical solutions based on the method of characteristics approximate analytical techniques using the perturbation method, and more recently numerical computer algorithms have been applied to integrate the differential equations for nearly horizontal flows. If we accept these equations as approximate models of the hydraulic system's behavior, all the above solutions contain various numerical errors. On the other hand, the reliability of the deterministic approach and the error bounds of such predictions are related to the various uncertainties in the dam destruction process (partial, total, slow, sudden) and the friction characteristics of the floodplain. These uncertainties can induce errors several orders of magnitude greater than the numerical ones. It is the objective of this paper to evaluate the influence of these uncertainties on the prediction of the flood wave propagation characteristics.

In view of the high potential damage of dam failure floods, evaluation of reliability in predicting the hydraulic consequences is very desirable. The emphasis in this paper is the stochastic description of the breach formation in an earthen dam and its evolution. Experimental and computational results show that in the case of earthen dams and flood levees, the mode of failure which is the most devastating is that of "partial breach" and not "complete overtopping." Since a complete dam failure is unlikely to occur in practice, an initial breach can be assumed. The size of the initial breach gradually increases because of erosion from water flowing through the breach with high velocities. The dimensions of the initial opening are considered as random variables having a probability distribution. Various probability density distributions such as the uniform, and the upper and lower triangular density functions are studied here. The influence of the assumed probability distribution is studied further in two steps. The first is to consider the stochastic characteristics of the discharge and stage hydrographs due to the dam failure. The second step is to consider the statistical moments of the computed inundation maps.

A Monte Carlo simulation is then used to compute the flooding characteristics in a hypothetical two-dimensional plain. The results indicate the influence of uncertainties in the input discharge hydrograph and that of the irregular topography of the flood area on the flood wave propagation. Furthermore, a sensitivity analysis is performed to evaluate the influence of the variation of the friction coefficient along the valley reach.

2. PHYSICAL CHARACTERISTICS OF THE FAILURE PROCESS

The case of accidental failure of an earthen dam or a flood levee is considered here. Various national and international committees have collected data from past accidents of total or partial failures of large dams. Information provided by ICOLD (International Commission on Large Dams, 1973) and by USCOLD (1975) indicates that in the case of a dam subjected to erosion, failure is gradual because the erosive action of the water discharging through an initial breach is progressive and because the failure process is three-dimensional. The breach is extended gradually to include the total mass of the dam. Each particular case has, of course, its own specific characteristics depending on the dam material and the geotechnical properties of the site. A systematic classification of the various causes that lead to the possible failure of a dam is given below.

2.1 Landslides

Past disasters show that periodic cycles of saturation of the porous material in the reservoir banks can lead to a loss of stability and produce major landslides. The sudden arrival in the reservoir of this material may produce a high solitary wave propagating against the earthen dam. An initial breach can form immediately in the dam and progressive erosion caused by the flowing water will lead to a partial or total failure of the structure.

2.2 Initial Breach Caused by Human Intervention

Although dams and flood levees are designed to resist extreme floods, abnormal precipitation together with incorrect operation of the reservoir gates can produce an initial breach and consequently a partial failure of the dam. This initial breach can also be caused by human intervention such as war actions.

2.3 Overtopping

In connection with an accidental failure of the dam spillway, an extreme flood can produce an overtopping of the dam and a progressive erosion of its crest as well as of downstream banks. Experimental and computational results indicate that this case is less critical than the case of an initial breach. This has a simple explanation. When overtopping, the body of water is distributed all along the crest length. Therefore, the local velocities and resulting scouring rates are less than in the case of an initial breach at one spot.

2.4 Earthquakes

An earthquake at the dam site can produce waves or landslides and consequently lead to a partial or total failure of the dam. Figure 1 shows the flow conditions in the case of the partial failure of an earthen dam. The top of the dam functions as a broad-crested weir. The flow conditions at the location of the breach are critical, with subcritical and supercritical flow upstream and downstream of the breach, respectively. A positive flood wave propagates downstream and a negative one moves upstream (Figure 1).

In a study of flood wave propagation, there are two ways to take into account the process of the dam failure. The first one is the more rigorous. The computation starts upstream of the dam site with the dam forming an "interior" boundary condition. Therefore, the negative wave propagating into the reservoir can be computed as well. The second procedure consists of separating the dam failure process from the rest of the upstream flow. Using the mass balance equation, the flood hydrograph can be computed first at the dam site and then used as input boundary condition for the downstream flood wave propagation. This technique is approximate because the influence of the negative wave, though usually small, is not taken into account. However, the method is simple and was adopted for the present study.

As far as the erosion process is concerned, Figure 2 shows the results of a recent experiment conducted at the Hydraulics Laboratory of the Aristotle University in Thessaloniki. Shown is the case of a two-dimensional flood levee with failure due to the combined action of overtopping and erosion. We can observe that erosion takes place at the downstream bank of the earthen structure where there was no protection.

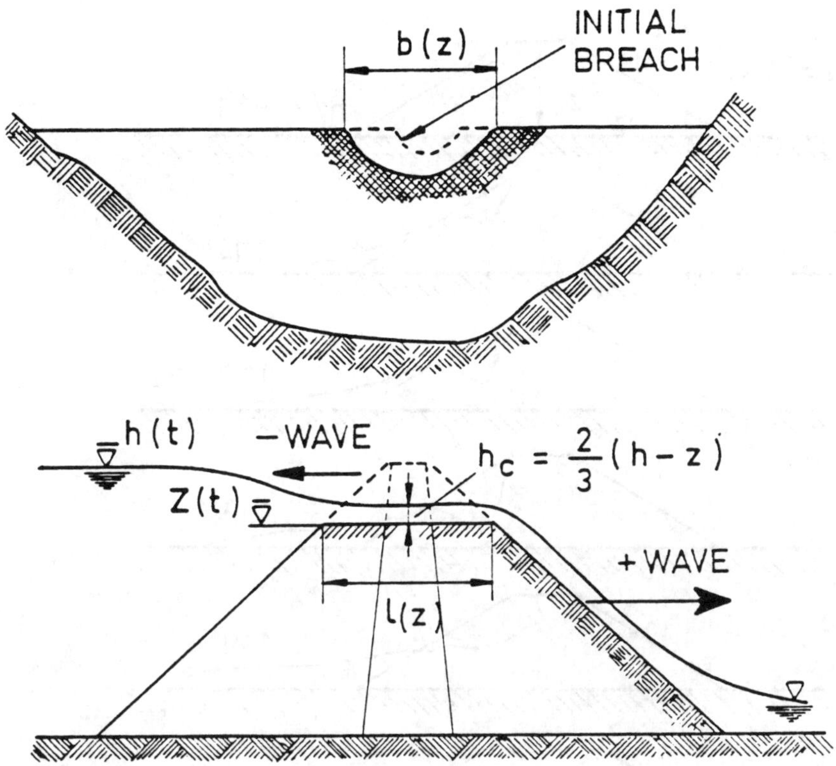

Figure 1. Flow conditions for dam failure due to initial breach

3. DETERMINISTIC FLOOD WAVE PROPAGATION STUDY

3.1 Discharge Hydrograph at the Dam Site

The basic parameters for a deterministic evaluation of the outflow discharge hydrograph at the dam site are the initial shape and size of the breach. The breach itself is assumed to form instantaneously at the onset of failure and either remain fixed or change thereafter depending on the rate of the erosion process. The two extreme cases to be studied usually correspond to the minimum and maximum size of the initial breach, respectively. They lead to a gradual or sudden breach of the dam.

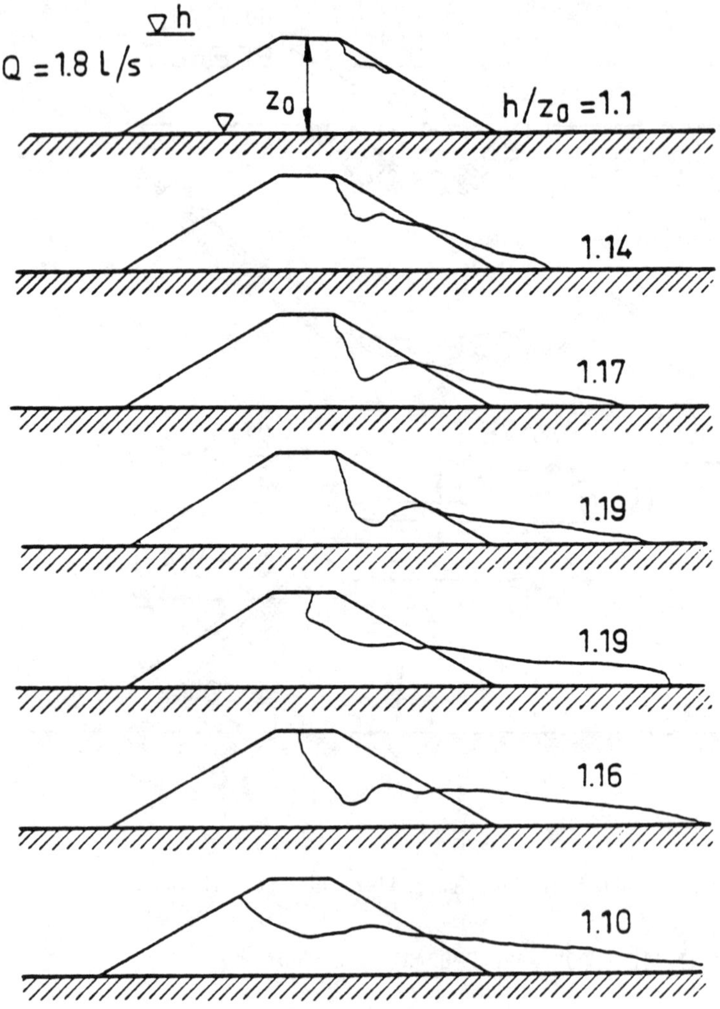

Figure 2. Laboratory experiment of a 2-D flood levee breach by overtopping.

For every initial shape and size of the breach, the outflow discharge and stage hydrographs can be computed using the following three relations:

(a) The energy equation relating the outflow rate $Q(t)$ at time t to the water head $h(t)$ at the reservoir entrance. This is usually a weir formula related to the geometric form of the breach.

(b) The mass balance equation expressing in differential form the continuity of the water entering and leaving the reservoir.

(c) The solid material mass balance equation, also in differential form. This relation is used when erosion of dam material is taken into account.

In the latter case, the flow through the breach can be considered to be flow over a broad-crested weir of geometric dimensions varying with time. Denoting the water level at the reservoir entrance as $h(t)$ and the ground elevation of the bottom breach as $Z(t)$ (Fig. 1), the outflow rate $Q(t)$ can be computed using the following expression:

$$Q(t) = K(t)[(h(t) - Z(t))]^n \tag{1}$$

where $K(t)$ depends on the time-dependent flow through the area as well as the shape of the breach, and n is a constant depending on the shape of the breach only.

In general we can assume that the top width b and the cross section A of the breach follow, at every time t, a monomial power law:

$$b = k(h - Z)^m \; ; \; A = k(h - z)^{m+1}/(m+1)$$

in which m and k are constants.

Special cases of this equation are: $m = 0$ for rectangular cross section of the breach, $m = 1$ for triangular and $m = 1/2$ for parabolic. In the case of a rectangular breach section, equation (1) takes the following form:

$$Q(t) = C \sqrt{2g}\, b(t)\, (h(t) - Z(t))^{3/2} \tag{2}$$

with $C \cong .38$ (broad-crested weir).

For a triangular breach such that the sides of the notch form an angle θ, equation (1) takes the form:

$$Q(t) = C' \sqrt{2g} \tan \frac{\theta(t)}{2} (h(t) - Z(t))^{5/2} \tag{3}$$

From topographic data the functional relation between reservoir volume V and water level h is usually known, so that:

$$V = V(h) \tag{4}$$

Letting $Q_I(t)$ be the flow rate entering the reservoir, the mass conservation equation is:

$$\frac{dV}{dt} = Q_I\ (t) - Q(t) \tag{5}$$

The time derivative can be transformed to yield:

$$(\frac{dV}{dh})\ (\frac{dh}{dt}) = Q_I(t) - Q(t)$$

or:

$$\frac{dh}{dt} = \frac{Q_I(t) - Q(t)}{(dV/dh)} \tag{6}$$

Equation (6) gives the time variation of the reservoir water level in terms of the inflow and outflow rate. The latter is calculated from equation (2) or (3), and the derivative dV/dh is computed at time t using equation (4).

Although most previous work (e.g. Cunge, 1970; Chamber and Katopodes, 1984) does not consider the effect of erosion at the breach location, an unpublished document (Cristofano, 1973) describes the relationship between the discharge rate through the breach and the erosion rate of the dam. In the present work, the Meyer-Peter formula is used to compute the solid flux rate, $q_s(t)$, per unit width of the breach:

$$q_s = \frac{(1/\beta(\tau - \tau_c))^{3/2}}{g\ \rho^{1/2}(\ \rho_s - \rho)} \tag{7}$$

where ρ_s is the solid material density, ρ the density of water, τ the shear stress and $\tau_c \cong .04\ (\rho_s - \rho)gd$, the critical shear stress on the solid surface of the breach related to the mean diameter, d, of the grains. As shown in Figure 2, the solid material mass balance equation must take into account the geometrical form of the eroded dam. Letting $\ell(Z)$ be the equivalent width of the eroded breach, equation (7) yields:

$$- \ell(Z)dZ = q_s(t)\ dt \tag{8}$$

or:

$$\frac{dZ}{dt} = - \frac{q_s(t)}{\ell(Z)} \tag{9}$$

Now, the computational procedure for the outflow hydrograph is as follows:

(a) Given the initial values of h(0) and Z(0) at time t = 0, the outflow rate Q(0) is computed from equation (2).

(b) New values of h(t) and Z(t) are computed by numerical integration of equations (6) and (9).

(c) The outflow rate is computed by introducing into equation (2) the new values of h and Z.

The Euler-Cauchy polygon or Runge-Kutta method can be used for the numerical integration of the ordinary differential equations (6) and (9). Shear stresses in equation (7) can be evaluated from:

$$\tau = \frac{\lambda \rho}{8} \mu^2 \tag{10}$$

where λ is the friction coefficient and μ is the characteristic velocity in the breach section, which is the critical velocity. Consequently, critical flow depth is assumed:

$$h_c = \frac{2}{3} (h - Z) \tag{11}$$

3.2 Flood Wave Propagation

The discharge hydrograph at the dam site is used as the boundary condition for the deterministic flood wave propagation study. Depending on the geometric shape of the river bed and the floodplain downstream, the mass and momentum balance equations can be written in one or two dimensional form.

In the case of one-dimensional modeling, using the water level h(X,t) and the flow rate Q(X,t) as variables, the Saint-Venant equations take the following form:

$$b(h) \frac{\partial h}{\partial t} + \frac{\partial Q}{\partial X} = 0 \tag{12}$$

$$\frac{\partial Q}{\partial t} + \frac{\partial}{\partial X} \left(\frac{Q^2}{A(h)}\right) = -g(A(h) \frac{\partial h}{\partial X} - g\, A(h)\, (I_f - I_o) \tag{13}$$

where b(h) and A(h) are the top width and the cross-section of the river bed, respectively.

If the flood wave propagates in the longitudinal (X) as well the lateral (Y) directions, a two-dimensional model must be used. Denoting U_X and U_Y as the velocity components in the X and Y directions, the flow rates per unit width q_X and q_Y are respectively:

$$q_X = U_X h \quad , \quad q_Y = U_Y\, h \tag{14}$$

where h is the water depth. Assuming hydrostatic pressure distribution along the vertical axis, conservation of mass and momentum leads to the following system of equations:

$$\frac{\partial h}{\partial t} + \frac{\partial q_X}{\partial X} + \frac{\partial q_Y}{\partial Y} = 0 \tag{15}$$

$$\frac{\partial q_X}{\partial t} + \frac{\partial}{\partial X}\left(\frac{q_X^2}{h}\right) + \frac{\partial}{\partial Y}\left(\frac{q_X\, q_Y}{h}\right) = - g\, h\, \frac{\partial h}{\partial X} - g\, h\, (I_{fX} - I_{0X}) \tag{16}$$

$$\frac{\partial q_Y}{\partial t} + \frac{\partial}{\partial X}\left(\frac{q_X\ q_Y}{h}\right) + \frac{\partial}{\partial Y}\left(\frac{q_Y^2}{h}\right) = - g\ h\ \frac{\partial h}{\partial Y} - g\ h\ (I_{fy} - I_{0y}) \quad (17)$$

I_{0X}, I_{0Y} and I_{fX}, I_{fY} are the X- and Y-direction bed and friction slopes, respectively. The Manning formula can be used to evaluate I_{fX} and I_{fY}. Here the inverse value of the friction coefficient n is used, namely K = 1/n.

$$I_{fX} = \frac{q_X^2}{K^2\ h^{10/3}} \quad , \quad I_{fY} = \frac{q_Y^2}{K^2\ h^{10/3}} \quad (18)$$

The system of equations (12) and (13) or (15), (16) and (17), together with the appropriate boundary and initial conditions, characterize in mathematical terms the response of the catchement area to a given input. To every flood hydrograph at the dam site there corresponds a different flood hydrograph for each location in the floodplain. In view of the irregular topography of the natural floodplain and the non-linearity of the system of partial differential equations, numerical integration is usually appropriate for obtaining deterministic solutions.

Over the past decade, various numerical methods, mostly based on finite differences, have been developed for unsteady shallow water flows (Mahmood and Yevjevich, 1975; Abbott, 1979; Cunge et al, 1981). Explicit or implicit numerical algorithms will induce various errors if only a limited number of terms in the Taylor series expansion are taken into account. Explicit algorithms (Ganoulis and Tolikas, 1983) suffer from so-called numerical diffusion. This leads to excessive attenuation of the input signal with an artificial diffusion coefficient related to the truncation errors. Implicit algorithms, such as the one used by Ganoulis and Koutitas (1980), introduce trailing effects because the harmonic components of the initial signal are propagated downstream with variable velocities.

For the one-dimensional model, the explicit algorithm is stable when the time step Δt and the space step ΔX fulfill the Courant condition

$$t \leq \frac{\Delta X}{C} \quad (19)$$

where $C = \sqrt{gh}$ is the flood wave velocity. Since the time scale of real dam-break floods is relatively small, e.g. of the order of a few hours, the small value of Δt computed from equation (19) does not imply a restriction on using an explicit algorithm. If a space grid is used and ΔX is carefully selected, this renders the computation simple without creating too much numerical diffusion.

Since analytical solutions are known for the one-dimensional case, they can be used to demonstrate the capability of the proposed explicit algorithm to control numerical errors due to artificial diffusion.

Figure 3 shows a comparison between the numerical computation and the analytical solution given by Ritter (1892). This is the classical case of the one-dimensional total break of a vertical dam. If friction is neglected, the method of characteristics can be applied to obtain the analytical solution:

$$\frac{h}{h_o} = \left(\frac{2-X/C_o t}{3}\right)^2 \tag{20}$$

where h_o is the initial water depth and $C_o = \sqrt{g\,h_o}$. From the results shown in Figure 3, we can conclude that even for this unfavorable case (no physical friction, shock front in the flood wave) the numerical algorithm can adequately predict the flow, especially after the first moments have elapsed and the front of the wave becomes smoother. Similar conclusions can be drawn from Figure 4 where the numerical method is compared with the method of characteristics for a trapezoidal channel. Values of $\Delta X = 1$ m and $\Delta t = 0.01$ were used for the numerical computations.

When the floodplain is wide and nearly flat, the non-linear terms appearing in equations (16) and (17) can be neglected. In this case the explicit algorithm can be successfully applied as shown later in a numerical example.

4. STOCHASTIC FORMULATION OF THE PROBLEM

4.1 Source of Uncertainties

Because of the various uncertainties involved in the description of the physical problem, several errors are introduced in the deterministic prediction. Uncertainties arise essentially from the incomplete information on the initial and boundary conditions. The numerical values of the physical parameters used in the mathematical model also are not correctly known. All these uncertainties can affect the reliability in predicting the severity of dam-failure floods and can be grouped in two principal groups.

4.2 Uncertainties Due to the Dam Failure Process

The form of the input hydrograph at the dam site, which is the input boundary condition for the flood propagation study, depends on the physical characteristics of the dam-break mechanism. It is difficult to describe it adequately because of the lack of available data. The latter is mainly due to the fact that a dam breach is a very rare event. Depending on the size of the initial gap, the discharge hydrograph can have different peak flow rates as well as different durations. If the initial opening is great, the discharge hydrograph has a large peak value and a short duration. On the other hand, for small initial gaps, the peak flow rate is small but has a long duration. In view of the multitude of possible causes for the breach, information should be given in terms of the lower and upper limits of the size of the breach. In the deterministic approach a conservative assumption will be used to

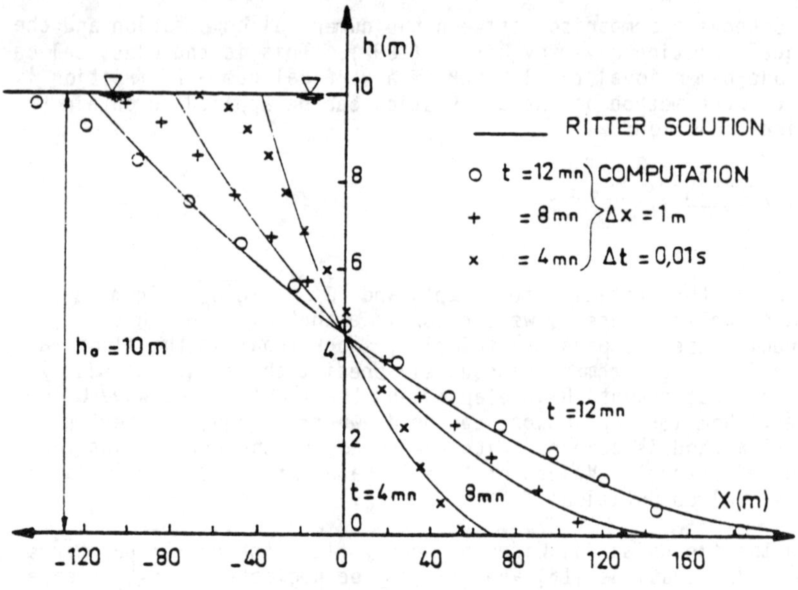

Figure 3. Dam-break flood in rectangular channel: comparison between numerical results and Ritter analytical solution.

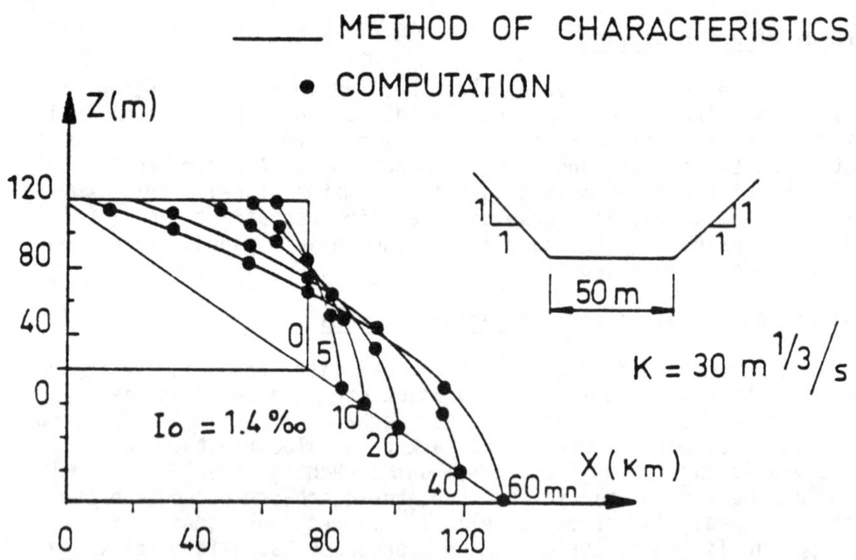

Figure 4. Comparison between numerical solution and method of characteristics for a trapezoidal channel.

predict the flood wave propagation, i.e. the largest possible gap will be assumed to exist. Thus, in the prediction of flood severity and damages, only the most unfavorable case is examined. Neither error bounds nor confidence intervals of these predictions are considered to be obtainable.

4.3 Uncertainties Due to the Inaccurate Description of Floodplain Characteristics

Apart from the limited precision of topographic data describing the floodplain morphology, a large inaccuracy will be introduced in describing the physical characteristics of the plain's surface. The latter information is necessary for evaluating the friction factors for the flooded area. There are two types of difficulties in doing so. The first is related to the great variation in land surface condition such as the distribution of cultured brush, forests, settlements. Natural drainage areas are generally heterogenous and vegetation conditions vary with the seasons. The second difficulty concerns our lack of knowledge in converting the above data into an appropriate "friction coefficient" such as the one appearing in Manning's formula.

By applying a "sensitivity analysis" the influence of the variation of friction coefficient has been studied in Ganoulis and Koutitas (1980). The technique consists of first obtaining the partial differential equations which describe the evolution in time and space of the so-called "sensitivities" of the system. These are defined as the derivatives of the physical variables Q(t) and h(t) against the K-coefficient in Manning's formula, i.e dQ/dK and dh/dK. These derivatives are proportional to the errors induced when calulating Q(t) and h(t) with error-laden K values.

Establishing the influence of such physiographic factors on the reliability of dam-break floods is beyond the scope of the present study and will not be discussed further.

5. STOCHASTIC SIMULATION OF DAM-BREAK FLOODS

As shown in Figure 5, the stochastic simulation of the problem is performed in two steps. Step 1 concerns the stochastic generation of the outflow hydrograph at the dam site (Fig. 5a). Step 2 is the computation of the statistical characteristics of the floods at every location, k, of the floodplain (Fig. 5b). In the above two steps the structure of the hydrologic process, i.e. the relation between the input and the output, is deterministic. The "system" consists of a set of differential equations simulating the dam breach process (step 1) and the floodplain (step 2). If the input is a random variable (as in step 1) or a random function (as in step 2), these differential equations become a set of stochastic differential equations.

Since erosion phenomena are considered, the mathematical model describing the dam-break process is non-linear. Therefore a Monte Carlo technique is appropriate for simulating the outflow hydrograph at the dam site. The initial top width of the breach b(0) is considered to be

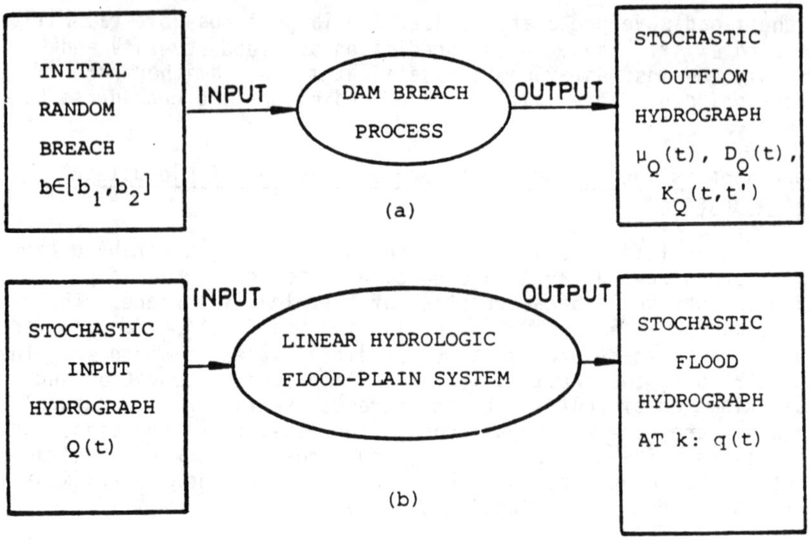

Figure 5. Stochastic dam-failure floods: outflow hydrograph (a) and flood hydrograph simulation (b).

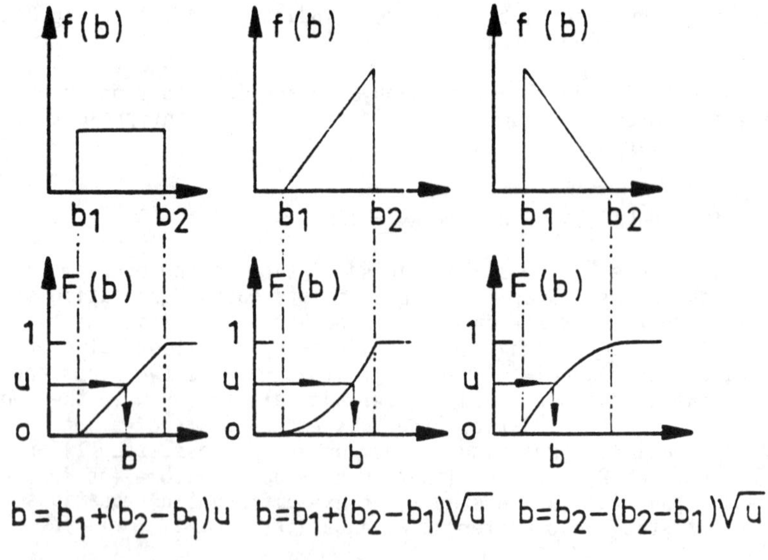

Figure 6. Monte Carlo simulation of samples.

a random variable which can vary between a minimum value b_1 and a maximum value b_2. As shown in Figure 6 either a uniform or an upper and lower triangular probability density function can be assumed for b(0) (Ang and Tang, 1984). If u is a uniform random variable between 0 and 1, then the Monte Carlo generation of initial top-width values, b = b(0), can be performed using the following expressions (Fig. 6).

<u>Uniform distribution</u>

$$b = b_1 + (b_2 - b_1)u \tag{21}$$

<u>Upper triangular distribution</u>

$$b = b_1 + (b_2 - b_1)\sqrt{u} \tag{22}$$

<u>Lower triangular distribution</u>

$$b = b_2 - (b_2 - b_1)\sqrt{u} \tag{23}$$

Usually several hundred values generated from equations (21), (22), (23) are sufficient to represent adequately the assumed probability distribution.

For every value generated for the initial top width, $b_i(0)$, the outflow discharge hydrograph $Q_i(t)$ is computed using the system of mass balance equations described earlier. Figure 7 shows schematically the different discharge hydrographs generated with a Monte Carlo simulation. As the total volume of water discharged at the reservoir is constant, the surface underneath the different curves is also constant. At every time t_k, a number N of flow rates:

$$Q_1(t_k),\ Q(t_k)\ldots,\ Q_N(t_k)$$

is generated which correspond to the realization of the random variable $Q(t_k)$. As t_k can vary, the random function Q(t), that is to say, the stochastic process Q(t), is then formed as an average as shown in Figure 7. This is a non-stationary stochastic process having the following statistical characteristics: the mean value (deterministic) function $m_Q(t)$, the dispersion $D_Q(t)$, and the covariance function $K_Q(t,t')$. Estimates of $m_Q(t)$ and $D_Q(t)$ for each time are calculated for consecutive values of t_k as follows:

$$m_Q(t_k) \cong \frac{1}{N} \sum_{i=1}^{N} Q_i(t_k) \tag{24}$$

$$D_Q(t_k) \cong \frac{1}{N-1} \sum_{i=1}^{N} (Q_i(t_k) - m_Q(t_k))^2 \tag{25}$$

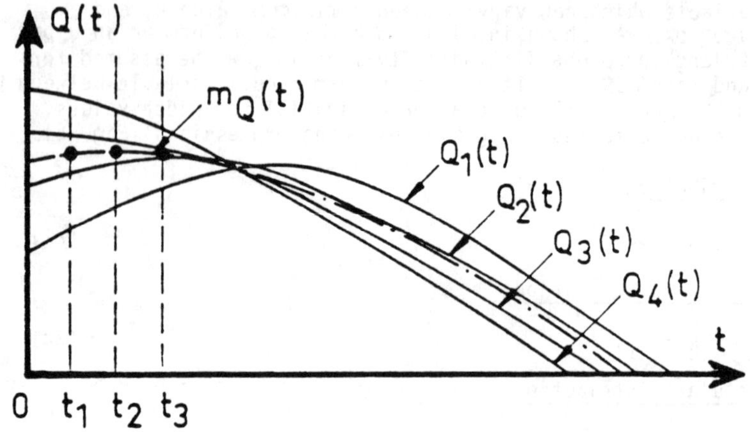

Figure 7. Computation of the stochastic outflow hydrograph.

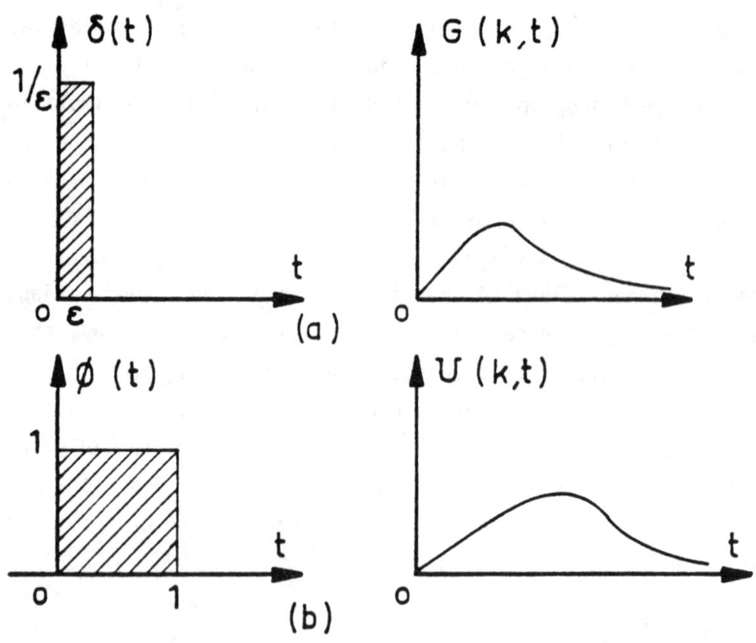

Figure 8. Green's function (a) and Unit hydrograph (b).

The covariance function, which is the correlation between the random variables $Q(t_k)$ and $Q(t_1)$, is given by the expression:

$$K_Q(t_k,t_1) \cong \frac{1}{N-1} \sum_{i=1}^{N} (Q_i(t_k) - m_Q(t_k))(Q_i(t_1) - m_Q(t_1)) \qquad (26)$$

The second step of the stochastic simulation (see Fig. 5b) can be analysed using the assumption of linearity of the floodplain system. This is the case of relatively shallow flood waves propagating over an almost flat surface. As the acceleration is very small, the zero inertia approximation can be used in the Saint Venant equations. For this quasi-linear hydrologic system, the principle of superposition applies. Knowing the response $G(k,t)$ to the unit impulse $\delta(t)$ at a given point k (Fig. 8a), the flood hydrograph $q(k,t)$ at the point k can be computed by means of the convolution

$$q(k,t) = \int_0^t G(k,t-\tau)\, Q(\tau)\, d\tau \qquad (27)$$

where $Q(t)$ is the input discharge hydrograph. Equation (27) can be written in discrete form. Denoting $t = n\Delta t$ and $\tau = i\Delta t$, we obtain:

$$q(k,n) = \sum_{i=1}^{n} Q(i) \int_{i-1}^{i} G(k,n-\tau)\, d\tau \qquad (28)$$

Introducing the new variable $\tau' = \tau-(i-1)$, the integral in equation (28) takes the form

$$\int_{i-1}^{i} G(k,n-\tau)d\tau = \int_0^1 G(k,n-i+1-\tau')d\tau' \qquad (29)$$

According to Figure 8b, the function $U(k,t)$ is the response at the location k to the unity pulse $\phi(t)$.

$U(k,t)$ is traditionally known as the unit hydrograph. The functions $G(k,t)$ and $U(k,t)$ are related as follows:

$$U(k,t) = \int_0^t G(k,t-\tau)d\tau \qquad t<1$$

$$U(k,t) = \int_0^1 G(k,t-\tau)d\tau \qquad t\geq 1$$

Define the discrete unit hydrographs as:

$$U(k,i) = \int_0^1 G(k,i-\tau)d\tau$$

Then equations (28) and (29) can be combined to give:

$$q(k,n) = \sum_{i=1}^{n} Q(i)\, U(k,n-i+1) \qquad (30)$$

Thus, the statistical characteristics of the response hydrograph can be computed. The mean value $m_q(k,n)$ and the autocovariance $K_q(n,n')$ are given by the expressions:

$$E(q(k,n)) = \sum_{i=1}^{n} U(k,n-i+1)\, E(Q(i)) \qquad (31)$$

$$K_q(n,n') = \sum_{i=1}^{n} \sum_{j=1}^{n'} U(k,n-i+1)U(k,n'-j+1)\, K_Q(i,j)\,) \qquad (32)$$

6. EXAMPLE OF APPLICATION

To illustrate the proposed methodology, an example of the computation is presented. The stochastic simulation describes a flood wave propagating in a two-dimensional, dry plain. The drainage area is about 20 km wide and 40 km long, laterally confined by mountains. Figure 9 shows two different perspective view of the area under study. Because of the irregular plain topography, a non-uniform flood wave propagation can be expected.

A series of examples using an initial top width of the breach b between 10 and 90 m were generated by a computer according to uniform, and then upper and lower triangular probability density functions. In Figure 10, the results of the simulation are presented for the case of a uniform probability distribution, with sample sizes of 20, 70, and 100. As can readily be seen, the precision of the simulation depends on the size of the sample. For 100 values, the law obtained is almost uniform. In the following, Monte Carlo simulation is applied to generate the outflow discharge hydrograph at the dam site. Figure 11 shows some computed dimensionless hydrographs of Q(t). Using the above information, the statistical measures $m_Q(t)$, $D_Q(t)$ and $K_Q(t,t')$ can be computed by using equations (24), (25) and (26).

The ability of the proposed explicit finite difference algorithm in producing inundation maps is illustrated in Figures 12, 13, 14 and 15. The perspective representation of surface elevation versus time after breach of dam, as given in Figure 12, should highlight the situation. As shown in Figure 13, lines of equal water depth give quantitative information on the inundated areas for various time steps. Despite the large amount of data and numerical results involved in these computations, two and three-dimensional computer graphics techniques adapted to an IBM-PC microcomputer were found to be very satisfactory. For every input discharge hydrograph, stage hydrographs can be computed at various locations of the plain. This is shown in Figures 14 and 15. For a unit impulse function at the dam site, the unit hydrograph can be computed numerically at every location of interest. Using equations (31) and (32), the mean values, the covariance and variance characteristics of the flood can be derived.

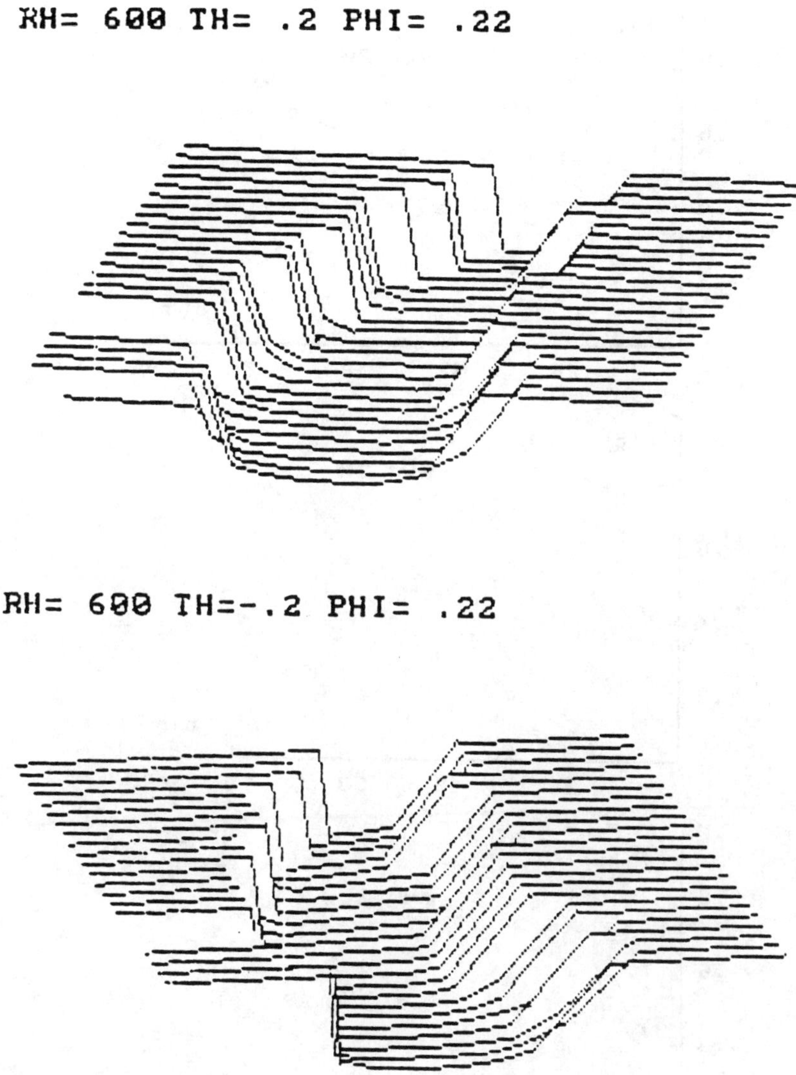

Figure 9. Two perspective views of the floodplain.

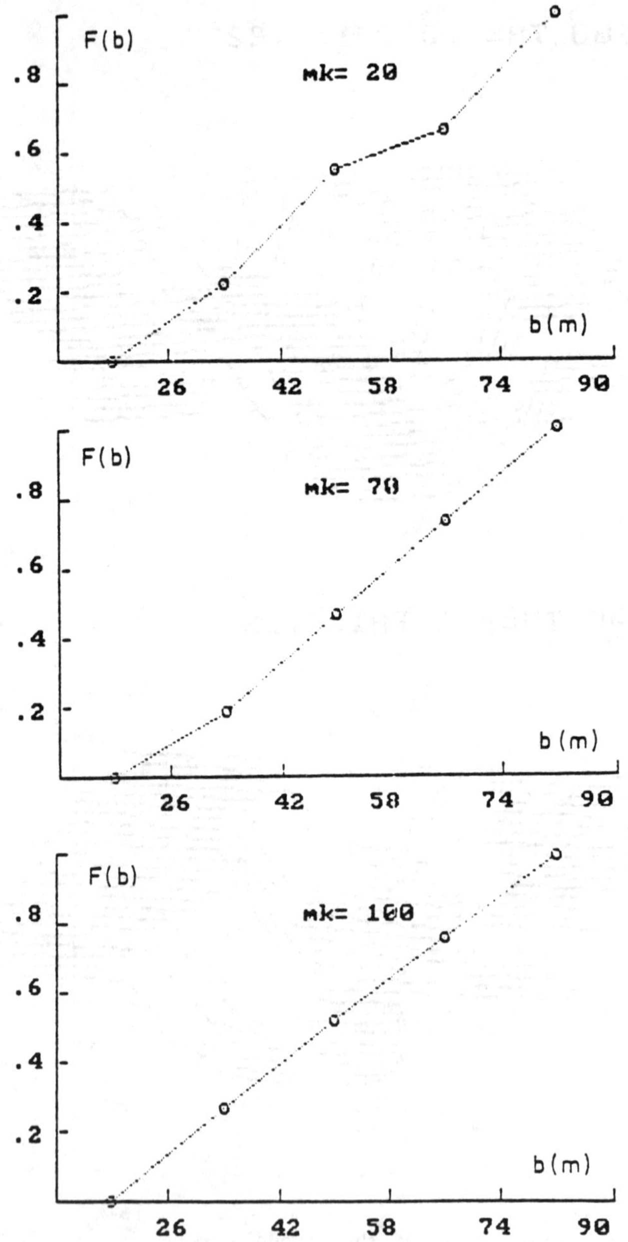

Figure 10. Monte Carlo generation of three samples of uniform probability density distribution.

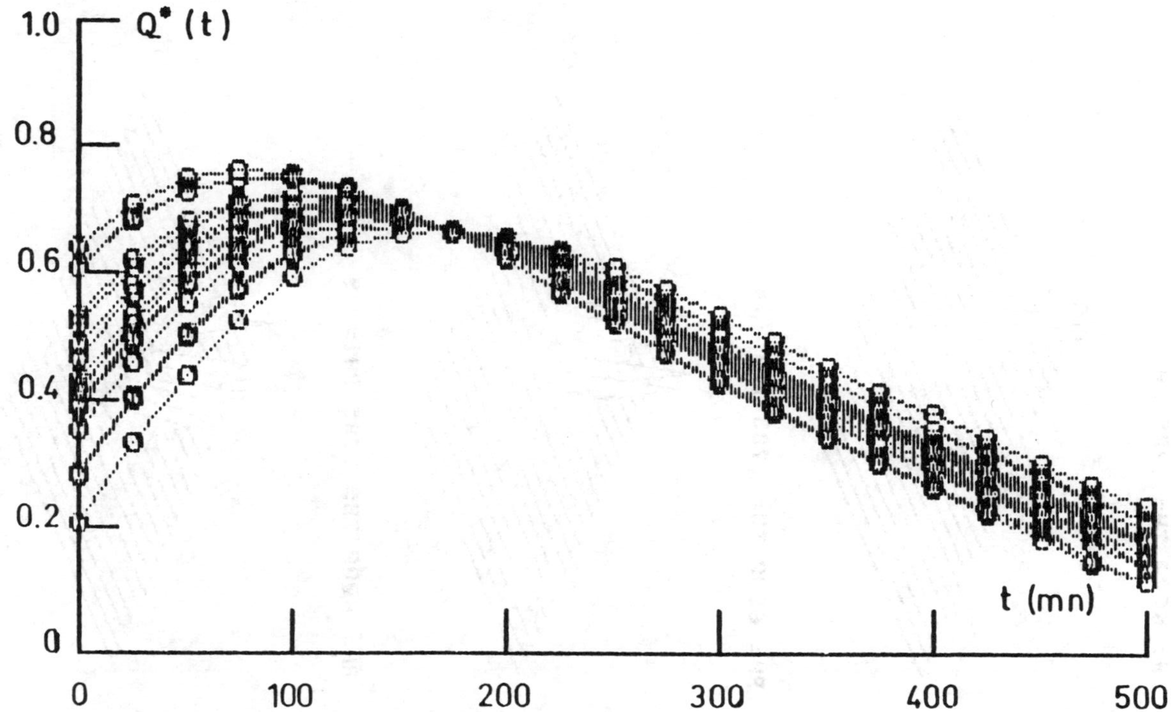

Figure 11. Monte Carlo simulation of the outflow discharge hydrograph at the dam site.

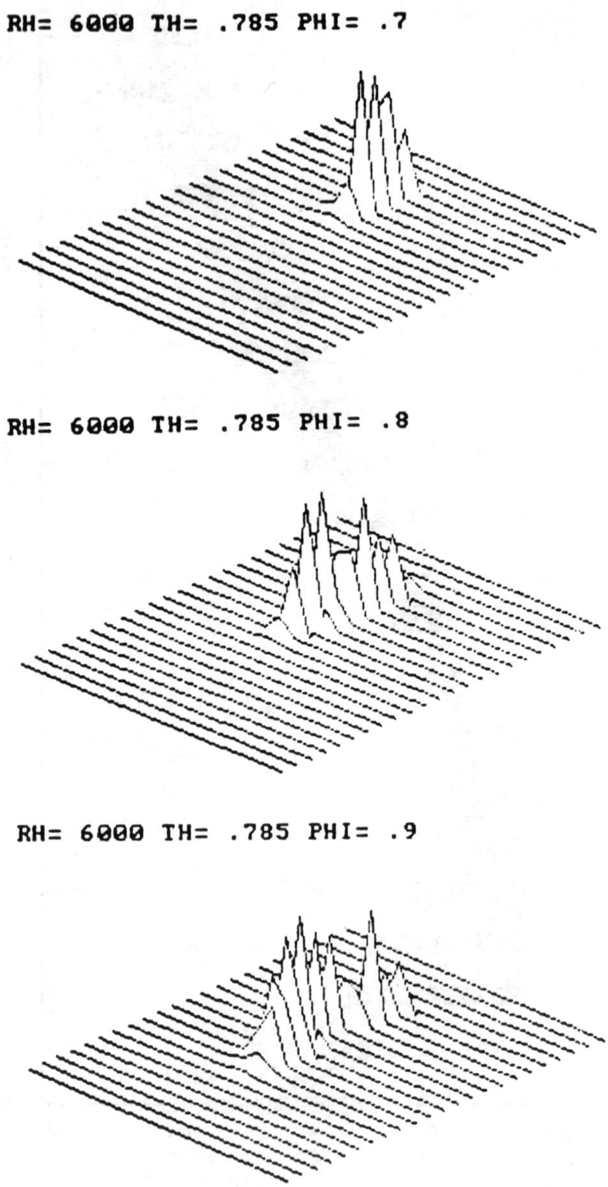

Figure 12. Perspective view of free surface elevation of dam-break flood versus time.

RH= 6000 TH= .785 PHI= .95

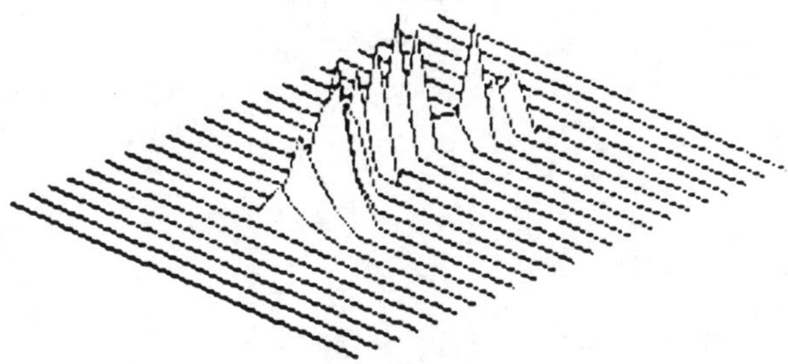

RH= 6000 TH= .785 PHI= 1.1

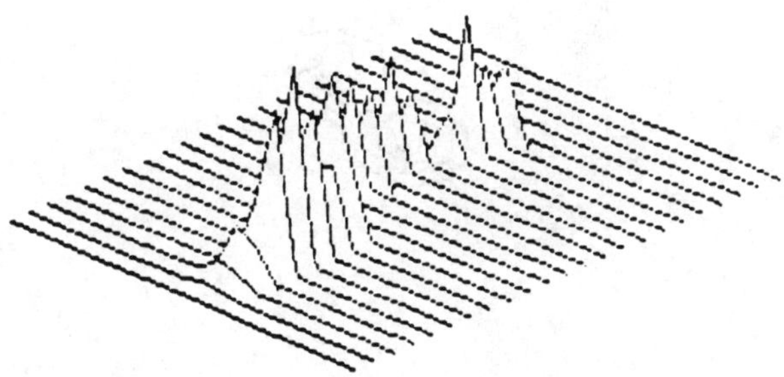

RH= 6000 TH= .785 PHI= 1.2

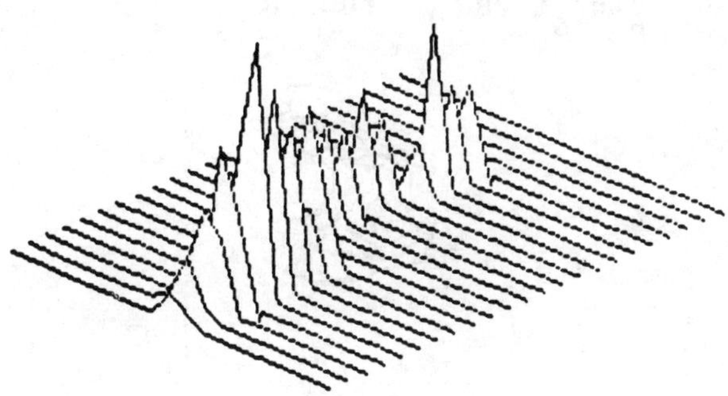

Figure 12--Continued.

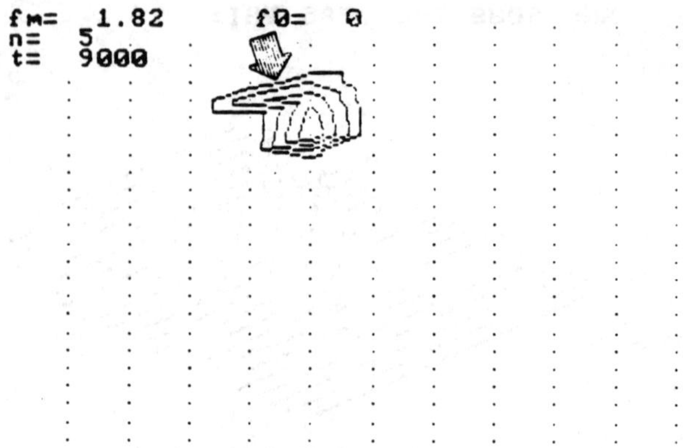

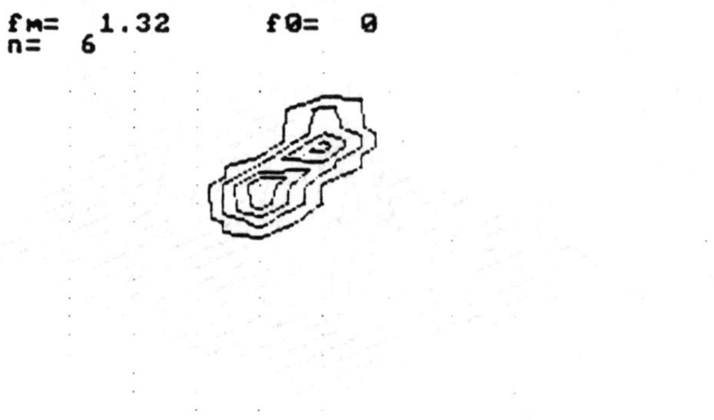

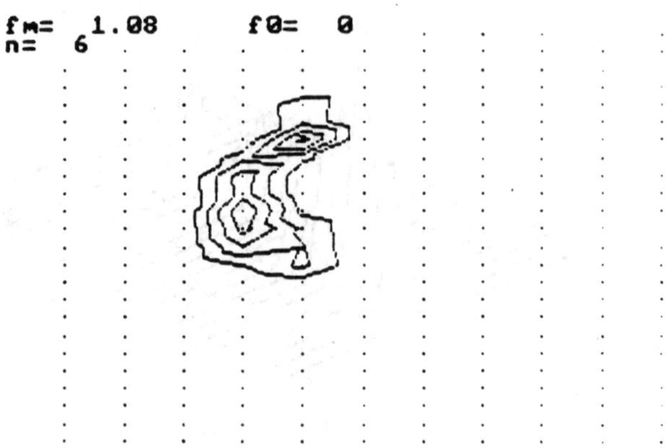

Figure 13. Lines of equal water depth versus time.

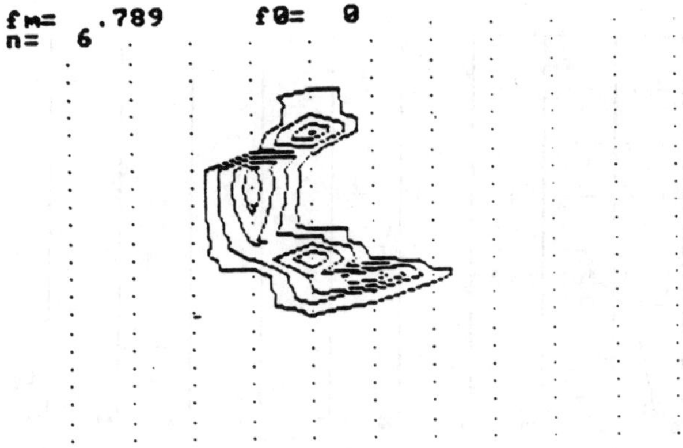

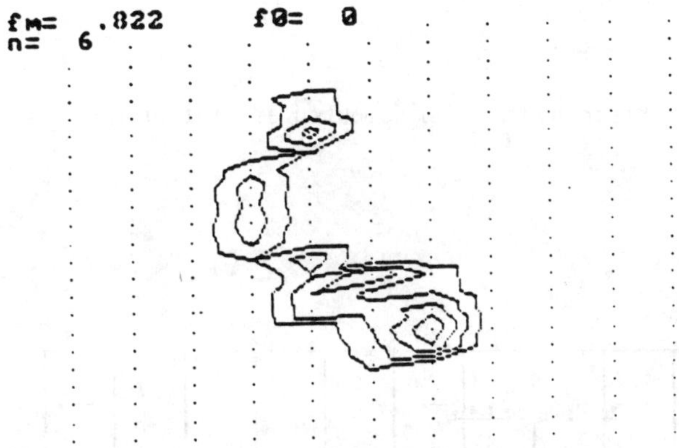

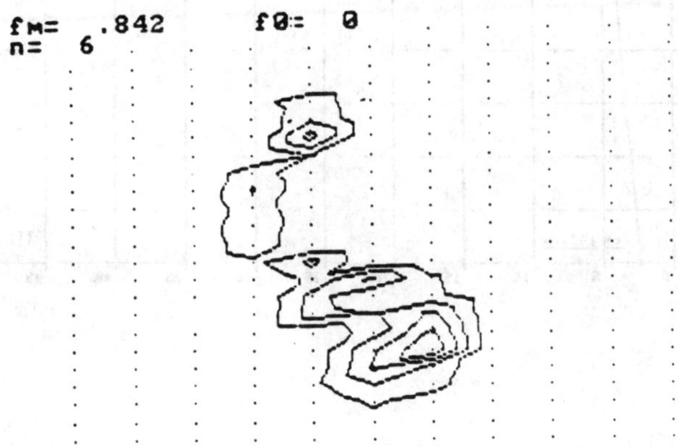

Figure 13--Continued.

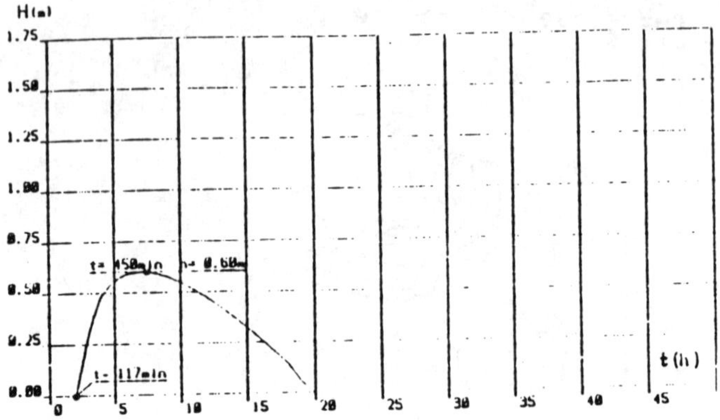

Figure 14. Stage hydrograph computed at station A.

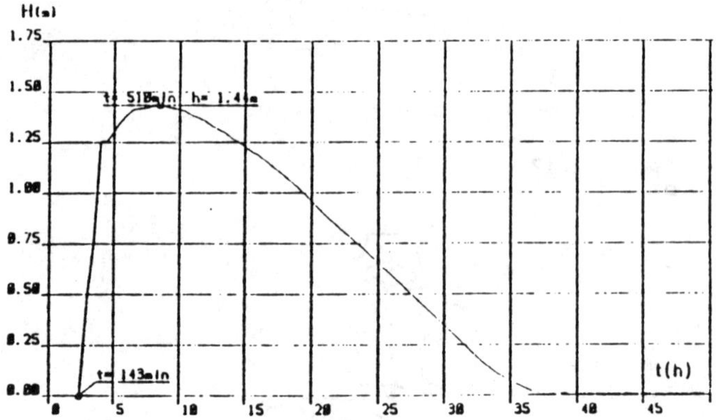

Figure 15. Stage hydrograph computed at station B.

7. CONCLUDING REMARKS

In view of the various uncertainties in predicting the dam break floods, a stochastic formulation of the problem is given which can provide reliability indices for the results obtained. By considering the size of the initial gap in the dam as a random variable, the outflow discharge hydrograph is computed at the dam site. Monte Carlo simulation and linear system analysis are applied to predict the statistical characteristics of the flood.

The same methodology can also be used in the case of another population of streamflows acting as input. Further research needs to be done to include the influence of the uncertainties resulting from variation in the friction coefficient. The method also needs to be extended to the case where non-linear effects are important.

ACKNOWLEDGEMENTS

The writer wishes to express his gratitude to Heinz-Peter Hadrys for carefully reviewing this manuscript.

REFERENCES

Abbott, M.B. (1979) Computational hydraulics, Pitman, London.

Ang, and W.H. Tang (1984) Probability concepts in engineering, Planning and Design, Vol. II, I. Wiley.

Chamber, D.R. and Katopodes, N.D. (1984) One-dimensional models for partially breached dams, Journal of Hydraulic Engineering, ASCE, 110, N 8, pp. 1086-1102.

Cristofano, E.A. (1973) Method of computing erosion rate for failure of dams. Unpublished Preliminary Report to US Bureau of Reclamation, Denver, Colo.

Cunge, J.A. Holley, F.M. and Verwey, A. (1981) Practical aspects of computation river hydraulics, Pitman, London.

Duckstein, L. and Plate, E. (1983) A system framework for hydrological reliability with application to the design of structures, Int. Assoc. Hydrol. Sciences, Symposium Hamburg (to be published in the J. of Hydrological Sciences).

Ganoulis, J. and Koutitas, C. (1980) Sensitivity of flood wave propagation model to errors in friction coefficient estimation, Int. Assoc. of Hydr. Res. (IAHR), Symp. on River Engineering and its Interaction with Hydrological and Hydraulic Research, paper D.25, Belgrad.

Ganoulis, J. and Tolikas, D. (1983) Influence du mode de rupture d'un barrage en terre sur la propagation de l'onde oval, XX IAHR Congrès, Moscow.

ICOLD (1973) Lessons from Dam Incident, Reduced edition.

Mahmood, K. and Yevjevich, Y. (Eds) (1975) Unsteady flow in open channels, Water Resources Publication, Co.

Plate, E. (1984) Reliability analysis in hydraulic design, 4th IAHR Int. Symp. on Stochastic Hydraulics, Univ. of Illinois at Urbana-Champaign, USA.

Ritter, A. (1892) Die Fortpflanzüng der Wasserwellen, Ver. Deutsch Ing. Zeitschr., 36, N 33, pp. 947-954.

USCOLD (1975) Lessons from Dam Incidents, Amer. Soc. Civil Eng., New York.

III. RELIABILITY AND RISK IN WATER SUPPLY SYSTEMS

III. RELIABILITY AND RISK IN WATER SUPPLY SYSTEMS

The concepts of risk and reliability can readily be applied to supply problems. In fact, one of the greatest advantages of reliability as a design tool is that it is equally well suited for many of different design problems, provided that the concept of failure or incident be defined appropriately. The concepts developed in the introductory paper by Duckstein et al appear to be broad enough to cover all cases of failure.

Section II was concerned mostly with structures whose usefulness is terminated once they have failed, therefore leading to definitions of reliability based on the time to failure. On the other hand, reliability in water supply problems refers to the probability of meeting given targets without having the structural part of the system affected by the failure. The target-related failure of a supply system is usually reflected in the resilience and the vulnerability performance indices (PI).

As defined in the introductory paper by Duckstein et al, the resilience of a simple system is a PI equal to its expected length of residence time in the non-failure mode. For a reservoir, which is in the failure mode when it cannot meet the target release, the resilience is the complement of the average time from the time of failure to the next time of full service. Inherent in this definition is that there is a return to normal. Reliability then is no longer the time to first failure but the average time between failures. Also, since failure to meet a target implies a rigid threshold, the word "failure" may be replaced by "incident". Structures that have this kind of reliability also exist. For example, if a flood levee along a river is destroyed by a flood, it will be repaired as quickly as possible. It therefore has a resilience which may be higher than that of a reservoir which has failed to supply the target. In other words, target-related failures can be treated in the same manner as structures except for the finality of the time to failure t for some of the structures and for the importance of estimating the distribution t for target-related reliability (so that quantities such as the variability of t may be ascertained.)

Supply problems with uncorrelated inputs are presented in Part III.1. A straightforward application of a level 3 design approach to water supply problems is given in the paper by Shamir, using the method described in the introductory paper of Section II. However, the integrations involved in the determination of the failure probability are often quite involved and are overcome, in Shamir's case, through the use of simplified forms of the relevant probability density functions. More complex distributions are suitably handled by applying simulation techniques, as illustrated in Schultz's paper in this section (Part III.1). The advantage of the simulation approach is that risk costs can easily be integrated into the PI and/or FM.

A second case treated in Schultz's paper involves the application of queuing theory, which represents an application of reliability analysis with correlated inputs. Such a case is very frequent in water supply problems because of the preponderant influence of the annual cycle of reservoir inflow on supply reliability. Since correlated data handled effectively by techniques used for uncorrelated data, simulation techniques with artificially generated correlated data are commonly used. Such techniques are illustrated in Part III.2. The first paper in this section, by Vogel, is of a fundamental nature; the failure probabilities of a reservoir of prescribed properties are determined by an input consisting of a Markov chain, whose parameters are systematically varied. It serves as an introduction to Part III.2, as it clarifies the role of correlation between neighboring time intervals in the determination of the time between failures of a reservoir (called PI^5 in the Duckstein et al paper). The use of simulation methods with situation-specific generation models of correlated random inputs is then illustrated by means of real-world examples. The three papers in Part III.2 by Schultz, Correia, and Benedini and Ciccioni, respectively, provide an overview of the broad scope of applications and conclude the presentation of current applications of reliability theory. The papers in the next (and last) section of the book cover more research-oriented aspects of reliability and risk analysis.

RELIABILITY OF WATER SUPPLY SYSTEMS

URI SHAMIR

Faculty of Civil Engineering, Technion - Israel Institute of Technology, Haifa 32000, ISRAEL

ABSTRACT

Reliability of water supply may be defined in a variety of ways. The paper begins with a discussion of the considerations which lead to selection of Figures of Merit (FM's) and Performance Indices (PI's) to measure supply reliability, then lists a number of possible definitions for PI's. The following sections are devoted to methods of analysis: the statistical-analytical method, fault tree analysis, and methods for analyzing the reliability of networks.

1. INTRODUCTION

A supply system is designed and operated to take water from sources, treat, store, convey and distribute it to consumers who are spread over an area, and whose demands change over time. The reliability of supply is affected by the availability of water in the sources, water quality, the performance of system facilities, and the random nature of the demands. We consider the reliability of supply and discuss: (a) measures of reliability, (b) sources of uncertainty which affect reliability, and (c) methods for analyzing the reliability of a supply system.

Our ultimate goal should be the ability to maximize the system's reliability for a given cost, or to minimize the cost for a given reliability -- so that efficient solutions for these two objectives can be identified.

2. MEASURES OF RELIABILITY

We can think of a very large number of Figures of Merit (FM's) and Performance Indices (PI's) for the reliability of a water supply system. PI's are measures of the physical (water) shortfalls. FM's are functions defined on the PI's, functions which express the "worth" of the PI's. Both are random variables, or can be defined as statistics of random variables (e.g., mean, variance, percentile) -- which themselves are random variables. Their selection is a compromise between the indicators we would like to have as PI's and FM's for reliability and what we can actually consider within the method of analysis. When the analysis is carried out by simulation, any desired PI and FM can be computed. If we wish to have an analytical method of analysis, then there are severe limitations.

Possible PI's for a water supply system are:

(a) Fraction of the demanded water which is supplied, computed over a specified period of time, such as a day, a month, a year.

(b) Fraction of time the consumer is connected to a source and has some supply (even if less than demanded).

(c) Number of failure events per unit time (say, one year).

(d) Number of failures per unit time which are longer than some minimum.

(e) Magnitude of the largest shortfall during the unit time.

(f) The number of customers who have at least one failure event, greater than some value, once or more during the time period.

Obviously we could define many more PI's. The FM's are expressed as functions of the PI's and reflect the losses resulting from failures and shortfalls in supply.

3. SOURCES OF UNCERTAINTY WHICH AFFECT RELIABILITY

Consider a supply system which contains sources, treatment plants, storage reservoirs, pumping stations, transmission and distribution pipes. Reliability of water supply, measured by any PI or FM, will depend on single or combined random events, some of which are:

(a) The amount of water in the sources (stored plus inflows) is insufficient.

(b) The quality of water in the sources is too low to be treated adequately by the treatment plants.

(c) Failure of part or the whole treatment plant.

(d) Failure of pumps.

(e) Failure of pipes.

(f) Failure of the power supply to pumps.

(g) Increase in demands beyond what was expected, and beyond what can be supplied by the system, including from storage.

4. METHODS FOR COMPUTING RELIABILITY

4.1 Overview

Barlow (1984) has given a historical perspective on the mathematical theory of reliability. Its modern history started in the 1950's, which he calls "the formative years". Life testing, electronic devices and missile reliability were the main topics. Statistical tools and experimental methods were developed. General system reliability was developed in the 1960's, and attention was given to the way in which individual components are linked into a system.

The dominant methodology of the 1970's was fault tree analysis, motivated in part by the requirements for designing nuclear plants. The method, which will be explained in some detail in a later section, computes the probability that two specified nodes will be connected. It uses a graph which describes the connectivity of the elements in the system, and uses element failure proabilities to compute overall system reliability. The method is computationally feasible for systems of modest size and complexity. The book by Henley and Kumamoto (1981) is a good reference on fault tree methods.

In the 1980's the emphasis has shifted to the reliability of networks, motivated by the interest in computer design, communications and computer networks. Agrawal and Barlow (1984) survey this field, considering only analytical methods and not stochastic simulation.

Much of the work that is relevant for water distribution systems is in the field of power systems and networks. Lie et al. (1977) deal with the concept of availability, which is defined as the ratio of the time the system is "up" to the total time. Availability may be defined at a specific instant, or may be expressed as an average over some time period (say, one year). Lie et al. review pdf's of inter-failure times and repair durations, and give equations for computing the availability of various system configurations.

Billinton et al. (1982) present algorithms for the loss of load expectation (LOLE) and duration and frequency (F&D) methods. LOLE is the expected number of days in a time period (say, one year) in which the peak demand exceeds the available generating capacity. The load is described by means of the load duration curve, which gives the percentage of time the load equals or exceeds any given value. The F&D method also deals with the number of days in which peak load exceeds capacity, but adds the evaluation of the cumulative probability and cumulative frequency of the difference between peak demand and capacity. In this analysis it is assumed that both demand and supply are lumped at one point (in space), and therefore location, transmission and distribution are not important.

In the field of water supply, the paper by Damelin et al. (1972) is an early attempt to define and compute the reliability of water supply. Reliability is defined as the ratio of quantity actually supplied to that which would have been supplied in the absence of failure (the demand), and computations for two real systems are carried out by stochastic simulation. Shamir and Howard (1981) derived analytical expressions for the probability distribution function of a reliability factor, viewing the supply and demands as lumped. Lvovsky (1984) used the same approach for computing the pdf of shortfalls in a more complex system. Howard and Associates (1984) used the same method for computing the pdf of a reliability measure defined as the ratio of supply to demand, and then converted this into an economic loss to the region being supplied, through an economic loss function.

Hashimoto et al. (1982) defined several measures of system performance which describe its operation under uncertainty: (a) reliability - the probability of meeting the demands, (b) resiliency - how rapidly the system recovers from a failure, and (c) vulnerability - how damaging is the failure when it occurs.

Tangena and Koster (1983) used fault tree analysis for a system of series and parallel elements. Hobbs (1985a,b) borrowed some of the procedures of LOLE from the power systems work and applied them to water supply systems.

Next, we review three methods for analyzing the reliability of water supply systems. These correspond to three models of the system, in increasing levels of detail:

(a) At the most aggregate level, the supply and the demand are each lumped at a few points in space, connected by a few links. The analysis is carried out for a single, typical, time period, such as a year. For such simple models it is possible to use LOLE and F&D methods (Hobbs, 1985a and b) and the statistical-analytical method (Shamir and Howard, 1981; Lvovsky, 1984; Howard and Associates, 1984). PI's of reliability are probability distributions of shortfalls or statistics of these distributions.

(b) At the next level of detail, the model can have more facilities, connections and demand points. Assumptions are made about the pdf's of times to failure and repair durations of each facility, and about the independence between them, and a fault tree type analysis is used to compute the availability of the demand nodes -- i.e., the probability that the node is connected to the source (Tangena and Koster, 1983).

(c) Finally, a model of the network is employed, including in it all the major facilities and links. Analytical methods for computing the availability for a given set of nodes becomes computationally impractical beyond a few tens of nodes and a similar number of links (Ball, 1980; Agrawal and Barlow, 1984). Stochastic simulation can be used, even for somewhat larger networks. It is, however, difficult to interpret the results to determine the reliability of the system being examined. It is even more difficult to use simulation as a prescriptive tool, to indicate how the system should be modified to achieve a desired level of reliability.

4.2 The Statistical-Analytical Method

A water supply system subject to random events -- such as fluctuations in quantity and quality of water in the sources, failure of facilities, changes in demands -- has a reliability PI which is a random variable. The approach taken by the statistical analytical method is to compute analytically the probability distribution of the chosen PI, given the probability distributions of all the random variables upon which it depends, and functional relationships between the PI and these variables. These computations turn out to be feasible only in relatively simple cases. But if the analytical expression can be obtained -- even at the cost of considerable simplifications -- there are important benefits: insight into the structure of the dependence of reliability PI's and FM's on variables and parameters, and the possibility of incorporating the resulting probability distributions explicitly into management models. Shamir and Howard (1981) defined two PI's for reliability of water supply. The first is reliability of discharge:

$$RC = 1 - (C/CT)^n \qquad (1)$$

where C is the lost capacity rate due to the failure, in units of discharge, out of a total rate required, CT. Values $n>1$ cause RC to decrease more rapidly as C approaches CT, while $n<1$ causes RC to drop more rapidly as C decreases. The second PI is reliability of volume:

$$RV = 1 - (V/VT) \qquad (2)$$

where V is the shortfall volume -- during a single failure event or for an entire time period (for example, one year) -- out of a total volume, VT, to be supplied.

For a single failure event:

$$V = C * D \quad \text{and} \quad VT = CT * D \qquad (3)$$

where D is the duration of the event, and thus V/VT = C/CT. The overall reliability factor is defined as:

$$RF = (RC + RV)/2 = 1 - [(C/CT)^n + (C/CT)]/2 \qquad (4)$$

C and V are the independent (functionally) random variables; RC, RV and RF are the dependent variables. For the special case n=2, the pdf of the reliability factor is:

$$f_{RF}(r) = \frac{2\ CT}{\sqrt{1+8(1-r)}}\ f_C \left\{ \frac{-CT\ [1- \sqrt{1+8(1-r)}]}{2} \right\} \qquad (5)$$

which holds for $0 \leq r \leq 1$. If the pdf of C is:

$$f_C(c) = [(m+1)/CT]*[1-C/CT]^m \qquad 0 \leq c \leq CT \qquad (6)$$

then the specific form of the pdf of the reliability factor is:

$$f_{RF}(r) = \frac{2\ CT^n}{CT^{n-1} + nC^{n-1}}\ \frac{m+1}{CT}\ \{ 1 - [C(r)/CT] \}^m$$

$$= \frac{2\ (m+1)}{\sqrt{1+8(1-r)}} \left\{ 1 + \frac{1 - \sqrt{1+8(1-r)}}{2} \right\}^m \qquad 0 \leq r \leq 1 \qquad (7)$$

Lvovsky (1984) applied the method to the water supply system shown in Figure 1. The random annual demand, whose pdf is given by

$$f_D(d) = B \qquad \underline{D} \leq d \leq \overline{D} \tag{8}$$

with $B = 1/(\overline{D}-\underline{D})$, is to be supplied from two sources: a reservoir of size V with annual inflow whose pdf is:

$$f_I(i) = \begin{cases} ai & i \leq i_o \\ b \exp(-ci) & i_o \leq i \end{cases} \tag{9}$$

and a groundwater aquifer with infinite capacity. The parameters a, b and c are selected such that $f_I(i)$ integrates to 1.0, and fits the observed inflow data. The pdf's of the demand and of the inflow are shown in Figure 2.

The two sources are connected to the demand area by facilities -- treatment plants, pipeline, pumping stations -- whose annual failure probability is known: p1 for the reservoir and p2 for the aquifer. Water is to be taken first from the reservoir, because it is cheaper, and only if it is not possible to meet the demands from that source -- either because there is not enough water in the reservoir or because the link has failed -- then an attempt is made to supply from the aquifer.

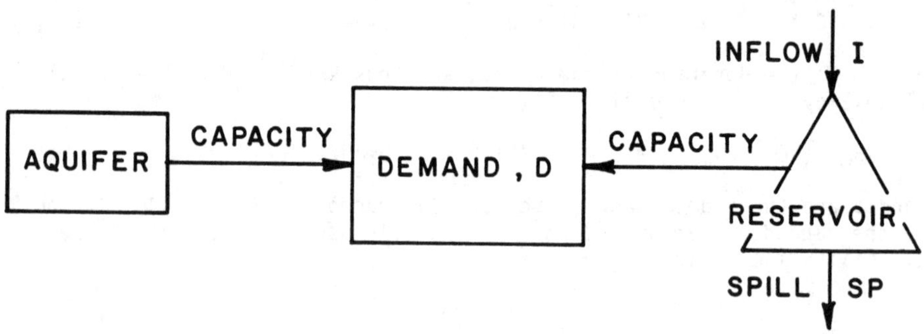

Figure 1. A Water Supply System

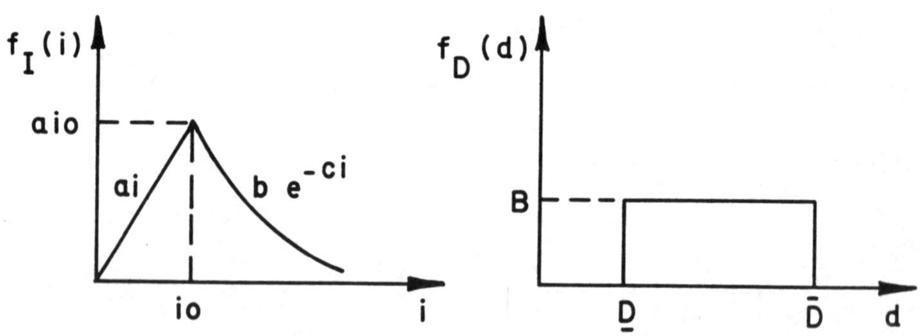

Figure 2. Probability Distributions for Inflow (I) and Demand (D)

The dependent random variables are: the amount supplied from each of the two sources, their sum, the volume in the reservoir at the end of the year, the spill from the reservoir, and -- the PI of interest for reliability -- the shortfall, S = (demand - supply). The pdf's of these variables are obtained by integrating the pdf's of the demand and the inflow into the reservoir over appropriate ranges, considering also the probabilities of failure of the links.

In order to carry out these integrations it is necessary to make several assumptions about the relative magnitudes of some of the variables. The following were used: (a) $Q2<V<Q1$; (b) $\underline{D}<S0$; (c) $Q1+Q2<D$; (d) $Q1-S0<io<Q1-S0+V$; (e) $io<Q1-\underline{D}$; (f) $\underline{D}+V-S0<\overline{io}$. Under these assumptions the pdf of the annual shortfalls, S, is:

$$f_S(s) = \begin{cases} (1-p1)B[Q1-\underline{D}-a(Q1-S0)^3/6+(1-p2)B(Q2-p1\underline{D}) & \text{for } s=0 \\ (1-p1p2)B & \text{for } 0<s\leq\underline{D} \\ B & \text{for } \underline{D}<s\leq\overline{D}-Q1-Q2 \\ B[(p1+p2-p1p2)+(1-p1)(1-p2)a(\overline{D}-Q2-S0-s)^2]/2 & \text{for } \overline{D}-Q1-Q2<s\leq D-Q1 \\ B[p1+(1-p1)(1-p2)a(\overline{D}-Q2-S0-s)^2/2 \\ \quad + (1-p1)p2a(\overline{D}-S0-s)^2/2] & \text{for } \overline{D}-Q1<s\leq\overline{D}-S0-Q2 \\ B[p1+(1-p1)p2a(\overline{D}-S0-s)^2/2] & \text{for } \overline{D}-S0-Q2<s\leq\overline{D}-Q2 \\ B[p1p2+(1-p1)p2a(\overline{D}-S0-s)^2/2] & \text{for } \overline{D}-Q2<s\leq\overline{D}-S0 \\ p1p2B & \text{for } \overline{D}-S0<s\leq\overline{D} \end{cases} \tag{10}$$

Equation (11) and Figure 3 give the specific form of this probability distribution for the following data (all physical variables are in a common unit of volume or volume/year): $\underline{D}=2$; D=28; Q1=16; Q2=8; V=12; S0=6; io=12; p1=p2=0.02.

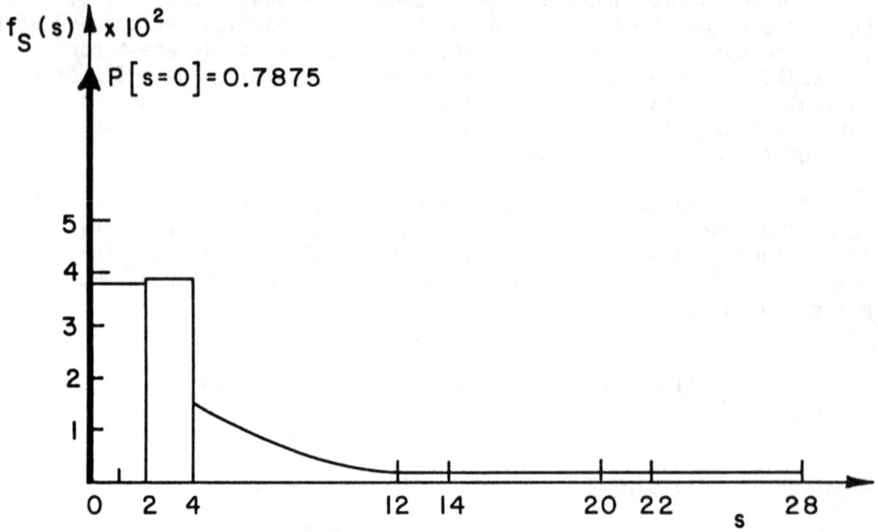

Figure 3. Probability Distribution of Shortfalls Eq. (4)

$$f^{S}(s) = \begin{cases} 0.7875 & \text{for } s=0 \\ 0.0384 & \text{for } 0<s<2 \\ 0.0385 & \text{for } 2<s\leq 4 \\ 0.0001s^2-0.0033s+0.0255 & \text{for } 4<s\leq 12 \\ 0.0001s^2-0.0034s+0.0282 & \text{for } 12<s\leq 14 \\ 0.000002s^2-0.00001s+0.0013 & \text{for } 14<s\leq 20 \\ 0.000002s^2-0.00001s+0.0012 & \text{for } 20<s<22 \\ 0.00001 & \text{for } 22<s\leq 28 \end{cases} \qquad (11)$$

A general equation has been developed for the expected value of the shortfall, which will not be given here for lack of space. For the data above it yields E[S]=0.95. Similarly, expressions for other moments or quantiles of the probability distribution can be developed, or, much more easily, can be computed for the specific form of the distribution after the values of the parameters have been inserted, as in Equation (11).

The analytical integration which leads to results such as (10) above can become quite complicated, and error prone. Still, if it can be carried out, then the results are extremely useful because they allow insight into the role of the various variables. The analytical expression can also be used for sensitivity analysis, as well as in an economic optimization of the system.

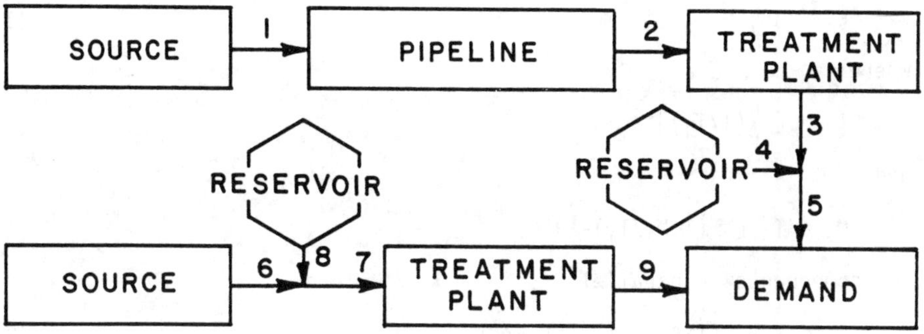

Figure 4. A Water Supply System with Two Sources

When the system is more complicated, and it is no longer possible to compute the probability distributions, it may still be possible to obtain analytical expressions for reliability measures, after making certain assumptions. Consider the system shown in Figure 4. Water is supplied from two sources, through separate systems. The facilities in the system are each subject to random failures, and the repair duration is also a random variable. If it is assumed that times between failures as well as repair durations of facilities are exponentially distributed independent random variables, then it is possible to compute reliability measures for the supply at the demand area.

For times to failures, t_w:

$$Pr[t_w \geq t] = \exp(-t/T) \tag{12}$$

with

$$T = E[t_w] = MTBF \tag{13}$$

and for the repair durations, t_f:

$$Pr[t_f \geq t] = \exp(-t/M) \tag{14}$$

with:

$$M = E[t_f] = MTTR \tag{15}$$

The availability (Henley and Kumamoto, 1981; Lie et al., 1977) of the component is defined as the average portion of time during which it is operational:

$$A = T/(T+M) \tag{16}$$

For components in series, the availability is:

$$A_s = \Pi A_i = T_s/(T_s+M_s) \tag{17}$$

where:

$$T = [\sum(1/T_i)]^{-1} \tag{18}$$

and

$$M_s = T_s [\Pi (1+M_i/T_i)-1] \tag{19}$$

For components in parallel:

$$A_p = 1 - \Pi (1-A_i) = T_p/(T_p+M_p) \tag{20}$$

where:

$$M_p = [\sum(1/M_i)]^{-1} \tag{21}$$

and

$$T_p = M_p [\Pi(1+T_i/M_i)-1] \tag{22}$$

For a system which contains a storage reservoir, the failure of components between that point and the source causes a shortfall only if the reservoir does not contain enough water to supply the demand while these components are being repaired. Denote by t_R the time to deplete the reservoir in supplying the demand, t_R = (volume in storage)/(demand rate), and by the subscript u that part of the system leading into the point where the reservoir is located, then the probability that the reservoir will be emptied before the failed components are repaired is:

$$(1-A_u)\exp(-t_R/M_u) \tag{23}$$

and the resultant availability of the system with the reservoir is:

$$A_R = 1-(1-A_u)\exp(-t_R/M_u) \tag{24}$$

The expected value of the time to failure is:

$$E[t_R] = T_R = (T_u+M_u)\exp(t_R/M_u)-M_u \tag{25}$$

For $T_u >> M_u$, which normally holds in practice:

$$E[t_R] = T_u \exp(t_R M_u) \tag{26}$$

The MTBF increases exponentially with the size of the volume which is in storage when the failure occurs.

4.3 Fault Tree Analysis

The fault tree method (Henley and Kumamoto, 1981; Lie et al., 1977; Tangena and Koster, 1983) uses the topology of the network and the probabilities of failures and repair durations of the individual components, and computes the availability at demand nodes. Availability is defined as the existence of a continuous link connecting the demand node to a source. This does not consider, therefore, the capacity of the remaining network (after a failure) to deliver water with adequate quantity and pressure.

To determine the availability, the minimal cut-sets of the network are identified. A cut-set = if all components in a cut-set fail then no water can reach the demand zone. It is termed "minimal" if the number of failed components is minimal. In Figure 4 , such minimal cut-sets are (5,9) and (1,4,7). For the supply from Source 1 alone, the following are minimal cut-sets: (5), (1,4). The second of these includes the case of Reservoir 1 emptying before Source 1 comes back on line. The availability of the system is closely approximated by:

$$A = 1 - \sum(1-A_i) + \sum(1-A_i)(1-A_j) \tag{27}$$

where the second term contains all pairs of i and j, and higher order terms have been neglected. The capacities of different supply paths can be used, together with their availabilities, to compute the shortfalls in meeting the demands.

The fault tree method is workable only as long as the system is not too complex, because identifying all the cut-sets becomes complicated and carrying out the computations is expensive.

4.4 A Case Study

Figure 5 shows demand projections to the year 2030 for Seattle, Washington (Howard and Associates, 1984). The mean was computed by projecting the population and per capita consumption. The distribution over the range, between E_1 and E_2, was developed on the basis of the variability observed in the historical demands - 48 years, from 1929 to 1976. It was assumed that the demand probabilities over this range are of the same shape as in the historical record, just scaled up to the projected demand level.

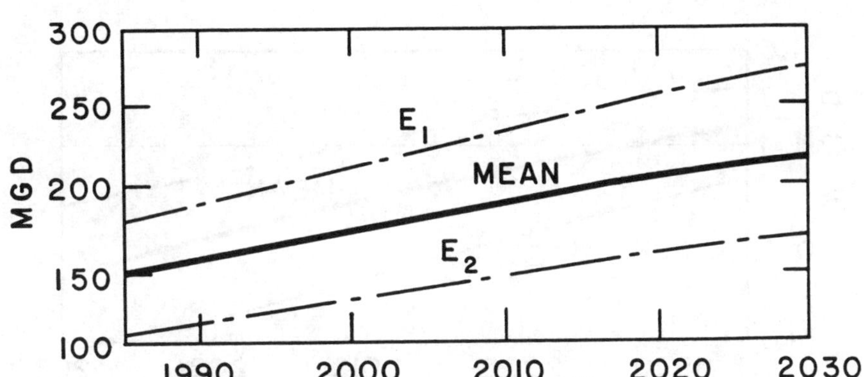

Figure 5. Average Annual Demand Projections

The histogram of the available annual supply in the 48-year record was used as the probability distribution for future supplies, if no additional sources or facilities are developed. Reliability is defined as the ratio of supply, S, to demand, D, i.e., R = S/D. For every future year it is possible to compute the histogram of reliability, R, from:

$$Pr[r_1 \leq R \leq r_2] = Pr[r_1 \leq S/D \leq r_2]$$

simply by counting the (relative) number of times that (S/D) falls in the considered range.

Costs due to water shortages were found to be greatest in the wet winter months (October to December), because in the summer, when demands are higher, consumers have more flexibility to reduce their consumption. Shortfalls will therefore occur in October and November, after the reservoirs have been depleted and many of the discretionary uses of water have been terminated. Thus the reliability and risk analysis was carried out for the winter months.

The projected mean reliability for the winter months, and its range, are shown in Figure 6 for the case where no additional sources or other facilities are developed.

An economic analysis of the region used historical marginal cost data for water supply and the concepts of consumer surplus to develop estimates for the costs of shortages at various levels of demand. Figure 7 shows this cost function, which is now used to assess future economic costs (losses) due to shortfalls.

Several proposed alternatives of capacity expansion were examined. For each, future reliability distribution was developed and converted into economic costs. These were added to the capital costs of the plan, all expressed in present values, so that the different alternatives can be compared. Figures 8 and 9 show the reliability and costs of shortfalls for one of these alternatives.

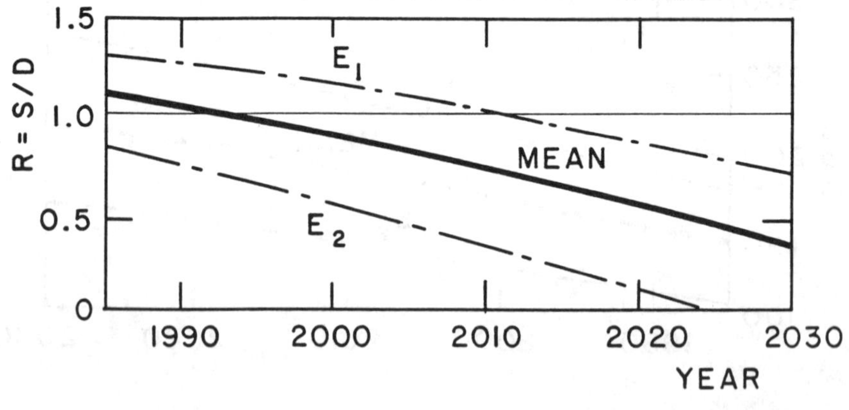

Figure 6. Seasonal (Oct-Dec) Reliability with No Source Development

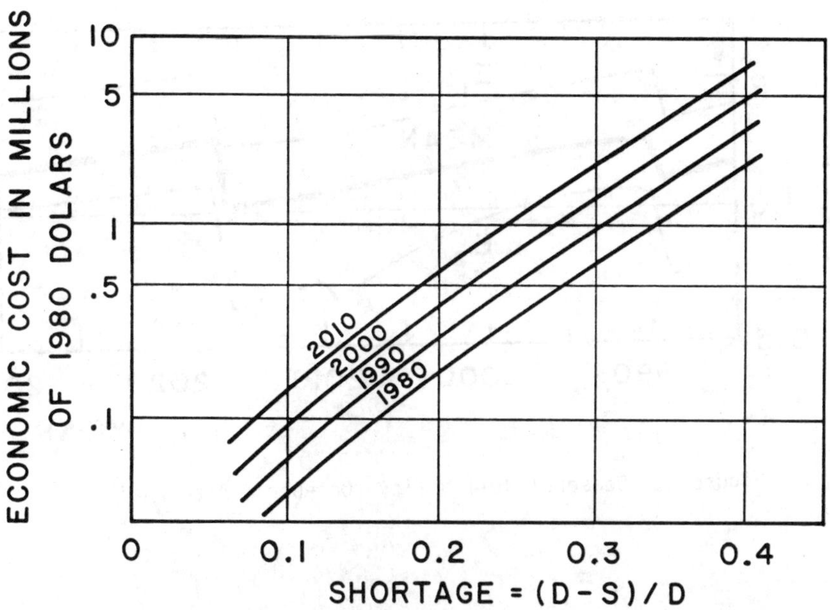

Figure 7. Economic Cost of Shortages, Wet Months (Oct-Dec)

4.5 Stochastic Simulation

Stochastic simulation can be used to conduct numerical experiments on a model of the supply system. Random events are generated from the probability distribution of each random variable. In a supply system, these could include random quantities and qualities of water at the sources, failures at treatment plants, in pumps and pipes, as well as variations in the demands. For example, if a random failure of a pipe has been generated, the capacity of the remaining system to supply water is evaluated by solving the network with the original demands imposed. The performance is evaluated, to see whether the demands can indeed be supplied at adequate pressures. Failures and shortfalls are recorded, so that at the end of the simulation run a statistical analysis can be conducted.

There is no conceptual difficulty in setting up a stochastic simulation model. To obtain a statistically significant sample, the model may have to be run for a long time. The main deficiency of simulation, however, is that it is merely descriptive, and not prescriptive. It tells us how reliable a specified system is -- as measured by any PI and FM -- but provides no indication what to do to improve the reliability in the most efficient way.

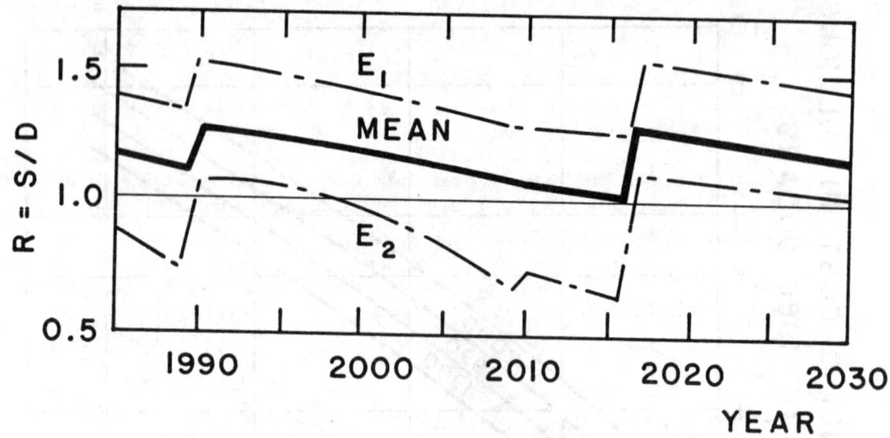

Figure 8. Seasonal Reliability (Oct-Dec), Alternative I

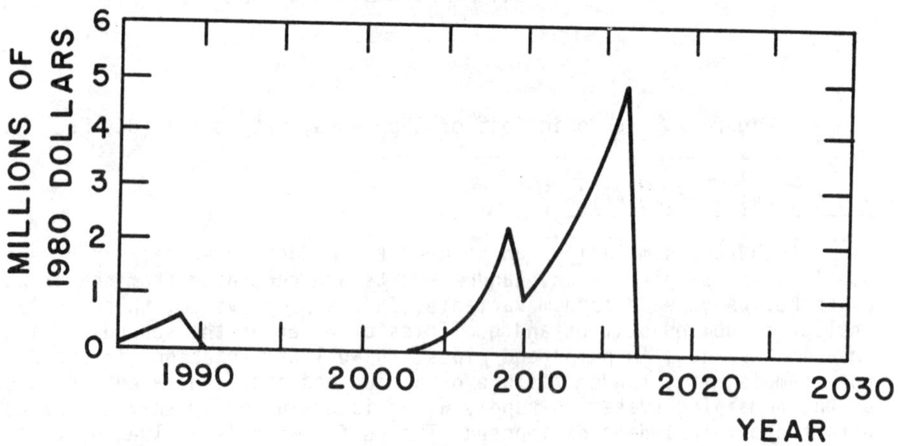

Figure 9. Risk of Seasonal Shortage (Oct-Dec), Alternative I

BIBLIOGRAPHY

Allan, R.N., R. Billinton and S.W. Lee. Bibliography on the Application of Probability Methods in Power System Reliability Evaluation, 1977-1982. IEEE Transactions on Power Apparatus and Systems, Vol. PAS-103(2) (1984).

Agrawal, A. and R.E. Barlow. A Survey of Network Reliability and Domination Theory. Operations Research 32(3) (1984) 478-492.

Ball, M.O. Complexity of Network Reliability Computations. Networks 10 (1980) 153-165.

Barlow, R. Mathematical Theory of Reliability: A Historical Perspective. IEEE Transactions on Reliability R-33(1) (1984) 16-20.

Billinton, R., C.L. Lee and G. Hamoud. Digital Computer Algorithms for the Calculation of Generating Capacity Reliability Indices. IEEE Transactions on Power Apparatus and Systems, PAS-101(1) (1982) 203-211.

Damelin, E., U. Shamir and N. Arad. Engineering and Economic Evaluation of the Reliability of Water Supply. Water Resources Research 8(4) (1972) 861-877.

Duckstein, L. and E.J. Plate. A System Framework for Hydrologic Reliability with Application to Design of Structures. International Association of Hydrological Sciences Symposium, Hamburg (August 1983).

Duckstein. L., E.J. Plate and M. Benedini. Incidents and Failures in Reservoirs and Dams: A Mathematical System Framework. American Geophysical Union Fall 1984 Meeting, San Francisco, (December, 1984).

Haimes, Y.Y. (Ed.). Risk/Benefit Analysis in Water Resources Planning and Management. Plenum (1981).

Hashimoto, T., J. Stedinger and D.P. Loucks. Reliability, Resiliency and Vulnerability Criteria for Water Resources Systems Performance Evaluation. Water Resources Research 18(1) (1982) 14-20.

Henley, E.J. and H. Kumamoto. Reliability Engineering and Risk Assessment. Prentice-Hall (1981).

Hobbs, B.F. Reliability Analysis of Urban Water Supply. To appear in "Computer Applications in Water Resources", ed. by H. Torno, ASCE, NY, (1985a).

Hobbs, B. Reliability Analysis of Water System Capacity. To appear in "Hydrology and Hydraulics in the Small Computer Age", ed. by W. Waldrop ASCE, NY, (1985b).

Howard and Associates, Ltd. Water Supply Reliability and Risk. Study for the City of Seattle. (1984).

Koenig, L. Optimal Fail-Safe Process Design. Journal WPCF 45(4) (1973) 647-654.

Lie, C.H., C.L. Hwang and F.A. Tillman. Availability of Maintained Systems: A State-of-the-Art Survey. AIIE Transactions 9(3) (1977) 247-259.

Lvovsky, M. Statistical-Analytical Models for Analysis of Water Resources Systems (in Hebrew). M.Sc. Thesis, Faculty of Civil Engineering, Technion - Israel Institute of Technology (1984).

Norrie, D.J.W. Optimal Reliability in Water Supply Planning: Risk Versus Cost of Supply Tradeoffs. M.Sc. Thesis, Department of Civil Engineering, MIT (1983)

Plate, E. Reliability Analysis in Hydraulic Design. Fourth IAHR Symposium on Stochastic Hydraulics, Urbana-Champaign, Illinois (1984).

Shamir, U. and C.D.D. Howard. Water Supply Reliability Theory. Journal of the American Water Works Association (July 1981) 379-384.

Shamir, U. and C.D.D. Howard. Reliability and Risk Assessment for Water Supply Systems. To appear in "Computer Applications in Water Resources" ed. by H. Torno, ASCE, NY, (1985).

Tangena, B.H. and P.K. Koster. Reliability of Drinking Water Supply Systems. Proceedings of the International Water Supply Association Specialized Conference on Security in Water Supply, Brussels (1983).

APPLICATION OF MODELS FOR RELIABILITY ASSESSMENT IN RESERVOIR OPERATION

Gert A. Schultz

Institute of Hydrology and Water Resources
Ruhr University Bochum
P.O. Box 102148
4630 Bochum, F.R. Germany

ABSTRACT

Within the framework of two related papers five different techniques are presented, all of which use reliability related performance indices or figures of merit (Duckstein et al, 1984) as decision criteria. In two cases these figures of merit are optimized. Each technique is presented along with an example of a real world project. This paper deals with two projects/techniques, both of which are subject to independent loading, while the companion paper discusses three projects/techniques that are subject to dependent loading.

The first technique discussed here represents a combination of flood frequency analysis and benefit/cost computations. The problem consists of evaluating the results of raising a dam for the purpose of (among others) mitigating flood hazards in the Ennepe valley of Germany. The performance index in this case is the reduction of flood damage costs due to the increased flood storage of the raised dam. The damage costs are calculated on the basis of expected present values of these costs.

The second technique presented in this paper is queuing theory. This technique is applied to the design of an extension of an existing water supply reservoir system in West Africa. Twelve probability transition matrices have to be computed based on the twelve probability distribution functions (pdf's) of the monthly inflows. In this way it is possible to compute for a given reservoir capacity and given withdrawal rates the corresponding reliability as a performance index (PI). By the iterative application of queuing theory for many combinations of future demands and reservoir capacities, it appears to be possible to construct storage-yield frequency curves for several prespecified PIs and thereby select the required storage capacity for the demand of the year 2000.

1. INTRODUCTION

The purposes of this paper and the accompanying paper (Schultz, 1986) are threefold:

a) presentation of some modern planning techniques,

b) proof that these techniques have been successfully applied and sold to decision makers in practice,

c) discussion of reliability related decision criteria in relation to and combination with other, equivalent performance indices or figures of merit.

The planning techniques discussed in these papers along with real world applications are queuing theory (an example of stochastic reservoir theory), optimization techniques such as dynamic programming and compromise programming (a multi-objective optimization technique), simulation, and a combination of economic and probability methods. The project examples were taken from the consulting work of the author in Europe and Africa. The performance indices (PI^s) or figures of merit (FM^s) used for the various projects differ considerably from project to project. They were, in each case, selected during or after long dialogs with the decision makers.

The two papers are organized such that a total of five planning techniques are presented and discussed, each with a different real world water project. Table 1 shows these techniques and projects together with their main specifications. In this paper two methods are presented which use independent loading, while the three methods presented in the other paper are based on dependent (correlated) loading as model input (Table 1).

2. FLOOD FREQUENCY ANALYSIS WITH COMBINED BENEFIT/COST RATIO ESTIMATION

The first technique to be discussed is No. 1 in Table 1. One of the oldest and simplest stochastic hydrological techniques is flood frequency analysis. The application of this method usually results in information on the relationship between peak discharge values and the corresponding probabilities of exceedance. If, for example, a flood levee has to be designed for a hundred-year recurrence interval (HQ_{100}) flood, this technique can be applied, providing the flood levee height itself is not considered as an independent stochastic variable (Plate, 1984).

In many cases, however, the decision maker is not content if the design is based on a simple PI alone. He wishes to get a FM indicating the ratio of the expected value of the present value of benefits (e.g. in the case of flood protection measures, avoided damage costs) to the construction costs (plus OMR cost).

In the project example presented here, a water authority (Ennepe Water Authority) in Germany had to decide whether or not an existing multipurpose dam was to be raised. Since most of the members of the water authority (besides a water supply company and the representatives of the cities involved) were factory owners, decision criteria had to be formulated in monetary terms.

Table 1. Specifications of Presented Techniques, Projects, Loading and Performance Indices.

Technique/ Project No.	Lecture No.	Methodology	Project Purpose	Hydraulic Structure	Performance Indices/ Figures of Merit	System: Loading	System: Resistance	Project Region	Contractor
1	I (Independent loading)	Flood frequency analysis combined with benefit/cost ratio estimation	Flood protection	Dam (to be raised)	Economic benefits	independent: flood volume	Reservoir flood storage	Ennepe River catchment Germany	Ennepe Water Authority
2		Queuing theory	Water supply in the year 2000	Dam (to be raised)	Water supply reliability	independent: water demand	Water supply capacity of reservoir	Cape Coast Ghana West-Africa	GWSC, Ghana
3	II (Dependent Loading)	Optimization (Dynamic programming)	Multipurpose: Hydropower & low flow augmentation	New dam	Bi-objective penalty function (to be minimized)	dependent: power demand & flow deficits	Water supply capacity of reservoir	Wupper River catchment Germany	Wupper Water Authority
4		Simulation	Multipurpose: drinking water supply and low flow augmentation	New dam	Target-related reliability for: - water supply - low flow augmentation	dependent: water demand & flow deficits	dto	Dhünn River catchment Germany	Wupper Water Authority
5		Multiobjective optimization (compromise programming)	Low flow augmentation (operation)	Existing reservoir system	Lp-metric (of compromise programming), (to be minimized)	dependent: flow deficits	dto	as example: Wupper reservoir system	--- (Research)

2.1 Problem Formulation

The existing Ennepe reservoir serves several purposes, the most important of which are

- drinking water supply

- flood protection

- low flood augmentation in order to improve the downstream water quality.

The existing dam fulfills its purposes at certain levels of reliability, with drinking water supply having the highest priority.

It would be possible to raise the existing dam considerably, thus increasing the reservoir capacity by 300% (from 12.6×10^6 m^3 to 35×10^6 m^3). In this way the reliability of the dam for meeting the various purposes would be increased. Decision criteria, however, include not only reliability, but also increased (monetary) benefits from the reservoir system.

Therefore, the water authority required from the consultant a comparison between the benefits from the existing system and those from the raised dam system for all the purposes which the reservoir must serve. The resulting benefits were compared to the respective costs, and the decision was made on the basis of maximum benefit/cost ratio.

Within the framework of this paper, the relevant computations will be shown only for the purpose of improvement of flood conditions downstream from the reservoir. Here again, three types of benefits were calculated

- reduced flood damage costs

- reduced maintenance costs of the downstream river reach

- increased prices of real estate.

Only the first type of costs will be discussed in this paper.

2.2 Technique Applied

Benefits are defined in this context as avoided flood damage costs. Then benefits increased by raising the dam are the differences of damage costs (raised dam vs. present conditions). Since downstream of the dam the Ennepe valley is highly industrialized, flood damage costs are a function of peak flood levels. Peak flood levels, however, represent a stochastic process, thus rendering damage costs stochastic. Therefore, it is only possible to compute expected values of damage costs on the basis of the probability distribution (pdf) of peak floods. Since two cases have to be compared, namely present dam vs. raised dam (i.e. increased flood storage in the reservoir), the pdf's of both present flood conditions and future flood conditions (i.e. with increased flood storage) have to be computed first and then compared. The project lifetime was assumed to be 50 years. In this way it became possible to compute the present value of the reduced damage costs for various interest rates.

2.3 Hydrological Computations

Since river gauges were neither at the dam site nor exactly in the area most endangered by floods, many hydrological computations were necessary before the flood frequency analysis could be made. Because these problems are, however, of no major interest for the general topic discussed here, it is assumed that reservoir inflow hydrographs can be constructed with the aid of pdf's of peak discharges combined with a standard dimensionless reservoir inflow hydrograph. Figure 1 gives the pdf of reservoir inflow peaks (Pearson III distribution and observed values). Altogether, nine different pdf's were computed and will be discussed later.

Next the standard dimensionless inflow hydrograph is multiplied by the flood peaks of specified probability, thus rendering flood hydrographs of specified probability. With the aid of a reservoir routing routine, these floods were routed through the reservoirs for two conditions:

a) present reservoir: no flood storage, bottom outlets closed; spillway length: 70.2 m; assumption: reservoir full at beginning of flood

b) raised dam: flood storage 3.5×10^6 m^3, bottom outlets closed; spillway length: 80.0 m; assumption: reservoir full up to lower limit of flood storage.

For the reservoir routing procedures it was assumed that the reservoir was full (case a) or only the design flood storage (3.5×10^6 m^3) was available.

This means that in terms of reliability theory only the loading (flood inflow volumes) was asumed to be stochastic, while the resistance (reservoir flood storage) was assumed to be deterministic, which is not true in practice since the available storage in the reservoir at the beginning of a flood is also a stochastic variable (Plate, 1982).

There were two reasons, however, why the resistance was assumed to be deterministic:

- major floods occur only during the wet season, when the existing reservoir is almost always completely full;

- the water authority wanted the benefit computations to be on the "safe side". This means that potential lower water levels in the reservoir at the beginning of a flood will reduce the flood damages even more, particularly for the raised dam. If this is compared to the present conditions, this means an additional extra benefit (probably small) from raising the dam.

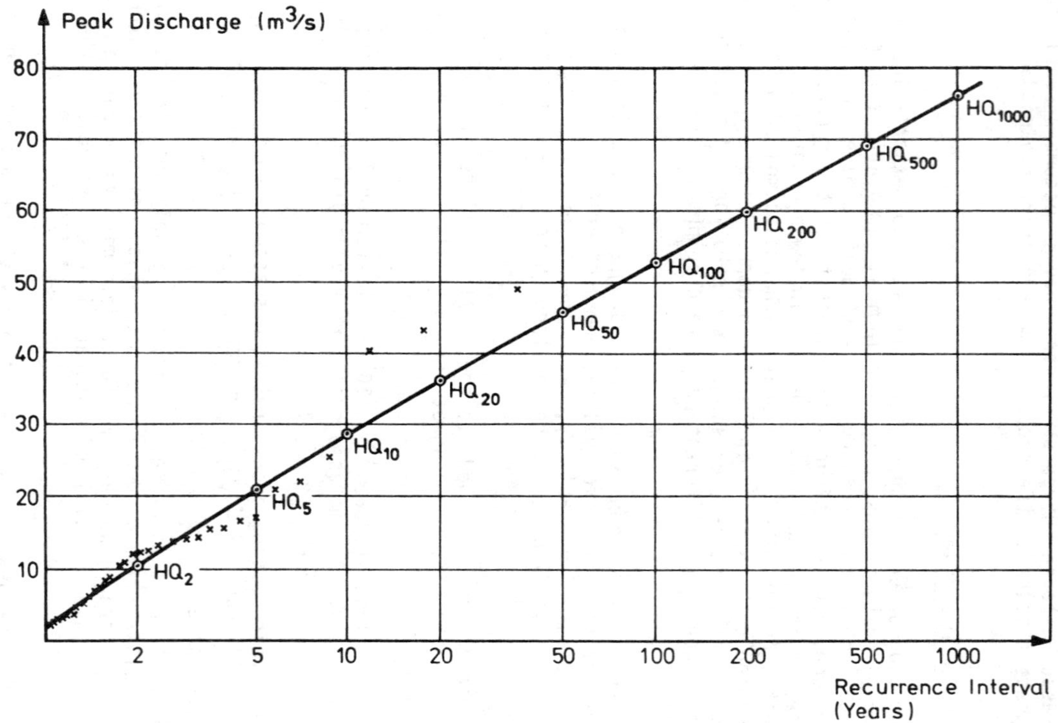

Figure 1. Flood Frequency Graph of Ennepe Reservoir Inflows (Pearson III-Distribution)

Figure 2 shows as an example the result of the reservoir routing procedure for project alternative a) present conditions and alternative b) dam raised for the 100-year flood. As can be seen, alternative a) (zero flood storage) yields almost no flood peak reduction while this is significant for alternative b).

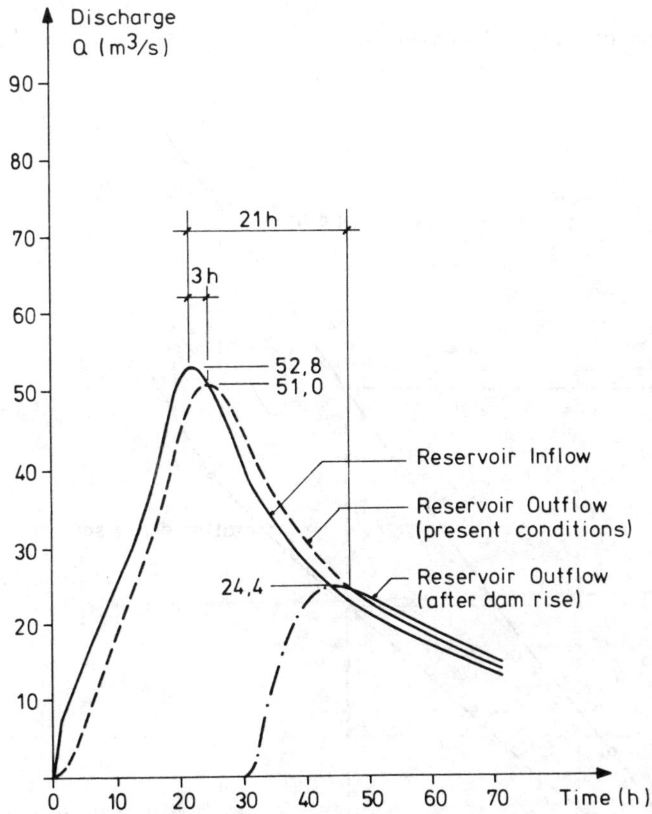

Figure 2. Eneppe Flood Hydrographs (100-year flood)

After combination of the routed floods of various probabilities of occurrence with the flows from other subcatchments, the flood peaks for the area of interest (further downstream) were computed for alternatives a) and b). Figure 3 shows the flood frequency curves for both alternatives as a function of the reservoir inflow floods. The influence of the dam raising is significant; the 100-year reservoir inflow flood is reduced from 142 m^3/s to 90 m^3/s.

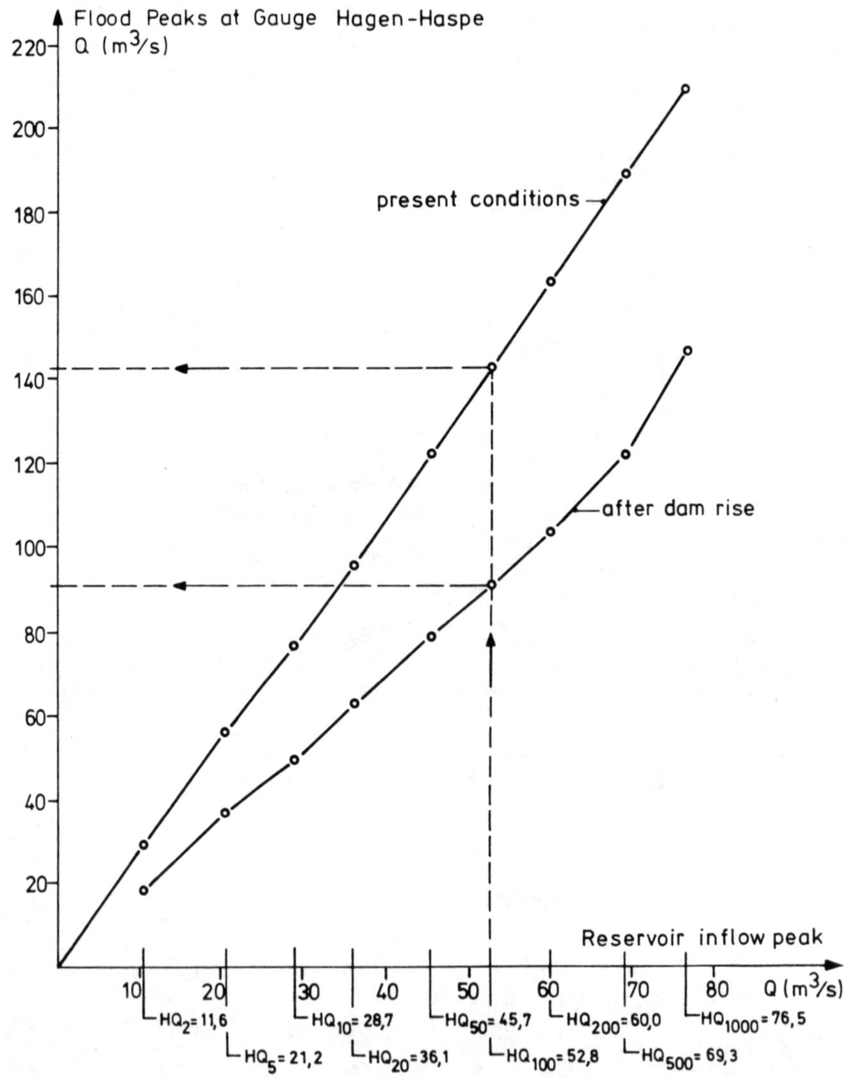

Figure 3. Peak Floods at Gauge Hagen-Haspe for Both Alternatives

2.4 Compution of Damage Cost

Since there have been large floods in the Ennepe valley during the last decades that have caused severe damage, the water authority has made monetary evaluations of those flood damages. These damage costs were plotted against the flood peak discharges to give the diagram of Figure 4.

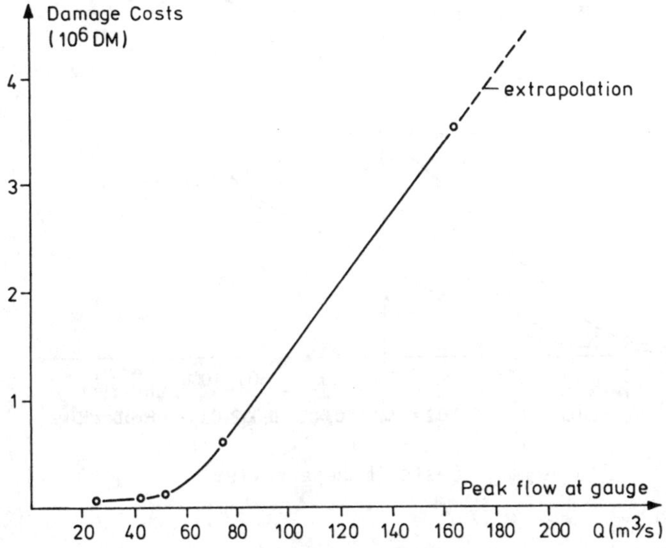

Figure 4. Monetary Flood Damage Function (Ennepe Valley)

Since the pdf's of peak discharges are known from Figure 3, it was possible to combine Figures 3 and 4 such that a diagram could be constructed representing flood damage costs as a function of probability of exceedance (Fig. 5) for both project alternatives.

By integration of the damage costs vs. probability of exceedance curve, the expected value of yearly damage costs can be computed. Figure 6 represents one curve of Figure 5 at a non-distorted scale showing how the integration can be executed at discrete steps

$$S = \sum_k s(\Delta p_k) \cdot \Delta p_k$$

where:

S = expected value of damage costs per year

s = damage costs

p = probability of exceedance

k = number of discrete interval (Fig. 6).

In order to compute the expected value of total costs of flood damage for the lifetime of the project (50 years), it is necessary to add up the discounted expected values of yearly damage costs, thus obtaining the present value of damage costs. The present value depends on the chosen interest rate (after deduction of the inflation rate). Table 2

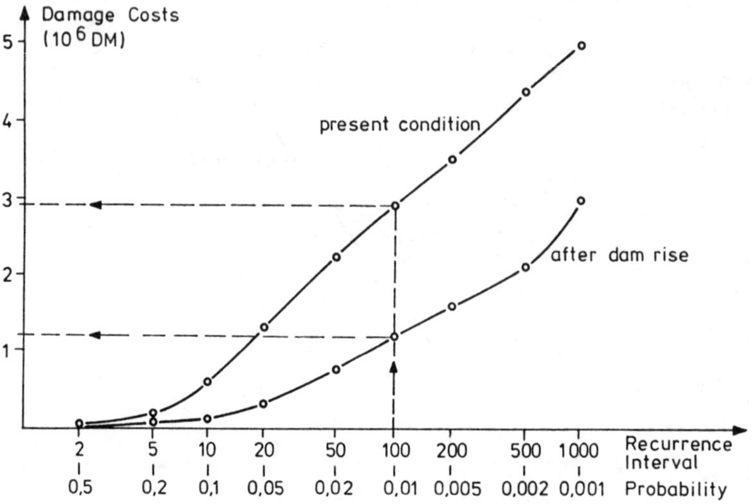

Figure 5. Pdf of Flood Damage Costs (Ennepe Valley)

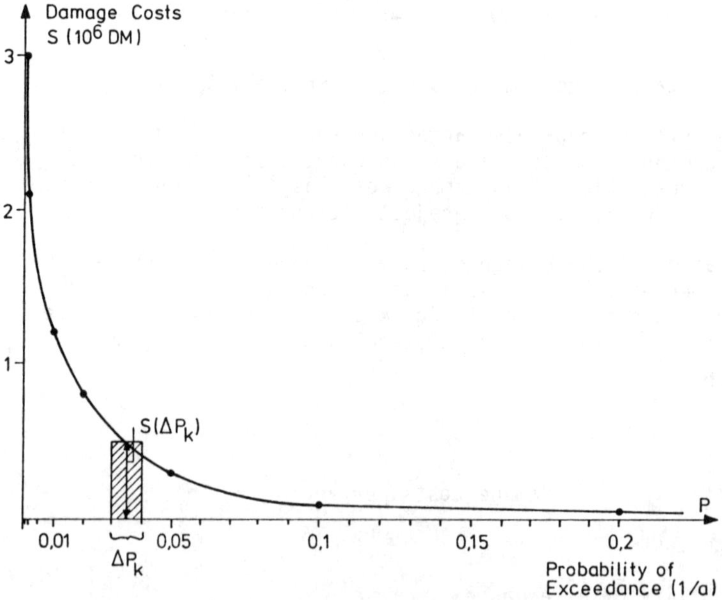

Figure 6. Determination of the Expected Value of Yearly Flood Damage Costs

shows these results for three different interest rates which are valid in Germany (Rudolph and Schultz, 1985). One can see that for the usual interest rate i = 3.5% per year, the total damage cost (present value) amounts to 6.86 million Deutsche Marks (for alternative 1), while after raising the dam these would be only 1.68 x 10^6 DM.

Thus the benefits of raising the dam would be the difference 6.86 - 1.68 = 5.18 millon DM.

Table 2. Expected Damage Costs and Present Values due to Reduced Flood Damages (Calculated for a Lifetime of 50 Years)

	Expected Value of Yearly Damage Costs	Present Values (10^6DM) i = 2%	i = 3.5%	i = 5%
Alternative a) Present Condition	0.2925	9.19	6.86	5.34
Alternative b) Raised Dam	0.0716	2.25	1.68	1.31
Damage Costs Reduction due to Dam Rise	0.2209	6.94	5.18	4.03

2.5 Influence of Chosen pdf of Flood Peaks on Damage Costs

Among hydrologists, the discussion of "which pdf is the right one for flood peaks" has been going on for decades. While the log Pearson III distribution is generally recommended in the US, others (Wallis, 1985) state that this distribution is highly unsuitable. This may be of importance for the design of the spillway length for a dam and other problems, but it is not necessarily important for all flood problems involving flood pdf's.

For the estimation of the present value of the expected value of flood damage cost reduction over the Ennepe project lifetime, nine different pdf's were used. It can be shown that the difference of these present values is rather small, i.e. in the range of the usual variations of interest rates as given in Table 2. The sensitivity of these results as a function of pdf's, interest rates, project lifetime, etc. will be presented in a later publication.

3. QUEUING THEORY APPLIED TO THE DESIGN OF A WATER SUPPLY RESERVOIR IN AFRICA

The characteristics of the water supply project in Ghana and the chosen method are given in Table 1 under Technique/Project No. 2. While the decision maker in the Ennepe project was not content with a reliability related PI alone but insisted on a FM in monetary terms, the decison maker in Ghana (GWSC = Ghana Water and Sewerage Corporation) was interested in a project design which also guarantees the water supply for the demand of the year 2000 with a prespecified (high) reliability. Thus, a technique was chosen which gives information about the dependence of required reservoir capacity on future demands and the corresponding supply reliability. For this purpose queuing theory appears to be suitable.

3.1 Problem Formulation

An existing water supply system based on an existing dam was going to become too small for the growing water demand of the increasing population within the region. The task was to decide whether the water demand of the year 2000 could be met by the existing reservoir if the dam were raised or whether a second reservoir had to be built.

The water demand is a function of population size. The Census Bureau had forecast a population growth of 2.5% per annum. This gives rise to an increase in water demand, shown in Figure 7. Furthermore, these data shown that the demand differs in January, February and March from the other months.

Thus, the first question to be asked was: what is the reliability of the existing water supply system for meeting the present water demand? Then the question had to be answered: what is the reliability of the existing system to meet future demands (e.g. for the year 1990 and 2000)? If the resulting reliability is not satisfactory from the point of view of the decision maker, then the reservoir capacity has to be increased such that the reliability again becomes acceptable. In the framework of these considerations, the point in time at which the existing system reliability becomes acceptable is of primary interest.

Usually such problems are solved with the aid of simulation techniques. In order to apply simulation it is necessary, however, to specify the reservoir capacity and the operating rule. Then simulation results will determine the probability of failure or some other PI. If, however, the reservoir capacity is not yet known and if the reliability is a desirable quantity for many potential project alternatives, simulation is very cumbersome since it has to be applied again and again until acceptable solutions are found by trial and error.

3.2 Queuing Theory

Queuing theory belongs to the field of stochastic reservoir theory (Moran, 1959; Lloyd, 1967). It was first applied to water reservoirs by Langbein (1958) and Fiering (1961). Examples of reservoir performance calculated on a large computer by use of this theory are given by Schultz (1973).

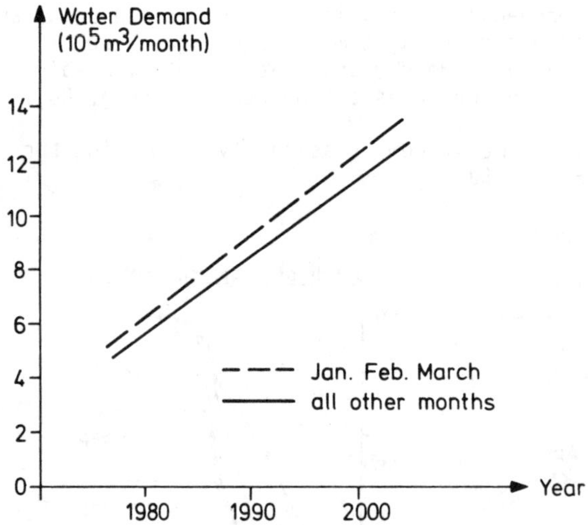

Figure 7. Future Water Demand in the Project Region (Cape Coast, Ghana).

Queuing theory allows for, given reservoir inflow pdf's and known demand (reservoir operating rule), computation of the probability of failure of the reservoir to meet the demand and also the probabilities of storage contents. This can be done for discrete time intervals (in our case chosen to be one month) and discrete storage increments.

In order to do this for each month, a transition probability matrix has to be calculated which gives one system of linear equations. Eventually all 12 such systems have to be solved simultaneously in order to obtain the desired probability distributions. It should be mentioned that queuing theory may be applied only if the operating rule can be expressed as a function of the storage content. Furthermore, this theory is unable to take into account the serial correlation of inflow values.

It is beyond the scope of this paper to introduce stochastic reservoir theory and queuing theory as a special case. Such information can be obtained from the available literature. After a very brief summary, the application of queuing theory will be presented and the results in terms of reliability related PI^s will be discussed.

The principle of queuing theory is as follows:

1. From hydrological measurements the reservoir inflow pdf's can be computed (see Fig. 8a for example). Such a pdf can be computed, e.g. for each month of the year.

2. A reservoir operating rule has to be specified as function of reservoir content (see Fig. 8b for example).

3. By formulation of probability transition matrices and simultaneous solution of the corresponding systems of equations (e.g. 12 such systems if monthly values are chosen), the pdf of reservoir contents (Fig. 8c) and the pdf of releases can be computed (Fig. 8d).

4. The reliability PI of the system (possibility of meeting the demand) can be read from Figure 8d.

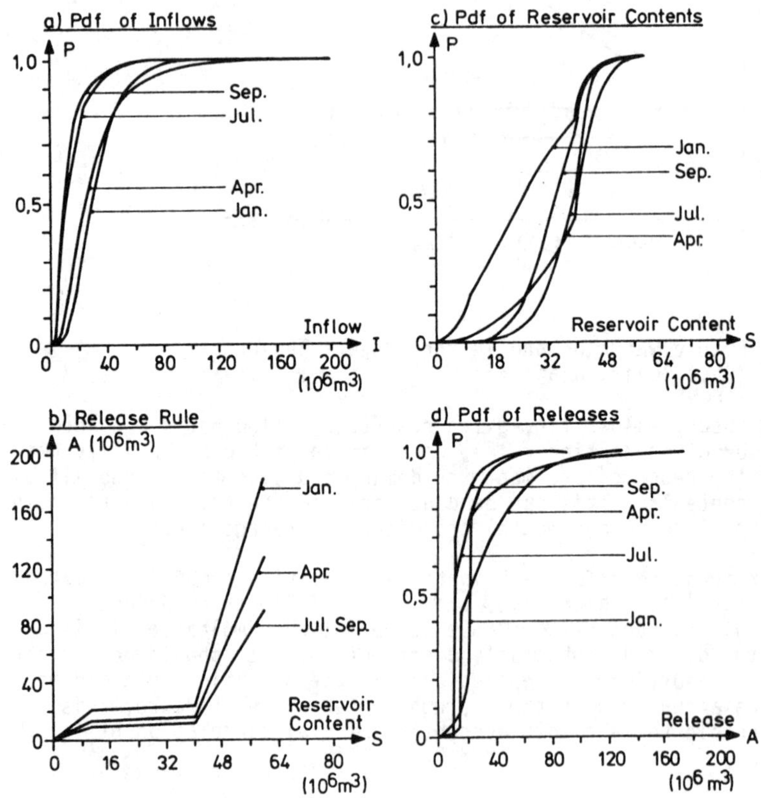

Figure 8. Probability Distributions for a Reservoir According to Queuing Theory

The construction of the probability matrices take into account the inflow pdf and the prespecified release rule. Single conditional transition probability $p(a_i,a_j)$ represents the probability that the reservoir content (state vector) S at time t (e.g. month) is $S_t = a_i$ given that $S_{t-1} = a_j$ in the previous time interval (month).

Therefore

$p(a_i,a_j) = p(S_t = a_i | S_{t-1} = a_j)$

The complete resulting transition probability matrix P is

$P = [p(a_i,a_j)],\ i = 1,\ldots,n;\ j = 1,\ldots,n$

In this case $S_t = a_1$ could mean reservoir empty and $S_t = a_n$ reservoir full.

In practice the non-conditional probabilities $p(S_t = a_i)$ are more important than the conditional probabilities $p(a_i,a_j)$. Thus, for time interval t the probability vector is:

$$p(S_t = a_i) = p_t(a_i) = \underline{p}_t \qquad i = 1,2,\ldots n \tag{1}$$

Now p_t can be computed with the aid of the transition probability matrix:

$$p_t(a_i) = \sum_{a_j=0}^{a_n} p\,(a_i,\, a_j)\,\cdot\, p_{t-1}\,(a_j) \tag{2}$$

and:

$$\underline{p}_t = P \cdot \underline{p}_{t-1} \tag{3}$$

A system of equations like Eq. (3) can be formulated for each month of the year considering that they are correlated by:

$$\underline{p}_{t-m} = P \cdot \underline{p}_{t-m-1} \tag{4}$$

Stable solutions for each $\underline{p}_t$ can be found by standard algebraic manipulations. With the aid of the known pdf's of reservoir contents for each month and the specified release rule, it is possible to estimate the resulting pdf's of release. From these the reliability of the system to meet the demand can be inferred (Fig. 8d).

3.3 Application of Queuing Theory to a Reservoir in Ghana

Queuing theory on the basis of monthly time increments and a suitable discretization of reservoir capacity was applied to the situation of Brimsu reservoir in Ghana (Schultz and Dafinis, 1977) according to the above mentioned "problem formulation." For the monthly inflow pdf's, the log Pearson III distribution was not rejected by a test of significance (Kolmogorov-Smirnov) on a 95% level. Therefore, this pdf was used (Fig. 9).

The operating rule is always such that the demand has to be met as long as there is sufficient water in the reservoir. If the storage volume in a particular month is less than the demand, only the available quantity is supplied.

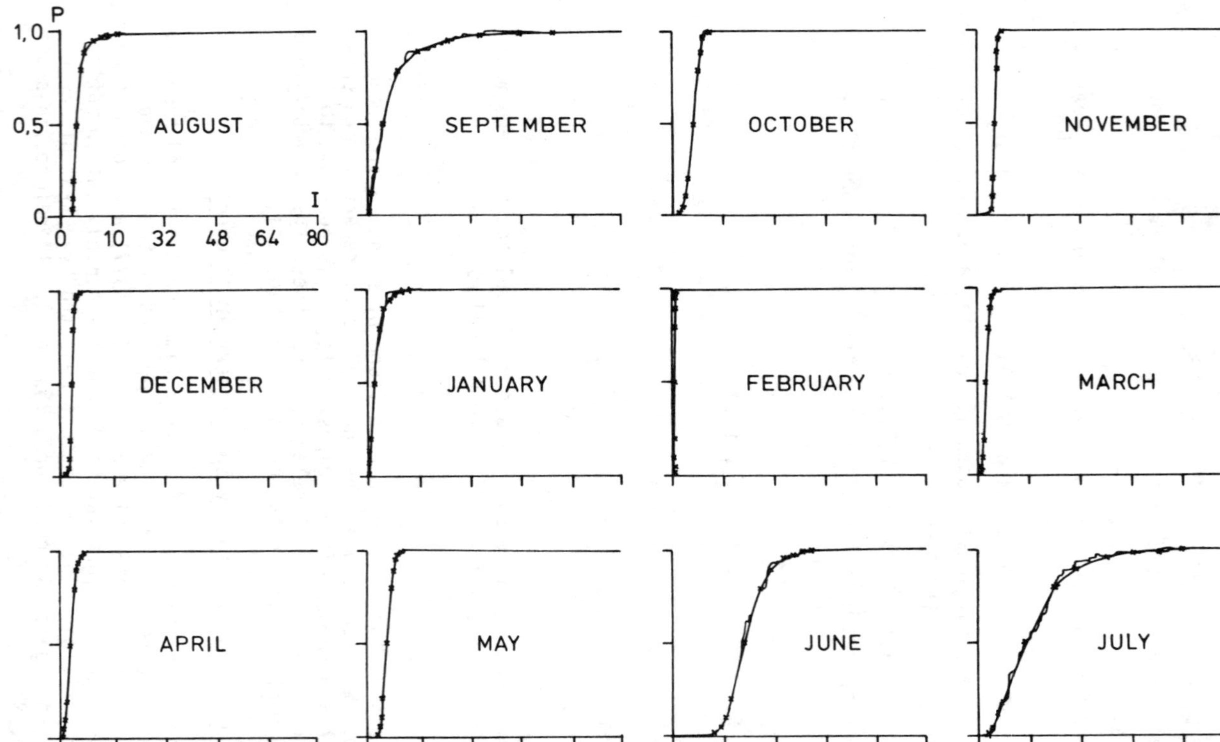

(p = 1 - probability of exceedance; I = monthly reservoir inflow 10^6 m^3)

Figure 9. Pdf of Monthly Reservoir Inflows (Log Pearson III)

According to this simplified operating rule, no compulsory release into the river downstream of Brimsu is provided and no permanent storage which may be released in order to preserve fish life.

In order to determine the required storage-yield-frequency relationships for Brimsu reservoir, it was necessary to compute about 50 times the reservoir performance by application of queuing theory. Here, only two representative examples will be given for the time horizons 1980 and the year 2000.

3.3.1 Reservoir performance in the year 1980. The demand for 1980 is 0.566 x 10^6 m^3/month for all months of year except January, February, and March, when it is 10% higher. The operating rule is given in Figure 10. The computations were carried out for various reservoir capacities. However, it seems appropriate to leave the present capacity at the same level because a rise of capacity is not necessary since the reservoir performance is very reliable. This can be seen for the probability density of reservoir contents (for the most relevant months) (Fig. 11) and for the probability distribution of reservoir release (not shown graphically). The demand is met with a probability of almost 100%, i.e. the risk of failure of the reservoir to meet the demand is almost zero.

3.3.2. Reservoir performance in the year 2000. The demand in the year 2000 was estimated to be 1.141 x 10^6 m^3/month for all months except January, February, and March when it is 10% higher. The operating rule is given in Figure 12. Again, the computations were carried out for various reservoir capacities, two of which are shown in the following graphs.

First the present reservoir capacity is used. Figure 13 shows the corresponding probabilities of releases (supply). Since there are non-zero probabilities of the reservoir not being full in February, March and April, during these months there is some risk of failure of the reservoir to meet the demand. Since this risk is rather high in March, another alternative from the computed ones is considered as a potential solution to the problem. In this alternative the reservoir capacity is increased from the present 2.08 x 10^6 m^3 to 3.41 x 10^6 m^3 by raising the dam by 1 meter. The release rule is given in Figure 14.

Figure 15 again shows the probability for reservoir releases.

The computed results show the reservoir to be more often at a higher level than in the previous case, and Figure 15 reveals that for the raised dam the supply is about 100% safe, i.e. the risk of failure of the reservoir to meet the demand is almost zero.

3.4 Results of the Application of Queuing Theory to Ghana

As a result of the numerous computations carried out with the aid of queuing theory, a storage-yield frequency graph was obtained for the Brimsu reservoir (Fig. 16).

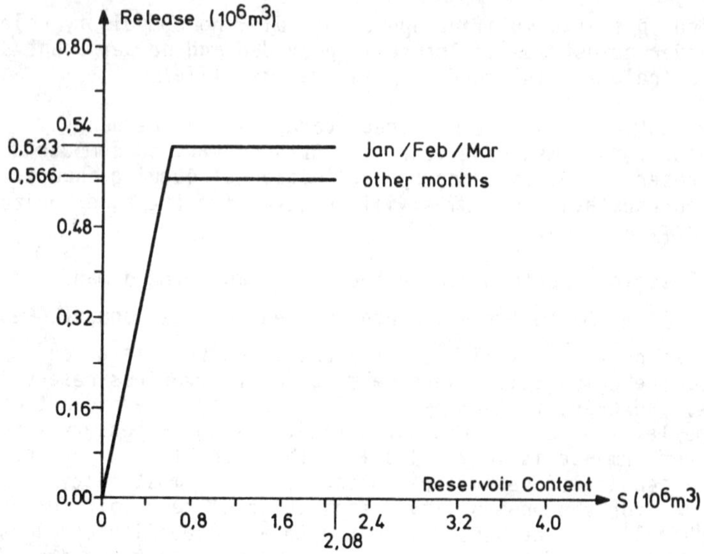

Figure 10. Reservoir Release Rule (Brimsu Reservoir Ghana)
Reservoir Capacity: 2.08 x 10^6 m^3; Demand of Year: 1980

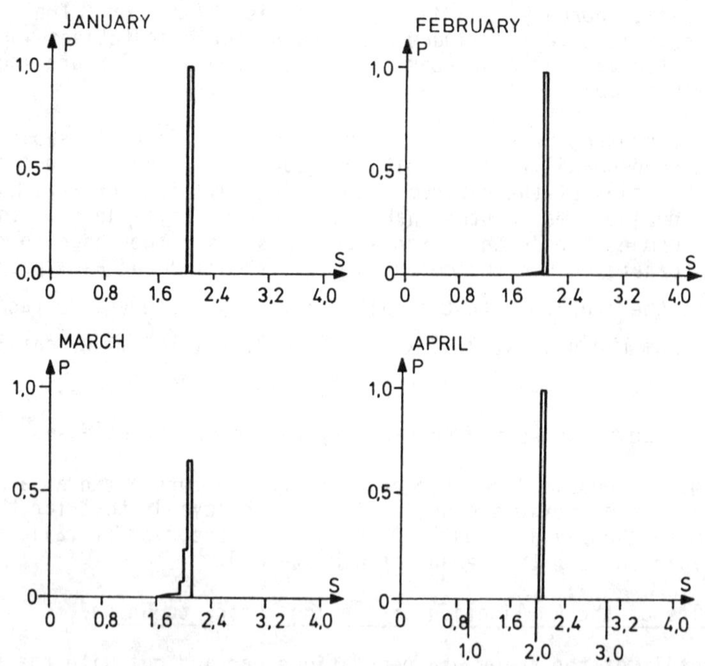

Figure 11. Pdf for Monthly Reservoir Content (Brimsu Reservoir)
Demand of Year 1980

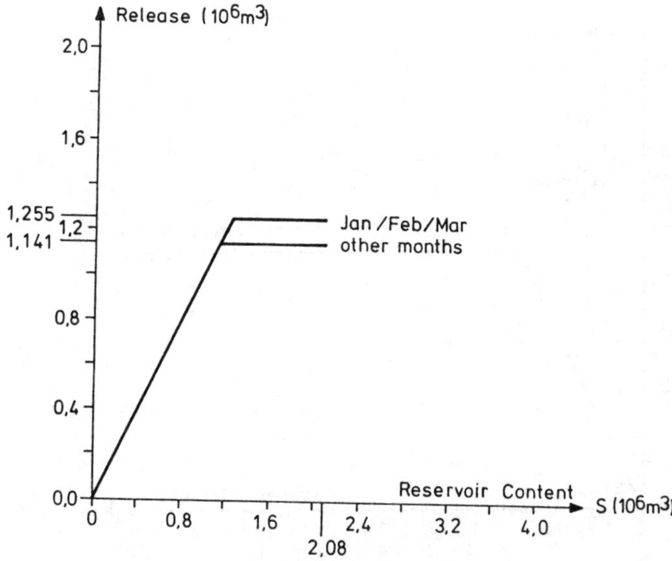

Figure 12. Reservoir Release Rule (Brimsu Reservoir)
Reservoir Capacity: 2.08 x 10^6 m^3; Demand for Year: 2000

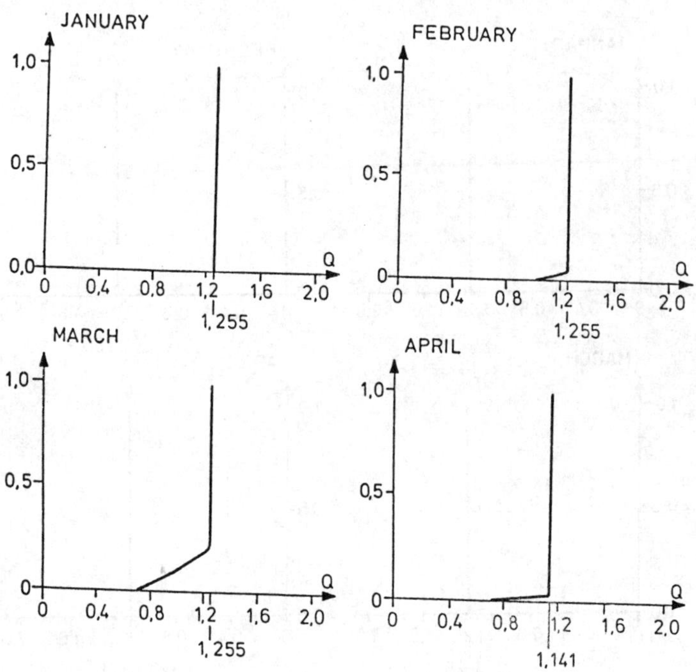

Figure 13. Pdf of Monthly Reservoir Releases (Brimsu Reservoir)
Demand of Year 2000

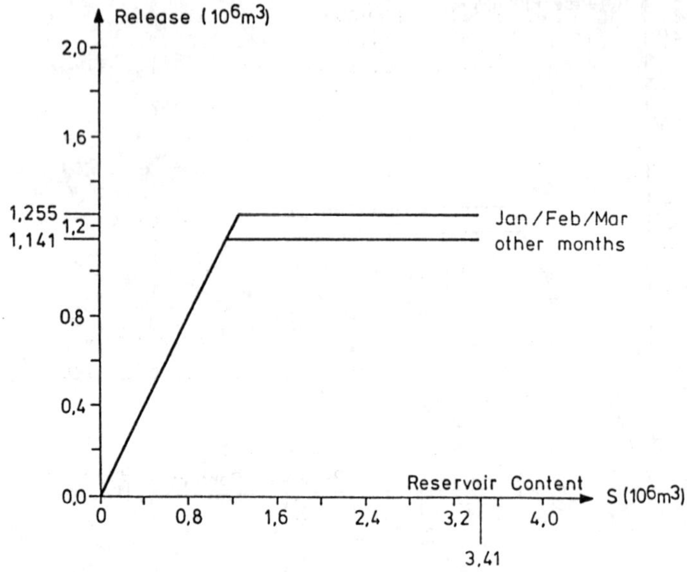

Figure 14. Reservoir Release Rule
Reservoir Capacity: 3.41 x 10^6 m^3; Demand for Year: 2000

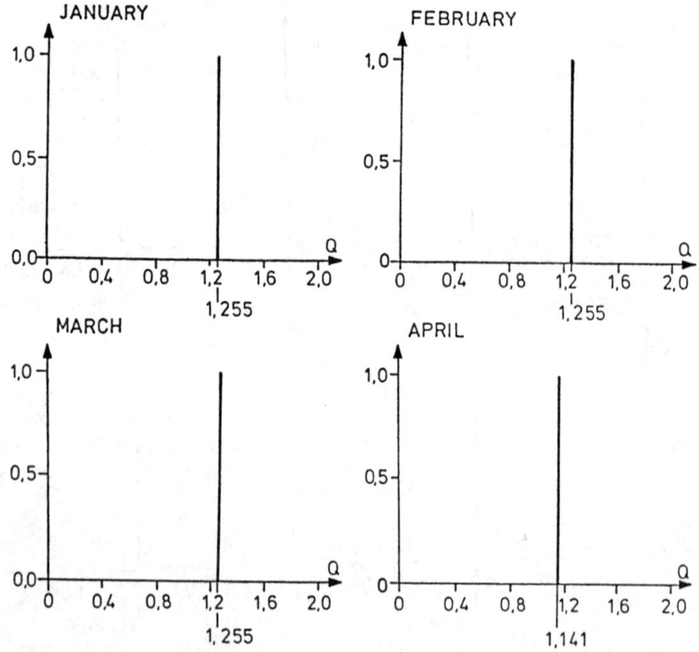

Figure 15. Pdf of Monthly Reservoir Releases (Brimsu Reservoir)
Demand of Year 2000. Dam Raised by 1 m.

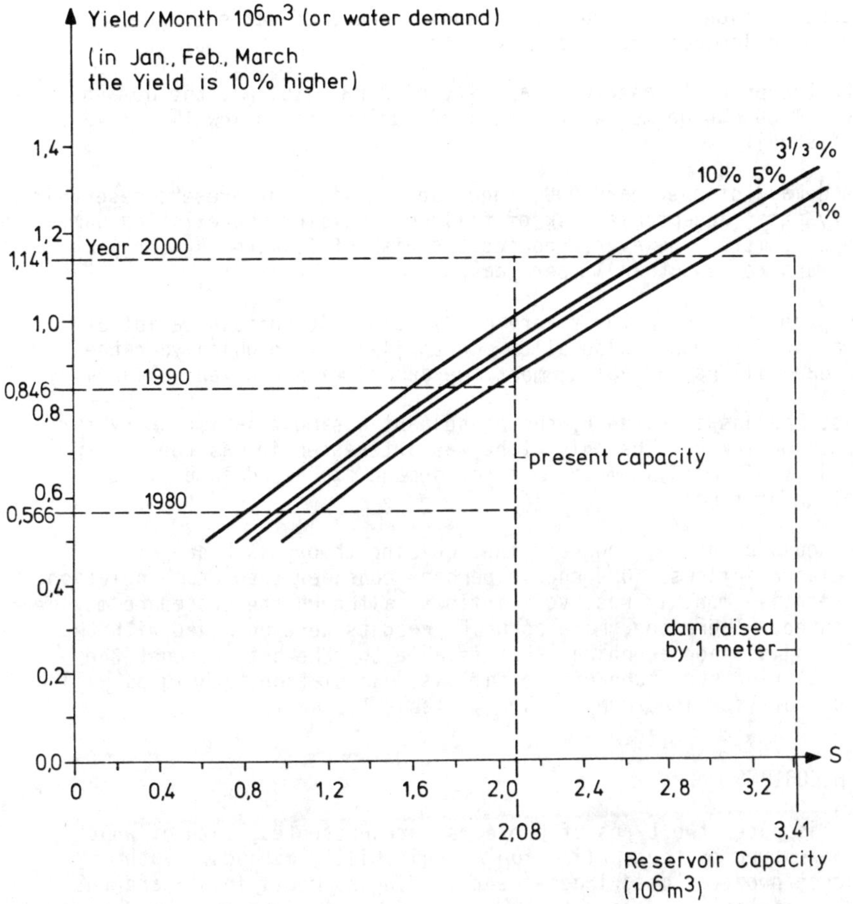

Figure 16. Preliminary Storage-Yield-Frequency Relationship Brimsu Reservoir, Ghana

The x-axis represents the variable reservoir capacity and the y-axis the yield, i.e the water withdrawn from the reservoir in order to meet the demand. The storage-yield relationship was computed for various probabilities of failure of the reservoir to meet the demand. The 1% curve indicates the risk of failure once in a hundred years (or 1% in any year); the 3-1/3% curve a risk of failure once in 30 years (or 3-1/3% in any year); the 5% line a risk of once in 20 years (or 5% in any year); and the 10% line a risk of once in 10 years (or 10% in any year).

Interpretation of the curves given in Figure 16 leads to the following preliminary results:

With the present reservoir capacity of 2.08 x 10^6 m^3, the demand of the year 1980 can be met with a risk of failure far below 1% per year (Line "1980").

The demand of the year 2000 cannot be met with the present reservoir capacity at an acceptable risk of failure. Raising the existing dam at Brimsu by 1 meter, however, reduces the risk of failure for the year 2000 demand to less than 1% per year.

The graph of Figure 16 is very handy, since it contains a lot of information. It would also allow one to identify in which year the raised dam will not be able to meet the demand with a given reliability.

Thus, the answer given by the consultant was most satisfactory for the decision maker. The only PI he was interested in was the reliability of the system to meet the demand expressed in terms of percentage of time.

It should be noted, however, that queuing theory is based on uncorrelated inflows. One should perhaps consider the autocorrelation of consecutive monthly reservoir inflows, although the procedure may be complicated. Therefore, more accurate results were obtained with the aid of a simulation technique that is able to take into account the autocorrelation structure of the inflows. Simulation techniques will be the topic of Technique/Project No. 4 (Table 1).

4. CONCLUSIONS

In this paper two types of projects were presented, both of which could be solved by the application of reliability methods. Both techniques/projects used independent loading as model input, and the hydraulic structure exposed to the loading was in both cases a dam which was to be raised.

In the second project (Ghana), the decision maker was content with a reliability related PI as basis for his decisions. In the first project (Ennepe), however, the decision maker was not prepared to decide on the basis of a reliability related PI alone. He requested a FM for which he wanted to compare the present values of expected values of benefits with construction costs (plus operation, maintenance, replacement). The computation of the benefits was based on the complete pdf's of damage costs which in turn were based on the pdf's of flood peaks.

A general evaluation of the five techniques/projects presented by the author will be given at the end of the companion paper (Schultz, 1986).

REFERENCES

Duckstein, L.; E. Plate, and M. Benedini (1984) Incidents and Failures in Reservoirs and Dams: A Mathematic System Framework. AGU Fall 1984 Meeting, San Francisco, December.

Fiering, M.B. (1961) Queuing Theory and Simulation in Reservoir Design. Journal of the Hydraulics Division, American Society of Civil Engineers, November.

Langbein, W.B. (1958) Queuing Theory and Water Storage. Journal of the Hydraulics Division, American Society of Civil Engineers, Vol. 84.

Lloyd, E.H. (1967) Stochastic Reservoir Theory. Advances in Hydroscience, vol. 4 (Academic Press, New York, London.

Moran, P.A.P. (1959) The Theory of Storage. Methuen Monographs (Methuen, London, Wiley, New York).

Plate, E.J. (1982) Design Flood and Hydrological Risk for Spillways of Reservoirs and Flood Protection Basins (in German). Wasserwirtschaft, vol. 72, no. 3., March.

Plate, E.J. (1984) Reliability Analysis in Hydraulic Design. Proceedings, IVth International Symposium on Stochastic Hyraulics, IAHR, Urbana.

Rudolph, K.U. and G.A. Schultz (1985) Project Evaluation for Present Conditions and Dam Rise of the Ennepe Reservoir (In German). Unpublished Report to the Ennepe Water Authority, Bochum.

Schultz, G.A. (1986) Reliability in Multipurpose Reservoir Operation Case Studied with Correlated Inflows. In Engineering Reliability and Risk in Water Resources, L. Duckstein and E. Plate, eds. Martinus Nijhoff Publishers BV, Dordrecht, The Netherlands.

Schultz, G.A. and A. Dafinis (1977) Hydrological and Water Resources Computations for the Cape Coast Water Supply Project. Unpublished Report to the Ghana Water and Sewerage Corporation, Karlsruhe.

Schultz, G.A. (1973, 1976) Water Resources Techniques For Reservoir Design (in German and Hungarian). Report no. 2, Institute of Hydrology and Water Resources, University of Karlsruhe, F.R. Germany. Published in 1973, revised in 1976.

Wallis, J.R. (1985) Scientific Procedures Applied to Planning, Design and Management of Water Resources Systems. IAHS Publication no. 147, Blackwell Scientific Publications, Oxford, UK.

THE RETURN PERIOD OF A RESERVOIR SYSTEM FAILURE

Richard M. Vogel

Department of Civil Engineering, Tufts University, Medford, MA 02155

ABSTRACT

It has become standard practice in the design of hydraulic structures to employ the average return period of a flood as the design event. An analogous index is introduced for the design of a water supply system: the average return period of a reservoir system failure, defined as the expected number of years until the occurrence of the first reservoir system failure. Other indices of the reliability of a water supply system are introduced. The use of these reliability indices is discussed in the context of a water supply reservoir design application.

1. INTRODUCTION

The storage-yield relation is the traditional tool used by water resource engineers to determine the required capacity of a storage reservoir to maintain a prespecified reservoir release. Although stochastic streamflow models have been available for a few decades, the most common approach in practice has been to base estimates of the required capacity of a storage reservoir upon application of the sequent peak algorithm (Thomas and Burden, 1963) or Rippl's mass curve (Rippl, 1883) to the historical streamflow record. While this approach is still advocated in recent textbooks on the subject (Clark et al., 1977; Steel and McGhee, 1979), it ignores the reliability associated with the resulting reservoir design capacity. Other textbooks (Linsley and Franzini, 1979; Linsley et al., 1982) discuss the application of stochastic streamflow models in conjunction with the sequent peak algorithm to generate the cumulative distribution function (cdf) of required reservoir storage, S, corresponding to a fixed planning period of length N years.

The cdf of S describes the relationship between the required storage capacity to meet a stated yield and the probability of failure-free reservoir operation, p, over an N-year planning period. Thus p is a measure of the reliability with which a reservoir of size S will provide failure-free operation over an N-year planning period. Duckstein et al. (1986) define p in a more general context where p is simply one performance index (PI^4) among a set of ten possible indices. Similarly Plate and Duckstein (1986) term p the project reliability for hydraulic design applications.

Engineers are often asked to convert statements of reliability over an N-year planning period to equivalent statements of annual reliability, R_a, or vice versa. Relationships between the annual reliability and the reliability over an N-year planning period associated with the design of flood control structures were developed by Thomas (1948) and further analyzed by Gumbel (1955) and Yen (1970) for independent events. These relations are in widespread use as evidenced by their inclusion in many textbooks on hydrology (Chow, pg. 8-34, 1964; Haan, pp. 70-75, 1977; and Linsley et al., pp. 349-350, 1982). This study develops analogous relationships between the annual reservoir reliability, R_a, and the probability of failure-free reservoir operations, p, over an N-year planning period.

Although use of stochastic streamflow models in conjunction with the sequent peak algorithm to estimate the cdf of S has been advocated by many authors since Fiering (1963), none of these investigators have evaluated which quantile, S_p, to choose in a design application. Similarly, when the sequent peak algorithm is applied to a single n-year historic streamflow record one is not sure which value of p, the probability of n=N year failure-free reservoir operation, to assign to the resulting estimate of the required storage capacity S. Here the length of the available historic streamflow sequence, n, is distinguished from the planning period or economic life of the proposed structure, N.

This study introduces a new index, T, the average return period of a reservoir system failure, analogous to the average return period of a flood event. Duckstein et al. (1986) also define the average return period, T, in a more general context where again T is simply one performance index (PI^5) among a set of ten possible indices. The relationships between T, p and N developed here are instrumental in choosing an appropriate quantile S_p for design purposes. Other indices are developed and their utility in choosing an appropriate design storage capacity is also considered.

2. RELATIONSHIP BETWEEN THE ANNUAL RELIABILITY OF A RESERVOIR SYSTEM AND THE PROBABILITY OF FAILURE-FREE OPERATION OVER AN N-YEAR PLANNING PERIOD

In this section, storage reservoir behavior is modeled, following Stedinger et al. (1983), concentrating upon both failure and regular (non-failure) years. In a given year a storage reservoir may be in either one of two states: (1) failure or (2) regular operation. Here a "failure" year is considered one in which the stated yield could not be met and a "regular" year is one in which the stated yield is provided or exceeded. While this analysis includes both the year in which the first reservoir system failure occurs as well as the duration of the failure, the actual magnitudes of the failures are ignored. For a more complete discussion of reliability measures associated with the frequency of failure years, failure durations and magnitudes of failures see Klemes (1979), Klemes et al. (1981) and Hashimoto et al. (1982).

Let the row vector $\underline{X}_y = (x_{1y}, x_{2y})$ specify the probability that a reservoir system is in either: (1) the failure state or (2) the regular (non-failure) state in year y. Also assume as did Stedinger et al. (1983), that $\underline{X}_y$, y = 1, ..., N forms a Markov chain with probability transition matrix

$$A = \begin{bmatrix} 1 - r & r \\ f & 1 - f \end{bmatrix} \tag{1}$$

where f is the probability that a failure year follows a regular year and r is the probability that a regular year follows a failure year. The two-state Markov model becomes

$$\underline{X}_{y+1} = \underline{X}_y A \tag{2}$$

Figure 1 depicts the two-state Markov model. The Markov assumption introduces memory into the process although the transition of the reservoir system in any given year y is only influenced by the reservoir system state in the preceding year y-1. More complex models are possible and have been considered by Klemes (1967, 1969). The simple model introduced here provides an approximation to the behavior of reservoir storage state transitions, and, most importantly, this model formulation leads to the simple and useful reliability indices developed in the following sections.

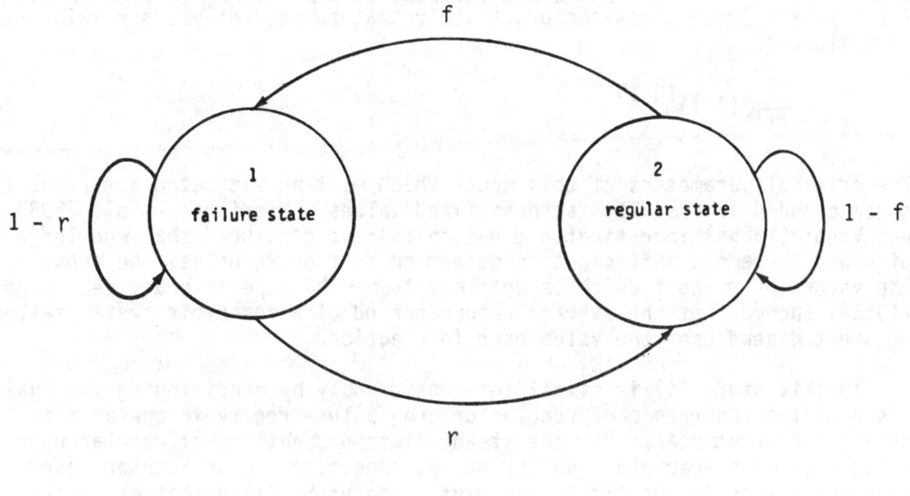

Figure 1 - The Two-State Markov Model of Reservoir System States

As y increases, $\underset{\rightarrow}{X_y}$ reaches a steady-state, and the solution to (2) becomes

$$\lim_{y \to \infty} \underset{\rightarrow}{X_y} = \left(\frac{f}{r + f}, \frac{r}{r + f} \right) \tag{3}$$

for $0 \leq |rf| \leq 1$. A derivation of this result may be found in Jackson (1975). Thus in the long run, the probabilities that the reservoir system will be in the failure or regular state are f/(r+f) and r/(r+f) respectively, regardless of the initial state of the reservoir system. Hence r/(r+f) is the steady-state probability of regular operation or the annual reliability R_a.

When the sequent peak algorithm is employed to determine the smallest storage reservoir design capacity required to assure regular or failure-free operation over an N-year planning period with probability p, then p becomes a steady-state probability. This is due to the fact that the sequent peak algorithm, as advocated by Thomas and Burden (1963), wraps the streamflow record around which generates the steady-state solution to the problem posed.

Now suppose we choose a reservoir capacity equal to S_p (the pth quantile of the distribution of S) using the sequent peak algorithm. Then the steady-state probability of regular (failure-free) operation over an N-year planning period, p, equals the probability of normal operation the first year r/(r+f) times the probability that subsequent years remain free of failures:

$$p = \frac{r}{r+f} (1-f)^{(N-1)} \tag{4}$$

The critical parameters of this model which must be estimated are r and f since p and N are usually taken as fixed values. Stedinger et al. (1983) and Vogel (1985a) investigated equation (4) and concluded that knowledge of p and N are not sufficient to determine r, f or R_a unless one knows the values of r and f which is unlikely to be the case in practice. Vogel (1985a) showed that the average return period of a reservoir system failure does not depend upon the value of r in practice.

In this study (4) is simplified considerably by conditioning the analysis upon the occurrence of regular or non-failure reservoir operations during the first year. Now the steady-state probability of regular operation over an N-year planning period, p, conditioned upon regular reservoir operations in the first year equals the probability that all years, subsequent to the first year, remain free of reservoir system failures:

$$p = (1-f)^{(N-1)} \tag{5}$$

Now due to this innovation p no longer depends upon r and (5) may be solved directly for f without resorting to a numerical algorithm as is required in the solution of (4).

$$f = 1 - p^{(1/(N-1))} \tag{6}$$

Furthermore f is completely specified by our knowledge of p and N and does not require assumptions regarding the value of r.

3. THE AVERAGE RETURN PERIOD OF A RESERVOIR SYSTEM FAILURE

In association with flood studies dealing with sequences of peak annual streamflows, Gumbel (1941) and Thomas (1948) defined the return period as the interval between flood events, where a flood event is defined as an annual peak flow above some threshold. Alternatively the return period may be thought of as the number of years until the occurrence of the first flood event. Since Thomas (1948), the meaning of the return period has changed. For example, Haan (1977, pg. 3) defines the return period as the average elapsed time between occurrences of a flood event. Thus, the return period was initially defined as the random time to an event, yet its meaning has changed to become the expected value of that random variable. This study distinguishes between these two definitions by using the terms return period and <u>average</u> return period.

Drawing an analogy to the average return period of a flood, the average return period of a reservoir system failure may be defined as the expectation of the return period, which is the number of years before the occurrence of the first reservoir system failure.

Let Z be the year in which the first reservoir system failure occurs. Then the steady-state probability of the first failure occurring in the Zth year, conditioned upon regular reservoir operation in the first year, equals the probability of Z-2 years of regular operation followed by a failure year. The probability mass function (pmf) of the time to the first failure is

$$P[Z = z] = \begin{cases} 0 & \text{if } z = 1 \\ f(1-f)^{z-2} & \text{if } z \geq 2 \end{cases} \tag{7}$$

Then the average return period of a reservoir system failure is

$$T = \mu_Z = \frac{1 + f}{f} \tag{8}$$

which may be combined with (6) to obtain

$$T = \frac{2 - p^{(1/(N-1))}}{1 - p^{(1/(N-1))}} \quad (9)$$

Similarly one obtains from (7) the variance of Z simply as

$$\sigma_z^2 = \frac{1 - f}{f^2} \quad (10)$$

Equations (7) through (10) provide a measure of the likelihood of future reservoir system failures. In particular, equation (9) is a simple relationship between T, p and N which provides a useful tool for determining which quantile, S_p, to choose in a design application. Equation (9) is the counterpart to the well known relationship between the average return period of a flood discharge, the planning period N, and the non-exceedence probability of that flood discharge documented in most textbooks on hydrology (see for example Chow 1964, Figure 9-61 and Linsley et al., 1982, Table 11-7). The simplicity of equations (7) through (10) results in large part from conditioning the entire analysis upon regular (or non-failure) reservoir system operation during the first year. If an unconditional approach is employed as was the case in the development of (4), equations (7) through (10) become much more complex because the expressions for P[Z = z], T and σ^2 include both r and f in addition to p and N, as is shown in Vogel (1985a). The equations developed here are recommended over those in Vogel (1985a) since the conditional approach makes physical sense and the resulting expressions are much simpler.

4. OTHER INDICES OF RESERVOIR SYSTEM PERFORMANCE

The average return period of a reservoir system failure, T, is simply the average number of years prior to the first reservoir system failure. Perhaps a more reasonable statistic would be to report the "likely recurrence interval" which is defined here to be that interval of time over which reservoir system failures are likely to occur, say 90% of the time.

The qth percentile of the distribution of the year in which the first reservoir system failure occurs, Z_q, may be obtained by choosing the largest value of Z_q such that

$$\sum_{z=2}^{Z_q} f(1 - f)^{z-2} \leq q \quad (11)$$

which may be solved for Z_q as follows

$$Z_q = \frac{\ln(1-q)}{\ln(1-f)} + 1 \quad (12)$$

where f is uniquely determined from (6) given values of p and N. Again equation (12) yields a very simple expression for a percentile of the distribution of return periods due in large part to having conditioned the entire analysis upon regular (or non-failure) reservoir system operation during the first year. Without resorting to this conditional analysis, the resulting expression for Z_q depends upon both r and f in addition to p and N and must be solved using a numerical algorithm as shown in Vogel (1985a).

5. SUMMARY

In the design of hydraulic structures it has become standard practice to employ the average return period of a flood discharge as the design event. This study developed an analogous index for the design of a water supply system: the average return period of a reservoir system failure. The resulting expressions are simplified dramatically by conditioning the entire analysis upon regular (or non-failure) reservoir operations during the first year. Percentiles of the distribution of the return period of reservoir system failures or simply the average return period of a reservoir system failure are readily estimated from the simple expressions developed here. These expressions are of particular value for the following reasons:

(1) The return period concept is a widely accepted index of reliability in the field of water resources engineering.

(2) The reliability indices developed here are simple to understand and easy to apply.

(3) Use of these indices provides a measure of the likelihood of future reservoir system failures which until now was unavailable in such a simple form. For an example of the use of the reliability indices developed here see Vogel (1985a, 1985b).

In a recent national assessment of our nation's water resources, the Water Resources Council (1978) concluded that 17 of the nation's 21 water resource regions have or will have a serious problem of inadequate surface-water supply by the year 2000. As increasingly marginal surface-water supply sites are pressed into service, target yields at both existing and proposed sites can only increase. In many instances, increased demands are being met by more efficient management and utilization of existing reservoir systems rather than by construction of new facilities (for an example of this recent phenomenon see Sheer and Flynn, 1983). Whether new facilities are envisaged or the existing reservoir system is to be operated more efficiently, the storage-reliability-yield relationship is a fundamental ingredient. The use of stochastic streamflow models in conjunction with the sequent peak algorithm may be used to develop the storage-reliability-yield relationship. The reliability indices developed here may then be employed to develop explicit statements regarding the likelihood of future reservoir system failures.

Past and recent research has identified weaknesses and potential problems with the traditional techniques for estimating the storage-reliability-yield relationship. Fiering (1967) documents important shortcomings associated with the strict use of the historic streamflow record. Stochastic streamflow models were developed to circumvent the shortcomings of the use of the historical required storage alone. Recent research indicates that stochastic streamflow models can be used to significantly improve the precision of estimates of the storage-reliability-yield relationship in comparison to the traditional approach of employing the historical streamflow record alone (Vogel, 1985a; Vogel and Stedinger, 1986) As is to be expected, the precision of estimates of the storage-reliability-yield relationship depends primarily upon the length of the available historic streamflow record. The precision of these estimates may be improved by employing streamflow record augmentation and/or extension procedures (Vogel and Stedinger, 1985). However, even for relatively long records, Vogel (1985a) and Vogel and Stedinger (1986) document substantial sampling variability associated with estimates of S_p. Given the short streamflow records available in most practical situations it has become evident that one should incorporate streamflow model parameter uncertainty into reservoir design and operations studies to obtain an honest account of the true likelihood of reservoir system failures (see Stedinger and Taylor, 1982 and Stedinger et al. 1985).

REFERENCES

Chow, V.T., ed., (1964) Handbook of Applied Hydrology, McGraw-Hill Book Co. Inc., New York.

Clark, J.W. , W. Viessman and M.J. Hammer, (1977) Water Supply and Pollution Control, Third Edition, Harper & Row Publishers, New York, 60-62.

Duckstein, L., E. J. Plate, and M. Benedini, (1986) Water Engineering Reliability and Risk: A System Framework, In: Engineering Reliability and Risk in Water Resources, L. Duckstein and E. Plate, eds., Martinus Nijhoff Publishers BV, Dordrecht, The Netherlands.

Fiering, M.B., (1963) Statistical Analysis of the Reservoir Storage-Yield Relations, In: Operations Research in Water Quality Management, H.A. Thomas, Jr., and R.P. Burden, eds., Harvard Water Resources Group, Cambridge, MA.

Fiering, M.B., (1967) Streamflow Synthesis, Harvard University Press, Cambridge, MA 1967.

Gumbel, E.J., (1941) The Return Period of Flood Flows, Ann Math. Statistics, Vol. 12, No. 2, 163-190.

Gumbel, E.J., (1955) The Calculated Risk in Flood Control, Applied Science Research, The Hague, Vol. 5A, 272-280.

Haan, C.T., (1977) Statistical Methods in Hydrology, The Iowa State University Press, Ames, Iowa.

Hashimoto, T., J.R. Stedinger and D.P. Loucks, (1982) Robustness of Water Resources Systems, Water Resour. Res., Vol. 18, No. 1.

Jackson, B.B., (1975) Markov Mixture Models for Drought Lengths, Water Resour. Res., Vol. 11, No. 1, 64-74.

Klemes, V., (1967) Reliability of Water Supply Performed by Means of a Storage Reservoir Within a Limited Period of Time, Jour. of Hydrol., Vol. 5, 70-92.

Klemes, V., (1969) Reliability Estimates for a Storage Reservoir with Seasonal Input, Jour. of Hydrol., Vol. 7, 198-216.

Klemes, V., (1979) Applied Storage Theory in Evolution, Advan. Hydrosci., Vol. 12, 79-141.

Klemes, V., R. Srikanthan and T.A. McMahon, (1981) Long-Memory Flow Models in Reservoir Analysis: What is their Practical Value?, Water Resour. Res., Vol. 17, No. 3, 737-751.

Linsley, R.K. and J.B. Franzini, (1979) Water Resources Engineering, McGraw-Hill Book Co., New York.

Linsley, R.K., M.A. Kohler and J.H.L. Paulhus, (1982) Hydrology for Engineers, McGraw-Hill Book Co., New York.

Plate, E.J. and L. Duckstein, (1986) Reliability in Hydraulic Design, In: Engineering Reliability and Risk in Water Resources, L. Duckstein and E. Plate, eds., Martinus Nijhoff Publishers BV, Dordrecht, The Netherlands.

Rippl, W., (1883) The Capacity of Storage-Reservoirs for Water-Supply, Minutes Proc. Inst. Civil Eng., Vol. LXXI, 270-278.

Ross, S., (1976) A First Course in Probability, MacMillan Publishing Co., Inc., New York.

Sheer, D.P. and K. Flynn, (1983) Water Supply, Civil Engineering, ASCE, Vol. 53, No. 6, 50-53, June.

Stedinger, J.R. and M.R. Taylor, (1982) Synthetic Streamflow Generation 2: Effect of Parameter Uncertainty, Water Resour. Res., Vol. 18, No. 4, 919-924.

Stedinger, J.R., B.F. Sule and D. Pei, (1983) Multiple Reservoir System Screening Models, Water Resour. Res., Vol. 19, No. 6, 1383-1393.

Stedinger, J.R., D. Pei and T.A. Cohn, (1985) A Condensed Disaggregation Model for Incorporating Parameter Uncertainty into Monthly Reservoir Simulations, Water Resour. Res., Vol. 21, No. 5, 665-675.

Steel, E.W. and T.J. McGhee, (1979) Water Supply and Sewage, McGraw-Hill Book Co., New York, 44-47.

Thomas, H.A., (1948) Frequency of Minor Floods, Journal of the Boston Society of Civil Engineers, Vol. 35, No. 1, 425-442.

Thomas, H.A., Jr., and R.P. Burden, (1963) Operations Research in Water Quality Management, Harvard Water Resources Group, Cambridge, MA, 1-17.

Vogel, R.M., (1985a) The Variability of Reservoir Storage Estimates, Ph.D. Dissertation, Cornell University, January.

Vogel, R.M., (1985b) Reliability Indices For Use in the Design of Water Water Supply Reservoirs, Jour. of the Water Resources Planning and Management Div., ASCE, submitted for review, February.

Vogel, R.M. and J.R. Stedinger, (1985) Minimum Variance Streamflow Record Record Augmentation Procedures, Water Resour. Res., Vol. 21, No. 5, 715-723.

Vogel, R.M. and J.R. Stedinger, (1986) The Value of Stochastic Stream-Streamflow Models in Reservoir Design Applications, Water Resour. Res., manuscript under preparation.

Water Resources Council, (1978) The Nations Water Resources, 1975-2000, Vol. 1, Summary, December.

Yen, B.C., (1970) Risks of Hydrologic Design of Engineering Projects, Jour. of the Hydraul. Div., ASCE, Vol. 96, No. (HY4), 959-966.

RELIABILITY IN MULTIPURPOSE RESERVOIR OPERATION: CASE STUDIES WITH CORRELATED INFLOWS

Gert A. Schultz
Institute of Hydrology and Water Resources
Ruhr University Bochum, P.O. Box 102148
4630 Bochum, F.R. Germany

ABSTRACT

Three techniques for the determination of reservoir system operating rules are discussed, all of which use reliability related performance indices (PI^s) or figures of merit (FM^s) as decision criteria. Two of the techniques were used in the context of consulting work for a river authority.

The methods applied for finding adequate reservoir operating rules are optimization and simulation. While in the first case (Wupper river system) optimization is used such that an objective function for two non-commensurate objectives is optimized, in the second application (Dhünn river system) simulation is used in order to obtain operating rules which fulfill reliability requirements according to specified PI^s for two purposes. In the first cast reliability PI^s are not primary (but secondary) decision criteria since the objective function does not contain reliability related parameters. In the second situation no optimum operating rule was obtained since the decision was made on the basis of reliability PI^s alone and no explicit optimization technique was applied.

The third technique discussed combines multi-objective optimization (compromise programming) with various PI^s, three of which are reliability related. In this way an optimum rule is determined such that all PI^s are combined to form an optimum compromise FM.

1. INTRODUCTION

In an accompanying paper by the author (Schultz, 1986), two techniques (queuing theory and a combination of flood frequency analysis with economic FM^s) were discussed. Both used independent loading as model input and both were concerned with reservoir design (raising of an existing dam) in combination with operation.

This paper presents two additional techniques using dependent loading which are applied to problems of reservoir operation. In the first case study (Technique/Project No. 3, Table 1) the task was the determination of the maximum possible performance of a reservoir system (Wupper system, Germany). The method applied was a dual-objective optimization technique.

Table 1. Specifications of Presented Techniques, Projects, Loading and Performance Indices

Technique/ Project No.	Lecture No.	Methodology	Project Purpose	Hydraulic Structure	Performance Indices/ Figures of Merit	System: Loading	System: Resistance	Project Region	Contractor
1	I (Independent loading)	Flood frequency analysis combined with benefit/cost ratio estimation	Flood protection	Dam (to be raised)	Economic benefits	independent: flood volume	Reservoir flood storage	Ennepe River catchment Germany	Ennepe Water Authority
2		Queuing theory	Water supply in the year 2000	Dam (to be raised)	Water supply reliability	independent: water demand	Water supply capacity of reservoir	Cape Coast Ghana West-Africa	GWSC, Ghana
3	II (Dependent Loading)	Optimization (Dynamic programming)	Multipurpose: Hydropower & low flow augmentation	New dam	Bi-objective penalty function (to be minimized)	dependent: power demand & flow deficits	Water supply capacity of reservoir	Wupper River catchment Germany	Wupper Water Authority
4		Simulation	Multipurpose: drinking water supply and low flow augmentation	New dam	Target-related reliability for: - water supply - low flow augmentation	dependent: water demand & flow deficits	dto	Dhünn River catchment Germany	Wupper Water Authority
5		Multiobjective optimization (compromise programming)	Low flow augmentation (operation)	Existing reservoir system	Lp-metric (of compromise programming), (to be minimized)	dependent: flow deficits	dto	as example: Wupper reservoir system	--- (Research)

The second case study (Technique/Project No. 4, Table 1) deals with the reservoir system in the Dhünn river catchment where the problem was to find an operating rule so as to satisfy two objectives. Here a simulation technique was applied.

2. DETERMINATION OF OPTIMUM PERFORMANCE OF THE WUPPER RESERVOIR SYSTEM

The Wupper water authority is responsible for the management of water quantity and water quality in the Wupper basin. A reservoir system has been constructed and is shown in Figure 1.

The multi-unit, multi-purpose reservoir system serves several purposes which include drinking water supply, low flow augmentation, hydropower, flood protection and recreation. The Wupper dam, which is presently under construction, is located in series with almost all other reservoirs and serves mainly the two purposes of hydropower generation and low flow augmentation with the aim of improving the rather low water quality in the downstream river reach.

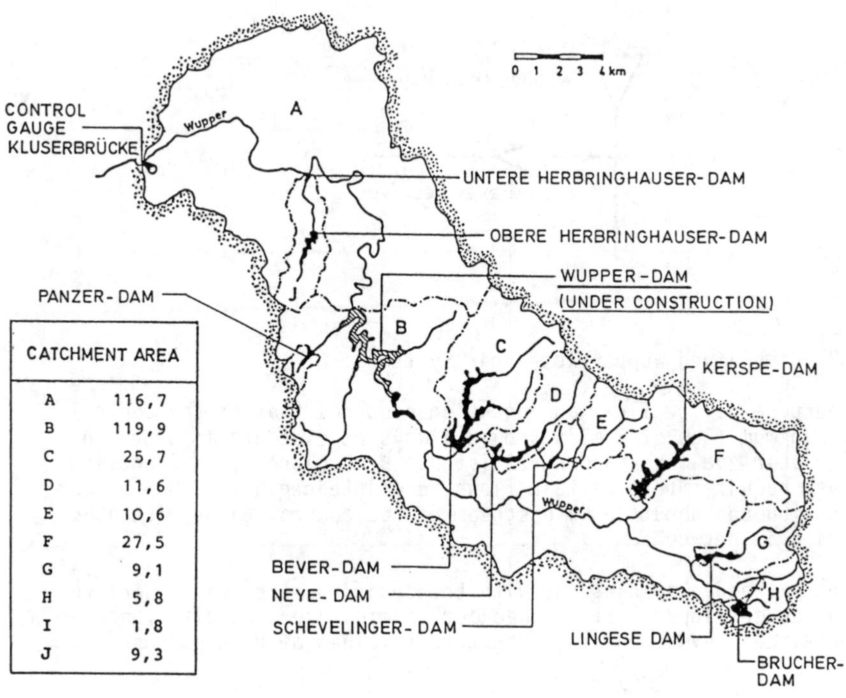

Figure 1. Wupper River Reservoir System

The water authority asked the consultant to determine the maximum possible performance of the Wupper reservoir (in conjunction with the other upstream reservoirs) for those two purposes. The first problem to be solved was the question of how such a "maximum possible performance" could be specified by a PI (performance index) or a FM (figure of merit), since the decision maker (water authority) himself did not have very clear ideas. It was decided that an objective function incorporating both (non-commensurate) objectives should be formulated and optimized on the basis of historical hydrological data available for 32 years.

2.1 Formulation of the Optimization Problem

The system for which the maximum possible performance is to be found is given in Figure 2.

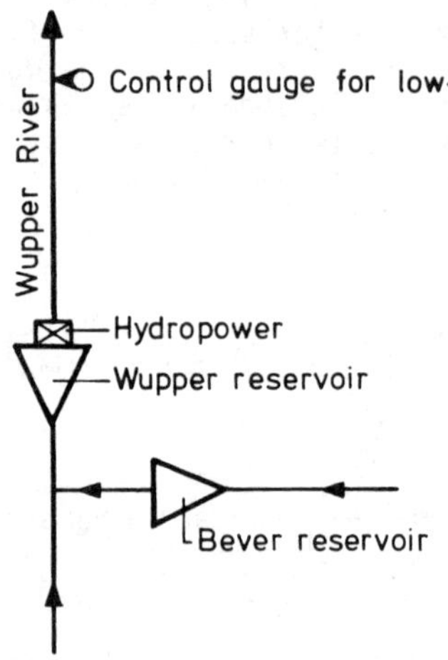

Figure 2. Simplified Wupper Reservoir System

The purpose of low flow augmentation requires that at the control gauge a minimum flow of 5 m^3/s should always be exceeded in order to maintain water quality standard II (stable β-mesoaprobic). Producing hydropower requires that certain flows be maintained at the Wupper dam which is situated about 20 km upstream of the control gauge with other tributaries in between.

During extensive discussions with the decision maker, an objective function was developed. It was decided to minimize a penalty function which penalizes deviations from the target values of both purposes:

$$\min Z = \sum_n e^{pe_n} \tag{1}$$

where:

Z = objective function to be minimized

pe = penalty points

n = number of time increments

The "penalty points" are dependent on the deficit discharges as compared to the target values of low flows as well as hydropower. Table 2 gives the relationship between the flow conditions achieved by reservoir operation and the corresponding penalty points as well as the resulting values of the objective function increments.

Table 2. Penalty Points and Penalty Function Values Depending on Low Flow (Deficits) and Turbine Discharge

Low flow augmentation low flows at control gauge, augmented by reservoir operation [m^3/s]	Hydropower Min. discharge through turbines at Wupper dam [m^3/s]	Penalty points pe [-]	Value of penalty function increment e^{pe} [-]
≥ 5.0	3.0	0	1.00
≥ 5.0	2.0	1	2.72
4.0	2.0	2	7.39
4.0	1.0	3	10.09
3.0	1.0	4	54.60
2.0	1.0	5	148.41
1.0	1.0	∞	∞

Table 2 can be interpreted as follows:

- If the flow through the turbines is Q_1 = 3m^3/s and the simultaneous discharge at the control gauge is not less than Q_2 = 5 m^3/s, the situation is satisfactory: no penalty points.
- If Q_1 = 1 m^3/s and also Q_2 = 1 m^3/s, the situation is not acceptable: ∞ penalty points.
- All combinations of Q_1 and Q_2 between these extremes are possible but not desirable and will be penalized with linearly increasing penalty points. Since these enter Equation 1 exponentially, the values of the objective function increase in a highly non-linear manner with decreasing flows.

This procedure of allocating penalty points to undesirable flows made it possible to combine two non-commensurate objectives into one objective function. In this way a multi-objective optimization problem has been solved. The technique used was Dynamic Programming, which allows one to deal with non-linear objective functions (Nemhauser, 1966; Hillier and Liebermann, 1967). Several constraints were formulated concerning reservoir contents, flood storages, minimum flows for ecological reasons, etc.

It was decided to use monthly values as time increments. Furthermore, it is necessary to know all reservoir inflows in advance in order to carry out the optimization procedure time step by time step. In this way the optimization was carried out on the basis of historical hydrological data, i.e. for n = 384 months (32 years). Since in real time reservoir operation it is not possible to know future reservoir inflows in advance, the optimization results represent the absolute maximum system performance which can never be achieved in real time. Equation 1 now reads:

$$\min Z = \sum_{n=1}^{384} e^{pe_n} \tag{1a}$$

and Belman's recursive equation:

$$f_n(S_n) = \min_{E_n}\{r_n(E_n) + f_{n-1}(S_{n-1})\} \tag{2}$$

where:

$f_n(S_n)$ = optimum value of the objective function for optimal decision sequence from time step n to 1

n = index for time step (counting backwards in time)

S_n = reservoir content at beginning of time step n

$r_n(E_n)$ = the term of the objective function value (e^{pe_n}) resulting from decision E_n at time n. See Table 2!

E_n = reservoir release at time step n.

Equation 2 is solved recursively beginning with n = 1 (last time increment) to n = 384, thus yielding the optimum result.

2.2 Optimization Results

The result of the optimization procedure consists of

a) the optimum value of the objective function (minimum of Z), and

b) the sequence of optimal decisions, i.e. the sequence of reservoir releases from which the optimum hydropower output and optimum low flow condition can be deduced.

According to the prespecified objective function, the optimum performance of the system (on the basis of historical data) is known and the problem is solved. It should be mentioned that by using this optimization procedure a FM was selected (implicitly) which is neither economical nor reliability related, since a physical objective function was chosen. Although the decision maker had developed the objective function in dialog with the consultant, he wanted more information on "what happened" after he had seen the results. This was the point where reliability came into perspective. Table 3 shows the optimization results along with information on how often desirable and undesirable situations occurred.

Table 3. Optimization Results, Wupper Reservoir System Operation

Decision		Penalty points	Penalty function	Optimization Results: Requirements fulfilled	
Low flow $[m^3/s]$	Hydropower $[m^3/s]$	pe	e^{pe}	no. of months in 32 years	% of time
5	3	0	1.0	268	69.8
5	2	1	2.7	25	6.5
4	2	2	7.4	75	19.5
4	1	3	20.1	5	1.3
3	1	4	54.6	11	2.9
2	1	5	148.4	0	0
1	1	∞	∞	0	0
			Sum:	384	100.0

Objective function*: $\sum_{n=1}^{384} e^{pe_n} = 1591.11$

*min possible absolute value: $\sum_{n=1}^{384} e^{0} = 384$

It is important to note that desirable results (0 penalty points) occurred about 70% of the time. The most undesirable result ($Q_1 = 0$; $Q_2 = 0$) did not occur during the historical 32 years. For all other relevant combinations of Q_1 and Q_2, the PI^s (i.e. probability of exceedance--percentage of time when this or a more desirable situation occurred) can be inferred from Table 3, thus giving reliability related PI^s for the optimum solution. Eventually the decision maker was satisfied with that solution (Schultz et al., 1983a).

The question may arise as to why the decision maker wanted to know the maximum possible performance of the system. The answer is easy: The decision maker (the Wupper water authority) was urged by the provincial government to provide water of standard II quality. This meant provision always of $Q \geq 5\ m^3/s$. From the optimization study the decision maker could prove that this requirement cannot be met, even under the (unrealistic) optimal conditions!

Looking back at the decision process, the following procedure was used:

a) Decision maker agrees to a priori specified (first) objective function (without reliability considerations).

b) Results are presented and decision maker disagrees on the basis of a posteriori shown reliability related PI^s.

c) Decision maker improves objective function.

d) New results are presented. Decision maker now agrees on the basis of reliability related PI^s.

This shows a new and interesting decision procedure incorporating a feedback between a priori and a posteriori decisions, which is not contained in most publications on multiobjective decision analysis (Goicoechea et al., 1982).

3. PERFORMANCE ANALYSIS OF THE DHÜNN RESERVOIR SYSTEM

The second case study deals with the application of a simulation technique in order to develop operating rules for the reservoirs in the Dhünn river system, Germany. The specifications are given in Table 1 under Technique/Project No. 4.

The water authority responsible for the Dhünn river catchment management asked the consultant to develop operating rules which meet the various purposes of the reservoirs in an adequate way for the multi-unit, multi-purpose reservoir system. The purposes are drinking water supply, low flow augmentation, flood control, and mandatory releases for fish and wildlife. The Dhünn reservoir system is shown in Figure 3.

Here again the question arises: what is meant by "operation rules which meet the various purposes of the reservoir system in an adequate way"? It would be possible to develop an objective function to be optimized analogous to the previous case study. This would be difficult, however, since optimization usually yields only an optimum sequence of releases if all (future) inflows are known, but not an operation rule (except for special cases like LDR (linear decision rule) or other simplifying assumptions). Due to this fact and the wish of the

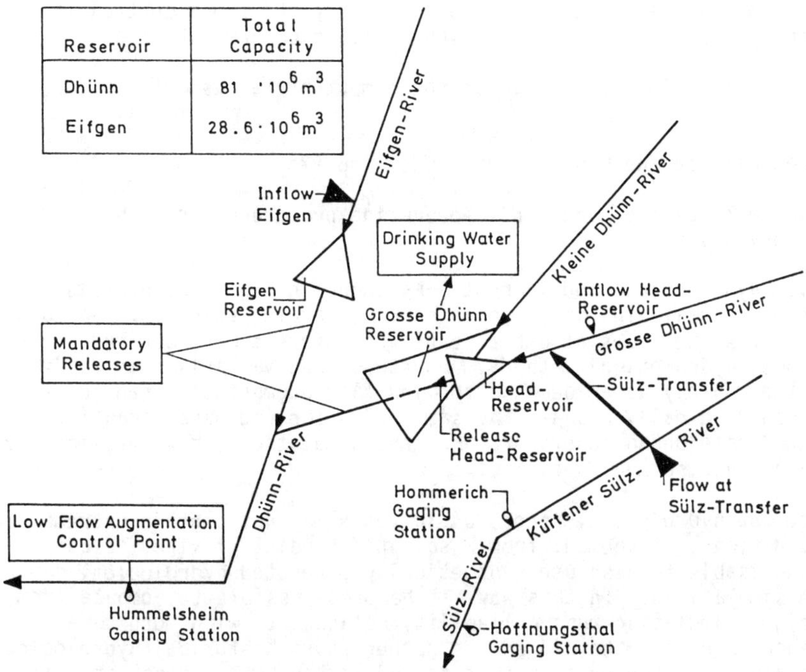

Figure 3. Dhünn River System

decision maker (water authority) to specify the PI^S or FM^S as reliability criteria alone, it appeared most appropriate to choose simulation as the method to be used.

In cooperation with the decision maker it was agreed that only two reservoirs (Dhünn and Eifgen dams) and only three purposes (drinking water supply, low flow augmentation and mandatory releases) had to be considered. Flood control was taken into account by constraints which keep a specified flood storage vacant in each month of the year.

3.1 Simulations of the Reservoir System's Operation

In simulation models (Hufschmidt and Fiering, 1966) the operating rules are given, but they can be changed parametrically until an acceptable solution is found. The simulation model is based on the reservoir mass balance equation for each reservoir (including evaporation losses), an operating rule as described later, a particular historic or synthetic hydrological record, and a set of water requirements for each possible low flow augmentation target (calculated on a daily basis). Priority is given to drinking water supply, then to low flow augmentation through the Dhünn reservoir, then to low flow augmentation through the Eifgen reservoir, and last to the mandatory releases and to the necessary releases to keep the flood control storage space empty (or spills in months with no flood control storage).

Before simulation can be run on the computer, two aspects have to be clarified:

- what will be the hydrological model input?
- what will be the conjunctive operating procedure for both reservoirs?

3.1.1 Hydrological model input. As shown in Figure 3, hydrological records were available for 30 years on four gauging stations. Based on these records the flows at other points of interest were derived using traditional hydrological methods. Due to strong variations in daily flows, the monthly water demands for low flow augmentation had to be determined on a daily basis. The same holds for the water transfers into the Grosse-Dhünn reservoir. For the simulations, however, monthly time increments seemed sufficient.

Since the hydrological record of 30 years may not contain extreme low flow conditions, it seemed, from a scientific point of view, to be highly advisable to also use synthetically generated hydrological data for the simulations. In this way, it becomes possible to compute from the various simulation runs reliability related PIs with much more confidence than with the aid of the rather short historical hydrological records alone. Although this fact is undisputed among scientists, it is always difficult to convince the decision maker of the necessity to use synthetic data for two reasons:

- he cannot recognize the advantage of using such abstract things as "synthetic data,"
- he is not prepared to pay considerably more money just for making a reliability related PI a bit more reliable.

In the case of the Dhünn reservoir project, the consultant succeeded in convincing the decision maker of this necessity. Therefore, in addition to the historical record, 10 cross-correlated reservoir inflow time series were used as input to the simulation model.

Although the simulation runs are based on monthly time increments, it was necessary to determine the low flow deficits on a daily basis. Therefore, a model had to be used which was also able to do this for generation of the cross-correlated data time series (inflow to the two reservoirs).

The method developed by Treiber (1975) was applied, in which one gauging station (Hoffnungsthal) was used as a pilot station. Other flows were determined using a regression analysis and adding some noise. Figure 4 shows as an example the results of the first synthetic record at the gauging station at Hummelsheim. The mean monthly values are similar to the observed ones, and the seasonal variation is identical. Other statistical parameters were reproduced with acceptable accuracy.

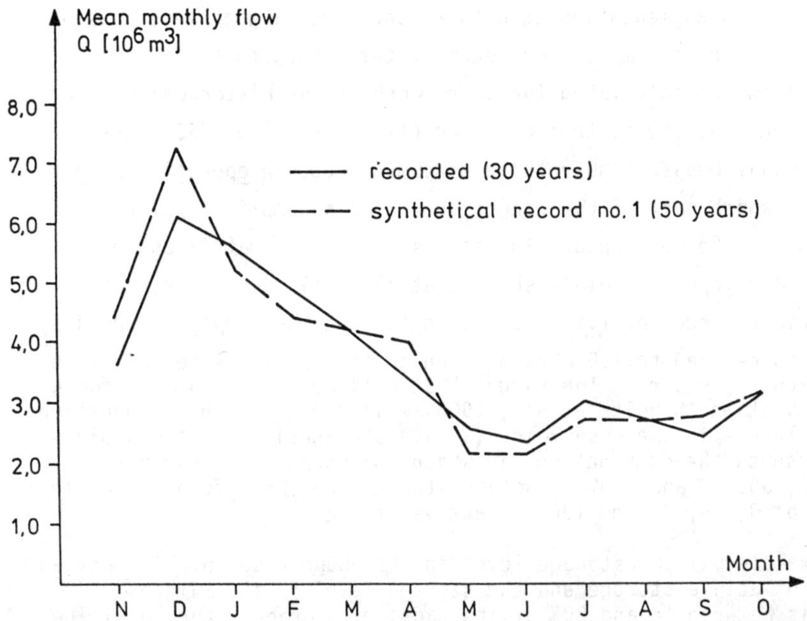

Figure 4. Comparison between Recorded and Synthetic Mean Monthly Flows at Hummelsheim Control Gauge

3.1.2. Reservoir operating procedure. The several objectives of the two reservoirs have to be accomplished in specified order. For instance, flood control is satisfied by keeping a certain capacity of each reservoir empty during flooding periods. This might require additional releases from each reservoir. Drinking water was supplied only from the Grosse-Dhünn reservoir directly, i.e. the water is pumped from the reservoir and is not included in the release. Low flow augmentation has to be provided at the downstream gauging station (Hummelsheim). The sum of the necessary releases from the reservoirs is obtained for each low flow augmentation target on a daily basis and then added to a monthly requirement. Small mandatory releases of 100 l/s (litre per second) have to be made from each reservoir. If the release requirement for low flow augmentation or for flood protection is larger than the mandatory release, no additional release is necessary. The water transfers from the Kürtener-Sülz River are such that at least a flow of 0.345 m^3/s is left in the river and a maximum of 25 m^3/s is transferred (capacity of the tunnel). The flood control storage for the Dhünn reservoir varies between 0 and 8.5 x 10^6 m^3 and for the Eifgen reservoir from 0 to 2.3 x 10^6 m^3. The drinking water supply is constant over the year and amounts to 42 x 10^6 m^3/a. Dead storages are 2 x 10^6 m^3 in the Dhünn reservoir and 0.5 x 10^6 m^3 in the Eifgen reservoir.

Three low flow augmentation targets at the control gage were chosen, namely 1.8, 1.5 and 1.0 m^3/s. For each of these targets the necessary additional flow was calculated for each month in the historical record (30 years long) and the synthetic record (11 records, each 50 years long) on a daily basis. Under these circumstances, an operating rule which states which part of the target should be accomplished by each reservoir has to be developed. For each simulation a matrix as in Table 4 was developed. Table 4 shows that for full reservoirs, both together have a target of 1.8 m^3/s and that for almost empty reservoirs, the target is reduced to 1.0 m^3/s or even to the mandatory release (100 l/s from each reservoir). The target that will be chosen in each month for the simulation (Schultz et al., 1983) will depend on the combination of storage levels in the reservoirs (active storages). As an example, Table 4 presents the combinations in which the Dhünn reservoir has limits of 0, 50, 90 and 100% of active storage and the Eifgen reservoir has limits of 0, 10, 80 and 100% of active storage.

As an example, if the storage level in the Dhünn reservoir is between 50 and 90% of active storage and the storage level in the Eifgen reservoir is between 10 and 80% of its active storage, the total low flow augmentation target will be 1.5 m^3/s (0.9 m^3/s for Dhünn and 0.6 m^3/s for Eifgen reservoir, respectively).

If natural flow between the reservoirs and the low flow augmentation control point is larger than 1.5 m^3/s, the target will be met with no additional release from the reservoirs.

3.1.3. Simulation results. During the simulations with a particular operating rule, the failures to meet drinking water supply, mandatory releases and/or low flow augmentation targets are recorded. At the end of each simulation the reliability of attaining each of these purposes is calculated. For the particular operating rule of Table 4, the results of all simulations are included in Table 5.

As can be seen, the reliabilities for mandatory releases and drinking water supply are very high (almost 100% in all cases). The reliability of a low flow augmentation target of 1.0 m^3/s is also very high (100% with historical record and 98.3% on the average with synthetic records). The higher targets have the lower reliabilities, i.e about 70% for 1.5 m^3/s and 13% for 1.8 m^3/s with synthetic records. Figures 5 to 7 show for this operating procedure the simulated reservoir contents of Dhünn (Figure 5) and Eifgen reservoirs (Figure 6), and of the augmented low flows in the Dhünn river (Figure 7) for a critical period in the historical data series.

Several other operating rules were simulated which differ mainly by the percentages of reservoir contents when the low flow augmentation targets were changed.

Table 4. Low Flow Augmentation Targets at the Hummelsheim Gaging Station as a Function of Reservoir Storage Levels

Storage Level(1) in Dhünn-Reservoir	Storage Level(1) in Eifgen-Reservoir 0%(6)	0%-10%	10%-80%	80%-100%(7)
0%(5)	D:(2) 0.1(4) E:(3) 0.1(4)	D: 0.1 E: 0.9	D: 0.1 E: 0.9	D: 0.1 E: 1.4
0%-50%	D: 0.9 E: 0.1	D: 0.7 E: 0.3	D: 0.9 E: 0.6	D: 0.9 E: 0.6
50%-90%	D: 1.4 E: 0.1	D: 1.4 E: 0.1	D: 0.9 E: 0.6	D: 1.4 E: 0.4
90%-100%(7)	D: 1.7 E: 0.1	D: 1.7 E: 0.1	D: 1.7 E: 0.1	D: 1.4 E: 0.4

(1) As percentage of active storage

(2) Low flow augmentation target for Dhünn Reservoir in m^3/s

(3) Low flow augmentation target for Eifgen Reservoir in m^3/s

(4) Only mandatory release of 0.1 m^3/s

(5) 0% is equivalent to dead storage (2×10^6 m^3) + drinking-water supply reservation (8×10^6 m^3) for Dhünn Reservoir

(6) 0% is equivalent to dead storage (0.1×10^6 m^3) and mandatory release reservation (2.9×10^6 m^3) for Eifgen Reservoir

(7) 100% is equivalent to capacity minus flood control reservation (variable).

3.1.4 Change of Priorities. While optimization models are usually quite rigid after being implemented, simulation models are very flexible. This feature could be used when a move started to change the priorities of the various purposes of the Dhünn reservoir system.

Table 5. Results of Simulations with Historical and Synthetic Records and Operating Rule of Table 4

Hydrological Record		Reliability According to Duration in % for:					
		Mandatory Releases		Drinking-Water Supply	Low-Flow Target at Control Gage[m^3/s]		
		Dhünn-Reservoir	Eifgen-Reservoir		1.0	1.5	1.8
Historical		100	100	100	100	74.4	17.5
Synthetic	1	100	100	100	99.2	81.2	17.8
"	2	99.5	100	99.5	98.2	67.2	12.7
"	3	99.3	100	99.5	97.7	71.8	9.0
"	4	100	100	100	98.8	67.5	12.8
"	5	100	100	100	98.5	66.2	9.0
"	6	98.5	100	98.5	96.0	66.0	13.8
"	7	99.7	100	99.5	98.2	50.5	2.3
"	8	98.8	100	98.8	96.7	68.2	12.2
"	9	100	100	100	100	70.8	11.0
"	10	99.3	100	99.3	98.2	76.6	22.2
"	11	100	100	100	99.5	80.2	23.5
Mean value of 11 simulations with synthetic records		99.6	100	99.6	98.3	69.7	13.3

While in the previous simulations the supply of 42×10^6 m^3/a drinking water to various cities had highest priority, later a discussion started as to whether:

- so much drinking water would really be needed in the future
- it would be optimal for ecological reasons to give the low flow augmentation of 1.8 m^3/s highest priority, because this way the river could be grouped into water quality standard II.

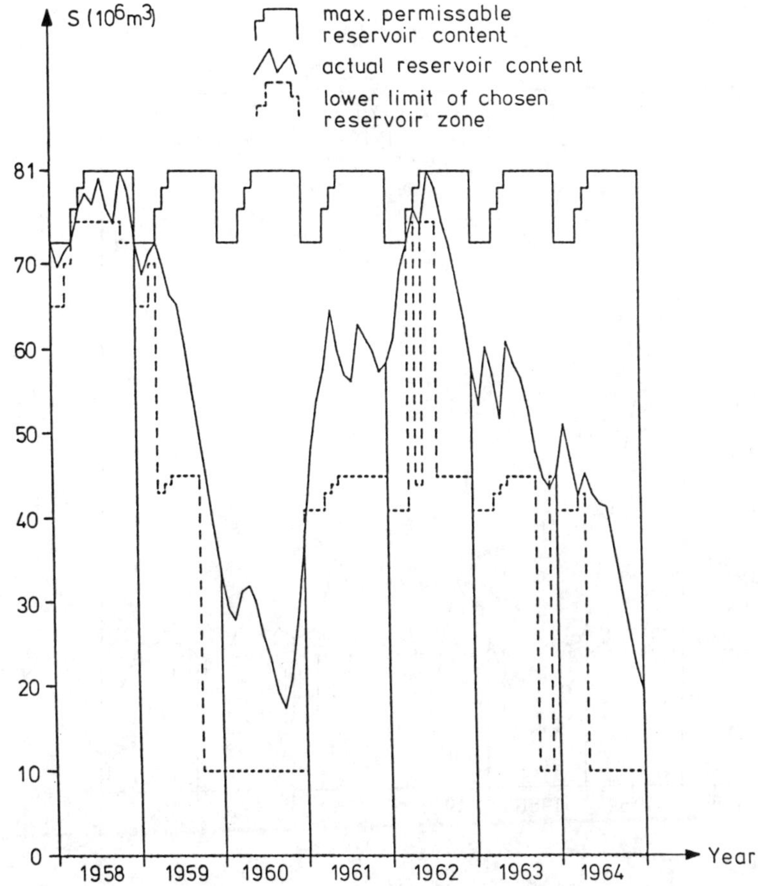

Figure 5. Reservoir Contents, Dhünn Reservoir (Simulated)

Under these new priorities more simulation runs were carried out with the same model. The results showed that it is possible to achieve the low flow target (1.8 m^3/s) with almost 100% reliability (~100% of the months) but the drinking water supply had to be reduced considerably from 42 x 10^6 m^3 to 31.5 x 10^6 m^3 (if still almost 100% reliability was required).

The results (Schultz et al., 1983b) were used as a basis for negotiations between the cities (representing the drinking water demand), the government agencies (representing the ecological demands), and the water authority (representing commercial interests). A compromise seems possible since obviously the future drinking water demand is less than 42 x 10^6 m^3 and the ecological pressure groups are

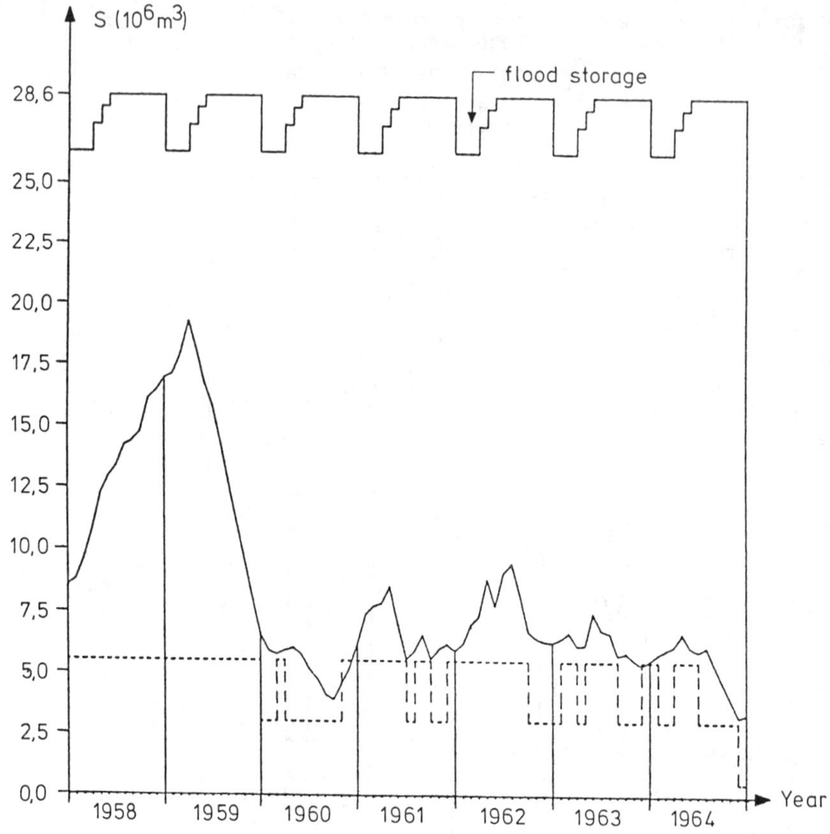

Figure 6. Reservoir Contents, Eifgen Dam (Simulated

getting more powerful, thus pushing improvement of water quality which means increased low flows.

The fact that the sudden change of priorities of purposes the water resources system has to serve could so easily be handled with the aid of the simulation technique again proves the high flexiblity of this approach.

4. OPTIMUM CHOICE AMONG VARIOUS PI^s AND FM^s

The last Technique/Project to be discussed is that of No. 5 in Table 1. Here the emphasis is put on the problem of how to make an optimum decision, if for each of the various possible project alternatives a whole set of PI^s and/or FM^s is offered to the decision maker. According to Duckstein et al. (1984), 10 PI^s and at least 2 FM^s are "on the market." These are only the reliability related ones, besides which there are several others not being reliability related.

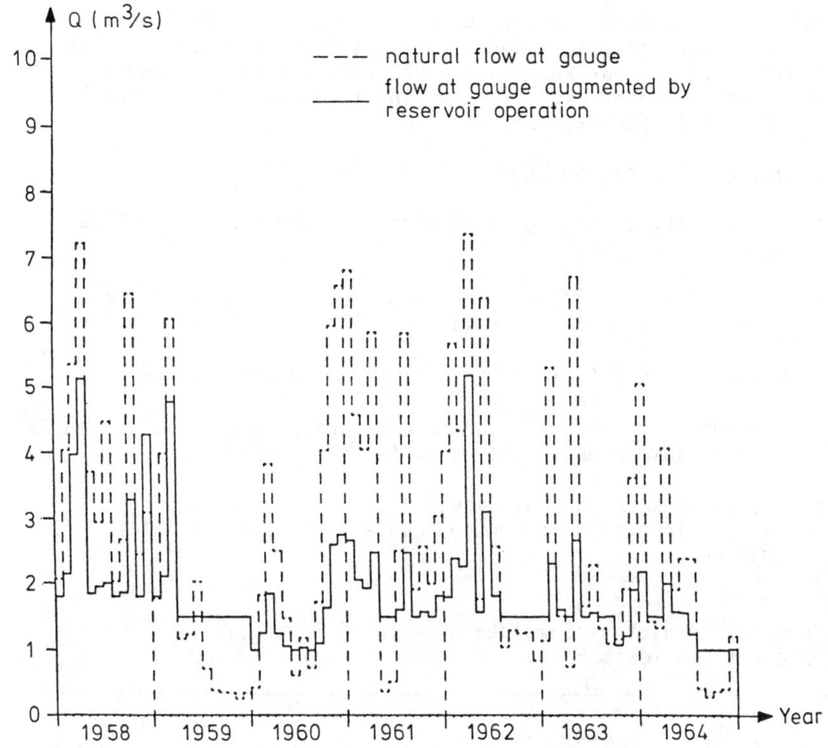

Figure 7. Discharges at Hummelsheim Gauge (Simulated)

If, for example, a decision maker is confronted with 5 project alternatives and (only) 5 PI^s, then he has to make his decision on the basis of 25 numbers, some of which are better, others worse for each project alternative. This information is more confusing than helpful and will certainly not be favourable for the acceptance of reliability theory in practice.

A solution would be to find a way to combine the PI^s in an optimum way which then would lead to an optimum decision, i.e. an optimum problem solution. If each PI is considered as an objective which has to be minimized (e.g. risk) or maximized (e.g. reliability), the problem would be amenable to the application of a multi-objective optimization technique. An example will now be presented where precisely this approach was chosen in order to find the optimum reservoir operation strategy out of 5 alternatives using 5 different PI^s, 3 of which were reliability related.

As an example, the Wupper reservoir system is chosen again. In contrast to Technique/Project No. 3, here a real-time operating rule has to be selected out of five potential rules.

Before this can be done a great number of simulations (Harboe, 1983) of alternative reservoir system operations has to be carried out with synthetic data, for each of which the 5 PI^s were evaluated. Table 6 gives the average value of each PI resulting from simulations with 10 synthetic data series for each alternative.

The following PI^s were computed:

- PI_1: Reliability according to duration for the target low flow augmentation level
- PI_2: Reliability according to duration for the reduced target (with switching factor F with $0 \leq F \leq 1.0$)
- PI_3: Lowest monthly flow at the control gauge over all 50 years
- PI_4: Percentage of months in which Bever reservoir is sufficiently full in summer months to allow for recreation
- PI_5: One-sided squared deviations, i.e., the sum of squares of target minus flow at control gauge, whenever the latter was smaller.

Table 6. Average Results of Simulations with 10 Synthetic Records (50 Years Each)

	PI_1	PI_2	PI_3	PI_4	PI_5
Operation rule j = Alternative No.	Mean reliability according to duration for: Target* [%]	(Target) F [%]	Average of lowest flow at control gauge [m^3/s]	% of months storage of Bever-Res. is in upper interval (average)	Mean yearly one-sided squared deviation [$10^6 m^3$]^2
1	89.56	99.46	1.924	25.09	29.18
2	86.68	99.56	2.124	27.26	37.31
3	86.69	97.89	1.132	24.90	17.51
4	92.07	99.28	1.279	24.12	22.97
5	94.26	94.26	0.582	23.53	16.54

*Target for Wupper reservoir = 5 m^3/s at control gauge

PI_1, PI_2, and PI_4 are reliability related. PI_3 represents the lowest flow during 50 years at control gauge (to be compared to the target of $5m^3/s$) while PI_5 represents a measure of flow deficits over the whole period of time (50 years).

A decision maker would have great difficulty in finding the optimum alternative from Table 6 since for most alternatives some PI^s are better and others are worse than for other alternatives. The best alternative would be that which satisfies all PI^s to a maximum degree, where the various PI^s could be given different weights. In this context, the five PI^s can be considered as measures of different objectives. Thus, the problem is one of decision under multiple objectives. A possibility for evaluation of these 5 alternatives with 5 objectives is given by the compromise programming technique (Goicoechea et al., 1982), in which a measure of distance to ideal values of the objectives is minimized. The measure of distance is the so-called L_p-metric which can be defined for alternative j as follows

$$L_p(j) = \left[\sum_{i=1}^{5} \alpha_i^p \frac{|f_i^* f_{ij}|^p}{|M_i - m_i|^p} \right]^{1/p}$$

where:

$L_p(j)$ = L_p-metric for alternative j (to be minimized over j to find optimal alternative)

α_i = weighting factor for objective i

f_{ij} = value of objective i (performance parameter) attained by alternative j (Table 8)

M_i = $\max_j f_{ij}$

m_i = $\min_j f_{ij}$

f_i^* = preferred value of objective i (can be chosen equal to M_i or m_i accordingly)

p = parameter which reflects the attitude of the decision maker

The corresponding values are included in Table 7. The values of α_i were chosen subjectively (Harboe, 1983) according to the relative importance of the different objectives. For example, since objective 4 (recreation) incorporates only one reservoir, the corresponding weighting factor was set very low. The ideal value for objective 5 is the lowest of all, namely $f_5^* = m_5$. In all other objectives the highest value is the preferred one.

Table 7. Parameters for Evaluation of L_p-Metric

Objective i	Definition of Objective	α_i	M_i	m_i	f_i^*
PI_1	Mean security according to duration for target	1.0	94.26	86.68	M_1 = 94.26
PI_2	Mean security according to duration for (Target).F	0.6	99.56	94.26	M_2 = 99.56
PI_3	Average of smallest flow at control gage	1.5	2.124	0.582	M_3 = 2.124
PI_4	% of months Bever-Res. is in upper storage interval in September	1.0	27.26	23.53	M_4 = 27.16
PI_5	Mean yearly one-sided squared deviation from target	0.5	37.31	16.54	m_5 = 16.54

The values of L_p for each alternative were calculated for several values of p as presented in Table 8. Among the 5 alternatives, the one with minimum value of L_p is alternative 1 (j=1) for all values of p ($p = 1,\dots,6$ and $p \to \infty$). Certainly other results can be obtained for different values of the weighting factors α_i. Sensitivity analyses showed, nevertheless, that for other α_i's the first alternative was still preferred for most of the p-values.

By application of compromise programming it became possible to select alternative 1 as optimum considering all 5 PI^s, which was not possible on the basis of the (confusing) information of Table 6.

5. CONCLUSIONS

After presentation of five modern planning techniques along with real world project examples in the two papers of the author (this paper and Schultz, 1986, Table 1), the following conclusions can be drawn:

1. All five techniques are related to probability, although in very differing ways.

2. Some of the models contain probabilistic elements as inherent model components (e.g. queuing theory, flood frequency analysis); others introduce stochasticity as a model input (e.g. economic analysis, simulation); others transform the model output into probabilistic information (e.g. optimization, simulation).

Table 8. Evaluation of Results through Compromise Programming (L_p-Metric)

Critical Storages with P_μ [%]	Switching Factor F	Alter-native j	$L_p(j)$						
			p=1	p=2	p=3	p=4	p=5	p=6	$p\to\infty$
70	0.60	1	1.189	0.720	0.650	0.630	0.624	0.622	0.620
80	0.60	2	1.500	1.118	1.040	1.015	1.006	1.003	1.000
70	0.80	3	2.239	1.403	1.238	1.168	1.128	1.103	0.999
60	0.60	4	1.378	0.886	0.832	0.822	0.819	0.818	0.818
0	1.00	5	2.200	1.619	1.531	1.510	1.503	1.501	1.500

3. Often planning techniques can be optimized by combination of two methods:

   - Stochastic models (e.g. queuing) with simulation (Technique/Project No. 2, Table 1), where queuing theory is used as screening model.

   - Optimization followed by simulation (Technique/Project No. 3, Table 1), where simulation was not discussed. Optimization in such cases is used as screening model.

   - Simulation followed by optimization (Technique/Project No. 5), where optimization is used to determine a FM of the simulation results.

   - Probabilistic models (e.g. flood frequency analysis) with economic models (e.g. benefit/cost analysis). Example: Technique/Project No. 1, Table 1.

4. Most important is the adequate choice of performance indices (PI^s) and/or figures of merit (FM^s) since they strongly influence the project decisions. Although in all the five techniques/projects presented probability was involved, reliability related PI^s are by no means the only decision criteria. Others are physical (e.g. minimum flow values), economical (benefit/cost ratio) or any type of objective function to be optimized (e.g. minimization of sum of squares of deviations from a target).

5. In many cases (e.g. if hydrological records are short) reliability related PI^s can be made more reliable by use of synthetic data (Technique/Project No. 4).

6. At present, it seems that too many reliability related PI^s are "on the market", some of which are highly correlated. The

variety of PI^S is more confusing than helpful to the decision maker. Only the significant ones should survive, and they could be combined into one single FM as shown in Technique/Project No. 5.

7. The choice of the PI^S is the job of the decision maker himself. Scientists can guide him and try to convince him if he is reluctant to accept modern reliability concepts.

8. It is hoped that the necessary distinction can be evaluated between:

   - recommended and less favourable planning techniques in water resources

   - useful and useless applications of reliability concepts taken from structural engineering that incorporate probabilities of system loading and system resistance

   - significant and less significant PI^S and FM^S in order to help the decision maker not become confused.

REFERENCES

Duckstein, L., E. Plate, and M. Benedini (1984) Incidents and failure in reservoirs and dams: A mathematical system framework. AGU Fall 1984 Meeting, San Francisco, December.

Goicoechea, A., D.R. Hansen, and L. Duckstein (1982) Multiobjective Decision Analysis with Engineering and Business Applications, John Wiley & Sons, New York.

Harboe, R. (1983) Use of critical storages in operating rules for reservoir system simulation, Proceedings, XIXth IAHR Congress, Moscow.

Hillier, F. and G. Liebermann (1967) Introduction to Operation Research, Holden Day, San Francisco.

Hufschmidt, M.M. and M.B. Fiering (1966) Simulation Techniques for Design of Water-Resources Systems, Harvard University Press, Cambridge, Mass.

Nemhauser, G.L. (1966) Introduction to Dynamic Programming, John Wiley & Sons, New York.

Schultz, G.A. (1986) Application of models for reliability assessment in reservoir operation. In: Engineering Reliability and Risk in Water Resources, L. Duckstein and E. Plate, eds., Martinus Nijhoff, Publishers BV, Dordrecht, The Netherlands.

Schultz, G.A., R. Harboe, W. Böhle, and R. Vogt (1983a) Optimization of reservoir systems operation: Wupper and Bever dam with and without Dörpe dam (in German), Unpublished report to the Wupper Water Authority, Bochum.

Schultz, G.A., R. Harboe, R. Vogt, and K. Schneider (1983b) Reservoir system Grosse-Dhünn and Eifgen dam: Reservoir simulations using historical and synthetic inflow time series. Unpublished report to the Wupper Water Authority, Bochum.

Treiber, B. (1975) Ein stochastisches Modell zur Simulation von Tagesabflüssen, Mitteilungen, Institut für Hydrologie und Wasserwirtschaft, Universität Karlsruhe, Heft 5.

ENGINEERING RISK IN FLOOD STUDIES USING MULTIVARIATE PARTIAL DURATION SERIES

Francisco Nunes Correia

Laboratório Nacional de Engenharia Civil, Av. do Brasil, 101
1799 Lisboa, Portugal

ABSTRACT

Flood risk analysis plays a crucial role in the design of many hydraulic structures and water resources systems. Traditionally, annual flood peak discharge is considered in the design process and flood risk assessment is based on the analysis of the annual maximum series. An alternative approach to the annual maximum series is the partial duration series. This method has proved to be particularly suitable when short records of data area available. In this method not only the annual maxima are considered, but all the peaks over a given threshold are taken into account. This threshold may have a physical meaning, such as the discharge value for which the submersion of a structure occurs or a river bank is overtopped.

The use of a technique based on the discharge values over a threshold makes it very attractive for taking into account not only the peak discharge but also the flood volume and the flood duration in the risk assessment process. In this paper these techniques are presented and their merits are discussed. Two Portuguese river basins (Corgo River at Ermida and Tera Creek at Pavia) are used for illustration.

1. INTRODUCTION

Assessment of flood risk plays an important role in the design and operation of many hydraulic structures and water resources systems.

For many decades hydraulic practice has concentrated on fitting a statistical distribution to the series of annual maximum floods. Fisher-Tippet type I and Log-Pearson type III are classic examples of distribution functions commonly used in engineering practice. Five parameter Lambda distribution and Wakeby distribution are examples of recent developments along the same lines. This approach, based on the annual maxima series, provides a very limited assessment of flood risk. In fact, risk is strictly formulated as the probability that the annual maximum peak discharge exceeds a given design value. Questions such as for how long the design value is exceeded and how frequently it may occur every year are completely excluded from this classic formulation.

In more recent years Todorovic and Zelenhasic (1970), Todorovic and Rousselle (1971), Cunnane (1973), Taesombut and Yevjevich (1978) and

Correia (1983), among other authors, successfully tried a different approach based on partial duration series.

This approach takes into consideration not only the maximum annual peak discharge but also all the values above a given threshold. A bivariate stochastic model is considered to describe the number of floods per year and the maximum discharge associated with these floods. Influenced by the traditional approach, in which only annual maxima are considered, it is very common that the partial duration series method is used to estimate the distribution of the annual maximum discharge. However, there is a much larger potential to be explored in this method. The consideration of flood volume, flood duration and flood peak in the process of assessing flood risk, may be an important step forward in improving the design of many water resources systems.

The purposes of this paper are the following

(i) to review briefly the use of partial duration series to characterize flood peaks;

(ii) to extend this formulation to the study of flood volumes and flood durations, under certain assumptions;

(iii) to illustrate the proposed methodology with the study of flood risk in two Portuguese watersheds.

2. PARTIAL DURATION SERIES FOR FLOOD PEAK CHARACTERIZATION

In this approach, any period of time during which the discharge is above a given threshold is considered a flood event. The threshold, or truncation level, may be established with the purpose of analyzing some flood consequences such as the submersion of a road or the overtopping of a river bank. The truncation level may also be established with the purpose of including a given average number of flood events per year.

Once the truncation level is defined, the flood events can be analyzed in terms of the number of events per year, peak discharges, durations, and flood volumes associated with the events. As mentioned previously, most of the research reported in the literature has been based on the consideration of two random variables, namely the number of events per year and the peak discharges associated with these events.

Zelenhasic (1970) and Todorovic and Zelenhasic (1970) based their formulation on the results obtained by Todorovic (1970) on the statistical distribution of the extreme value of a random number of random variables. According to these authors, the distribution function of the maximum exceedance during a time interval (o,t) is given by

$$P_t(x) = \sum_{k=0}^{\alpha} \left\{ [H(x)]^k \cdot P(E_k^t) \right\} \qquad (1)$$

where H(x) is the distribution function of all exceedances in the interval (0,t) and $P(E_k^t)$ is the probability that k exceedances occur during this time interval. This expression is valid under two basic assumptions:

(i) successive exceedances ξ_1, ξ_2, ..., are mutually independent random variables with common distribution H(x), i.e.

$H(x) = P[\xi_p \leq x]$

(ii) the distribution of ξ_1, ξ_2, ..., is independent of the number of exceedances that may occur in the time interval (0,t).

For practical purposes, the problem is not if these assumptions are fully verified but whether some degree of violation is significant from the point of view of practical applications. With the purpose of guaranteeing the independence assumptions, it is common to select a truncation level such that the average number of exceedances per year is less than 4 or 5. It is also common to "filter" exceedances that are very close in time to avoid a clustering effect that may occur, mainly in small and torrential watersheds.

Zelenhasic (1970) and Todorovic and Zelenhasic (1970) derive the distribution function given by Eq. (1) for the time interval of one year, when the distribution of exceedances is a gamma distribution and the number of exceedances per year is a Poisson stochastic process. A special case of the gamma distribution, corresponding to the exponential distribution, is used by these authors, which leads to the following distribution function of the largest annual exceedance

$$P(x) = \exp(-\lambda\, e^{-\beta x}) \tag{2}$$

where λ is the parameter of the Poisson stochastic process corresponding to the average number of exceedances per year, and β is the parameter of the exponential distribution, corresponding to the inverse of the mean magnitude of the exceedances. This approach was applied to the Greenbrier River at Alderson and to the Susquehanna River at Wilkes-Barre, in the U.S.A., where 72 years of daily flows were available. Results showed a good agreement with data, and the partial duration series performed better than the classical use of the Gumbel distribution applied to the annual maxima series.

Todorovic and Rousselle (1971) extended the previous model for nonidentically distributed exceedances on the assumption that only those exceedances during a particular season may be considered identically distributed. This extension was successfully applied to the Greenbrier River at Alderson.

After this pioneering work, many other authors were attracted by the potential of this approach and compared results with the traditional methods based on the annual maxima series.

Cunnanne (1973) proved that under commonly used assumptions and for return periods greater than about 10 years, the partial duration series estimate of the annual maximum discharge for a given return period,

Q(t), has smaller sampling variance than the annual maxima series estimate if an average number of exceedances larger than 1.65 is considered.

NERC (1975) used the Poisson and the negative binomial distributions to model the number of annual exceedances and the exponential distribution to model the magnitude of exceedances. In this study, data from 26 hydrometric stations from all the United Kingdom were used to compare the use of partial duration series and annual maxima series. It was found that, in most cases, the Poisson distribution fits reasonably well the number of annual exceedances and the exponential distribution fits the distribution of the magnitude of exceedances. However, these distributions cannot be taken for granted as they were clearly rejected for some stations. As a final and somewhat prudent recommendation, NERC (1975) suggests that for small samples ($N \leq 10$) the approach based on partial duration should be used to estimate the mean annual maximum discharges. Taesombut and Yevjevich (1978) conducted a very extensive study based on data from 17 stations in the United States. The Poisson, the mixed Poisson, the Hyper-Poisson, the negative binominal, and the mixed geometric distributions were used to model the number of annual exceedances for an average number of exceedances from 1 to 5. The exponential, the gamma, the Pearson type III, the Weibull, the three-parameter lognormal, and the mixed exponential distributions were used to model the magnitude of exceedances. These authors concluded that in the range of truncation levels with an average number of exceedances from one to four, the independence assumptions underlying the use of partial duration series can be accepted. The mixed Poisson or the Poisson distributions, and the mixed exponential or the exponential distributions, gave the best results in modeling the annual number of exceedances and the magnitude of exceedances. The authors proved that the estimates of annual flood peaks of given return periods, Q(T), from the partial flood series have smaller sampling variances than the corresponding estimates from the annual flood series when the average number of exceedances per year is larger than 1.65. This result was confirmed with synthetic data generated by a model of daily flows. As a basic conclusion, Taesombut and Yevjevich (1978) state that "when the model of partial flood series is developed with assumptions for its derivation supported by data for low truncation levels, the partial flood series is more efficient or more useful in estimating annual flood peaks than the annual flood series, especially in the case of small sample sizes."

Correia (1983) applied the partial duration series method to daily flow records from 12 stations in Portugal representing the several hydrologic regions of the country. Truncation levels corresponding to an average number of annual exceedances from 0.5 to 4.0 were considered. Estimates of the annual peak discharges for return periods from 5 to 1,000 years were compared with results obtained from the Gumbel, Log-Gumbel, Lognormal, Pearson type III, Log-Pearson type III, and the general extreme value distributions applied to the annual flood series. The independence assumptions of the partial duration series were easily accepted for the Portuguese data. The Poisson and exponential distributions were found adequate to model the number of annual exceedances and the magnitude of exeedances, respectively.

It was found that the partial duration series with an average number of annual exceedances from 2 to 3 gave the best results in fitting the observed annual maximum daily flows and were significantly more efficient in estimating the annual peak discharges for the return periods considered in the study. The use of partial duration series was recommended principally when short records of data are available.

In all the studies mentioned previously, only the peak discharge of every flood event is considered and flood risk is simply characterized in terms of the annual flood for a given return period. The flood duration and volume are not considered, although they are important aspects to be taken into account in the design of many flood control systems.

Research that is currently being done on the modeling of flood events as multivariate partial duration series is reported in the following sections. These results should not be considered complete nor final, but simply an attempt to introduce the consideration of flood duration and volume into the process of flood risk analysis.

3. MULTIVARIATE PARTIAL DURATION SERIES IN FLOOD RISK ASSESSMENT

A flood event is defined as a sequence of discharge values above a given truncation level. The peak discharge, the duration, and the volumes of the flood are denoted by X, Y and Z, respectively.

It is physically understandable, and data confirms, that there is a much stronger dependence between flood volume and duration, and flood volume and peak, than between duration and peak. For this reason peak and duration will be studied first and volume will be considered later.

If flood peak, X, and duration, Y, are independent, the joint probability density function is given by

$$h(x,y) = h(x) \,.\, h(y)$$

and the distribution function is given by

$$H\,(x, y) = \int_0^x \int_0^y h(x)\, h(y)\, dx\, dy \qquad (3)$$

where h(x) and h(y) are the marginal density functions of flood peak and flood duration.

The joint distribution function of X and Y when the time interval of one year is considered may be given by

$$P\,(x, y) = \sum_{k=0}^{\infty} [H(x, y)]^k \,.\, P(E_k) \qquad (4)$$

where $P(E_k)$ is the probability that k flood events occur in one year. F(x, Y) represents the probability that in one year x and y are simultaneously not exceeded

$$F(x, y) = P(X \leq x \text{ and } Y \leq y)$$

If the number of flood events per year is Poisson distributed, Eq. (4) may be written as

$$P(x, y) = \sum_{k=0}^{\infty} [H(x, y)]^k \cdot \frac{e^{-\lambda} \lambda^k}{k!}$$

$$= e^{-\lambda} \sum_{k=0}^{\infty} \frac{[\lambda H(x, y)]^k}{k!}$$

$$= e^{-\lambda} e^{\lambda H(x, y)}$$

and finally

$$P(x, y) = e^{\lambda H(x, y) - \lambda} \tag{5}$$

If the marginal distribution function of flood peak and duration are both exponential, with parameters α and β respectively, Eq. (3) becomes

$$H(x, y) = \left(\int_0^x \alpha e^{-\alpha u} du\right) \left(\int_0^y \beta e^{-\beta w} dw\right)$$

$$H(x, y) = (1 - e^{-\alpha x}) (1 - e^{-\beta y}) \tag{6}$$

Substituting Eq. (6) in Eq. (5) we get

$$P(x, y) = \exp\left[\lambda (1 - e^{-\alpha x}) (1 - e^{-\beta y}) - \lambda\right] \tag{7}$$

For any given flood peak x and flood duration y, Eq. (7) gives the probability that these values are simultaneously not exceeded during one year. The probability that x or y are eventually both exceeded is 1-F(x,y), and the return period associated with this event is given by

$$T(x, y) = \frac{1}{1-P(x, y)} \tag{8}$$

For design purposes it may be important to consider the probability that both x and y are exceeded. In this case the return period is given by

$$T^1(x, y) = \frac{1}{1-F_M(x) - F_M(y) + F(x, y)}$$

where $F_m(x)$ and $F_m(y)$ are respectively the marginal distribution functions of X and Y.

In these expressions the parameters λ, α and β are easily estimated for any given truncation level. In fact λ is the average number of annual flood events, α is the inverse of the mean flood peak discharge, and β is the inverse of mean flood duration.

Independence between flood peak and duration may not be realistic, principally when the truncation level is low and the annual number of flood events is relatively high. If X and Y are not independent, their joint density function is given by

$$h(x,y) = h(x|y)\ h(y) \tag{9}$$

and the distribution function is

$$H(x, y) = \int_0^x \int_0^y h(x|y)\ h(y)\ dx\ dy \tag{10}$$

Let us assume that the relationship between X and Y is modeled adequately by a linear model. The value of X is then given by

$$X = AY + B + \varepsilon$$

If we assume that the residual ε is normally distributed

$$\varepsilon \sim N\ (0,\ \sigma^2)$$

The conditional density function $h(x|y)$ is also normal

$$h(x|y) \sim N\ (\mu(y),\ \sigma^2)$$

with

$$\mu(y) = Ay + B$$

If the flood duration is exponentially distributed Eq. (9) becomes

$$h\ (x, y) = \frac{1}{\sqrt{2\pi}\ \sigma}\ e^{\frac{-[x - \mu(y)]^2}{2\sigma^2}}\ .\ \beta\ e^{-\beta y} \tag{11}$$

This function can be numerically integrated to get $H(x,y)$ for any pair of values x and y. Once $H(x,y)$ is obtained, Eq. (5) provides the nonexceedance probability for these values and Eq. (8) the corresponding return period.

The parameters of this model are λ, the average number of annual exceedances; β, the inverse of the mean flood duration; and A, B and σ^2, the regression parameters of X on Y.

For practical purposes it is possible to characterize flood risk in terms of annual maximum peak discharge and annual maximum flood duration if the assumptions of Eq. (7) and Eq. (11) can be accepted. In fact it was shown that for any pair (x,y) it is possible to find the annual exceedence probability and the corresponding return period.

The problem of assessing the flood volume must be addressed now. It is very common that the flood volume associated with a flood event of duration Y and peak discharge X can be computed, with reasonable accuracy, as the area of the triangle with base Y and height X.

$$Z = XY/2 + e \tag{12}$$

If the residual e is normally distributed with zero mean and variance σ_e^2, it is possible to consider the (1-α) confidence interval for the flood volume Z.

$$Z \varepsilon \left| \frac{XY}{2} - \sigma_e \ \phi^{-1} \ (\frac{\alpha}{2}) \ , \frac{XY}{2} + \sigma_e \ \phi^{-1} \ (\frac{\alpha}{2}) \right| \tag{13}$$

If the assumptions of Eq. (13) are supported by data, it is possible to establish not only the return period associated with any pair of peak discharge and flood duration values, but also the confidence interval of any size for the flood volume associated with this pair of values.

A more rigorous and general formulation to include peak discharge, flood duration, and flood volume in the flood risk assessment process will be tried in future research work. Nevertheless, the formulation presented in this paper seems to be a fruitful step forward, if data support the necessary assumptions.

4. CASE STUDY: CORGO RIVER AT ERMIDA AND TERA CREEK AT PAVIA

4.1 Location and Main Characteristics of the River Basins

Corgo River at Ermida is a watershed located in a mountainous and wet region of northern Portugal. The area of the watershed is 258 km^2. The altitude of the hydrometric station is 120 m. Twenty-one years of daily flow data are available at this station.

Tera Creek is a watershed with 610 km^2. It is located in a much flatter and arid zone. The altitude of the hydrometric station is 224 m. Twenty years of daily flow data are available at this hydrometric station.

4.2 Marginal Characterization of Peak Discharge, Duration, and Volume of Flood Events

Truncation levels corresponding to average number of flood events per year equal to one, two, and four, were considered. The hypothesis that the number of flood events per year is Poisson distributed was tested using an approximation to the normal distribution recommended by NERC (1975) and Cunnane (1979). This test is based on results derived by Sukhatme (1938) and uses the Chi-square statistic. Significance levels for which the hypothesis can be accepted are presented in Table 1.

The hypothesis that the marginal distributions of the peak discharge, the duration, and volume of the flood events are exponential was checked with a chi-square test. The significance levels of acceptance for the two river basins and for the three truncation levels are also presented in Table 1.

The hypothesis that successive peak discharges, successive flood durations, and successive flood volumes are serially independent were tested. For each case, the lag one autocorrelation coefficient was computed and tested for not being significantly different from zero. A test based on the student distribution, recommended by Yevjevich (1972) was used. The significance levels for which the hypothesis can be accepted are presented in Table 1 for the two rivers and the three truncation levels.

Analysis of Table 1 shows that only the hypothesis that the flood volume is exponential for Tera Creek at Pavia for a Lambda equal to two and four are clearly rejected. All the other hypotheses can be accepted with significance levels from 1% to 99%, with most values between 10% and 90%.

The truncation levels corresponding to the average number of flood events per year equal to one, two and four are presented in Table 2.

The mean flood peak discharge, the mean flood duration, and the mean flood volume corresponding to the inverse of the exponential parameters are also presented in Table 2.

4.3 Dependence Structure of Flood Characteristics

Linear correlation coefficients between peak discharge and duration, volume and peak discharge, and volume and duration were investigated for the two rivers and the three truncation levels. Results are presented in Table 3.

The hypothesis that the correlation coefficient is equal to zero is rejected for all the cases concerning volume versus peak and volume versus duration. For the peak versus duration cases, the hypothesis is rejected for $\lambda = 4.0$ for both rivers. For $\lambda = 1.0$, the hypothesis is not rejected with a significance level of 48% for the Corgo River and 8%

Table 1. Significance levels for which the Poisson, the exponential and the serial independence hypothesis can be accepted

River Basin	λ	Poisson distribution	EXPONENTIAL DISTRIBUTION			INDEPENDENCE		
			Peak discharge	Flood volume	Flood duration	Peak discharge	Flood volume	Flood duration
Corgo	1.0	39%	35%	70%	1%	14%	11%	73%
at	2.0	96%	86%	88%	70%	24%	64%	17%
Ermida	4.0	7%	99%	24%	59%	73%	26%	25%
Tera	1.0	58%	8%	64%	51%	28%	44%	32%
at	2.0	52%	9%	0.5%	76%	12%	28%	82%
Pavia	4.0	66%	71%	0.5%	99%	45%	35%	5%

Table 2. Truncation levels and mean peak discharge, duration, and volume of flood events

River Basin	Truncation level (m^3/s)	Poisson Parameter λ	MEAN FLOOD CHARACTERISTICS Peak (m^3/s)	Duration (days)	Volume (10^3 m^3)
Corgo	110.	1.0	48.45	1.70	5,581.44
at	72.	2.0	51.21	2.40	7,629.72
Ermida	32.	4.0	54.19	5.15	13,187.18
Tera	86.	1.0	74.89	1.69	10,140.29
at	51.	2.0	54.50	2.21	8,232.67
Pavia	6.7	4.0	55.27	7.27	14,160.17

Table 3. Linear correlation coefficients between flood characteristics

River Basin	λ	Peak versus duration	Volume versus peak	Volume versus duration
Corgo	1.0	0.175	0.771	0.456
at	2.0	0.284	0.713	0.787
Ermida	4.0	0.552	0.734	0.889
Tera	1.0	0.522	0.818	0.855
at	2.0	0.372	0.814	0.697
Pavia	4.0	0.681	0.800	0.861

for the Tera Creek. For $\lambda = 2.0$, the hypothesis is accepted with a significance level of 8% for Corgo and 4% for Tera. The conclusion is that flood volume is highly correlated with peak discharge and flood duration.

The peak discharge and flood duration can be considered linearly independent for $\lambda \leq 2.0$ and linearly dependent for $\lambda = 4.0$.

The equations presented in a previous section for the independent and linear dependent cases were both applied to all truncation levels. The reason for this procedure is that in most cases the rejection of a zero correlation coefficient is not very clear. In fact, only for Corgo River with $\lambda = 1.0$ the significance level is 48%. For all the other cases, the significance level is smaller than 10%. Moreover, the computation of the nonexceedance probabilities of pairs of peak discharges and flood durations, for both the independence and dependence assumptions, allows for the comparison of the final results in both cases.

The parameters of the linear regression models for the two rivers and the three truncation levels is presented in Table 4.

The hypothesis that the flood volume can be computed by Eq. (12) was accepted. The correlation between flood volumes computed by this equation and observed flood volumes are presented in Table 5. In this table the standard deviations of the residuals appearing in Eq. (12) σ_e, are also presented.

Table 4. Parameters of the linear regression models of peak discharge versus flood duration

River Basin	Discharge threshold (m^3/s)	Poisson parameter λ	Mean flood duration (days)	Λ	β	σ^2
Corgo	110.	1.	1.7	- 7.54	61.27	1,727.724
at	72.	2.	2.4	6.67	35.24	1,830.720
Ermida	32.	4.	5.15	5.10	27.91	1,717.783
Tera	86.	1.	1.69	25.08	32.45	2,630.972
at	51.	2.	2.21	11.67	28.69	3,012.790
Pavia	6.7	4.	7.27	5.64	14.23	1,954.076

Table 5. Correlation coefficients between computed and observed flood volumes and standard deviation of the residuals appearing in Eq. (12)

River Basin	λ	ρ	σ_e (10^3 m^3)
Corgo	1.0	0.954	1.156
at	2.0	0.951	2.469
Ermida	4.0	0.921	11.185
Tera	1.0	0.945	3.499
at	2.0	0.989	1.875
Pavia	4.0	0.902	24.623

4.4 Final Results for the Corgo River at Ermida and Tera Creek at Pavia

The models presented for the assumptions of independence and linear dependence were both used for all the truncation levels. Results are displayed in Figures 1 through 12. The volumes computed by Eq. (13) for λ = 2.0, for given durations, and for 90% confidence intervals are displayed in Figures 13 through 16. Naturally, other truncation levels and other flood durations can be considered.

The marginal distributions of annual maximum peak discharge and annual maximum flood duration are presented in Table 6 and 7.

These values were obtained under the assumption that peak discharge and duration of all flood events are exponentially distributed and Eq. (2) can be applied with the adequate parameters.

It can be seen in Figures 1 through 12 that for short durations there is an asymptotic behavior with respect to the return period.

The return period is related by Eq. (8) to the nonexceedance probability of two variables considered simultaneously, namely flood peak discharge and duration. For any given duration, y, the discharge, x, may increase indefinitely and the return period will still be limited by the marginal distribution of y, as can be seen in Eqs. (6) and (7) for the case of independence and in Eq. (11) for the case of linear dependence.

It can also be seen in Figures 1 through 6 that for large durations the distribution of peak discharge given a flood duration tends to the marginal distribution of peak discharge as might be expected.

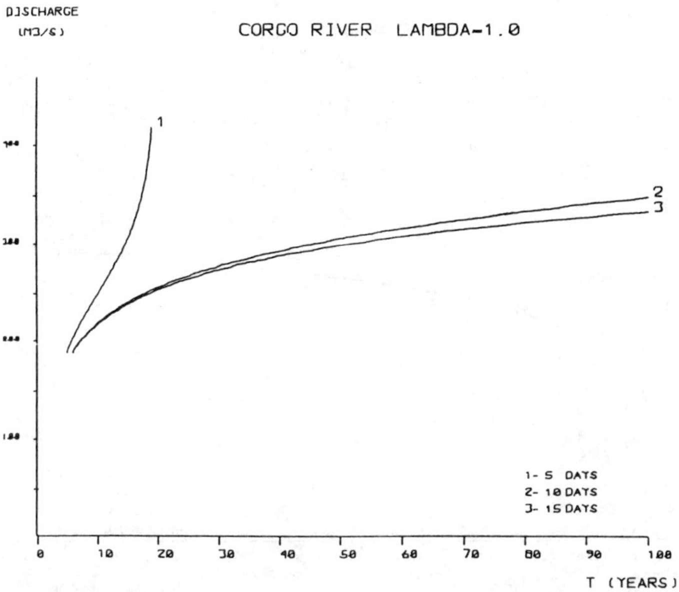

Figure 1. Flood peak versus return period for given durations at River Corgo with λ = 1.0 and assuming independence between flood peaks and durations.

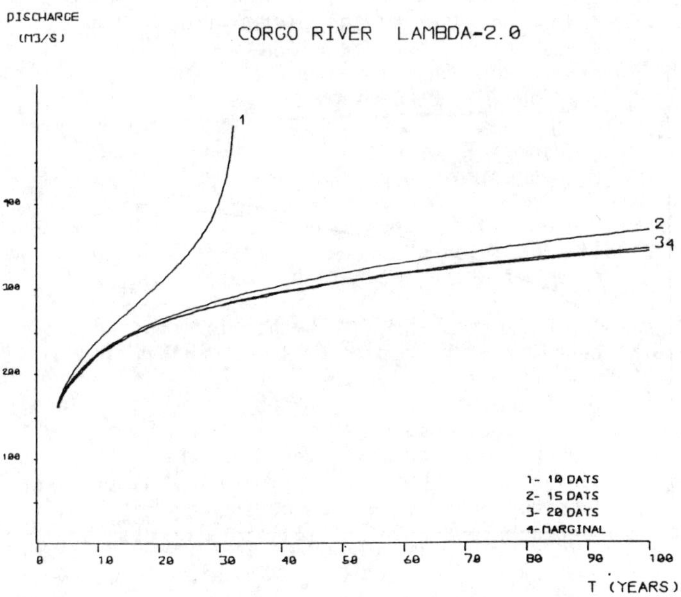

Figure 2. Flood peak versus return period for given durations at River Corgo with λ = 2.0 and assuming independence between flood peaks and durations.

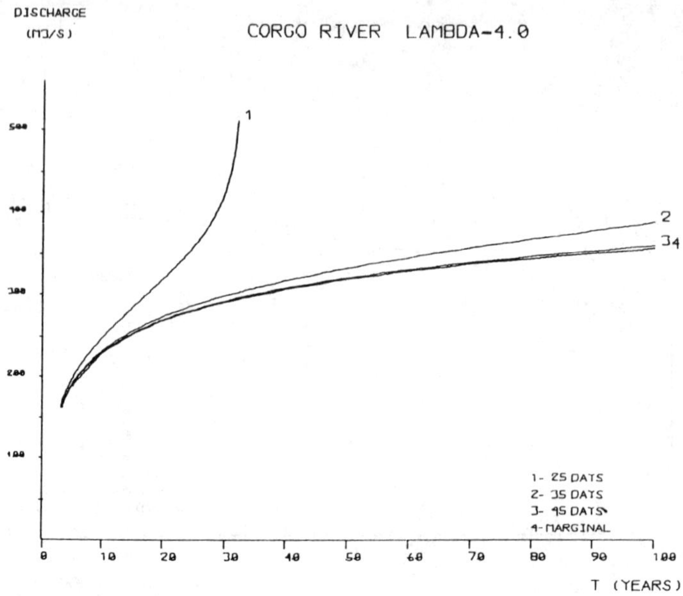

Figure 3. Flood peak versus return period for given durations at River Corgo with λ = 4.0 and assuming independence between flood peaks and durations.

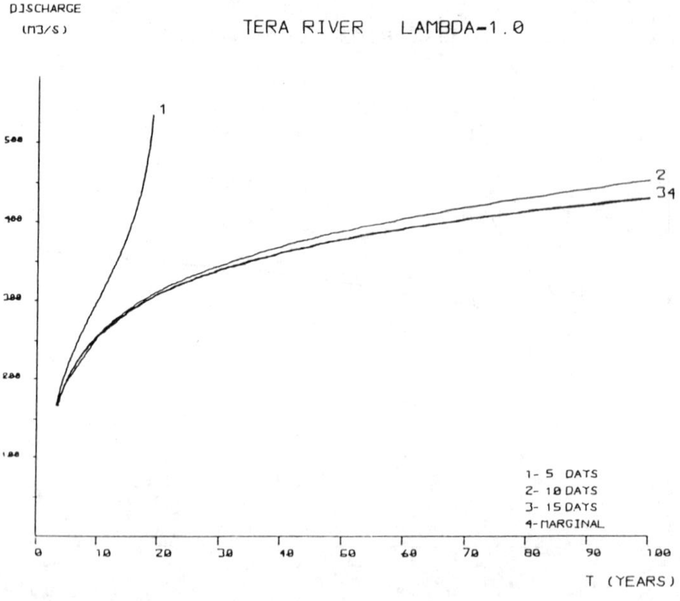

Figure 4. Flood peak versus return period for given durations at Tera Creek with λ = 1.0 and assuming independence between flood peaks and durations.

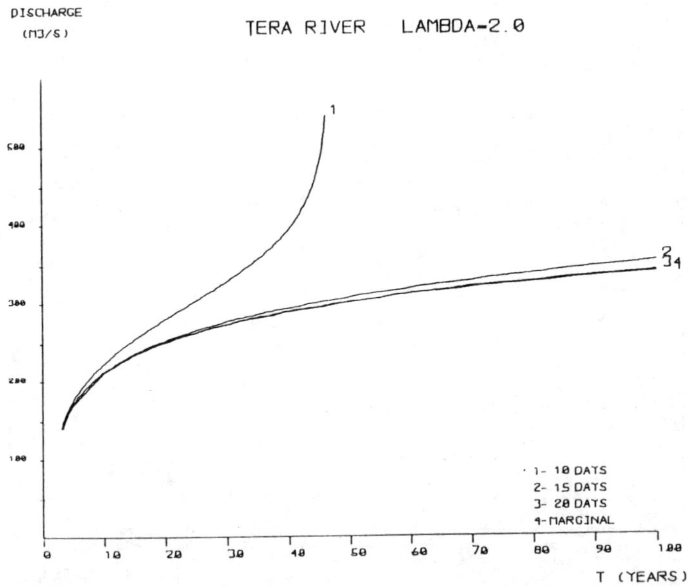

Figure 5. Flood peak versus return period for given durations at Tera Creek with λ = 2.0 and assuming independence between flood peaks and durations.

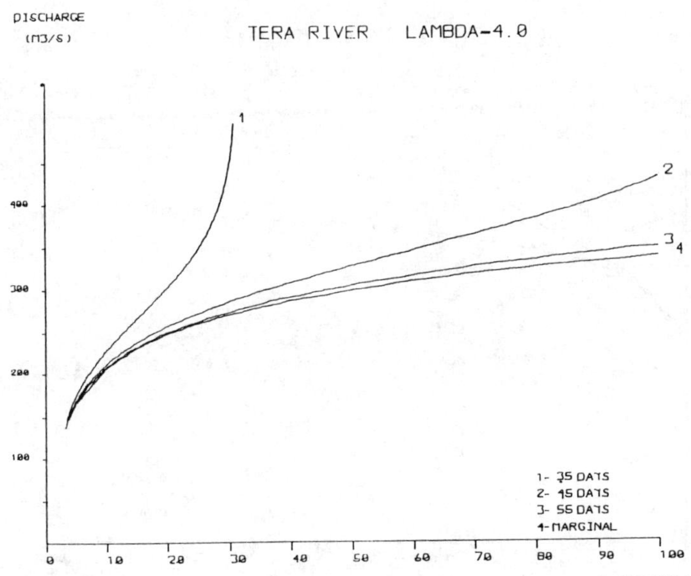

Figure 6. Flood peak versus return period for given durations at Tera Creek with λ = 4.0 and assuming independence between flood peaks and durations.

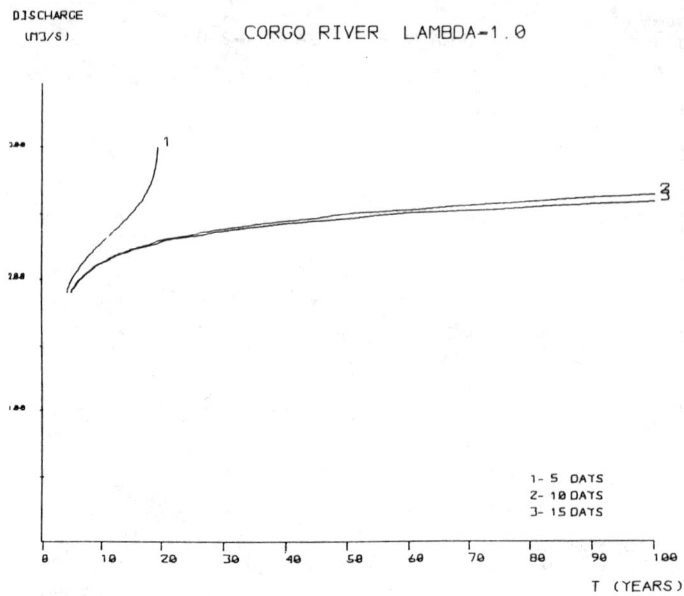

Figure 7. Flood peak versus return period for given durations at River Corgo with λ = 1.0 and assuming a linear dependence between flood peaks and durations.

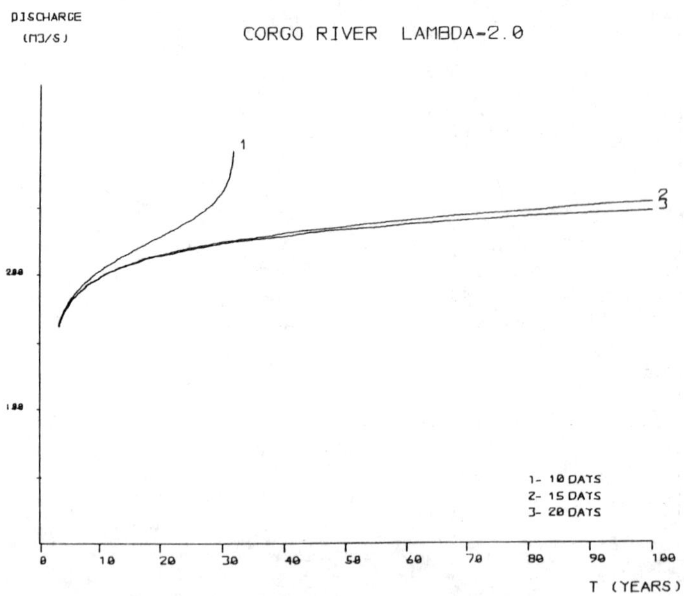

Figure 8. Flood peak versus return period for given durations at River Corgo with λ = 2.0 and assuming a linear dependence between flood peaks and durations.

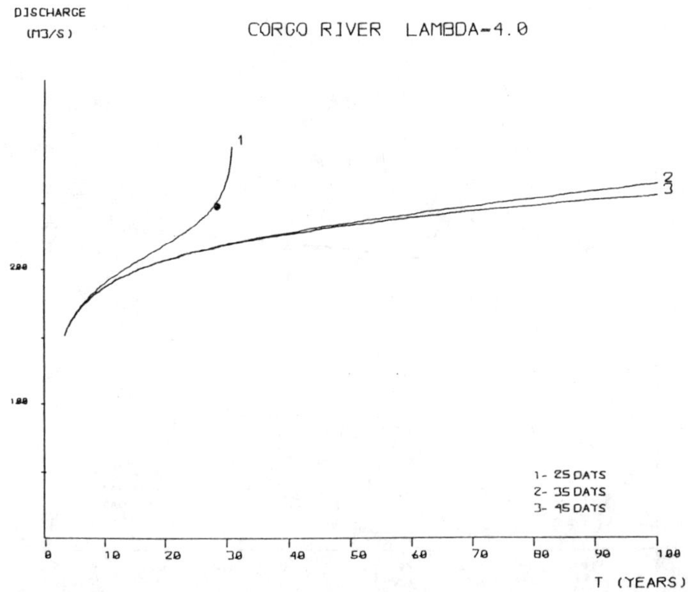

Figure 9. Flood peak versus return period for given durations at River Corgo with λ = 4.0 and assuming a linear dependence between flood peaks and durations.

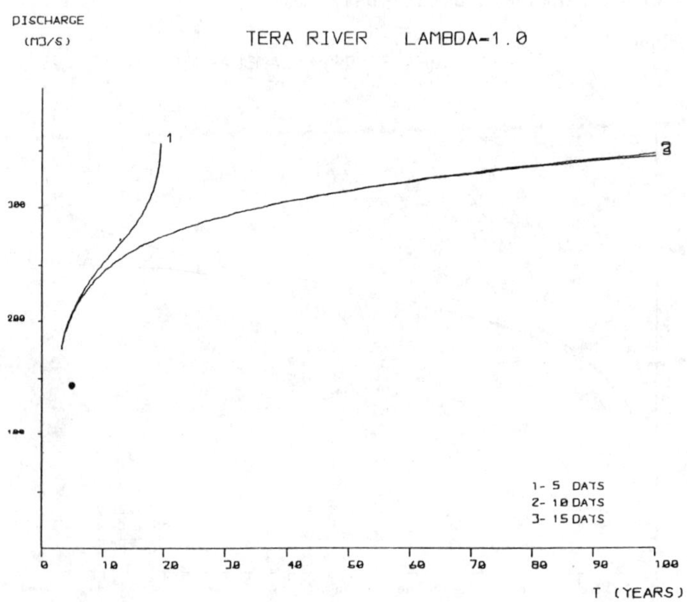

Figure 10. Flood peak versus return period for given durations at Tera Creek with λ = 1.0 and assuming a linear dependence between flood peaks and durations.

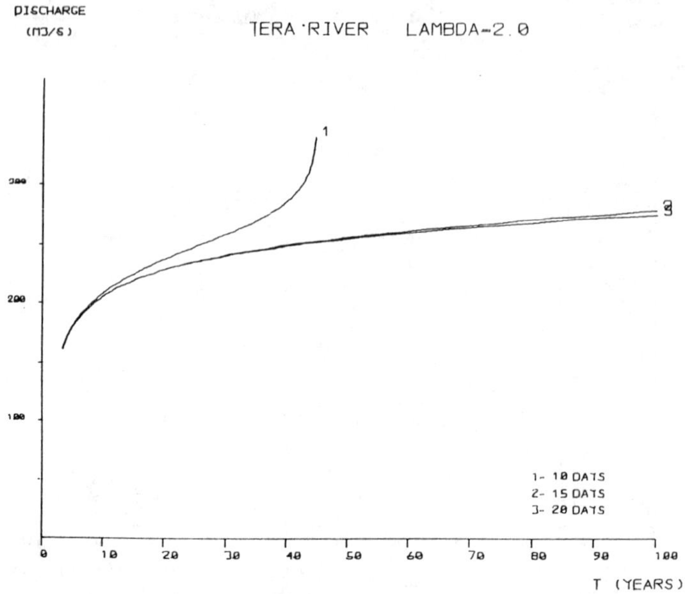

Figure 11. Flood peak versus return period for given durations at Tera Creek with λ = 2.0 and assuming a linear dependence between flood peaks and durations.

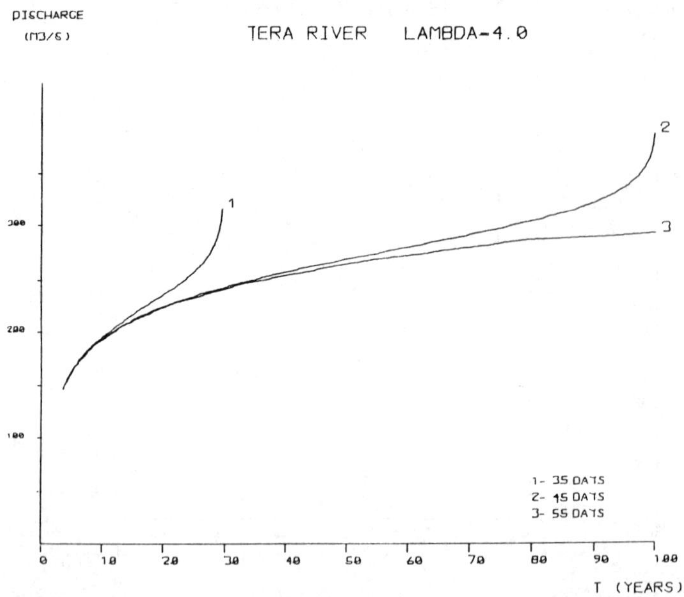

Figure 12. Flood peak versus return period for given durations at Tera Creek with λ = 4.0 and assuming a linear dependence between flood peaks and durations.

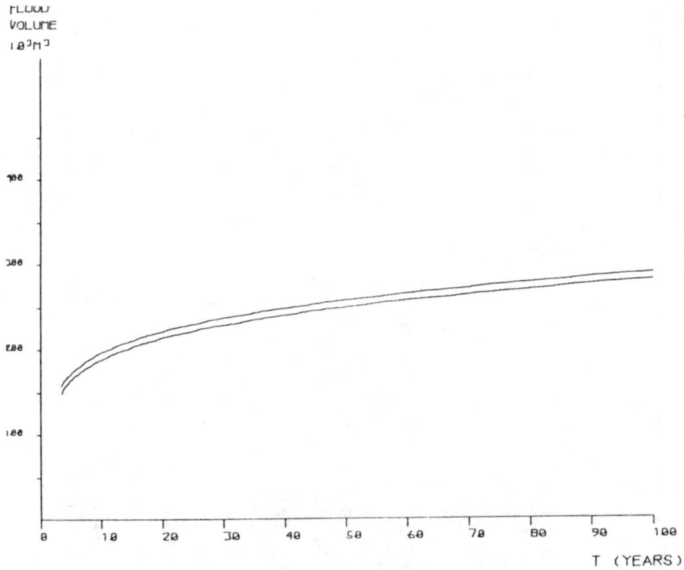

Figure 13. Flood volume 90% confidence interval for λ = 2.0 and 15 days flood duration at River Corgo, assuming independence between flood peaks and durations.

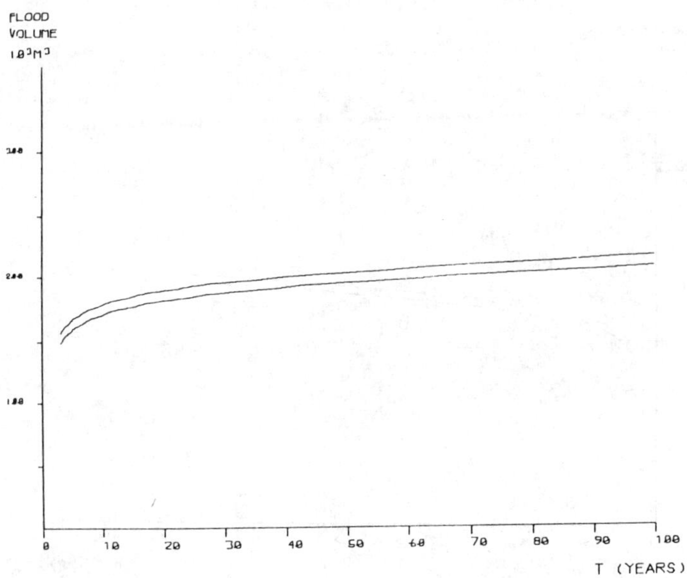

Figure 14. Flood volume 90% confidence interval for λ = 2.0 and 15 days flood duration at Tera Creek, assuming independence between flood peaks and durations.

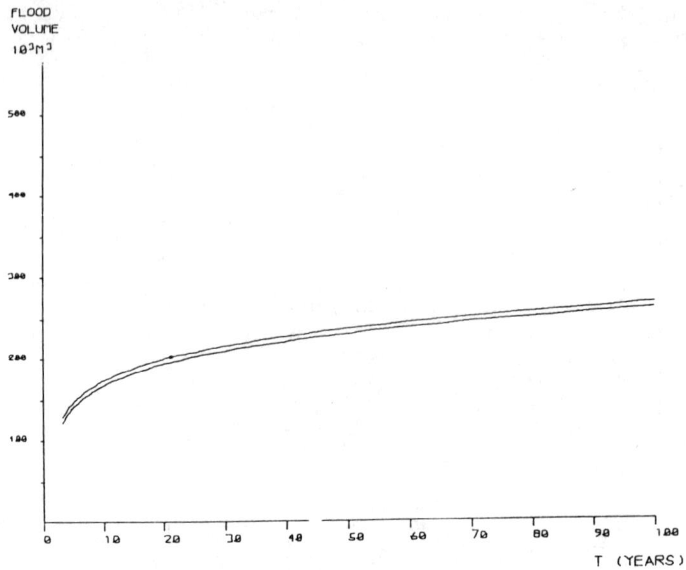

Figure 15. Flood volume 90% confidence interval for λ = 2.0 and 15 days flood duration at River Corgo, assuming a linear dependence between flood peaks and durations.

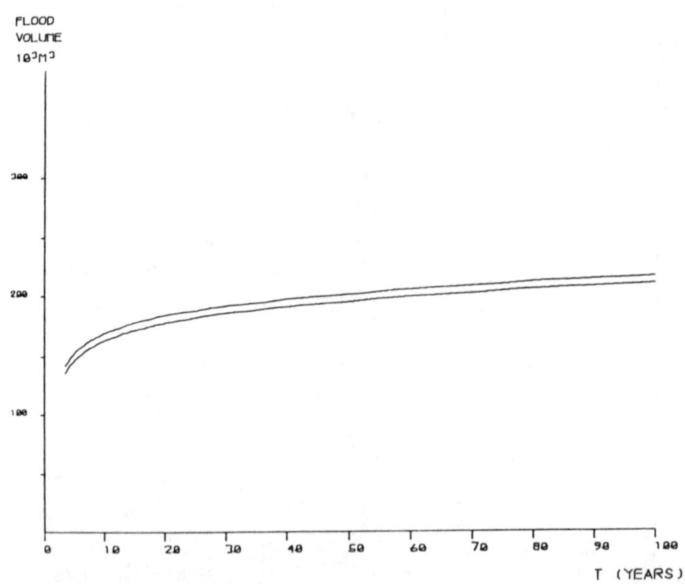

Figure 16. Flood volume 90% confidence interval for λ = 2.0 and 15 days flood duration at Tera Creek, assuming a linear dependence between flood peaks and durations.

Table 6. Marginal distribution of annual flood discharges in m^3/s

River		RETURN PERIOD (YEARS)					
Basin	λ	5	25	50	100	500	1,000
Corgo	1.0	182.67	264.97	299.05	332.88	411.05	444.66
at	2.0	184.31	271.29	307.31	343.07	425.69	462.22
Ermida	4.0	188.41	280.45	318.57	356.41	443.84	481.43
Tera	1.0	198.33	325.54	378.22	430.51	551.34	603.28
at	2.0	170.52	263.10	301.43	339.48	427.42	465.22
Pavia	4.0	166.22	260.10	298.98	337.57	426.75	465.08

Table 7. Marginal distribution of annual maximum flood duration in days

River Basin	λ	RETURN PERIOD (YEARS) 5	25	50	100	500	1,000
Corgo	1.0	2.55	5.44	6.63	7.82	10.56	11.74
at	2.0	5.26	9.34	11.03	12.70	16.58	18.24
Ermida	4.0	14.86	23.61	27.23	30.83	39.14	42.71
Tera	1.0	2.53	5.41	6.59	7.77	10.50	11.67
at	2.0	4.85	8.60	10.16	11.70	15.26	16.80
Pavia	4.0	20.98	33.33	38.45	43.52	55.25	60.29

It is interesting to note that the annual peak discharge for any given return period, Q(t), is larger under the assumption of independence between flood peaks and durations, and smaller when a linear dependence model is considered.

CONCLUSIONS

The use of partial duration series in flood risk analysis has the advantage of easily allowing for the characterization of floods in terms of peak discharges, flood durations, and flood volumes.

In this paper results for the joint distribution of annual flood peak discharge and annual maximum flood duration are obtained for the case of independence between flood peaks and durations, and for the case of linear dependence between these flood characteristics.

The results were applied to two Portugeuese rivers, the Corgo River at Ermida and the Tera Creek at Pavia. More research is needed before final conclusions can be drawn. The use of multivariate partial duration series may be considered, however, a very promising technique for the assessment of engineering risk in flood studies.

ACKNOWLEDGEMENTS

Mrs. Maria Alzira Santos, researcher at LNEC, is gratefully acknowledged for the fruitful discussions on the topic of this paper.

This research was done under Project 62/11/8490 financed by PIDAC (Central Administration Investment and Development Plan). Data were provided by Direcção Geral dos Recursos e Aproveitamentos Hidraulicos.

REFERENCES

Correia, F.N. (1983) Métodos de Análise e Determinação de Caudais de Cheia (Methods for the Analysis and Determination of Flood Discharges). Lisboa, Laboratório Nacional de Engenharia Civil.

Cunnane, C. (1973) A Particular Comparison of Annual Maxima and Partial Duration Series Methods of Flood Frequency Prediction. Journal of Hydrology 18.

Cunnane, C. (1979) A Note on the Poisson Assumption in Partial Duration Series. Water Resources Research, Vol. 15, No. 2.

NERC (1975) Flood Studies Report. London, Natural Environment Research Council.

Sukhatme, P. (1938) On the Distribution of Chi-Square in Samples of the Poisson Series. Journal of Royal Statistics Society, Vol. 5, Supl. 1.

Taesombut, V., V. Yevjevich (1978) Use of Partial Duration Series for Estimating Distribution of Maximum Annual Flood Peak. Hydrology Paper No. 97, Fort Collins, Colorado State University.

Todorovic, P. (1970) On Some Problems Involving Random Number of Random Variables. Annuals of Mathematical Statistics, Vol. 41, No. 3.

Todorovic, P., E. Zelenhasic (1970) A Stochastic Model for Flood Analysis. Water Resources Research, Vol. 6, No. 6.

Todorovic, P., J. Rousselle (1971) Some Problems of Flood Analysis. Water Resources Research, Vol. 7, No. 5.

Yevjevich, V. (1972) Probability and Statistics in Hydrology. Fort Collins, Water Resources Publications.

Zelenhasic, E. (1970) Theoretical Probability Distributions for Flood Peaks. Hydrology Paper No. 42. Fort Collins, Colorado State University.

CONJUNCTIVE USE OF SURFACE AND GROUNDWATER IN A PROBLEM OF ENVIRONMENTAL PROTECTION: A CASE IN SALENTO PENINSULA IN SOUTHERN ITALY

M. Benedini and Gb. Cicioni
Water Research Institute
Via Reno 1 -00198 Rome, Italy

ABSTRACT

The Ionic area of Apulia, in Southern Italy, has a very peculiar surface hydrology. The average annual rainfall is merely 500 mm, which gives rise to very little runoff. Groundwater resources are extensive and are favoured by natural recharge through the fractured soil. The groundwater system consists of a lens-shaped freshwater layer floating on top of intruded sea water. A multifaceted economic development plan has been formulated for this area by local and national authorities. However, the problems of environmental protection are related to the construction and operation of urban and industrial wastewater treatment plants. Extracted groundwater is mixed with brackish water from a coastal spring that has a flow of about 2 m^3/s and with diversion of 2.4 m^3/s from a large aqueduct system, starting from a reservoir more than 200 km away. This quantity of water is requested for irrigating an area on the order of 20,500 hectares. For this system an 'incident' is any situation in which a discharge of contaminated water, either onto the topsoil or below ground, is characterized by quality parameters that exceed acceptable limits. A secondary set of incidents can be defined if there is not enough water available to meet the demand for the most common uses. The situation in which the farmers have no water for irrigation is particularly serious, especially when the level of the water table is lowered or the piezometric surface of the saline water has risen.

Several formulations of the "incident modes" are examined and discussed, given the available data and the peculiarity of the system, and a "figure of merit" is proposed for use in the decision making process.

1. INTRODUCTION

This paper describes a problem that can be studied by means of reliability analysis and refers to a typical system of combined surface and groundwater resources. Although an existing situation is considered, many simplifications have been introduced in order to emphasize the methodological aspects. Consequently, none of the decisions and management policies included in this paper necessarily commit the authorities responsible in the described area, but reflect exclusively the needs of a scientific exercise.

A typical example of the conjunctive use of surface and groundwater is the case of the Ionic area of Apulia, in southern Italy. This region of the "Heel of Italy" has in fact a very characteristic surface hydrology due to an annual average rainfall of less than 500 mm (the corresponding value for the whole country is 1,000 mm), which yields very low surface runoff. Wherever they exist, river beds have no water for most of the year.

Conversely, groundwater resources are quite extensive and are favoured by natural recharge through a highly permeable and fractured soil. The existing aquifer belongs to a lens-shaped layer that floats on top of the intruded sea water, which originates from the opposite Adriatic coast.

The peculiar geological conditions of this area have favoured the natural upwelling of groundwater in several places in the form of springs. The latter are located both inland, in the vicinity of the shoreline, and on the sea bottom in shallow submarine vents.

This area (Fig. 1) comes under the purview of an intensive economic development plan fostered by local and national authorities. The portion examined in this exercise is part of a very large irrigation project some 140,000 hectares in area, with a total demand of 323.4×10^6 m^3/year of water; the irrigation season is concentrated into 175 days (from April 8 to September 30). The specific demand is therefore 2,310 m^3/ha or 13.20 m^3/ha d.

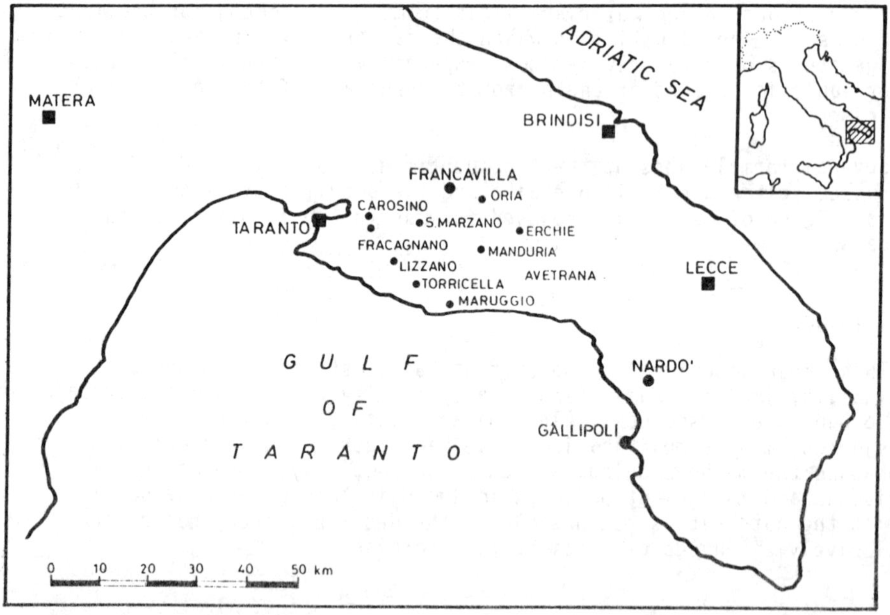

Figure 1. Francavilla and surrounding towns in the Ionic area of Apulia.

The problems of environmental protection have become very important, especially since the promulgation of national law "319/76", which governs water quality control. This law primarily makes provision for building and operating urban and industrial wastewater treatment plants. In the area under consideration, particularly with regard to the inland settlements, the final discharge of treated water encounters several difficulties because there are no surface water bodies capable of receiving and diluting the discharged water in accordance with environmental requirements.

The problem is aggravated by the industrial settlements, which in this area are mostly olive mills and other agricultural processing plants. Although operating only seasonally and for only very short periods, these plants discharge a huge amount of highly polluted wastewater, which cannot be accepted by the sewerage and conveyed to the urban treatment plants. Up to the present, mill owners have been forced to discharge untreated water directly onto the topsoil, which very often results in percolation and groundwater pollution.

The amount of wastewater is summarized in Table 1.

Table 1. Estimated wastewater from urban zones.

TOWN	POPULATION	URBAN DISCHARGE (l/s)	INDUSTRIAL DISCHARGE (l/s)	TOTAL DISCHARGE (l/s)
AVETRANA	7,772	24.1	8.9	33.0
CAROSINO	5,600	17.4	3.3	20.7
FRAGAGNANO	5,342	16.6	4.0	20.6
LIZZANO	8,765	27.2	15.8	43.0
MANDURIA	30,608	94.9	35.7	130.6
MARUGGIO	5,067	15.7	5.5	21.2
MONTEPARANO	2,344	7.3	1.4	8.7
PULSANO	9,167	28.4	5.9	34.3
SANMARZANO	8,033	24.9	5.0	29.9
TORRICELLA	3,821	11.8	4.0	15.8
ERCHIE	8,097	21.1	12.7	33.8
FRANCAVILLA	32,912	102.0	56.6	158.6
ORIA	14,812	45.9	16.5	62.4
Total values	142,340	437.3	175.3	612.6

- An eleven-hour working time is assumed

Urban and industrial water demands are met by long-distance transfer of water using several facilities, some of which (the famous Apulian Aqueduct) are several decades old. A new transfer scheme, the "Sinni Aqueduct," is supplied by a reservoir located more than 200 km away and carries a flow of about 12 m^3/s, which is used partly for agricultural purposes.

The resources locally available consist, first of all, of several pumping wells withdrawing from the freshwater aquifer. One of the main available resources is the Chidro Spring, in which a perennial flow of about 2 m^3/s has been ascertained. Unfortunately the discharging water is characterized by high salinity (more than 3,000 g/m^3) and cannot be used directly on the crops. It first has to be mixed with freshwater in order to reduce the salinity to an acceptable level.

2. PRELIMINARY INVESTIGATION

Because all the possibilities of economic development are in practice related to the amount of available water, several efforts have been made to identify the size of the aquifer and its behaviour in relation to the natural recharge, the interaction with the underlying saline strata, and the effects of withdrawals (Cotecchia, 1975). These aspects can now be described and examined by means of some mathematical models that have been tested through repeated applications. A simulation model can be used to describe the shape of the water table, both in natural conditions and under exploitation. The same model can also map the characteristics of the interface between fresh and saline water, showing the "intrusion cones" caused by intensive pumping (Sirangelo and Troisi, 1980; Benedini et al, 1982; Benedini et al, 1983).

Other aspects can be also investigated by means of this simulation model: e.g., the mutual interaction among the pumping wells. This is very useful for the purpose of ascertaining (Cicioni et al, 1982), by means of a linear programming model, the best way to exploit the available groundwater if some constraints are imposed in order to keep the drawdown of the water table (and, correspondingly, the rise of the interface) within acceptable limits. The programming model can also assess the maximum quantity of water that can be withdrawn for irrigation and other purposes.

3. ENVIRONMENTAL CONSTRAINTS

A thorough investigation has been conducted in order to identify the areas where a discharge could be performed without hampering the aquifer, taking into due account the local geological conditions as well as the existing and future exploitation of groundwater. Following this investigation, the whole zone has been split into (1) the areas where all discharge is forbidden, both on the topsoil and underground; (2) the areas where discharge is allowed in the topsoil and in the underground, through dispersion holes and natural karst formations; and (3) the areas where discharge will be possible only after appropriate preliminary tests.

The amount of wastewater discharged from the urban settlements can be determined, as usual, as a fraction of the total quantity of potable water supplied to the population. The quantity of industrial wastewater can be evaluated after the number of employees for the various industrial activities is known, assuming an appropriate per-capita volume.

4. SYSTEM DEFINITION

The system studied in this exercise (Fig. 2) consists of a large portion of the area previously outlined and is identified by the urban zones of Sava, Maruggio, Torricella, and Manduria; and the surrounding farmland including also a part of the farmland of Francavilla Fontana, Avetrana, Grottaglie, Torre Santa Susanna, Lizzano, Fagagnano, and San Marzano. The total irrigable area is about 20,450 hectares, inside which the following partition has been proposed:

- 3,900 hectares to be irrigated by means of extracted groundwater or reused urban and industrial wastewater

- 8,000 hectares to be irrigated by means of water stored in the Manduria reservoir

- 8,550 hectares to be irrigated after mixing water from the Chidro Spring and Sinni Aqueduct.

The facilities planned for supplying this area, beside the single pumping wells run directly by the farmers, consist of:

a) a pumping rig at the Chidro Spring

b) the high-pressure "intake tower" (G) at the Sinni Aqueduct main

c) the Manduria reservoir built above the ground level with a "crown" structure, having a total capacity of 16×10^6 m^3

d) the "mixing tank N.1" (H)

e) the "distribution node" (A)

f) the pumping plant and the "re-emission tower" at the downstream section of the Sinni pipeline

g) the facilities for collecting and conveying the treated wastewater

h) the wastewater treatment plant

i) the distribution network in the fields.

In order to make the most rational use of available water and to avoid any possible waste, the system has the capability of storing water temporarily unused by farmers in the Manduria reservoir and also of returning it to the Sinni Aqueduct if and when required by some downstream utilization. This will occur particularly at night, when the farmers prefer to stop all irrigation activity.

The Regional Plan for Water Protection, laid down in accordance with the above-mentioned Law "319/76", provides for the urban discharges of Maruggio and Manduria, which are the only settlements that already have working treatment plants, to be disposed of directly underground, partly through artificial boreholes and partly through natural caves and sinkholes. At present, discharge takes place directly onto the ground.

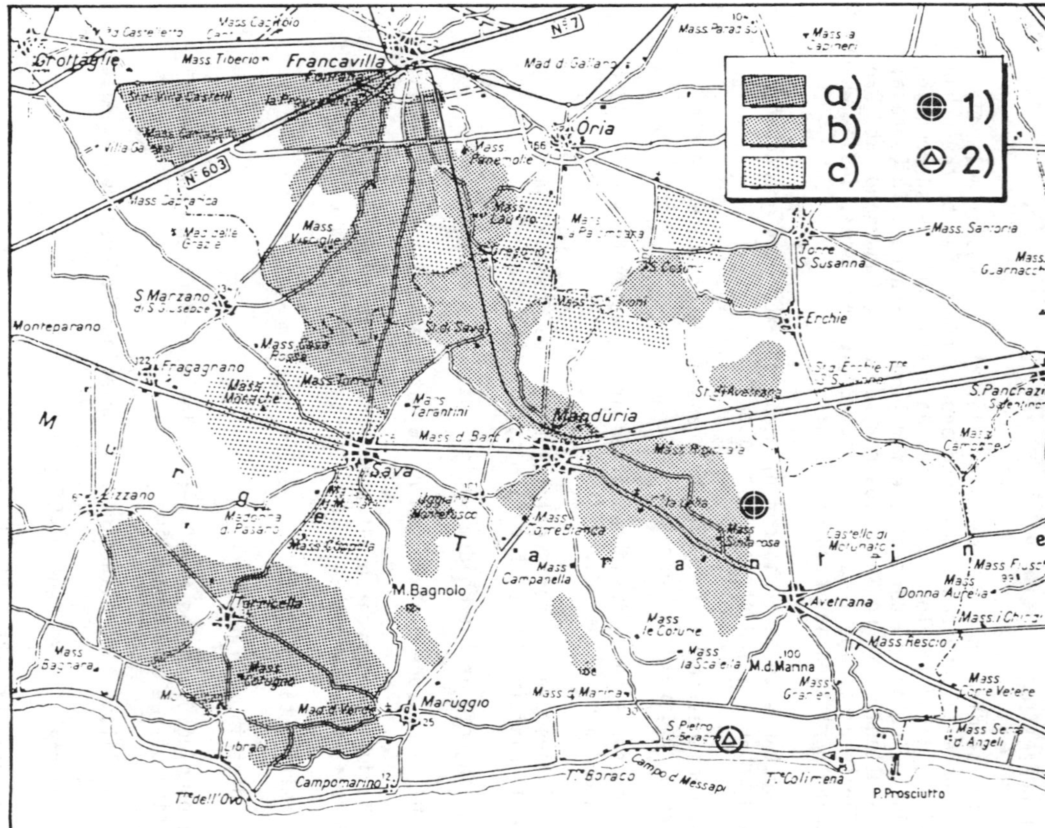

Figure 2. Areas irrigated by means of water from (a) Chidro Spring and Sinni Aqueduct, (b) Manduria reservoir, and (c) extracted groundwater and treated wastewater.

New plants have been designed for Sava and Torricella. In the case of these plants it has been assumed that the final discharge of treated wastewater will take place underground as in the case of the surrounding villages.

Discharge underground obviously will be carried out during the winter wet season, with the secondary purpose of restoring the aquifer for the following dry season. In summer, treated wastewater can be directly diverted to the irrigation network, with the aim of helping the farmers by increasing the quantity of water available.

The simplified system adopted for the exercise is shown schematically in Figure 3.

During daylight hours, when irrigation is in progress, brackish water from Chidro Spring is pumped continuously and at a constant rate. From the "distribution node" (A) it is then pumped into "mixing tank N.1" (H), where an appropriate quantity of freshwater from Sinni Aqueduct is added. A diluted brackish water is thus produced which is then delivered through the pipeline (M). Before reaching the irrigation plots, this water is mixed with water coming from the pumping wells, the resulting salinity always being kept under control at the cross-section b (water coming from wells may also have a high salt content). Further downstream it is mixed with treated wastewater, with any unacceptable concentrations of residual pollutants being abated by dilution. Finally, the water is delivered in an acceptable quality through the outlet N to the farmers, to meet their requirements.

Direct delivery of water from pumping wells through P, as well as from Manduria reservoir through R and from the treatment plant outlet T, occurs whenever possible "outside the system"; however the quantities handled must be constantly monitored, as they contribute to decreasing the demand on the System.

Working overnight and on rainy days, when irrigation is not being performed, water from Chidro Spring is conveyed to the Manduria reservoir for storage, or if necessary, straight into the Sinni Aqueduct for a possible utilization downstream (some balancing reservoirs, not included in the system, are provided). From the Manduria reservoir water can either be delivered to the crops through R or discharged into the Sinni Aqueduct.

Such a system can be described daily by the following state variables;

i-the discharge in the connection pipes

ii-the volume stored in Manduria reservoir

iii-the salt concentration of the transported water, as a result of diluting the brackish water of Chidro Spring with the freshwater of Sinni Aqueduct

iv-the final quality after adding the treated wastewater

v-the salinity of water from pumping wells.

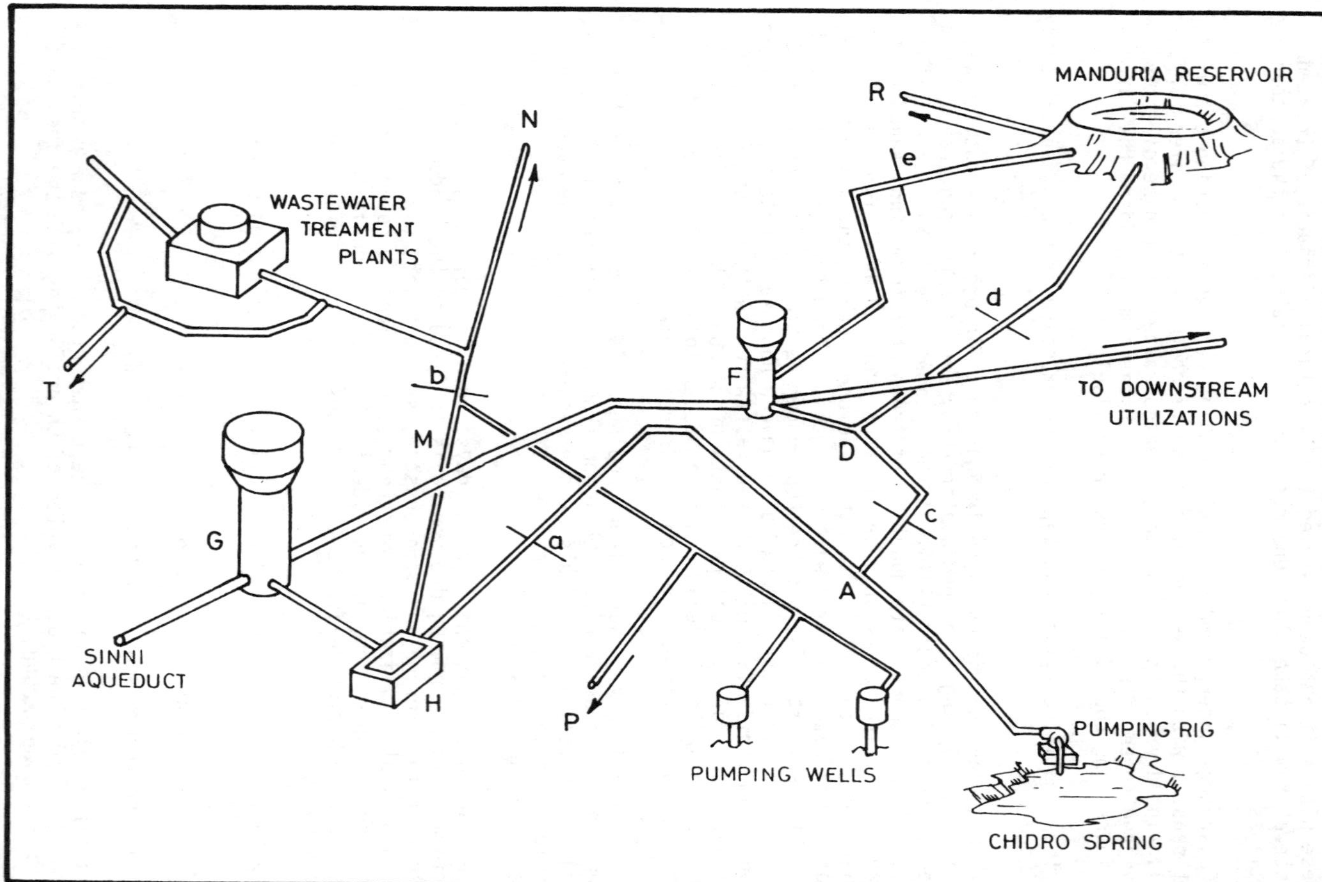

Figure 3. Schematic diagram of the system.

5. THE SYSTEM'S INPUT

The first input is due to the hydrology, which is uncontrollable. Several rainfall gauging stations have been operating in this area for many years. Every year, on the average, there are from 50 to 80 rainy days, with an almost dry summer.

The following sets can be identified:

i-expected rainfall, uncontrollable

ii-demand for water by farmers, controllable

iii-quantity available at the outlet of the treatment plan, uncontrollable

iv-discharge at the Chidro pumping rig, constant for all the irrigation period

v-quality of treated wastewater, controllable.

6. THE OUTPUT

The economic structure in this area, as in the rest of the country, is based on free enterprise and a free market and is related to several persons and bodies who can benefit from the correct operation of the system. This makes it difficult to determine an output in economic terms.

The only appreciable output is therefore the quantity of water available to farmers. As the flow coming from Chidro Spring is a continuous free resource, the best solution is to exploit it as is, at a constant, maximum rate. Also the treated wastewater is free, and thus the need to avoid any environmental harm requires that it be removed as quickly as possible from any potential source of contamination. Also this water should be exploited in its maximum available quantities.

Any decision regarding the optimal operation of the system is therefore confined to the utilization of the flow withdrawn from Sinni Aqueduct and that extracted by means of pumping wells.

As far as the quantity withdrawn from the Sinni Aqueduct is concerned, a working rule should be defined every day to regulate the valves at the "intake tower" G. In the case of the pumping wells, it would be most desirable to have an operation rule for each one of them. However, since this is unrealistic, the allotment of water quantities to groups of wells aggregated in limited zones seems to be the only possible solution.

The output trajectory must therefore be defined as the quantity of water required in order to meet the farmers' demand, namely:

i-from Sinni Aqueduct

ii-from the underground

iii-from Manduria reservoir

7. THE "INCIDENTS"

As mentioned above, the average quantity of water to be delivered to the crops in the irrigation season has been evaluated to be in the range of 2,300 m^3 per hectare. Combining the 0.80 m^3/s of Chidro Spring with the 2.40 m^3/s from Sinni Aqueduct and the 0.613 m^3/s of treated wastewater, a total flow of 3.61 m^3/s can be achieved. This flow will serve 12,450 hectares and, in order to assure the quantity required during the 175 irrigation days, it must be delivered for an average daily period of about 11 hours.

To irrigate the remaining 8,000 hectares, about 105,000 m^3 are required from Manduria reservoir, with a flow of 2.58 m^3/s. The total flow involved in this experiment is therefore 6.59 m^3/s, to be considered an average value for 11 hours a day. This is indeed the optimal situation and corresponds to a specific value of 0.32 l/s per hectare.

The salt concentration of water from Chidro Spring is about 3,000 g/m^3. After mixing its flow of 0.80 m^3/s with the 2.40 m^3/s derived from the Sinni Aqueduct, the concentration is decreased to 750 g/m^3. This value is tolerated by most crops and does not cause any harm to the soil.

If this water is in turn mixed with that coming from the wells, which, as seen in Table 2, has an average salinity of 1,000 g/m^3, the resulting concentration will be further decreased to a value that will obviously depend on the quantities involved.

The total quantity of treated wastewater, up to 0.61 m^3/s, has a negligible content of salts and, from the viewpoint of salinity, can be safely delivered to the crops.

In line with the above, one of the possible incidents to be expected is the shortage of water from the Sinni Aqueduct. The overall priority for the freshwater transported is in fact for potable needs. This demand can be particularly pressing during the driest summer, taking into consideration also the growing needs of developing tourism. This sometimes means that more water must be released towards the utilizations downstream and that the pre-established value of 2.40 m^3/s at the "intake tower" must be reduced.

For the purposes of this methodological exercise, let us suppose, first of all, that the whole flow of 2.40 m^3/s must be diverted downstream.

Table 2. Characteristics of the pumping wells in the area.

IDENTIFICATION NUMBER	AVERAGE DISCHARGE (l/s)	DRAWDOWN (m)	AVERAGE SALINITY ($\times 10^3 g/m^3$)
1	20	0.5	1.8
2	30	1.0	1.4
3	15	2.5	0.9
4	5	2.0	0.8
5	20	0.5	1.5
6	20	3.0	0.5
7	20	1.9	0.7
8	30	1.7	0.9
9	5	3.5	0.5
10	5	5.0	0.6
11	10	1.8	1.0
12	5	3.0	0.4
13	--	---	---
14	10	1.5	0.6
15	25	0.2	1.2
TOTAL DISCARGE	220	TOTAL SALINITY	1.0

At this stage, the farmers should be induced to increase the rate at the pumping wells to meet the demand for irrigation and also for the purpose of diluting the saline water of Chidro Spring. However, over-exploitation of the wells will cause the water table to drop to a level which may be detrimental to the conservation of the freshwater layer inside the aquifer.

On the other hand, keeping the amount of extracted groundwater to its maximum of 0.20 m^3/s and mixing it with the water from Chidro Spring will give rise to practically unacceptable salinity values. If the total available wastewater is assumed to be mixed with the total available groundwater, the final salt concentration will reach the value of 2,600 g/m^3, which is still unacceptable for the crops. Greater concentrations can be tolerated, but only on occasion and for short times. It is therefore necessary to reduce also the quantity of water taken from Chidro Spring, and the total quantity of water consequently becomes too small to satisfy the overall demand.

To conclude, a cutoff at the Sinni Aqueduct will reduce the quantity of usable water much more than would be expected at a first glance. Water from Manduria reservoir is thus the last available resource, but it is not supposed to be always sufficient for an area exceeding the 8,000 hectares for which the reservoir was designed.

Another incident that has occurred several times so far is that wastewater treatment temporarily becomes ineffective. This may be due to the delayed installation of facilities or to bad management of the existing ones (in too many places urban and industrial treatment plants, often built in accordance with the most advanced technology, cannot be operated due to bureaucratic reasons mainly related to the hiring of qualified personnel).

As, for the time being, there are no possibilities of transporting the untreated wastewater outside this area (indeed, a feasibility study for an emergency collecting canal discharging into the sea is being carried out), the first solution to the problem is to discharge the untreated wastewater directly onto the topsoil or into the aquifer. Such processes fall easily within the reach of the juridical authorities who are now particularly wary about health safeguards and environmental protection. One of the most common measures taken by them is to stop some industrial activities deemed to be the major sources of pollution, no matter how economically promising they are. This type of incident becomes evident when the water quality parameters exceed the imposed limits of acceptability.

A small quantity of contaminated water spilled on the ground may cause only a little temporary harm to the environment and human health. Conversely, a large quantity of untreated water, like that coming from the outlet of urban or industrial sewers, discharged on dry ground, first generates a stagnant and evil smelling pond and then favours percolation through the soil and consequent contamination of the underlying aquifer.

The risk of causing harm to the environment can be decreased by mixing this contaminated water with fresh and saline water drawn from Chidro Spring and Sinni Aqueduct. If such an incident occurs during the wet season, Chidro and Sinni water can be used freely.

During the dry season the water resulting from such a mixing may be used for irrigation only if dilution has decreased the concentration of those pollutants harmful to the crops and, at the same time, if the salinity is kept within acceptable limits. Otherwise the water must be discharged on authorized sites under the supervision of the health authority, thus foregoing a useful contribution to irrigation.

It is expected that the use of freshwater and saline water for diluting polluted wastewater may have higher priority for sanitary and health reasons, so a conflict with the farmers will arise. In such a situation the farmers can rely only on the water withdrawn from pumping wells and on that stored in the Manduria reservoir.

The possibility of insufficient water being stored in Manduria reservoir also exists.

It is not unrealistic to couple the two incidents described above, since a shortage at Sinni Aqueduct might occur at the same time as a treatment plant breaking down. This will necessitate diverting all the flow of Chidro Spring and resorting to wells also for the purpose of diluting a contaminated discharge.

8. THE TIME SCALE

All the events characterizing this system occur in a short time interval. Rainfall and other hydrological events, although lasting only a few hours, are always expressed as daily characteristics.

A daily time scale is also useful for considering the working conditions of the sewage, even though using average values, in accordance with some specific singularities like the availability of water supplied to the community, the efficiency of treatment plants, and the occurrence of highly concentrated pollutants in the waste water.

In order to conduct an accurate analysis of the system, a very long temporal horizon is necessary so that enough events are included for there to be a good probability of including incidents and singularities. At any rate, as the system is supposed to work only during the irrigation season which begins April 8 and ends September 30, the "time vector" is made up of 175 elements corresponding to the whole sequence of irrigation days.

9. THE SYSTEM EXPERIMENT

Let Q(i,j) be the farmers' demand at site i and day j. Under the assumptions formulated above, the local index i can represent the various irrigation districts in the area considered. For the sake of simplicity, following the partition indicated in Section 4, this area can be split into only three sub-areas, irrigated respectively by means of extracted groundwater and treated wastewater (i=1; 3,900 hectares), by means of water stored in the Manduria reservoir (i=2; 8,000 hectares), and by means of water mixed from Chidro Spring and Sinni Aqueduct (i=3; 8,550 hectares). The following balance equations can be written

$$
\begin{aligned}
Q(1,j) &= U(j) + W(j) \\
Q(2,j) &= M(j) \\
Q(3,j) &= C(j) + S(j)
\end{aligned}
\tag{1}
$$

where

U(j) is water withdrawn from underground

W(j) is treated wastewater

M(j) is water released from Manduria reservoir

C(j) is water taken from Chidro Spring

S(j) is water derived from Sinni Aqueduct

The three sub-areas identified above are not necessarily independent of one another but, in accordance with the operation rules defined in the previous sections, they can be partly or totally interrelated for a certain time. For this reason it is better to refer to the total demand of the area by means of the following global equation

$$Q(j) = U(j) + W(j) + M(j) + C)(j) + S(j) \qquad (2)$$

Equations (1) are therefore a special case of equation (2).

10. SOME "INCIDENT MODES"

Equations (1) can be used to express the incidents. In order to conduct the analysis, some "mode" functions have to be defined.

If the available resources are temporarily in a position of not fulfilling the farmers' demand, the right-hand sides of (1) are less than the Q's, or

$$\frac{Q(1,j) - (U(j) + W(j))}{Q(,j)} = \lambda(j)$$

$$\frac{Q(2,j) - M(j)}{Q(2,j)} = \mu(j)$$

$$\frac{Q(3,j) - (C(j) + S(j))}{Q(3,j)} = \nu(j)$$

where λ , μ and ν are continuous variables in the field 0-1. The value 0 means the farmers' demand is completely fulfilled, the value 1 means zero water availability.

The analysis can therefore be conducted simply by considering these terms, and it seems quite correct to assume they are "incident modes" for the shortage of water.

Such "modes" are functions of other terms which can either be stochastic or deterministic.

The state of the system's failure is assumed when $\lambda = \mu = \nu = 1$.

11. PERFORMANCE INDICES

Water is necessary to the crops throughout the irrigation season, except for the rainy days (6 days on the average). For a few other days water may not be necessary, depending on the farmers' working calendar (ploughing, harvesting). Water delivery can therefore be stopped at most for 20 days, which means the "Grade of Service" is assumed equal to 88%.

If not enough water is available to satisfy the demand, farmers can rotate delivery to crops on a daily basis, taking care that a single crop is not left without water for more than a couple of days. In this way the crops can remain productive and suffer less damage. This is not to be considered an incident and, as shown in Figure 4, a daily term δ (j) can be introduced to quantify this reduction of water; the daily "quality of service" can therefore be expressed as

$$1 - \delta (j)$$

A weighted mean of this term can be taken for the total irrigation period. For the purpose of this exercise, this term becomes δ - 1.18 m^3/s. Therefore, it may be written that

$$\frac{Q - 1.18}{Q} = \frac{6.59 - 1.18}{6.59} = 0.82$$

The limit above which water delivery can be considered satisfactory is 82%.

In accordance with the above, a delay in delivering the required quantity may not be harmful. A two-day lag is assumed for the hottest and driest weeks, but as far as water quality control is concerned, all possible discharge of wastewater must be immediately stopped. The "speed of response" is therefore confined to one day.

Figure 5 shows the salinity that results from mixing water from Chidro Spring and Sinni Aqueduct. If the salt concentration is above a certain limit, the resulting water cannot be used and is constantly $\nu = 1$. As soon as the acceptability threshold is reached, water can be accepted by the crops and ν can be assumed equal to zero. Note that the variation from 1 to 0 is assumed to be discontinouous: in practice a continuous variation band could be considered, since there is a "transition zone" in which salinity can be partly tolerated by some crops for a short time.

A situation which is more relevant to environmental considerations is shown in Figure 6, where water from pumping wells is supposed to be used in order to abate pollution caused by untreated wastewater having a BOD content of 50 g/m^3 (the maximum acceptable value stated by the Italian legislation is 40 g/m^3). To achieve acceptable values, at least 20% groundwater is necessary. Also in this case a discontinuous variation of ν is considered.

A quite different situation is shown in Figure 7 regarding the Manduria Reservoir. A minimum available flow is identified, below which there is not enough water for irrigation. Above this flow it is possible to deliver water to some parts of the area or to the crops more sensitive to water shortage; the more water available, the larger the portion of the total area that can be irrigated. The "incident mode" is therefore expressed as a continuous function of the water level in the reservoir.

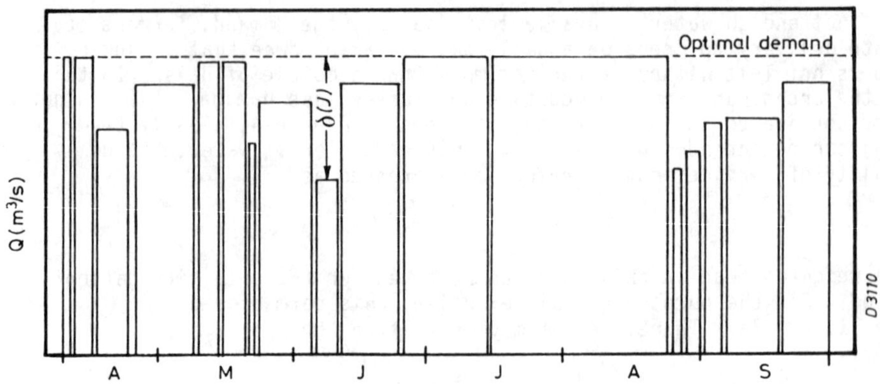

Figure 4. Seasonal variation of water demand.

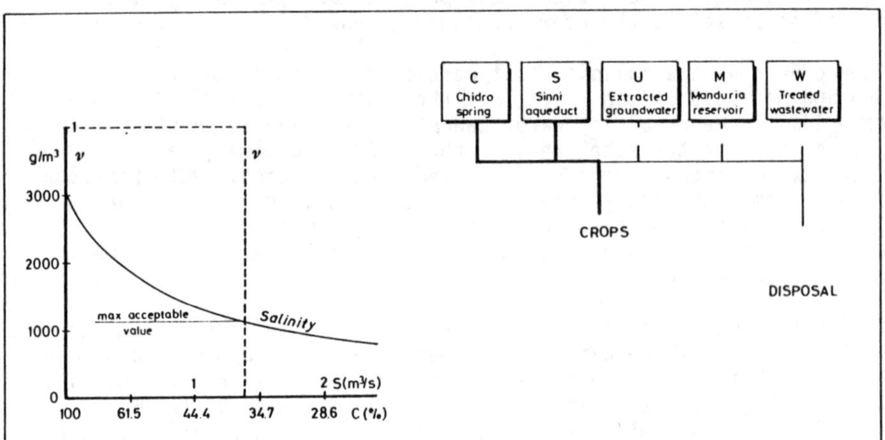

Figure 5. Combined use of water from Chidro Spring (C) and Sinni Aqueduct (S) with the purpose of reducing the salinity, and the resulting "incident mode" ν.

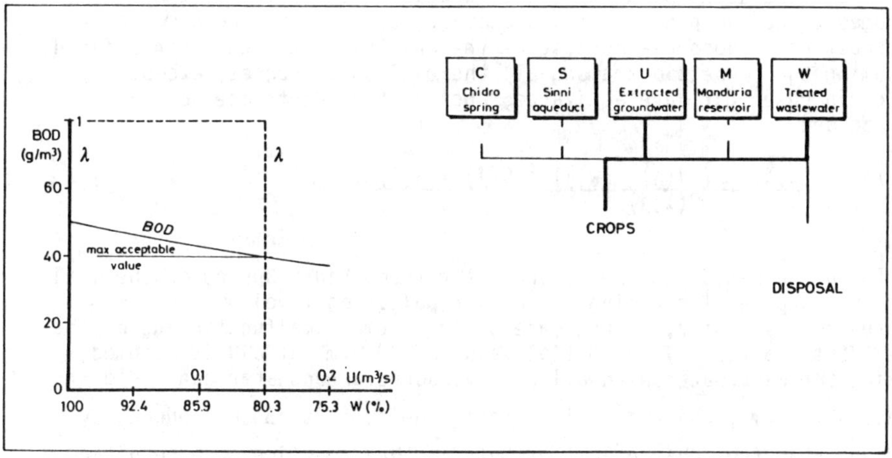

Figure 6. Combined use of extracted groundwater (U) and untreated wastewater (W) having 50 g/m of BOD, with the purpose of abating the pollution, and resulting "incident mode" λ.

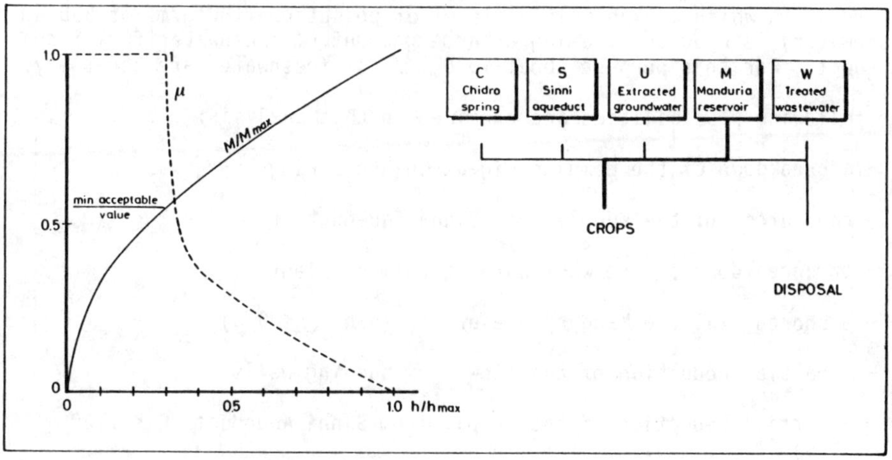

Figure 7. Evaluation of the "incident mode" μ at the Manduria reservoir.

In the preceding paragraphs the various parts of the system are assumed to be independent of one another, but in practice a mutual interaction is possible because of the way the system has been defined. Assuming the connections among all the available sources, except the Manduria reservoir, are activated, another "incident mode" can be introduced

$$\lambda' = \frac{Q(1,j) - (V(j) + W(j) + C(j) + S(J))}{Q(1,j)}$$

In such a situation the saline water from Chidro Spring can be used for the purpose of reducing the higher pollution level due to the untreated wastewater, in the case of an assumed malfunctioning of the treatment plants. If an initial value of 70 g/m^3 of BOD is assumed, mixing the wastewater with all the extracted groundwater can yield at least 52.7 g/m^3, which is still unacceptable; this can be reduced by adding water from Chidro Spring, but, at the same time, the resulting salinity increases. Figure 8 shows the possible behaviour, together with an interval of acceptability for both pollutants and salinity. As a consequence, the "incident mode" λ', initially equal to 1 owing to the high level of pollution, falls sharply to 0 after the level of acceptability is reached and returns to 1 beyond an increased value of salt concentration.

Another combination of different sources of water is shown in Figure 9, in which a high initial level of pollution (150 g/m^3 of BOD in wastewater) is reduced by using a large amount of freshwater from Sinni Aqueduct. For this purpose about 1.70 m^3/s of freshwater are necessary.

The following possibilities are examined in this analysis:

- a breakdown of the pumping rig at Chidro Spring
- the cutoff of the supply from Sinni Aqueduct
- the breakdown of the wastewater treatment plant
- a shortage at the Manduria reservoir ($h/h_{max} < 0.3$)
- a partial reduction of the flow from pumping wells
- a partial reduction of the supply from Sinni Aqueduct ($Q < 1.00$ m^3/s)

The "incident modes" described in Section 10 have a different probability of occurrence. Table 3 shows an "interaction matrix" in which the various possible combinations have been examined. The first column of this matrix contains the incident's frequency (how many times during the irrigation period considered), the second contains the incident's duration in days, and the third the probability. The "boxes" of the matrix contain the joint probability, p(m,n), resulting from the product of the simple probabilities defined respectively in column m and row n ($m \neq n$).

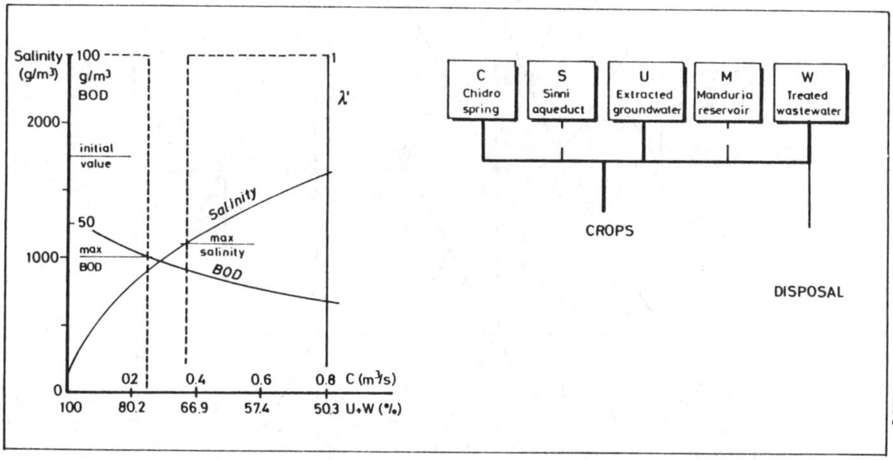

Figure 8. Combined use of water from Chidro Spring (C), extracted groundwater (U) and untreated wastewater (W) with the purpose of reducing the pollution level and at the same time maintaining an acceptable salinity. The "incident mode" λ' is also shown.

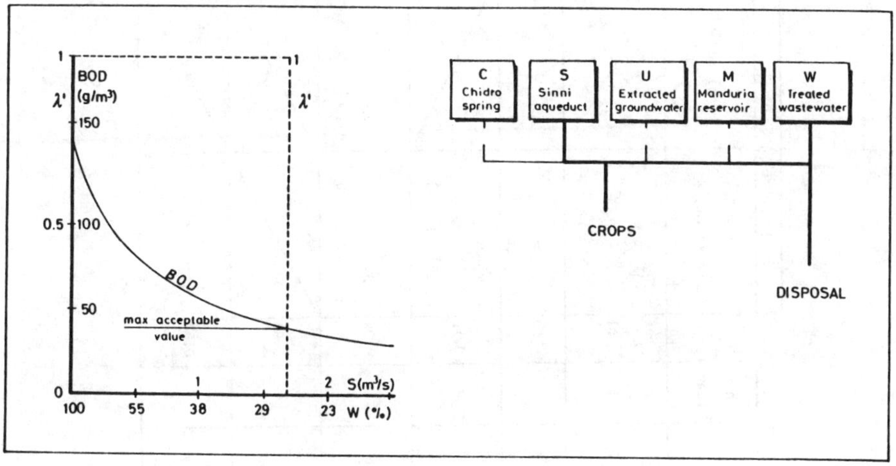

Figure 9. Combined use of water from Sinni Aqueduct (S) and untreated wastewater (W) having initially 150 g/m of BOD, with the purpose of abating pollution, and the corresponding "incident mode" λ'.

Table 3. Interaction matrix.

Incident	F	D (day)	P	1 Chidro Breakdown	2 Sinni cut-off	3 Sinni partial reduction	4 Wastewater treat_ ment Breakdon	5 Manduria shortage	6 Underground partial reduction
Chidro Breakdown (C = 0)	1	2	0.01	$\nu = 1$					
Sinni cut-off (S = 0)	8	2	0.09	$\nu = 1$ $P(1.2) = 9.10^{-4}$	$\nu = 1$				
Sinni partial reduction ($S \leq 1\ m^3/s$)	10	2	0.11	$\nu = 1$ $P(1.3) = 11 \cdot 10^{-4}$		$\nu = 1$			
Wastewater treat- ment Breakdown ($BOD > 70 g/m^3$)	60	1	0.34	$\lambda = 1$ $\lambda' = 1$ $P(1.4) = 3.4 \cdot 10^{-3}$	$\lambda' = 1$ $P(2.4) = 4.1 \cdot 10^{-3}$	$\lambda' = 1$ $P(3.4) = 4.1 \cdot 10^{-2}$	$\lambda = 1$		
Manduria shortage ($h/h_{max} \leq 0,3$)	5	1	0.03					$\mu = 1$	
Undeground partial reduction ($U \leq 0.1\ m^3/s$)	4	1	0.02				$\lambda = 1$		

The resulting probability, P, of all the incidents which can occur to the system is the probability of the summation of all the events yielding $\lambda' = \lambda = \mu = \nu = 1$; that is, of the six initial incidents and the five incidents resulting from a combination of the former ones. In accordance with Venn algebra, such a probability is the summation of the probabilities of the single events minus the "combined probabilities", which in this case have a very small value and can be neglected.

Since the reliability is the complementary value of the resulting probability, it follows that

$$IR = 1 - P$$

and, with the numerical values considered, IR = 0.327. In other words, the system is expected to work 57 days during the irrigation period without any incident.

Table 3 also shows that the total probability of having $\nu = 1$ is 0.22 and that this occurs as a combination of effects having the maximum composed probability $p(1,3) = 12 \times 10^{-4}$.

Determination of other indices following the general theory (Hashimoto et al, 1982; Duckstein and Plate, 1983; Bredehoeft and Young, 1983) encounters more difficulty, and in some cases it is a meaningless task. The "index of mission reliability" and the "index of availability" are already included in the recommendation that the system should provide water under zero rainfall conditions, i.e., for 169 days during the irrigation period.

The occurrence of a breakdown at the wastewater treatment plant on a rainy day is not an incident, as, with no demand for agriculture, the total available water can be used to dilute a polluting discharge.

Similarly, the recovery from an incident or a failure is immediate, i.e., as soon as a pump or a valve has been repaired. This system is supposed to have full resilience.

On the other hand, the vulnerability would be a more interesting aspect. However, although not involving the system directly, it does affect the crops, which, after a certain number of days without water, lose their quality and their economic return to the farmers.

12. OPERATION POLICIES

The occurrence of any of the incidents described above is a peculiarity of this system. An operation rule, suitable for adoption in management activities, will depend simply on the possibility that one or more sources of water supply are temporarily missing. The managers of such a system--in practice the people who operate the various hydraulic devices and the farmers--have a clear idea of what to do in order to overcome an event, even of the most serious kind, e.g., the ineffectiveness of a wastewater treatment plant.

From this exercise they must acquire the capacity to assess the quality of the working facilities, especially when a question arises about the possibility of growing a certain crop that has an economic return dependent on irrigation. In other words, these people should rely on the answer to the following questions:

a) Given that many incidents cannot be completely avoided, is there any possibility of reducing their number and their frequency, at the same time increasing the reliability of the system?

b) Suppose an intervention along these lines is possible and effective, how can the improved situation be evaluated, also in comparison with other possible alternative solutions?

Obviously there are very few chances of reducing the possibility of mechanical breakdowns and power outages, even though the most sophisticated equipment has been installed. This possibility is related to the complexity of the system and is expected to increase in the future as a result of wear and age.

The only possible effective action is to complete the treatment plants as quickly as possible and to reduce the probability of polluted water being discharged to the same level that would exist with an efficiently operating treatment plant.

The role of environmental control is therefore very important also for the high performance and good productivity of an irrigation scheme.

Once such a solution has been achieved, the probability of a failure of the wastewater treatment plant can be compared with the overall probabilities of the system's failures. Starting again from Table 4, the following relation can be written

$$r = P/PTP$$

PTP being the probability of a breakdown at the treatment plant.

The value of r will increase as the probability of a breakdown at the treatment plant approaches 0. This term may therefore be considered a "figure of merit" of the system.

It is easy to show that improved efficiency of the treatment plant capable of reducing the breakdown probability to 0.20 will raise the "figure of merit" to 3.23 from the present value of 1.90: this term therefore appears very useful in emphasizing the effect of any intervention on the system.

13. FINAL CONSIDERATIONS

The application of risk analysis theory to a system having these features can prove a helpful tool for pointing out some management rules, in a way which could easily be put in the hands of a responsible authority.

The fundamental aspects considered in the preceding sections point to the need for reviewing and summarizing the kind of information that is necessary and the level of detail required in order to apply this methodology. The most essential data are the following:

- typical hydrological pattern
- water demand
- threshold of acceptability and effects of water shortage
- nature and frequency of incidents

If such information is available, the analysis can be carried out in great detail, for instance emphasizing the effects of a partial reduction of water. This would mean that only selected crops could be irrigated while the others would be left without water. With respect to the environmental problems, a more detailed analysis could help focus on the impact caused in more restricted zones, where the spreading of untreated (or partially treated) wastewater can cause serious consequences.

There is still some doubt as to the possibilites of bringing these methods into the purview of field engineers. Pursuing this analysis seems to be the primary task for the purpose of showing the real possibilities and the limits of the methods through the solution of some specific numerical examples. This is the goal for the continuation of this exercise.

The most interesting aspect will be to consider also the economic impact (Simonvic and Orlob, 1984), particularly in relation to the costs involved in running the various plants and recovering from incidents. In this respect it will be advisable to define some objective functions, in order to minimize total costs of the system's operation.

ACKNOWLEDGEMENT

For the interest shown in this exercise and the essential information made available, the Authors are indebted to the Southern Italy Development Fund, the Regional Administration of Apulia and the Land Improvement Consortium of Arneo.

REFERENCES

Benedini, M., G. Giuliano, B. Sirangelo, and S. Troisi (1982) Preliminary results from a methodological study in the field of groundwater resources management, Proceedings of the Symposium on Mathematics and Computing Simulation, Southampton.

Benedini, M., B. Sirangelo, S. Troisi, and M. Vurro (1983) Mathematical model of a coastal aquifer subject to seawater intrusion: the Nardo' aquifer (Italy) as an example, Proc. of the 8th Salt Water Intrusion Meeting, Bari.

Bredehoeft, J.D., and R.A. Young (1983) Conjunctive use of groundwater and surface water for irrigated agriculture: Risk aversion, Water Resour. Res., Vol. 19, No. 5.

Cicioni, G.B., G. Giuliano, and S. Troisi (1982) Formulation and application of a linear programming model of aquifer management, Proc. of the Internat. Conf. on Modern Approach to Groundwater Resources Management, Int. Assoc. for Hydraulic Research, Capri (Italy).

Cotecchia, V. (1975) Studies and investigations on Apulian groundwaters and intruding seawaters (Salento Peninsula), Quaderno IRSA (Italian Water Research Institute), No. 20.

Duckstein, L. and E.J. Plate (1983) A system framework for hydrologic reliability with application to design structures, Proc. of the Symposium of the International Association for Hydrological Sciences, Session 7, Hamburg.

Hashimoto, T., J.R. Stedinger, and D.P. Loucks (1982) Reliability, resiliency and vulnerability criteria for water resources system performance evaluation, Water Resour. Res., Vol. 20, No. 6.

Simonovic, S.P., and G.T. Orlob (1984) Risk-reliability programming for optimal water quality control Water Res. Res., Vol. 18, No. 1.

Sirangelo, B., and S. Troisi (1980) A methodological study of new management techniques of water resources management, 1: The hydrodynamic simulation model IDROSIM. Quaderno IRSA (Italian Water Research Institute), No. 50 (in Italian).

IV. RELIABILITY AND RISK AS FACTORS IN DECISION MAKING

IV. RELIABILITY AND RISK AS FACTORS IN DECISION-MAKING

In this section, reliability and risk analyses are embedded into various models of decision-making under uncertainty. In fact, the third part of this section tends to show that, as suggested at the end of the Introduction paper, reliability and risk are integral parts of a multiple objective analysis.

Consider the elements of a decision problem using the two parallel examples of levee design and reservoir release, as given in Table 1. The three parts of this section may be considered as dealing with various aspects of Table 1. The first part, composed of three papers, begins with Sorooshian's study of the uncertain features of hydrologic input or generalized load. Thus, Sorooshian's paper contributes to the understanding of part 2a of the state of nature S (discharge Q). For example, complex rainfall-runoff models may be quite "unreliable" when calibrated with data that do not trigger certain interactions; then whenever the real world provides the right combination of inputs, large errors in the hydrologic output variable, which are propagated through the calculation of the generalized load λ to the overall reliability estimate, may result.

The second paper, by Klemes, deals with both parts a) and b) of the state of nature. Specifically, Klemes reinforces Sorooshian's assertions about the unreliability of hydrologic models that are used for assessing hydrologic reliability. Either empirical or causal models may be inadequate because of non-stationarity of processes or non-homogeneity of extreme events, as described by their probability distribution functions. Klemes then provides examples and suggestions for improving what may be considered as the state of nature S (elements 2a and 2b of Table 1).

In the third paper, Bernier proposes (implicitly) that the difficulties raised in Sorooshian's and Klemes' papers be modeled via a Bayesian approach. Such a Bayesian decision theoretic approach covers all the elements of Table 1; however, the consideration of model uncertainty (entry 2b) is only mentioned in Bernier's paper.

Part IV.2, Applications and Advances, deals with more complicated systems and techniques than Part IV.1. One paper refers to groundwater control (Bogardi et al), another to surface water allocation schemes (Krzystofowicz) and the third to advanced Bayesian methodology applied to minimization of the consequences of extreme events, namely floods (Bernier). Referring again to Table 1, it appears that all three papers possess the essential elements of a decision problem with, however, the following variations:

(i) Bogardi et al use a combination of fault tree and Bayesian analysis to estimate the figure of merit "implementation reliability" (see the Introduction paper). Both resistance and loading are taken as

Table 1. Elements of Two Typical Hydrologic Decision Problems

	Levee design	Reservoir Release
1. Decision space A	Height h of levee	Released volume v
2. State of nature S:		
a) Natural	a) Flood stage $\underset{\sim}{x}$	a) Net hydrologic input
b) Probability model uncertainty	b) Class of possible probability distribution function (DF) $\{F(x\|\theta)\}$ of $\underset{\sim}{x}$	b) Class of $\underset{\sim}{Q}$ possible DF of $\underset{\sim}{Q}$: $\{F(q\|\lambda)\}$
c) Parameter uncertainty	c) Space $\{\theta\}$ of the parameter vectors θ of F $(x\|\theta)$	c) Space $\{\lambda\}$ of the parameter vectors λ of $F(q\|\lambda)$
d) Resistance	d) Same as a) b) c) for resistance ρ	d) Same as a) b) c) for resistance ρ
3. Probability distributions P	a) Model $F(x\|\theta)$ b) Prior distribution $\pi(\theta)$ ...	a) Model $F(q\|\lambda)$ b) Prior distribution $\pi(\lambda)$ ...
4. Performance index, figure of merit, utility or loss function L (single or multi-objective)	Loss L $(H,\underset{\sim}{x})$ or risk function $g(h,\theta)$	Loss L $(v,\underset{\sim}{q})$ or risk function $g(v,\lambda)$

random, with probability distributions subject to parameter uncertainty. Model uncertainty (2b in Table 1) is not considered.

(ii) Krzystofowicz emphasizes the expected utility approach and relates supply reliability to utility-type performance indices. Model and sample uncertainty are not considered.

(iii) Bernier uses all the elements of Table 1 with the performance index represented by a physical quantity, that is, peak flood discharge.

Table 2 summarizes the specific elements of decision theory considered in the above three papers.

Part IV.3, Multicriterion and Conflict Analysis, has four papers emphasizing the point made earlier that incident and failure-related criteria are often in conflict with one another, and also with economic or social criteria. Two papers deal with agricultural problems in different climatological conditions (Hiessl, Bogárdi and Bárdossy), one with reservoir control (Guariso et al), and one with conflict resolution in international river basin management (Hipel and Fraser).

Generally speaking, multicriterion decision-making is an emerging branch of water resources management, as may be seen in sampling the recent literature and examining the bibliography of several papers in this book, especially in Sections III and IV. Just as hydraulic engineers were advised in the introduction to Section II to modify design practices in the direction of performing reliability analysis under random load and resistance, water resources engineers have much to gain by becoming acquainted with multiple objective and conflict analysis. For example, the trade-offs between cost and reliability, or reliability and resilience, can then be made explicitly. Furthermore, in case of conflicts between interest groups, a systematic approach for dialogue is provided. Table 3 compares the essential features of the four papers in Part IV.3. It may be noted that the first three papers account for hydrologic input uncertainty and use numerical criteria, while the paper by Hipel and Fraser treats a case where non-numerical criteria are present.

Table 2. Papers in Section IV.2 Versus Elements of a Decision Problem

	Bogardi et al	Krzysztofowicz	Bernier
1. Decison space A	Elements of protection system against inrush hazard	Water supply target	Estimates of parameter ρ of DF maximum flood discharge $F(q\|\rho)$ and of occurrence rate λ of floods
2. State of nature S	Loading L capacities C parameters of DF of L and C	Actual water supply V	q, ρ, λ
3. Probability distribution	Models for L and C given the parameters	DF of V (no parameter uncertainty)	$\pi(\lambda)$, DF $F(q\|\rho)$, priors $\pi(\rho)$, $\pi(\lambda)$
4. PI or FM	Implementation reliability FIR	Expected utility (or reliability versus height)	Maximum annual flood flood q

Table 3. Comparison of Four Multicriterion Papers

Author	Type of Problem	Criteria or Objectives in Conflict	Solution Techniques
Hiessl	Irrigation under uncertain supply in semi-arid regions	-- Crop yield -- Reliability -- Resilience -- Vulnerability	-- Mathematical progamming (simulation - optimization) -- ε-Constraint technique
Bogardi and Bardossy	Agricultural watershed management under rainfall uncertainty	-- Economic (income, investment) -- Environmental (phosphorus and sediment loading) -- Hydrologic (water yield)	-- Composite programming
Guariso, Orlovski, Rinaldi and Soncini-Sessa	Management of a multi-purpose reservoir with inflow forecasting	-- Lake floods -- Evaporation -- Supply to users	Compromise programming with $p = \infty$ (minimax)
Hipel and Fraser	Conflict between interest groups regarding a large scale engineering project	U.S. & Canadian -- Political groups -- Environmental groups -- Engineering	Metagame analysis

THE IMPACT OF CATCHMENT MODELING ON HYDROLOGIC RELIABILITY

Soroosh Sorooshian

Department of Hydrology and Water Resources, University of Arizona, Tucson, AZ 85721

ABSTRACT

Plate and Duckstein (this proceedings) advance the notion of "hydrologic reliability" which interacts with two other quantities, namely "hydraulic reliability" and "structural reliability" to give a measure of total system reliability. The hydrologic variable, discharge, which is the output of a selected catchment model, is the main ingredient used in the evaluation of the hydrologic reliability. The purpose of this paper is to discuss some of the factors that influence the accuracy of discharge estimates by catchment models. It is argued that as long as the accuracy of forecasts generated by such models is questionable, we may wish to exercise caution in accepting the computed value for the hydrologic reliability which is used to give us the overall reliability of a water resources system.

1. INTRODUCTION

In their paper entitled "Reliability in Hydraulic Design", Plate and Duckstein (this proceedings) present an approach for calculating the total reliability of a water resources project. They define a quantity called "hydrologic reliability", RE_G, which interacts with two other quantities, namely hydraulic reliability (RE_H) and structural reliability (RE_S) to give the total system reliability. The procedure for computation of RE_G is quoted from Plate and Duckstein:

> "The general input vector X consists of the continuous functions $x_i(t)$ describing discharge or rainfall, etc.--in any event, it is usually a time series of a continuous natural process which has been measured. From these inputs, a hydrologic primary variable, for example a discharge Q, is calculated by means of a model, which could be as simple as a measured stage discharge relation or as complicated as a simulation model of runoff, in which the catchment model is subjected to time functions which have been generated artificially from the input functions $x_i(t)$. The hydrologic reliability is then defined in the traditional manner by comparing the output variables from the hydrologic model, for example the peak discharges of floods Q, with a permissible discharge, for example a design discharge Q_E."

The hydrologic variable $Q(_t)$ is transformed through hydraulic calculations into a design variable (e.g., stage reading). The information regarding the design variable(s) is subsequently used in selecting the systems structure. Such information is used to obtain RE_G, RE_H, and RE_S and subsequently to generate the estimate for the total system reliability. The part that is of interest to this author is the role and the importance which the calculated hydrologic variable $Q(_t)$ plays in the computation of systems reliability. If such an estimate is inaccurate, it is likely that the accuracy of the computed total system reliability, R_E, will be questionable. Therefore, the purpose of this paper is to discuss some of the factors that influence the accuracy of discharge estimates by catchment models. It is argued that as long as the accuracy of forecasts generated by catchment models is questionable, then we may wish to exercise caution in accepting the computed value as a measure of reliability of our water resources system.

MODELS OF THE CATCHMENT PROCESS

For a long time, hydrologists have labored to understand the physics underlying the various subprocesses observed to occur in the hydrologic cycle. Their work, both experimental and theoretical, has resulted in sophisticated representations of the behavior of soil-water interactions, and this accumulated knowledge has been instrumental in the development of various types of mathematical models. With the advent of digital computers, a new generation of hydrologic models, known as catchment models (e.g., Stanford model [Crawford and Linsley, 1966] and Sacramento model [Burnash et al., 1973]) was developed to simulate the internal description of a watershed with the primary purpose of forecasting river flows.

Most catchment models available today recognize the presence of different vertically stratified zones of soil in the ground. The soil mass is most often conceived to have two layers, or "zones", the upper and the lower, connected through vertical percolation. Each zone is commonly modeled using reservoirs (usually linear). Channel inflow consists of baseflow, interflow, and overland flow, and consideration is made for evapotranspiration depletion. The set of operations that are performed by most conceptual models can be summarized as follows: (1) interception of rainfall, (2) direct runoff from impervious areas, (3) infiltration of water from surface to upper zone, (4) surface runoff of water not accepted by the upper zone, (5) percolation of water from upper to lower zone, (6) interflow (lateral subsurface flow) from upper zone, (7) baseflow (lateral groundwater outflow into river), (8) evapotranspiration from all storages to the atmosphere, and (9) routing of river inflow to downstream gaging point. Catchment models usually accept lumped precipitation and potential evapotranspiration data input, at fixed intervals of time, which represents equivalent uniform depths of precipitation and potential evapotranspiration over the catchment area. The model output is usually the stream stage or flow volume but may also consist of evaporation water loss data and/or volume of subsurface water storage.

In spite of the relative sophistication of catchment models, their potential for providing accurate streamflow forecasts has not been fully realized, leading many engineers to question their usefulness and cost effectiveness. A major problem appears to be that catchment models contain structural peculiarities which make the application of model identification/parameter estimation methods and other useful systems theoretic concepts extremely difficult (ref: Sorooshian and Gupta, 1983; Gupta and Sorooshian, 1983). Before providing some specific examples of the above problems, let us briefly describe the model identification process.

General Identification Problem

Consider a catchment system, $\mathcal{P}$, for which various characteristics such as inputs 'u' (e.g., precipitation, thermal radiation), states 'x' (e.g., various soil moisture components), and outputs 'z' (e.g., streamflow, potential evapotranspiration) can be observed. The general identification problem can be broadly stated as that of finding a model 'M' whose input-state-output behavior is as close as possible to that of the catchment. The solution of this problem involves two major stages: (1) identification of a suitable structure (the mathematical functions relating the inputs, states, and outputs) for the model, and (2) calibration of the model parameters. This notion is geometrically demonstrated in Figure 1. Let $\mathcal{U}$ represent the universal set. If $\mathcal{M}$ represents the set of all mathematical models ($\mathcal{M} \subset \mathcal{U}$), then structure selection is equivalent to choosing a specific subset M ($\underline{\theta}$) of $\mathcal{M}$, where $\underline{\theta}$ represents the parameters of the specified model set ($\underline{\theta} \varepsilon \overline{\Theta}$, Θ = feasible set). Having done this, we must then select particular values for the parameters (say, $\underline{\theta} \varepsilon\ \Theta$) such that the model M (θ) is, in some sense, "closest" in its behavior to the catchment $\mathcal{P}$. Clearly, we would like to choose those parameter values for which the model exactly reproduces the behavior of the catchment $\mathcal{P}$ under all circumstances. Referring to Figure 1, we see that this is only possible if the catchment $\mathcal{P}$ ($\mathcal{P} \subset \mathcal{U}$) is contained within the restricted model set M (θ), $\underline{\theta} \varepsilon\ \Theta$ (e.g., M_1 = $\mathcal{P}_1$ in Figure 1). Given the extreme complexity of the catchment process, it is clear that $\mathcal{P}$ will not in general be contained in the set M ($\underline{\theta}$), (e.g., $\mathcal{M}$ = $\mathcal{P}_2$ in Figure 1), and our interest therefore lies in finding that model M ($\underline{\theta}$) whose behavior is closest to that of $\mathcal{P}$ (e.g., M_2 is "closest" to $\mathcal{P}_2$). As the geometric interpretation of Figure 1 suggests, the more realistic (less restricted) the model set M ($\underline{\theta}$), the "closer" our identified model will be to catchment $\mathcal{P}$. Note that, in general, the "closeness" of M ($\underline{\theta}$) to $\mathcal{P}$ will be evaluated using some suitably chosen estimation criterion that measures the difference between the model and catchment outputs. The success of the identification then clearly depends on appropriate treatment of both of the stages discussed above. For the purpose of simplicity, let us first assume that a suitable model structure has been successfully identified. Let us go so far as to assume that the selected model set actually contains the catchment process, $\mathcal{P}$, in which we are interested. Therefore, our problem is to establish a calibration methodology which, based on available information such as historical precipitation and streamflow observations, will enable us to find those parameters $\underline{\theta}$ for which the model M ($\underline{\theta}$) is identical in behavior to $\mathcal{P}$. This entails:

(1) Specification of a measure of "closeness" (called estimation criterion) between the model and the catchment. This is usually defined in terms of the differences between the model and catchment outputs when both are subjected to the same inputs; and

(2) Selection of a method for identifying those parameter values which "optimize" (minimize or maximize as appropriate) the chosen estimation criterion. Since the estimation

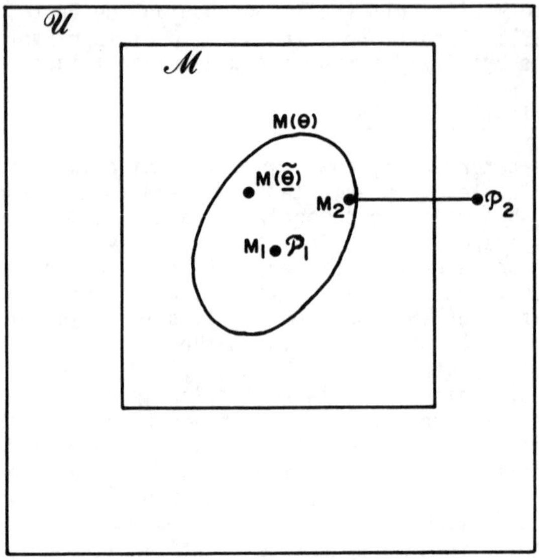

Fig. 1. Geometric interpretation of the identification problem.

criterion is usually nonlinear in the parameters, this usually involves implementation of an iterative optimization algorithm.

The mathematical representation of the above technique may be given as follows:

$$\text{optimize } g(E) = g[Q - F(I, \theta)] \tag{1}$$

where $Q^T = [q_1, \cdots, q_n]$ is the vector of single output variables (streamflows), I is the multivariable vector of input variables (usually precipitation and potential evapotranspiration records, F( ) is the deterministic model of the catchment response, θ is the vector of model parameters whose values must be estimated, E is the stochastic time series of the additive errors, and g( ) is the selected estimation criterion.

It is interesting to note that in many of the reports of studies involving catchment models, aside from a brief mention of the calibration procedure employed, there is rarely any discussion of the merits of the calibration results. The main reason for this omission is probably the lack of agreement or understanding as to what exactly is to be achieved in the calibration stage (in view of the assumptions accompanying catchment models). It is important to ask whether the purpose of calibration is (1) to obtain a "unique and conceptually realistic" parameter set which closely represents our understanding of the physical system, or (2) to obtain a (any) parameter set which gives the best possible fit between the model simulated and the observed hydrograph (i.e., for the calibration period). From the point of view of physical modeling, it is clear that a method which emphasizes both aspects is desirable. Unfortunately, as Sorooshian and Dracup (1980) pointed out, most of the commonly employed calibration methods tend to emphasize the latter aspect. The danger in this case is, as they discussed, that the resulting parameter set, though producing a close reproduction of the observed hydrograph during calibration, may result in poor performance when used for forecasting. This conjecture was strongly supported by the results reported by Sorooshian, Gupta, and Fulton (1983). Certainly, much of the blame for poor forecasting ability can be placed on the inability to obtain unique and conceptually realistic parameter sets. In the context of the calibration methodology discussed above, some researchers have looked into the causes of our inability to find accurate parameter estimates. Ibbitt and O'Donnell (1971) and Johnston and Pilgrim (1976) list the following features of catchment models and their calibration methodology as the primary reasons:

(1) interdependence between model parameters;
(2) indifference of the objective function to the values of "inactive" (threshold-type) parameters;
(3) discontinuities of the response surface; and
(4) presence of local optima due to the nonconvexity of the response surface.

Sorooshian and Gupta (1983) presented a detailed analysis of the reasons for the above properties.

Identifiability Problems in Catchment Models

Essentially, there are three elements which are responsible for the response surface problems mentioned above and which, in turn, hinder the accurate calibration of catchment models. They are: (1) model structural representation, (2) data and their associated measurement errors, and (3) imperfect representation of the physical process by the model.

Model structural representation. Let us consider the case of the simple two-parameter model depicted in Figure 2. (Note that the following simple linear reservoir model is a subprocess commonly used in conceptual R-R models where upper and lower zone storages are modeled). The two parameters are M and K, where M (unit: length L) represents the threshold parameter limiting the depth of the reservoir, and K (unit: time inverse T^{-1}) is the recession rate coefficient. Notice that surface runoff R_t (unit L) is equal to zero as long as storage level X_t (unit L) does not exceed the value of the threshold parameter M.

First, let us hypothesize that this model is a perfect representation of the physical process. This enables us to assume a true set of parameters and, for a hypothetical sequence of inputs, to generate the true watershed output. Under these ideal circumstances, the calibration of this should be rather simple. The problem with most such models is, however, that even under such perfect conditions, they are poorly identifiable.

To clarify the above point, let us denote the true values of M and K as M^* and K^*, respectively. Under our assumed ideal conditions, we can select and optimize a fitting criterion in order to estimate M^* and K^*. The

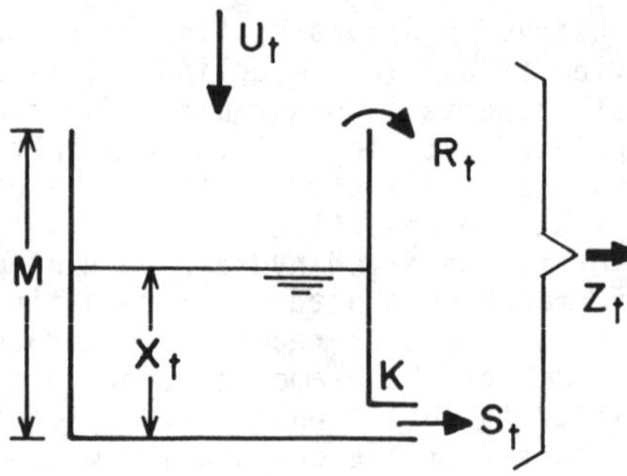

Fig. 2. A simple two-parameter linear reservoir model with threshold parameter M.

general shape of the response surface contours which are influenced by the mathematical relationship of the parameters would be similar to that shown in Figure 3 for a criterion such as the ordinary least squares (OLS) criterion. This shape may be a little different for different criteria (e.g., narrower contours, etc.) but, nevertheless, the interaction direction and the shape of the contours in the inactive zone for M should generally remain the same. In this case, it should be clear that the termination point of the optimization is dependent on the choice of initial values of the parameters, especially M. If the initial value of M is selected so that the starting point is in the region of the elongated valley, any search algorithm, even though partially successful in the K direction, will

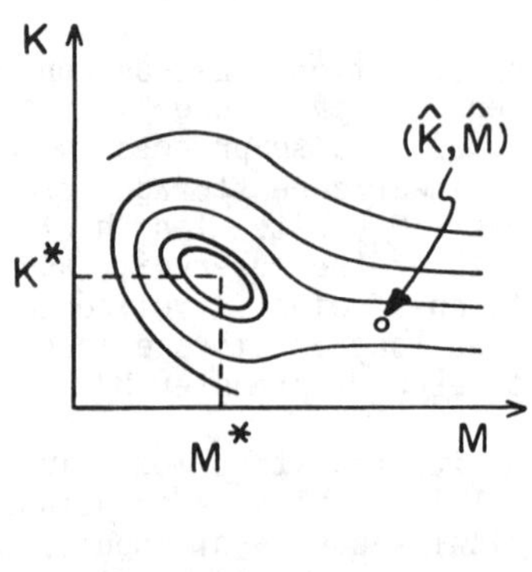

Fig. 3. Response surface contours for model of Figure 2.

make little progress (if any) in the M direction. As a result, the search will terminate at a nonoptimal point such as $\hat{M}$, $\hat{K}$.

For this simple problem, the obvious solution is to choose a small initial value of M so that the search remains in the region of elliptical contours and progresses towards the optimum (K*, M*). In a full-scale CRR model, however, the structural equations are far more complicated, and the choice of appropriate initial parameter estimates is not as obvious.

For example, in our work (Sorooshian and Gupta, 1983) with the U.S. National Weather Service's Soil Moisture Accounting Model (SMA-NWSRFS), we observed some peculiarities on the response surface of the parameter space which made it impossible to obtain a unique "best" parameter set for the model during calibration. In particular, we observed the existence of an extremely long and extended valley in the plane formed by two of the parameters of the percolation equation of the model using OLS as the estimation criterion (see Figure 4).

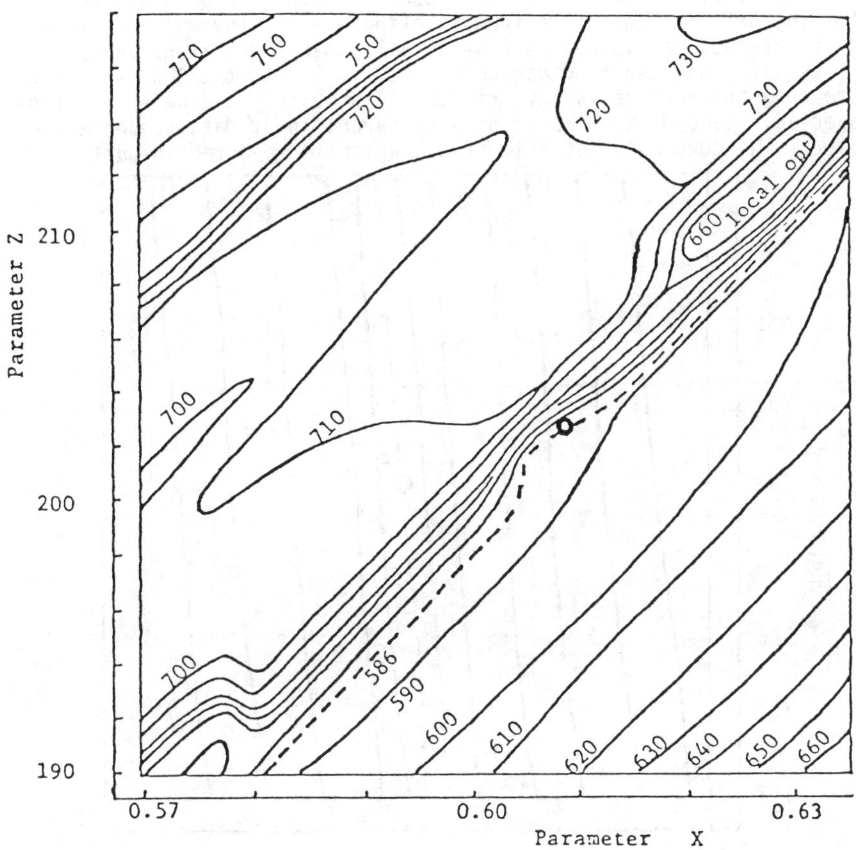

Fig. 4. OLS criterion contour map in region of OLS calibration best parameter set (Z versus X).

The above peculiarity prompted concern regarding the reason(s) for such behavior. A controlled study using synthetic data was conducted. The model parameters were set at prespecified values. The same rainfall data were used to generate a sequence of synthetic true streamflow values, using the model. Thus, the model becomes a true representation of the process. The two percolation parameters (z versus x) contour plots for the OLS criterion are presented in Figure 5a. Clearly, the response surface shows a long interactive valley which makes the presence of a unique pair of x and z meaningless. The fact that the model was assumed perfect in this study and yet the response surface showed such behavior is strong evidence that the problem was a consequence of the chosen structural representation of the percolation subprocess.

In order to ensure that the problem (i.e., the presence of the inclined and extended valley on the response surface) was indeed a structural one, two other studies were conducted. First, a different estimation criterion was employed to see if the response surface shape would be different. The criterion was a maximum likelihood estimator which filters the effect of heteroscedastic errors (for details, see Sorooshian, Gupta, and Fulton, 1983). The resulting contours were similar to the OLS case, and the same long interactive valley oriented in precisely the same direction was observed. Second, the impact of a different input data sequence was simulated. The conclusion of this study (which is reported in Gupta and

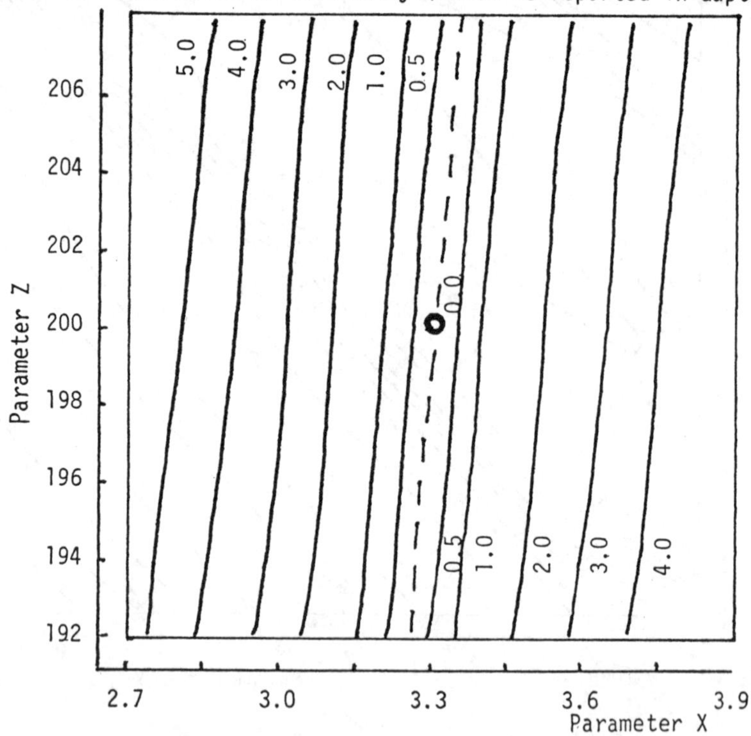

Fig. 5a. OLS criterion contour map for Z versus X using synthetic data.

Sorooshian, 1983) was that the extent of the valley was unaffected, and the only detectable difference was a progressive change in the angle of orientation of the valley (Figure 5b).

Let us now reflect on the consequences of the above observations. The most important one is that the existence of an extended valley implies an inability to obtain a unique parameter set for the model using either the manual or automatic calibration procedures. In the case of the automatic procedure, the search will terminate at the first point at the bottom of the valley that it encounters. Thus, the results of the calibration are highly dependent on the choice of parameter values used to initiate the search. Recognizing that each pair of Z and X along the valley results in the same value of the objective function, one might presume that it does not matter which pair is chosen. This presumption would be false for two basic reasons.

To understand the first reason, let us pretend that the true values of the parameters are known. Due to the nature of the calibration data being used, there will exist a valley (say, R_1) on the Z -X response surface oriented in some direction, this direction being dependent on the data (see Figure 6). Let us assume that the search technique selects a point (shown by the open circle) in the valley at some distance from the true optimum

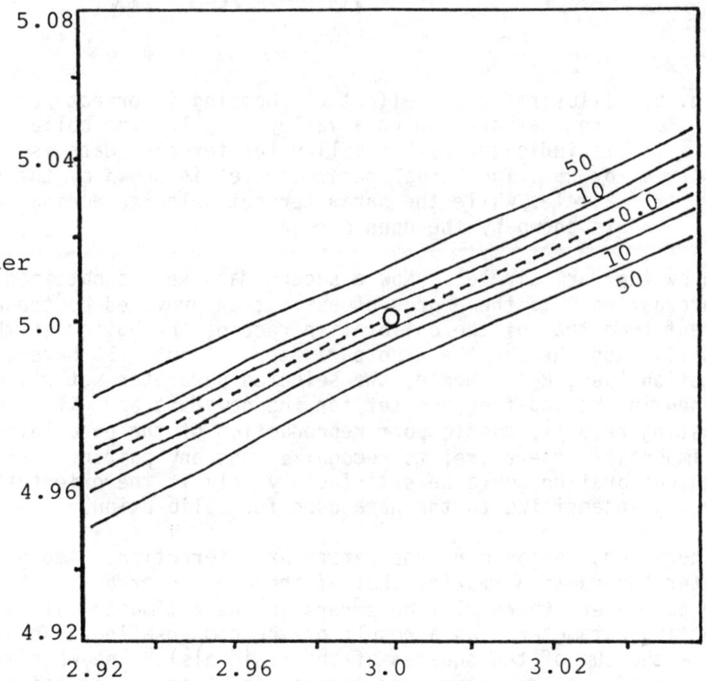

Fig. 5b. OLS criterion contour map for Z versus X using synthetic data different than the sequence used in the case in Fig. 5a.

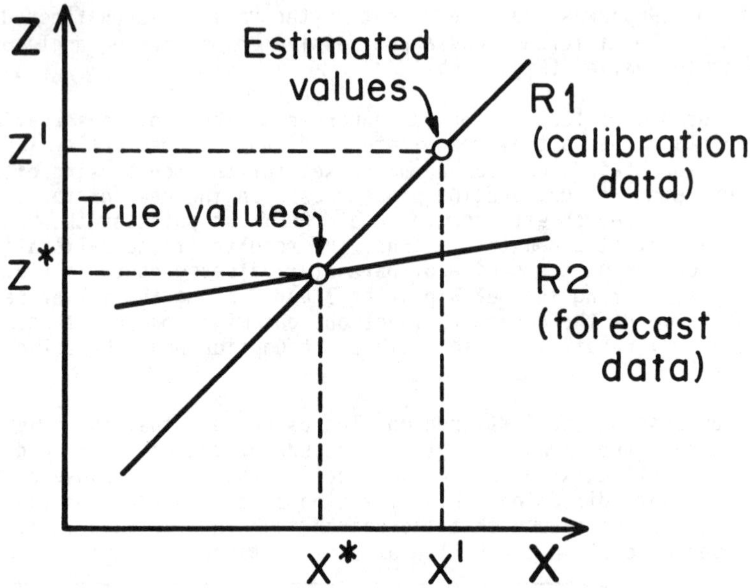

Fig. 6. Illustration of effect of choosing incorrect parameter set from calibration data valley. Valley for calibration data is indicated by R_1; valley for forecast data is indicated by R_2. The "true" parameter set is shown by the dark circle, while the parameter set selected during calibration is shown by the open circle.

(shown by the dark circle). Now a second data set is obtained to be used for forecasting. If the degree of activation provided by these data is different from that of the calibration record, the bottom of the valley will still pass through the true parameter set but will have a different orientation (say, R_2). Hence, the selected parameter set will no longer be contained in the indifference set for the new data and will give nonoptimal forecasting results, due to poor reproduction of the percolation process. It is important, therefore, to recognize that any pair of Z and X obtained through calibration would be satisfactory only if the orientation of the valley was insensitive to the data used for calibration.

The second reason concerns parameter interaction. Among other things, parameter interaction implies that if there is an error in the estimation of one parameter, there will be errors in the estimated values of the other interacting parameters (as a result of the compensating changes required to minimize the sum of the squares of the residuals). The magnitude of these errors depends on the extent and importance of the interaction.

Fortunately, some progress has been made to resolve the type of non-identifiability problem discussed above. Gupta and Sorooshian (1983) reported on a reparameterization approach which satisfactorily eliminated the nonidentifiability of the parameters Z and X of the SMA-NWSRFS percolation subprocess. The possibility of using a reparameterization approach can be explored only after a nonidentifiability problem has been detected. In the case of the percolation equation of SMA-NWSRFS, two-dimensional plotting of the response surface confirmed the existence of the problem. Such an analysis would become overwhelming and impractical as the number of model parameters grows. In our most recent work (Sorooshian and Gupta, 1985), we have presented the details of a methodology through which the local identifiability of model parameters can be examined. Application of the proposed tests can identify those parameters that are most likely to give us trouble in the estimation phase. Once the presence of such parameters is verified, one can proceed to examine the severity of the problems through response surface contour plotting and perform some form of reparameterization similar to the one reported by Gupta and Sorooshian (1983). In the view of this author, the estimation problems caused by model structure representation are the most critical aspect of the modeling process. More research is needed before any marked improvement in model accuracy can be achieved.

Calibration Data and the Effect of Measurement Errors

Calibration data are another factor which can influence the precision of a model. Note that the objective of the model calibration phase is to transform the information contained in a given data set into estimates for the model parameters. There are at least two aspects of the calibration data which have direct influence on the precision of the parameters: measurement errors and the type (range and frequency of recording, etc.).

Measurement Errors

Errors in any of the three main data requirements, namely precipitation, potential evapotranspiration, and streamflow, will result in poor estimation of the parameters unless they are accounted for and their effects filtered out through the estimation process. Unfortunately, no theoretical work has been reported that can tackle the errors associated with the inputs (precipitation and potential evapotranspiration) of nonlinear deterministic models such as CRR models.

There has been some work done in the area of accounting for streamflow errors. Ibbitt (1970) tested the effects of random and systematic errors in the streamflow data and found them to significantly influence the accuracy of the parameter estimates. Errors which are systematic (correlated) and have inhomogeneous variance (heteroscedastic) are more serious and are known to exist in most streamflow data records. Sorooshian (1978) and Sorooshian and Dracup (1980) proposed appropriate methods to account for each of these error types. The new procedures, which are based on maximum likelihood theory, have been shown to exhibit superior performance when compared to some of the more commonly employed criteria, in particular, the ordinary least squares (OLS) [Sorooshian and Dracup, 1980; Sorooshian, Gupta, and Fulton, 1983]. In both of these studies, the authors stressed the fact that the selected objective function plays a major role in forming the shape of the response surface. If the objective function is selected to properly account for the stochastic properties of the additive model errors (streamflow errors included), then the response surface is more likely to have an elliptical shape, making it easier for the optimization method to search for the best parameter values.

Evidence of this fact is presented in Figure 7, which should be compared with Figure 4. Note that the estimator which filtered out the effect of heteroscedastic errors (HMLE) resulted in improved response surface conditions and enhanced the likelihood of recovering the best parameter values.

Another interesting observation was that good parameter estimates were obtained using only one year of data; use of longer periods such as two and three years served only to marginally improve the estimates. Some of the same conclusions have recently been confirmed by independent researchers (Lemmer and Rao, 1983; Ibbitt and Hutchinson, 1984; and Delleur et al., 1984). For the benefit of the readers, the ML estimators are presented in the Appendix.

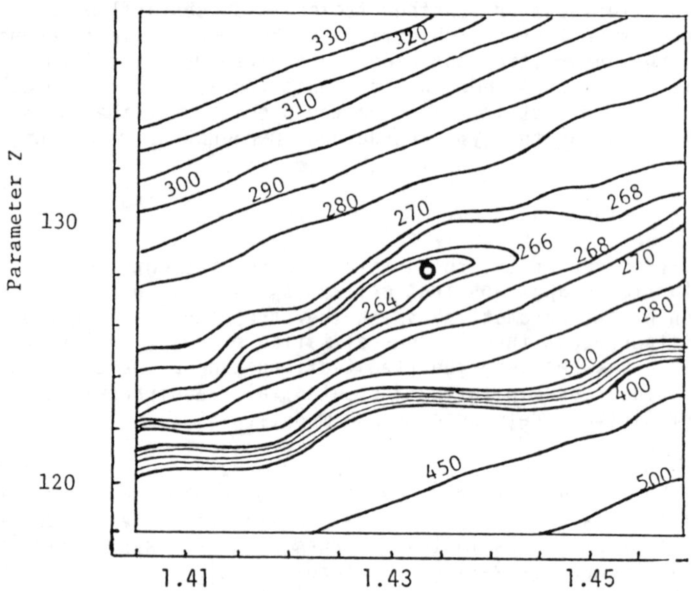

Fig. 7. HMLE criterion contour map in region of HMLE calibration best parameter set (Z versus X).

Type of Data

The second aspect of data which influences the identifiability of the model parameters relates to the type and the quality of the data. It has often been suggested or implied in the literature that the data used should be "representative" of the various phenomena experienced by the catchment. Many researchers have attempted to satisfy this requirement by using as large a data set as possible without demonstrating superior results. Sorooshian, Gupta, and Fulton (1983) have pointed out that, rather than length, it is the quality of information contained in the data which is important. They also stated that the data sequences which contain greater "hydrologic variability" are more likely to activate the various operational modes of the model sufficiently to result in reliable parameter estimates. The issue of how to measure "hydrologic variability" has recently been presented by Gupta and Sorooshian (1985a,b). A theoretical investigation into the relationship between the data, the model structure, and the precision of the parameter estimates was conducted. Some interesting results, particularly related to threshold parameter identifiability (see parameter M in Figure 2) were obtained. It was shown, for example (Gupta and Sorooshian, 1985a), that the precision to which the threshold parameter M can be identified is (1) directly proportional to the number of times the reservoir switches from the nonoverflow mode ($R_t = 0$) to the overflow mode ($R_t \neq 0$), and (2) is independent of the duration of the overflow mode. This has interesting implications to the choice of calibration data since it implies that a data set containing a few large storm events may be less informative than a data set containing many storms of moderate size.

The above theoretical results have been numerically confirmed by us using simulation studies. Figures 8a-d depict the ordinary least squares (OLS) response surface in the K-M parameter space for four different data sets of the same length but with different numbers of runoff-generating storms. The figures clearly show that the precision of the parameters improves directly as a function of the number of reservoir spills (overflow events). Notice that with no spill (Figure 8a), the response surface shows a horizontal extended valley in the direction of parameter M. This insensitivity results in the nonidentifiability of parameter M when no spills occur. The identifiability improves as the number of spills increase. This is evident by the degree of improvement and the concentricity and steepness of the response surface. With improved concentricity and steepness of the response surface in a multiple spill case, our parameters are estimated with improved precision.

This simple example relates to a very basic two-mode switching structure, and the result in this case indicates that identifiability is mainly a function of one characteristic of the data, namely the switching frequency. Typical CRR models have more complex structures and several different modes of operation. Consideration of the relationship between different vertical storages and lateral flow components of a more complex operational model will require more detailed analysis. In these cases, it seems reasonable to expect that other characteristics of the data, such as storm duration, interstorm time-interval, variation in storm intensity, etc., will become relevant. Research in this area is lacking at the present time.

Response Surface Contour for k-M Parameters

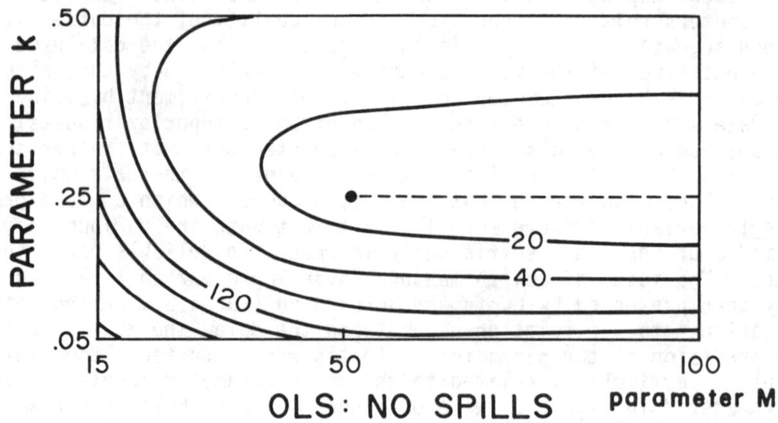

Response Surface Contour for k-M Parameters

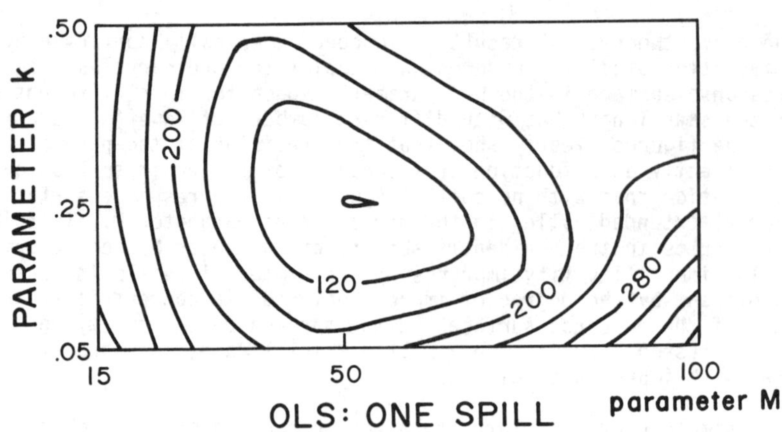

See figure caption on next page.

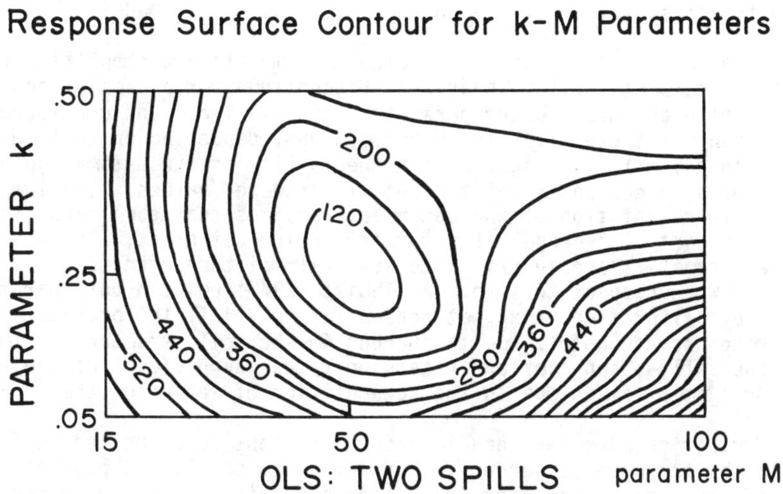

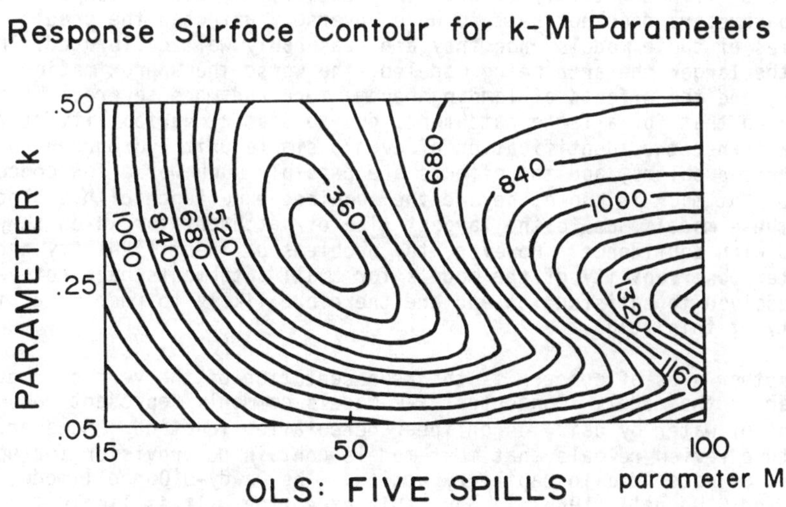

Figs. 8a-d. Ordinary least square response surface contour plots in the K-M parameter space of the linear reservoir model of Fig. 2 for various spill conditions. (a) Condition with no spill shows the insensitivity and the nonidentifiability of the M parameter. (b) through (d) show the relative improvement in the identifiability of parameter M as the number of spills increases. (Note: True parameter values K* = 0.25, M* = 50).

Imperfect Representation of the Physical Process by the Model

In general, most existing conceptual R-R models are simplified nonlinear, lumped parameter, time-invariant, discontinuous representations of a complex nonlinear, distributed parameter, time-variant, and continuous system. Many of these simplifying assumptions, depending on their degree of seriousness, will adversely affect the ability of the estimation procedure to find a unique and realistic set of parameter values. The lumped parameter representation of the catchment requires that each parameter assume some sort of average value for the entire catchment. The assumption that the parameters are invariant to seasonal moisture changes has a similar effect with respect to time. As Ibbitt (1970) pointed out, parameters obtained by calibration using wet season data may fail to forecast dry season response adequately due to changes in the soil structure. Similarly, the fall season overland recession rates may differ from their spring or summer values due to the accumulation of leaves on the ground.

A further problem that has long been apparent concerns the lumping of data used by the model in order to correspond to the lumped parameter nature of the model. The data from different rain gauges are aggregated in some manner and converted to average equivalent uniform depth of rainfall over the catchment. The effects of this lumping of spatially-distributed data are still not entirely clear. In practice, the use of lumped data and model parameters does not seem to have adversely affected the prediction abilities of these models once they are reasonably well calibrated. However, the larger the area being modeled, the worse the approximation becomes, and the effects of lumping become more and more severe. It is well known that for a large catchment, due to spatial variability of rainfall patterns, the identification of even a simple unit hydrograph poses uniqueness problems, and the effects are possibly much worse for conceptual models. Not much is known, beyond the acquired experience of hydrologists using these models, as to the largest size of catchment for which they can be used with confidence. However, the problems of identifiability and parameter observability of the models for small catchments have not yet been resolved to satisfaction and are therefore likely to receive a higher priority at this stage.

Another area of concern is the representation of the vertical movement of water in the model. Conceptual R-R models commonly represent downward movement of water by using a nonlinear percolation function. However, literature review reveals that most models contain no provision for upward movement of water due to capillary action. The Dawdy-O'Donnell model (Dawdy and O'Donnell, 1965) is a notable exception. It is likely that a number of other such deficiencies exist in most models that make them far less generally applicable than we would like.

Finally, it should not be overlooked that certain components in conceptual models are likely to better represent their associated physical subprocesses than do other components in the model. Remember that most conceptual models are viewed as consisting of an upper zone, a lower zone, and an interconnecting percolation process. As Gupta (1982) argues, the lumped, linear reservoir model of the lower zone is a comparatively more accurate representation of its corresponding physical subprocess than is the lumped, linear reservoir representation of the upper zone. Hydrologists have long recognized that baseflow into a stream can be fairly accurately modeled as the outflow from one or more linear reservoirs (e.g., Linsley et al., 1975). However, recent research (Parker, 1977) has indicated that as the size of the catchment increases, the runoff and interflow components (upper zone) are better represented by nonlinear relationships. In the same vein, another interesting point is the fact that the upper zone of the model has a much faster response to inputs than has the lower zone. Hence, the upper zone is far more likely to be sensitive to spatial variability in rainfall patterns. This problem is likely to become more severe as the size of the catchment becomes larger. On the other hand, the slower response of the lower zone and its relative homogeneity makes it less sensitive to spatial variability of inputs. It should be recognized that a model is esentially only as good as its weakest link. However, parameter estimation techniques that properly weight the more accurate parts of the model could help in obtaining improved parameter estimates. This concept deserves to be examined more thoroughly. Last but not least are the difficulties resulting from the presence of threshold parameters, as abrupt thresholds rarely exist in a real system.

Much of the discussion provided so far has been the result of our experience over the last five years with the soil moisture accounting model of the U.S. National Weather Service's river forecast system (SMA-NWSRFS). Other operational models have similar problems, and as long as these issues are not addressed, it would be difficult to place much faith in the forecasts generated by these models.

Model Selection

So far, we have discussed some of the problems which have been determined through various studies and are known to be partly responsible for the unreliability of conceptual catchment models. The reality of the situation, however, requires that "some" model must be used (reliable or not) to provide planners and decision makers with answers required in their design and operational planning. Therefore, the question is: What type of model should one choose? This issue was briefly discussed by this author recently (Sorooshian, 1985) and the opinion expressed in that paper follows.

The literature of the last two decades reveals two clear trends in the area of on-line streamflow forecasting. On the one hand, there has been the development of "conceptual" type models which purport to be "physically-based". The second trend has been towards developing "systems theoretic" models (Sorooshian, 1983), i.e., linear or quasi-linear input-output (rain-fall-runoff) models that are designed around the powerful theoretical base of linear systems theory (for examples of this type of models, see Wood, 1980). As discussed above, even though conceptual models are believed to be inherently more accurate in their representation of catchment behavior, calibration difficulties often limit their usefulness. In contrast, systems theoretic models (e.g., time-series models, state-space form relations, etc.) are usually easier to construct and calibrate but have been strongly criticized as employing unrealistic assumptions about the nature of the physical system (e.g., ignoring the nonlinear dynamics of the watershed process). It has sometimes been argued that, from an engineering point of view, the usefulness of a catchment model (be it conceptual or systems theoretic) need not depend on its conceptual realism so much as on its capability to reproduce input-output behavior. Some researchers have attempted to compare certain conceptual and systems theoretic models from this point of view. More often than not, the published results have supported the systems theoretic models. The reasons for this are probably quite varied and have not been clearly discussed in the literature. Sorooshian (1983) mentioned two important points, however. First, the state-of-the-art of parameter estimation in conceptual models has not been adequately refined, whereas the solution techniques available for systems theoretic models are comparatively highly efficient. The second reason is that the comparisons are rarely carried out under conditions which would highlight the inadequacies of either type of model. As suggested by Linsley (1982), the most important property of a model (and the least often tested) should be its inherent accuracy, i.e., it should not be a question of prediction accuracy under average or slowly changing conditions, but one of model credibility under extreme or rapidly varying conditions. For example, Kitanidis and Bras (1980) found that under rapidly changing hydrologic conditions, the conceptual SMA-NWSRFS model performed significantly better than an ARMAX (Autoregressive Moding Average with Exogeneous Inputs) linear stochastic model with on-line adaptively estimated parameters and states. Though the ARMAX model was found to be more reliable in forecasting the most important features of the hydrograph, such as the beginning of the rising limb, the time-to and height-of peak, and the total water volume. Similar results have been noted by Todini and Wallis (1978), Andjelic and Szollosi-Nagy (1980), and O'Connell and Clarke (1981).

The problem of whether to employ a systems theoretic or a conceptual model is really just a smaller facet of a larger problem, i.e., how to decide on a level of complexity of model structure appropriate to the modeling of a given catchment. Of the various issues related to this problem, one that has not received the attention it deserves, is that of system scale. Consider the conceptual diagram presented in Figure 9 which illustrates the fact that the variability in the hydrologic output of a catchment system is influenced by two major factors: (1) variability in the inputs, and (2) properties associated with the physical structure of the catchment (e.g., expansion and compression of time scales, damping and attenuation, non-linearities, input-dependent system modality, losses, etc.). It seems entirely possible that, in certain catchments, most of the variability in the outputs can be related, in a relatively simple manner, to variability in the input. That is, the influence of input variability on output behavior predominates over the effects of system structure, so much that influences of the latter are difficult to separate from those of the former. An example might be a small catchment whose response is dominated by surface runoff (due, for example, to minimum infiltration loss or ground-water flow contribution, etc.). At the other extreme, we might have a catchment in which the effects of the input variable are completely dampened by the system. In such a case, the characteristics of system structure obviously control properties of the output hydrograph. An example might be a very large catchment with little or no surface runoff (due to high infiltration and/or channel loss rate, etc.). The former catchment seems to be a prime contender for the systems theoretic modeling approach while, with the latter, we might benefit more from a conceptual approach.

In the above examples, we have referred to the spatial aspects of scale. In a similar manner, the time scale at which the catchment is to be

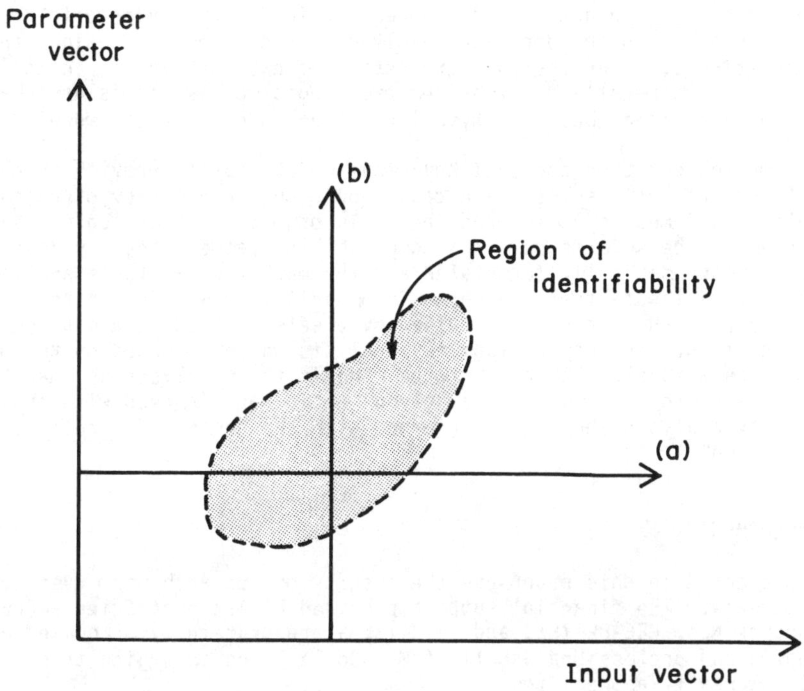

Fig. 9. Region of identifiability in the parameter/data space

(a) Identifiability studies in the input space conditioned on a chosen parameter set.

(b) Identifiability studies in the parameter space conditioned on a chosen input data set.

modeled will also have an important role in the selection of an appropriate model structure. It is felt that this area of research deserves a great deal of attention. If we can establish some means of identifying the relative importance of the hydrologic inputs versus the system structure (with respect to simulating system outputs), the problem of selecting an appropriate model (conceptual or systems theoretic or something in between) should be much easier to deal with.

Summary

Our dear colleague, Vít Klemeš, (this proceedings) addressed this workshop by saying:

"The reliability of performance of water resource projects as well as of individual structures interacting with water depends, among many other things, on an adequate representation of the hydrologic processes relevant to the decision involved in the planning, design, and operation.......As the characteristics of extreme hydrologic conditions are generally derived from hydrologic models, it is desireable to have an idea about the hydrological adequacy of these models."

In my presentation above, I have made an attempt to provide an overview of some recent results which casts doubt on the adequacy of conceptual rainfall-runoff models as forecasting tools of extreme hydrologic events. There seems to be some degree of incompatibility between the way most models are mathematically formulated and the mathematical tools available to properly calibrate them. Such an incompatibility will definitely affect the accuracy of the forecasts provided by a selected model, and hence impact the results of any reliability analysis which is based on the outputs of such a model. The point to be made is that analysts who use the output of a hydrologic model as a "given" entity and proceed with their reliability analyses should be concerned with the degree of unreliability of the "given".

ACKNOWLEDGMENTS

Ideas presented in this paper are the results of research work over the past 10 years. The financial support provided by National Science Foundation grants Nos. CEE-8217823 and INT8413539 are gratefully acknowledged. The superb and professional skill of Ms. Corla Thies in typing this manuscript is greatly appreciated.

APPENDIX

Description of Maximum Likelihood Estimators

1. The Maximum Likelihood Estimator for the Autocorrelated Error Case (AMLE)

$$\underset{\theta,\rho,\sigma_\nu^2}{\text{Min}}\ \text{AMLE} = \frac{n}{2}\ell n(2\pi) + \frac{1}{2}\ell n \frac{\sigma_\nu^{2n}}{1-\rho^2} - \frac{1}{2}\rho^2\sigma_\nu^2\varepsilon_1^2$$

$$+ \frac{1}{2\sigma_\nu^2}\sum_{t=2}^{n}(\varepsilon_t - \rho\varepsilon_{t-1})^2 \tag{1a}$$

where

$$\sigma_\nu^2 = \frac{1}{n}\quad - \rho^2\varepsilon_1^2 + \sum_{t=2}^{n}(\varepsilon_t - \rho\varepsilon_{t-1})^2 \tag{1b}$$

and ρ is estimated from the implicit equation

$$\varepsilon_1^2 - \sum_{t=2}^{n}\varepsilon_{t-1}^2\ \rho^3 + \sum_{t=2}^{n}\varepsilon_t\varepsilon_{t-1}\ \rho^2 + \sigma_\nu^2 - \varepsilon_1^2 + \sum_{t=2}^{n}\varepsilon_{t-1}^2\ \rho$$

$$- \sum_{t=2}^{n}\varepsilon_t\varepsilon_{t-1} = 0 \tag{1c}$$

where $\varepsilon_t = q_{t,obs} - q_{t,sim}$ = residual at time t and $q_{t,obs}$, $q_{t,sim}$ are the measured and the simulated flows at time t, respectively, θ is the set of parameters to be estimated, n is the number of data points, σ_ν^2 is a constant variance term, and ρ is the first lag autocorrelation coefficient. The above estimator is developed based on the assumption that the output errors are Gaussian with a constant variance and correlated according to a first lag autoregressive scheme [for details, see Sorooshian and Dracup (1980)].

Note that, in the case ρ = 0, the estimates obtained using AMLE are equivalent to those of the Simple Least Squares (SLS) criterion:

$$\underset{\theta}{\text{Min}}\ \text{SLS} = \sum_{t=1}^{n}\varepsilon_t^2 \tag{2}$$

2. <u>The Maximum Likelihood Estimator for the Heteroscedastic Error Case (HMLE)</u>

$$\underset{\theta,\lambda}{\text{Min}}\ \text{HMLE} = \sum_{t=1}^{n} w_t \varepsilon_t^2 \left[n \prod_{t=1}^{n} w_t \right]^{1/n \quad -1} \tag{3a}$$

where w_t is the weight at time t, and it is computed by

$$w_t = f_t^{2(\lambda-1)} \tag{3b}$$

where f_t is the expectation of $q_{t,true}$ (either $q_{t,obs}$ or $q_{t,sim}$) and λ is the unknown transformation parameter which stabilizes the variance. The implicit expression to estimate λ is

$$\sum_{t=1}^{n} \ell n\,(f_t) \sum_{t=1}^{n} w_t \varepsilon_t^2 - n \sum_{t=1}^{n} w_t \ell n(f_t) \varepsilon_t^2 = 0 \tag{3c}$$

Briefly, the HMLE estimator is derived based on the assumption that the errors are Gaussian with mean zero and covariance matrix V, where $V_{t,t} = \sigma_t^2$ and $V_{t,t+s} = 0$ for $s \neq 0$.

Stabilization of the variance is attempted through the use of the Box-Cox [Box and Cox, 1964] power transformation which relates the variance of each error to its associated output value [see Sorooshian and Dracup, (1980)].

In this study, $f_t = q_{t,obs}$ was used in the computation of the weights (the original procedure reported in the aforementioned papers used $f_t = q_{t,sim}$). Fulton (1982) has shown that this results in a more stable estimation scheme. It is interesting to note that if the variances of the additive errors are homogeneous (independent of time or magnitude of the associated flows), then the procedure will automatically select the value of $\lambda = 1.0$. This results in $w_t = 1$ for all t and the estimation criterion reduces to the SLS. If, however, the variance of the errors is proportional to a power function of the magnitude of the flows, then the procedure will select a value of $\lambda = 1.0$. Pertinent to our problem is the case where the error variance increases as the flow values get larger [Sorooshian and Dracup, (1980) have in great length discussed the underlying reasons]. In this situation, the ML estimate of λ will be less than unity, and this ensures that, in the estimation criterion, the errors associated with lower flows (which contain more reliable information) are weighted more heavily.

REFERENCES

Crawford, N.H., and R.K. Linsley, 1966, Digital simulation in hydrology: Stanford watershed model Mark IV, Dept. of Civil Engineering, Tech. Report 39, Stanford University, Stanford, California.

Burnash, R.J.C., K.L. Ferral, and R.A. McGuire, 1973, A generalized streamflow simulation system-conceptual modelling for digital computers, Report, Joint Fed. State River Forecast Center, U.S. National Weather Service and California Dept. of Water Resources, Sacramento, California.

Sorooshian, S., and V.K. Gupta, 1983, "Automatic Calibration of Conceptual Rainfall-Runoff Models: The Question of Parameter Observability and Uniqueness," Water Resources Research, 19(1):260-268.

Gupta, V.K., and S. Sorooshian, 1983, "Uniqueness and Observability of Conceptual Rainfall-Runoff Model Parameters: The Percolation Process Examined", Water Resources Research, 19(1):269-276.

Sorooshian, S., and J.A. Dracup, 1980, Stochastic Parameter Estimation Procedures for Hydrologic Rainfall-Runoff Models: Correlated and Heteroscedastic Error Cases, Water Resources Research, 16(2):430-442.

Sorooshian, S., V.K. Gupta, and J.L. Fulton, 1983, "Evaluation of Maximum Likelihood Parameter Estimation Techniques for Conceptual Rainfall-Runoff Models - Influence of Calibration Data Variability and Length on Model Credibility," Water Resources Research, 19(1):251-259.

Ibbitt, R.P., and T. O'Donnell, 1971, Fitting methods for conceptual catchment models, J. Hydraulic Div. Amer. Soc. Civ. Eng., 97(HY9), pp. 1331-1342.

Johnston, P.R., and D. Pilgrim, 1976, Parameter optimization for watershed models, Water Resources Research, 12(3):477-486.

Sorooshian, S., and V.K. Gupta, 1985, "The Analysis of Structural Identifiability: Theory and Application to Conceptual Rainfall-Runoff Models," Water Resources Research, 21(4):487-495.

Sorooshian, S., 1978, "Considerations of stochastic properties in parameter estimation of hydrologic rainfall-runoff models," Ph.D. Dissertation, University of California, Los Angeles, California.

Lemmer, H.R., and A.R. Rao, 1983, Critical Duration Analysis and Parameter Estimation in ILLUDAS, Water Resources Research Center, Tech. Rep. #153, Purdue University, West Lafayette, Indiana.

Ibbitt, R.P., and P.D. Hutchinson, 1984, "Model parameter consistency and fitting criteria," Proceedings of International Federation of Automatic Control (IFAC) 9th World Congress, Budapest, Hungary, July 2-6, 1984, Vol. IV, pp. 153-157.

Delleur, J.W., J.M. Bell, M.S. Blumberg, M.H. Houck, H.R. Lemmer, L.E. Ormsbee, H.R. Potter, A.R. Rao, and H. Schweer, 1984, Problem-oriented Evaluation of Institutional Decision Making and Improvement of Models Used in Regional Urban Runoff Management: Application to Indiana, Water Resource Research Center, Tech. Report #164, Purdue University, West Lafayette, Indiana.

Sorooshian S., and V.K. Gupta, 1985a, "The Automatic Calibration of Conceptual Watershed Models Using Derivative-Based Optimization Algorithms," Water Resources Research, 21(4):473-485.

Sorooshian S., and V.K. Gupta, 1985b, "The Relationship Between Data and the Precision of Parameter Estimates of Hydrologic Models," Journal of Hydrology, 81:57-77.

Ibbitt, R.P., 1970, Systematic parameter fitting for conceptual models of catchment hydrology, Ph.D. Dissertation, University of London.

Dawdy D.R. and T. O'Donnell, 1965, Mathematical models of catchment behaviour, J. Hydraul. Div. Am. Soc. Civ. Eng., 91(HY4), pp. 123-137.

Gupta, V.K., 1982, Calibration of Conceptual Rainfall-Runoff Models: Problems Caused by Model Structure, M.S. Thesis, Case Western Reserve University, Cleveland, Ohio, 134 p.

Linsley, R.K., M.A. Kohler, and J.L.H. Paulhus, 1975, "Hydrology for Engineers", 2nd Edition, McGraw-Hill, New York.

Parker, R.S., 1977, Experimental study of drainage basin evolution and its hydrologic implications, Hydrol. Paper 90, Colorado State University, Fort Collins, Colorado.

Sorooshian, S., 1985, "Synthesis of Hydrologic and System Sciences in the Development of Rainfall-Runoff Models", Revue Internationale des Sciences de l'eau, Vol. 1, No. 1/4, pp. 21-28.

Wood, E.F., 1980, Recent Developments in Real-Time Forecasting/Control of Water Resource Systems, Pergamon, New York.

Sorooshian, S., 1983, "Surface Water Hydrology: On-Line Estimation," Review of Geophysics and Space Physics, 21(3):706-721.

Linsley, R.K., 1982, Rainfall-Runoff Models: An Overview, In: V.P. Singh (Editor), Rainfall-Runoff Relationship, Proc. of the International Symposium on Rainfall-Runoff Modeling, Mississippi State, Mississippi, May 18-21, 1981, WRP, Littleton, Colorado, 3-22.

Kitanidis, P.K., and R.L. Bras, 1980, "Real-Time Forecasting with a Conceptual Hydrologic Model, 2: Applications and Results," Water Resources Research, 16(6):1034-1044.

Todini, E., and J.R. Wallis, 1978, "A real-time rainfall-runoff model for an on-line flood warning system," In: Proc. AGU Conference on Application of Kalman Filtering Theory to Hydrology, Hydraulics and Water Resources (ed. by Chao-Lin Chiu), Department of Civil Engineering, University of Pittsburgh, Pittsburgh, Pennsylvania, pp. 355-368.

Andjelic, M., and A. Szollosi-Nagy, 1980, "On the use of stochastic structural models for real-time forecasting of river flow on the River Danube," In: Hydrological Forecasting (Proc. Oxford Symposium, April, 1980): IAHS Publication No. 129, pp. 371-380.

O'Connell, P.E., and R.T. Clarke, 1981, "Adaptive hydrological forecasting: a review," Hydrological Sciences Bulletin, 26(2):179-205.

EMPIRICAL AND CAUSAL MODELS IN HYDROLOGIC RELIABILITY ANALYSIS

Vít Klemeš

National Hydrology Research Institute, Environment Canada, Ottawa, Ontario, Canada

ABSTRACT

The reliability of performance of water resource projects, as well as of individual structures interacting with water, depends, among many other things, on an adequate representation of the hydrologic processes relevant to the decisions involved in the planning, design and operation. Especially important is the representation of extreme hydrologic conditions, since it is the response of a facility to the extremes which affects its reliability most. Because the characteristics of extreme hydrologic conditions are generally derived from hydrologic models, it is desirable to understand the hydrological adequacy of these models. This adequacy is especially important for drawing inferences regarding probabilistic aspects of extreme hydrologic events, since the concept of reliability of a facility is usually related to the probability of some kind of failure in its functioning. Questions concerning this adequacy are discussed with reference to common practices in the estimation of probabilities of extreme floods and water shortages.

1. INTRODUCTION

Hydrologic reliability is usually understood as a probability that the provision of a specific service will not have to be interrupted or stopped because of hydrologic reasons. Thus, the problem central to its analysis is the probability of hydrologic events, with an emphasis on probabilities of extreme events. What do we really know about these probabilities and how can we estimate them?

In theory, these two questions should be related approximately as follows: One should first assemble the available hydrologic evidence and knowledge, incorporate them into a probabilistic hydrologic model, and use the model to estimate the desired probabilities. In practice, however, the situation is quite different. The difference is that the current models used for the computation of hydrologic probabilities are based largely on arbitrary assumptions about the probabilistic behaviour of hydrologic processes rather than on empirical evidence or theoretical knowledge and understanding of these processes. Thus, rather than being probabilities of hydrologic events, these estimates are numbers whose relevance to hydrologic probabilities is more a matter of wishful thinking than one of substance. This probably will be flatly denied by those who produce or use these "probabilities" for various theoretical manipulations in the optimization of design or operation of water resources systems or hydraulic structures. It will be argued that since the models are based on actual hydrologic records and are designed to "preserve" their basic statistical parameters, they give an approximation of the true probabilistic picture despite some assumptions which do not originate in hydrology. It will be further pointed out that some assumptions are necessary to simplify the reality in order to make the analysis mathematically manageable and some are needed exactly because we do not know what the reality is. So, the argument will continue, in spite of the fact that these assumptions introduce a certain element of uncertainty into the models, "such approximate models are sufficiently realistic that their use improves the planning process significantly" (Fiering and Jackson, 1971). The crown of the argument is usually a statement that the engineer needs at least an approximate number since a project cannot wait until enough data and knowledge are accumulated, which may take many years.

The flaw in this argument is that it redefines the problem. Rather than addressing the adequacy of the probabilistic hydrologic models, it defends a point which has not been contested, i.e., the engineer's need for some rationale for arriving at design (or operation) parameters from hydrologic data. In other words, the argument confuses the problems of "scientific truth" and the "concepts of expediency" (Yevjevich, 1968). The main reason for this confusion is that often hydrologists are hydraulic engineers by background and, consciously or not, identify hydrology with the above-mentioned engineering rationale for obtaining design parameters for water-related projects.

It has also been argued (Klemeš, 1983; Plate, 1984) that a better hydrologic foundation of hydrologic models does not automatically translate into better designs of water resource projects. This is particularly true in the sense that a specific project need not be equally sensitive to all aspects of a hydrologic process; in other words, hydrologic uncertainty must not be equated with decision uncertainty which is project specific (Klemeš, 1983). However, knowing the hydrologic content of a model will help the engineer determine the weight which a modelling result should carry in a given case, thus providing him with a better basis for making intelligent choices. Strengthening the scientific basis of water management decisions is becoming increasingly important in the present era, because the rapid development of new mathematical formalisms and explosion of computer technology exert irresistible pressure to compute whatever is computable, regardless of whether it has some substance or merely represents a modern version of Potemkin villages. The situation is now much worse than it was a decade ago when M.B. Fiering characterized it as follows:

> The engineering literature is replete with mathematical models, optimization techniques, Bayesian analyses, exotic formulations for synthetic flows, and all manner of computer studies. We seek optimal plans, optimal operating policies, optimal estimates of parameters, optimal anything. We are swept up in a litany of automatic computation, sensitivity analysis, and model making. It has become a new religion...(Fiering, 1976).

The purpose of this paper is to examine the hydrologic content of some common probabilistic concepts presented as "sufficiently realistic" descriptions of the hydrologic truth and regarded as the basis for the sophisticated formal structure of reliability analysis of water resource systems.

2. EMPIRICAL AND CAUSAL HYDROLOGIC MODELS

A detailed discussion of the differences between empirical and causal models was given elsewhere (Klemeš, 1978, 1982). Here, it will suffice to say that these two kinds of models do not contradict each other but represent two hierarchical steps in the description of some apparent regularity in the physical prototype. Empirical models simply describe the regularity, or pattern, as it manifests itself in the observations, ie., on the basis of empirical evidence. Causal models go one step further by tying the pattern not only to the observed facts but also to a physical theory that explains them. In other words, while empirical models are only descriptive, causal models are both descriptive and explanatory. This difference usually is not too important if the model is used merely for an interpolation of the observed pattern or as a summary of the data. However, it becomes very important if it is to be used for extrapolation, i.e., for drawing inferences about events beyond the range of observation and thus beyond the coverage of an empirical model. Since hydrologic models are used chiefly for extrapolation and hydrologic reliability analysis rests exclusively on such extrapolation, it is important to recognize the difference in the quality of extrapolation obtained from the two kinds of models.

This problem is rather clearly understood in connection with essentially deterministic models, but it tends to be blurred if the model involves probabilistic aspects beyond the description of random measurement errors. For example, discharge measurements in a given river cross-section may yield values which show an approximate straight line pattern in a stage-discharge plot. However, it is well recognized that there is no inherent reason why extrapolation of a fitted straightline far beyond the range of the data points should be believed and nobody would seriously suggest that such extrapolation becomes more reliable with a greater mathematical refinement in the fitting of this straight line. Yet, such practice is common in present attitudes towards probabilistic models fitted to hydrologic data. As if by magic, the introduction of probabilistic aspects releases a hydrologic problem from all its earthly connections and elevates it to the astral heights of the calculus of probability and mathematical statistics. Actually, the only substantial difference between the transformation of hydrology into probability theory and the transformation of an egg into a rabbit in the magician's hands seems to be in the magic formula employed - in the latter case it is "abracadabra" while in the former it is "let's assume": Let's assume the streamflow is a stationary and ergodic stochastic process, let's assume the maximum annual flood peaks are identically distributed random events, let's assume the observed flood peaks are a sample from a Gumbel distribution, etc. In other words, let's forget the hydrological reality and let's play mathematical games. After all, isn't is more fun to make your own rules than trying to decipher nature's rules?

It seems that many analysts have missed the following point. The role of hydrology is not to provide the decision maker with an ornamental epithet "hydrologic" for whatever mathematical construct he may choose to represent his musings about water-related inputs to a decision model, but rather to examine the compatibility of the various "let's assumes" with the available hydrological evidence and, eventually, to replace arbitrary stochastic and statistical models with models based on a hydrologically sound theory. It may take many years before such a comprehensive physically sound theory of stochastic hydrologic modelling is available. But such a theory can never be developed merely by piling one "let's assume" on top of another and without a critical analysis of the weaknesses of the current stochastic models.

3. FLOODS

The basis of probabilistic formulation of reliability as related to flood protection is the so-called flood frequency analysis. To analyze the frequencies of floods is a legitimate objective. The problem is that, as a rule, flood frequency analysis does not do this.

Firstly, flood frequency analysis is usually not concerned with floods as such but only with their peak discharges. Frequency of peak discharges of different magnitudes is of course a legitimate object of study, especially from a purely descriptive point of view, regardless of its practical relevance. However, it would be equally legitimate to study the frequency of maximum flood stages, flow velocities, the frequency of flood volumes or flood durations above a given stage or discharge. From the point of view of reliability analysis, any of these aspects can be the most relevant depending on the specific concern under consideration. For example, during the flood on the Danube River in the summer of 1965, the levees in the Czechoslovak section collapsed in many places, not because of overtopping but because of a long exposure to seepage under stages not exceeding the "safe" range and because of the resulting gradual washout of material from the levees or the base. In Canada, river ice is an important factor in reliability against flood damage, while in California it is the sediment and debris transported by the flood. Thus, the common practice of relating the reliability of flood protection to the frequency of peak discharges alone may introduce a large element of uncertainty that differs not only with regard to the object of concern but also from one river to another depending on how the various hydrologic aspects of floods are correlated in each case.

Leaving aside the problems which flood frequency analysis does not address, what does one learn from it about the problem it claims to address, namely the frequency of flood peak discharges? In general very little, and much of that may be completely wrong, especially when it comes to the probabilities of extremely high peaks on which rests the whole concept of reliability against flood damage. The natural frequencies of flood peaks in the historic series are, in fact, almost never analyzed. We do not learn whether there seems to be any pattern such as clustering of high or low peaks, trend, or some other feature, nor any indication of some hydrological, geophysical or other context that could shed light on the historic flood record. What happens is that the actual time sequencing is completely ignored and the flood record is declared purely random. The ostensible reason for this assumption is to "simplify the mathematical treatment." This, however, is rather amusing when one sees how the laudable resolve to keep things simple is then hastily abandoned and the use of the most advanced theories is advocated for the treatment of this artificially random sample on the pretext that the "greatest amount of information" must be extracted.

In this manner, the difficult problem of recurrence of flood peaks higher than a given flow Q is first reduced to two oversimplified parameters of its average return period T_Q and a probability $P_Q=1/T_Q$ with which the flow Q can be exceeded in any given year or (in case of a "partial" series) at any given time. Then the concern for simplicity is suddenly abandoned and an equally strong concern for mathematical finesse in the estimation and extrapolation for flows beyond the range of observations is advocated.

Figure 1 illustrates the general relevance of the average return period and the corresponding exceedance probability. It can be seen that during the 61-year period 1920 to 1980 the incidence of high flows exhibited features which would be difficult to reconcile with pure randomness. Thus, all the really high peaks occurred in the second half of the record; the whole decade 1930-1940 was dry, while the whole following one was wet; the variability during the last twenty years was extremely high, while during the first twenty years it was low. However, this information is not reflected in the values of return periods that would, for all practical purposes, represent the flood record.

For example, an annual maximum peak flow higher than about 820 m^3/s would be assigned a return period of slightly under 4 years, although this figure was never exceeded during the first 28 years of the record; or a peak flow higher than 600 m^3/s which would be classified as a 2-year flood (one that occurrs, on average, every other year) was not reached in 17 consecutive years, from 1924 to 1940, and then was exceeded in all but one of the following 12 years; or, a 15-year flood, determined from the first 30 years of the record, would be about 1000 m^3/s while it would be about 2000 m^3/s if estimated from the last 30 years of data.

The one good thing about the average return period is that at least it is what it says it is -- an average computed from the duration of the actual return periods. The fact that it may have no meaning for the purpose for which it has been intended, i.e., as a measure of the central tendency of the actual return periods, is a different matter. To presume such a meaning is a great leap of faith. However, an even greater leap of faith is to regard its inverse value P_Q as a probability of exceedance. Here, even the name may be misleading and the number may have little to do with probability. For this (now standard) probabilistic interpretation to be valid, there must be a good reason to believe that each of the recorded floods could have occurred equally likely in any year of the historic record. For the record shown in Figure 1, this seems very unlikely. To claim that any rearrangement of the record would be equally likely would be to dismiss the empirical evidence for long-term fluctuations of precipitation, temperature and other aspects of climate over large areas. If there were no evidence one way or the other, the assumptions of complete randomness of flood occurrences from one year to the next could perhaps be justified. But to force this assumption against strong empirical evidence to the contrary, and without being able to offer any plausible reason for it, is hardly a scientific approach.

Because of these reasons, the question of the relative merits of empirical and causal models for frequency analysis of historic floods does not even arise. The random model is neither empirical nor causal, but teleological -- it is dictated solely by a stubborn determination to use a simple probability calculus for manipulating the numerical values of flood peaks in the belief that such manipulation would bestow the desired probabilistic meaning upon the resulting numbers.

The general point to be emphasized is similar to the one mentioned in connection with the average return period. As the average is a measure of central tendency only if such tendency exists, so a frequency is a measure

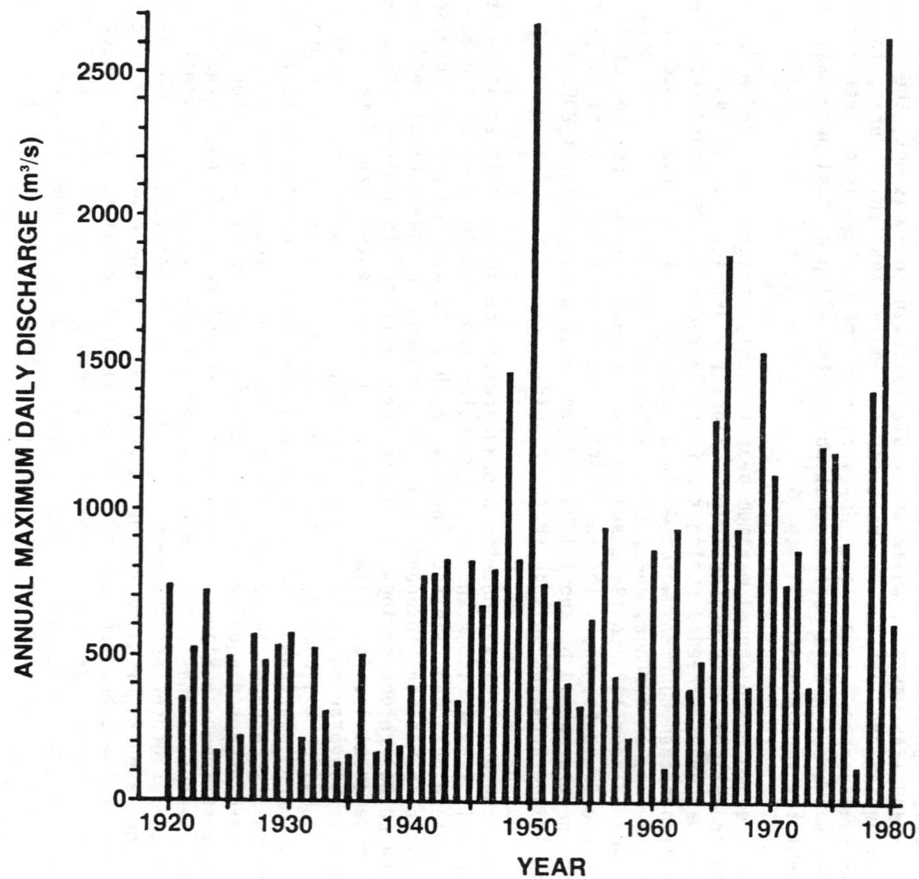

Fig. 1 Annual maxima of daily discharge for Red River at Emerson, Manitoba, Canada.

of probability only if the entity concerned has a probabilistic nature. Obviously, if a population statistics shows that the per capita frequency of entering marriage is 0.6, this does not mean that, given a life expectancy of 70 years, a man marries on the average once in 108 years, nor that there is a 0.86% probability that he will marry in any given year of his life.

The meaning of flood frequencies becomes especially important when frequency distributions are fitted with mathematical models of probability distribution functions and these are extrapolated to obtain estimates of rare floods, typically the 100-year or 1000-year flood, used for design purposes. While an ordered sequence of historic peak flows is a statement of fact summarizing the frequencies of peak flows during a specific period of time, fitting it by an analytically specified distribution function does not automatically make it valid for any other period, nor does it strengthen the case for its commonly used probabilistic interpretation. Least of all does its extrapolation give necessarily meaningful or reliable information on the actual probability of large floods. On the contrary, such a fit and its extrapolation are valid only to the extent to which the assumptions are met.

From the empirical point of view, the main difficulty is this: The existence of return periods shorter than the historic record is a matter of fact, and the degree of legitimacy of the assumption that the inverse values of their averages can be interpreted as probabilities of exceedance can be judged on the basis of the record's compatibility with a stationary random series. However, this line of reasoning cannot be applied to probabilities of extreme floods, since the empirical evidence for long return periods is not available in the relatively short historic records. One would need a stationary period on the order of 1000 years long to have a reasonable empirical basis for believing that a 100-year flood is not a vacuous concept (as it would be if, for instance, all the 10 or so highest floods occurred at the very beginning or end of the record.).

For much rarer floods, e.g., a 10,000-year flood, such empirical evidence is not conceivable even hypothetically since it would require a stationary record on the order of 100,000 years. This would extend at least 80,000 years into the last glacial period, which would make the record nonstationary be definition. Thus, for very rare floods, the probabilistic interpretation of their frequencies cannot be derived from empirical evidence for flood return periods. And to extrapolate this interpretation from its plausibility for small floods is quite arbitrary and difficult to justify on physical grounds. It is very different to say that a relatively low flood can be exceeded in every year with a probability of 10% than to say that an extremely high flood can be exceeded in every year with the same probability of say 0.01%. It may well be that an extreme flood is a result of some long-memory process, for example, a volcanic eruption that has released forces accumulating for hundreds or thousands of years, so that its annual probability slowly increases and, after the occurrence, suddenly drops down only to start growing again. Such a mechanism would be easily compatible with a short-memory or memory-less process for small floods.

The fact that we do not know a solution to a given problem should not mean that we can substitute for it a solution for a different problem. Unfortunately, this is done quite frequently. For example, even a markedly nonhomogeneous flood sample is often treated as homogeneous because, if it were split into two or three subsamples, some or all of them may be "too small to allow a realistic fit." The only really realistic conclusion --

that is, if a realistic fit cannot be made, then it simply is not made -- is seldom reached. This is because in "flood frequency analysis" the main objective is to do a "frequency analysis", not to be realistic. Any hydrological problem that interferes with the fitting of a simple mathematical distribution model and its interpretation as a probability distribution is simply swept out of the way and hydrological reliability or credibility never enters the picture. Even if the problem of uncertainty is considered, it is only a statistical uncertainty concerning inferences from a small random sample about the properties of its parent probability distribution. The fundamental problem of uncertainty about the relevance of the assumptions assigned to the flood regime is studiously avoided because of the intolerable risk that its study may reveal emptiness behind the ostensibly scientific façade of the whole flood frequency analysis.

To summarize, the empirical modelling of flood probabilities is still at a level where stochastic hydrology as a whole was about 50 years ago and their causal modelling has barely been recognized as a legitimate problem.

4. DROUGHTS

It is probably fair to say that stochastic hydrology owes its existence to concerns about reliability of water supply during periods of drought. This concern did lead to a genuine analysis of sequential patterns in time series of hydrologic records while its flood-related counterpart did not because of an early awareness that droughts can extend over several years and that it is not possible effectively to alleviate the resulting water shortages if each dry season is considered in isolation from the neighbouring ones, as it is usually possible with flood damage mitigation.

The basic tool for providing adequate water supply during droughts has always been water storage in reservoirs. Engineers realized more than a hundred years ago that considering the water deficit in the driest year on record was not sufficient for estimating adequate reservoir size, as had been customary before. The old practice was associated with the fact that most of the old reservoirs were built for low levels of development where the amount of water supplied even during the driest season was less than the lowest annual amount of the runoff so that the reservoir could be refilled every year. This is analogous to the practice in the design of flood control storages where it is assumed that the storage can be emptied between each two consecutive floods. However, as the demands for water supply grew, the above assumption became invalid. This was first pointed out by Rippl in 1883 and led to an interest in the study of the patterns in which dry years occur, in the search for periodicities and time symmetries, and eventually to application of the theory of stochastic processes to the modelling of hydrologic time series.

It is interesting to note that, as is still the case with flood series, the series of annual flow totals (which are the backbone of reservoir reliability estimation) were originally also regarded as purely random, without any attempt to verify this assumption. It was only in the 1940s that the assumption of randomness was superseded by the assumption of persistence, for which there was an empirical evidence in the existence of a linear correlation between flows in successive seasons and years (Kritskiy and Menkel, 1940). However, a genuine analysis of the sequential structure of hydrologic time series started only in the 1960s when their first empirical stochastic models started to appear.

Once again, as pointed out in connection with flood frequencies, the numerical characteristics of streamflow time series, such as the moments, autocorrelation function, crossing properties, ranges of the residual mass curve, etc., are statements of empirical facts. What remains open to question is their probabilistic interpretation. For the common interpretation, the basic assumption is that the series is a realization of a stationary and ergodic stochastic process. Without it, the theory of stochastic processes would have little to offer and all the stochastic hydrologic models, as well as the standard interpretations of reservoir and water supply reliability based on them, would be meaningless.

Given the pivotal role of these assumptions, it would seem not only logical but imperative that their validity in the hydrologic context be carefully analyzed. This is not and never has been the case. If there is an assumption that can be made with any confidence, then it is that there is no stochastic process for which a given flow series would be one of a multitude of equiprobable realizations. The assumption of a stochastic process implies the existence of a mechanism capable of generating sample series with the same structure, either simultaneously or in sequence. Examples include sample paths of microscopic particles moving in a fluid where the generating mechanism is represented by the movement of the fluid molecules; error characteristics of a number of self-guiding radars of a given make following a target over a prescribed path; distortions of a given signal by a set of amplifiers of the same kind; growth patterns of individual plants of the same species planted simultaneously on a plot with homogeneous soil; sets of measurements carried out under controlled stationary conditions in consecutive days, etc.

Because of the intrinsic impossibility of a stochastic process related to streamflow series at a given gauging station, the concept has to be saved by invoking assumptions of stationarity and ergodicity, which make it possible to conceive of a stochastic process where the equiprobable sample paths are represented by segments of a single time series. However, the stationarity assumption is often in direct contradiction to empirical evidence, even on a short time scale and more so on a long one. With regard to ergodicity assumption, there is no basis on which to judge it since it is defined only in the context of an ensemble of realizations. The very fact that only one realization exists makes it at the same time meaningless and valid by definition, since only an ergodic stochastic process can be "composed" from segments of one stationary series.

Without the benefit of these assumptions, the probabilistic interpretation of numerical characteristics of a time series is severely limited. The historic frequency distribution of annual flows no longer has to define their probabilities in individual years, the reservoir reliabilities no longer have to have the connotation of probabilities of failure-free operation, etc. In the modelling context, the assumption of stationarity is vital since without it the modeller's aim to "preserve" the historic parameters is completely misguided and population parameters become elusive.

Another important assumption which is almost never tested, although it is always implied in the stochastic flow models, is that only linear relationships exist within the series. There is no a priori reason why this should always be so. The only reason why this assumption is not tested seems to be the fear that it may be invalid. In such a case, the impressive mathematical tools now being employed in stochastic modelling would have to be considered irrelevant.

However, even if a model were designed to account for all empirically verified features of a historic flow record, and if the fit of the model were perfect, its extrapolation beyond the range of values given by the record is not justified without physically based reasons. In other words, the extrapolation must be plausible hydrologically because an extrapolation of a mathematical pattern beyond the range of data used to develop it is not hydrology. The numbers produced by it are ***unreliable***, even meaningless, and it would be paradoxical to use them as a basis of ***reliability analysis*** of real-life situations. This is more so now than it ever was in the past because the most compelling piece of empirical evidence is the fact that hydrologic conditions are changing due to man-made causes. Therefore, the relevance of historic records and of empirical models "preserving" the historic patterns and parameters are becoming potentially less and less relevant to the future to which our reliability analyses are supposed to apply. Consider, for instance, the "long-term mean" which is so basic to all stochastic hydrologic models, and to all our thinking about hydrologic and climatic processes, that it does not even occur to us to question its meaning. Yet, in the context of a given time series the mean is nothing more than the "best" horizontal straight line fitted to it in the least-square sense. But why should we cling so strongly to the concept of fitting (and extrapolating) such a straight line to a series which we claim to be unpredictable (Klemeš, 1974)? Is it to imply a belief that the development of its future central tendency will proceed in this simplest deterministic manner? Moreover, what is the empirical evidence for this assumption that the more remote past values we include in the computations the more accurate an estimate of future developments we will get.

5. CONCLUSIONS

It is perhaps the greatest paradox of the stochastic theory of hydrological processes (and of the reliability analysis based on it) that, while justifying its necessity by their stochastic nature, i.e. their unpredictability, we postulate the existence of a perfectly determined and intrinsically immutable system which generates them. This is nothing else but a disguised "Laplacean illusion" of a perfectly deterministic world, this time interpreted as a set of dice that God has created in the beginning and has been using ever since. This rigid "deterministic picture of stochasticity" is now being increasingly challenged. The stochasticity of nature is seen as irreducible, as if the set of the dice were continuously being perfected and modified by a playful and unpredictable God. Scientists now talk more about "becoming" than "being", about the options of the development being open and unpredictable (Bohm, 1957; Čapek, 1961; Prigogine and Stengers, 1984). From this point of view, our cherished concepts of "populations", "marginal distributions", etc. are just chimeras which exist only in our minds; in the real world, they are perpetually in the state of formation ("becoming") with no fixed design being followed. Estimating them becomes an intrinsic impossibility, an ill-posed problem. The only thing we can attempt is to project the course of the most immediate developements.

No doubt these views reflect not only the modern findings in physics (irreversible and dissipative processes), cosmology (evolutionary theory of the universe), and mathematics (catastrophy theory, theory of unstable dynamic systems), but also the increasing instability and accelerated changes in the social order and the impact of man on the environment. For hydrology, the consequences are very tangible indeed, not only theoretically but in the most direct practical sense through the environmental changes which are already taking place. Examples of these changes are changes in basin responses due to changes in land use, in water chemistry due to industrial emissions and waste disposal, in climate due to the changing balance of trance elements in the atmosphere, etc. This highlights the necessity for relating hydrological processes to other geophysical processes and trying to estimate their development from the forseeable changes in the latter ones. It is a reliability analysis based on these inescapable perspectives that the decision maker needs, not probabilistic games played with past records according to rules which are as relevant to reality as are the rules of ancient caligraphy to computerized word processing.

At the present stage of understanding of hydrological and other geophysical processes, it is very likely that the reliability-related questions of interest to the engineer and decision maker cannot be answered in a quantitative way. While this may be disappointing, there is nothing wrong with it. What is wrong, however, is to cover up this unsatisfactory situation by pretending that solutions artificially concocted from historic hydrologic data give the desired answers to real-world problems just because we give them names of the desired things like probability, risk, reliability.

The only consolation is that the decision maker in water resource development does not face this apalling situation only on the hydrologic side of reliability analysis. The other main basis on which his "optimal" decisions rest -- economics -- seems to be no more scientific and no less pretentious than stochastic and statistical hydrology. The following pronouncements of some leading economists closely parallel the message of this paper.

> An examination of economics as a discipline reveals that it is based on an epistemology, or method of establishing the validity of its knowledge claims, that runs counter to the norms of science. The prevalent view among economists is that formal (mathematical) proofs are not just helpful or indeed even necessary but also sufficient - and thus that empirical proofs can be dispensed with altogether. The result has been the development of a body of economic theory which is based entirely on axiomatic reasoning, without empirical support for the key propositions upon which the theory rests.
>
> The core of economic theory consists of four theoretical constructs which ... are derived from two more fundamental relationships. One is a 'utility' function ... the other ... is a production function (Eichner, 1985).

> Neither the utility function U = U(x) nor the production function O = O(z) are functions in the mathematical meaning of that term. That is, no rule is provided by which, given U(x), the value of U can actually be computed, and similarly for O and O(z) ... Why are problems of pure economic theory stated in such a peculiar, incomplete fashion? The reason is not far to seek. The utility function U(x) is not given explicitly because it cannot be given. It is an artificial construct of the theorist's mind and does not correspond to anything in the real world. ... Exactly the same is true of the production function O(z) ... All of this is not the application of mathematics to the economic problems of the real world. Rather, it is the application of highly precise and elaborate mathematics to an entirely imaginary and fanciful economic cloud cuckoo land (J. Blatt as quoted by Eichner, 1985).

> Page after page of professional economic journals are filled with mathematical formulas leading the reader from sets of more or less plausible but entirely arbitrary assumptions to precisely stated but irrelevant theoretical conclusions.
>
> Nothing reveals the aversion of the great majority of the present-day academic economists for systematic empirical inquiry more than the methodological devices that they employ to avoid or cut short the use of concrete factual information ... Year after year economic theorists continue to produce scores of mathematical models and to explore in great detail their formal properties; and the econometricians fit algebraic functions of all possible shapes to essentially the same sets of data without being able to advance, in any perceptible way, a systematic understanding of the structure and the operations of a real economic system (Leontief, 1982).

We only need to substitute "stochastic hydrology" for "economics", "flood probability distribution function" for "utility function", etc., and have a very fitting closure of this paper.

REFERENCES

Bohm, D. (1957) Causality and chance in modern physics. Routledge and Kegan Paul, London.

Čapek, M. (1961) The philosophical impact of contemporary physics. D. Van Nostrand, Princeton, N.J.

Eichner, A.S. (1985) The lack of progress in economics. Nature, Vol. 313, 7 February, pp. 427-428.

Fiering, M.B. (1976) Reservoir planning and operation. In: Stochastic Approaches to Water Resources, Vol. II, edited by H.W. Shen, pp.. 17-1 to 17-21, H.W. Shen, Fort Collins, Colorado.

Fiering, M.B. and B.B. Jackson (1971) Synthetic streamflows. Amer. Geophysical Union, Washington, D.C.

Klemeš, V. (1974) The Hurst phenomenon - a puzzle? Water Resources Research, Vol. 10, No. 4, pp. 675-688.

Klemeš, V. (1978) Physically based stochastic hydrologic analysis. Advances in Hydroscience, Vol. 11, pp. 285-356.

Klemeš, V. (1982) Empirical and causal models in hydrology. In: Scientific Basis of Water Resource Management, pp. 95-104, National Academy Press, Washington, D.C.

Klemeš, V. (1985) Hydrological uncertainty in water management context. XVIIIth General Assembly IUGG, Hamburg, 1983, IAHS Publ. No. 147, pp. 175-186.

Kritskiy, S.N. and M.F. Menkel (1940) A generalized approach to streamflow control computations based on mathematical statistics. (In Russian) Gidrotekhnicheskoye Stroitelstvo, No. 2, pp. 19-24.

Leontief, W. (1982) Academic economics, Science, Vol. 217, 9 July, pp. 104-107.

Plate, E.J. (1984) Reliability analysis in hydraulic design. Fourth IAHR Internat. Symp. on Stochastic Hydraulics. U. of Illinois at Urbana-Champain, Ill.

Prigogine, I., and I. Stengers (1984) Order out of chaos. Bantam Books, New York, N.Y.

Yevjevich, V. (1968) Misconceptions in hydrology and their consequences. Water Resources Research, Vol. 4, No. 2, pp. 225-232.

ELEMENTS OF BAYESIAN ANALYSIS OF UNCERTAINTY IN HYDROLOGICAL RELIABILITY AND RISK MODELS

Jacques M. Bernier

Laboratoire National d'Hydraulique - 6 quai WATIER - 78400 Chatou-France

ABSTRACT

Bayesian methods dealing with uncertainties in hydrology are described. Examples of application concerning estimation of reliability parameters are given. These examples allow to be explored the relationship between hydrological data, hydrological models and consequences of their use in the decision processes involved.

I. INTRODUCTION

Several types of uncertainties are involved in decision processes concerning the management and design of water resources systems :

- natural uncertainty conditioning the environment of the system due to the random nature of such state variables of the system as : river discharge, precipitation, water demand, resistance of soil or concrete... This uncertainty is described in terms of distribution functions and is taken into account in performance indices or figures of merit characterizing the system (Duckstein et al., 1984).

- technological uncertainties (due to measurement, sampling, model choice) generated by the failure of the decision maker to estimate fully the correct hypotheses, adequate model or estimation to be used for making adequate decisions. Classical approaches deal with these uncertainties by means of sensitivity analysis which separates the statistical problem of choosing the correct hypothesis or estimate, from the decision problem for management or design.

Statistical Decision Theory is concerned with these problems of technological uncertainties and their direct influence on the consequences of decisions when the decision-maker knows some statistical information such as hydrological data. Bayesian analysis is a part of Statistical Decision Theory where the technological uncertainties are taken in account in terms of probability distributions not only in the case of statistical quantitative information but also in the case where information is delivered as qualitative expert advice.

This paper is devoted to a presentation of the elements of these theories applied to reliability and risk problems in water resources systems studies.

II. A RELIABILITY MODEL

Let us consider time-varying state variables of a given system, so that the state trajectories enter at random epochs into specific critical domains corresponding to a failure mode. Figure I shows the case of a single state variable (a river discharge for instance) and critical domains defined as intervals. The upper interval allows us to take the flood failure mode into account and the lower interval allows us to accommodate critical drought periods. States variables could be defined in terms of reservoir content as well.

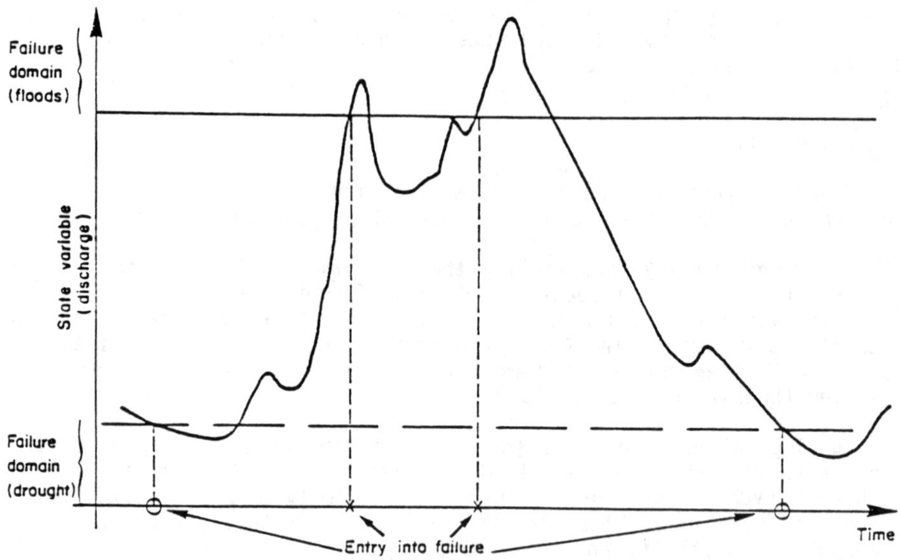

Figure I : Chronological curve of the state variable

For illustration purposes, consider reliability problems initiated by the entry into failure modes, for which the following quantities are defined :

- the failure rate $\lambda(t)$ such that :

$\lambda(t)\,dt$ is the probability of failure in dt at time t, given that the system was under control at time t (i.e., not in a failure mode).

- the reliability R(t) which is the probability that the failure has not occurred for a given period of length t :

$$R(t) = 1 - F(t) = \exp\left[-\int_0^t \lambda(u)du\right] \qquad (1)$$

where the failure rate depends only on time and F(t) is the cumulative distribution function of the continuous duration of life t of the system without failure.

The failure rate $\lambda(t)$ permits a process classification which is of interest for operational purposes. For simple, natural or man-made systems in stationary conditions, the function $\lambda(t)$ is usually constant. For more complex systems or in periods outside the stationary domain, the failure rate can depend on time, on the observed past behaviour of the process such as the last time duration life at time of failure (the so-called renewal hypothesis), or on external variables such as precipitation in a hydrological context.

Let the information for evaluating system reliability be obtained by direct observation of the system itself and neither by inspection of the components of the system nor by computation of the combined probabilities of failure as is common in industrial complex systems.

Thus the available information is given in terms of observed failure entries during a given time period T (periods of failure being disregarded or not being accounted for if they are small compared with no failure periods).

Given the simple hypothesis of constant λ, the complete information is contained in the sufficient statistics N : total number of observed failure entries on T, the probability of which is :

$$P(N, \ ) = e^{-\lambda T} \left(\frac{\lambda T}{N!}\right)^{N} \qquad (2)$$

This probability, called the likelihood function, entirely describes what it is necessary to know about the available information on the unknown value of the failure rate.

At this time two kinds of technological uncertainties appear :

- the sampling errors due to the random nature of the information N ; a measure of the resulting uncertainty is linked with the total duration T.
- the model uncertainties due, in this example, to imprecise identification of the factors affecting the reliability of the system : these uncertainties can make the failure rate non-constant with time or with varying conditions.

III. DECISIONAL ASPECTS

Given the damaging consequences of failure, a protection decision must be made in reliability studies. Such a decision, labelled a, belongs to a given set A. Generally the decision can be represented by the chosen residual failure rate (after protection action) so labelled a. So the consequences of a decision a can be economically assessed as :

- a protection cost (investment), $C(a,\lambda)$, decreasing with the residual failure rate a (see Figure II).
- a damage cost, $D(a,\lambda)$, increasing with the residual failure rate a.
  Note that these costs also depend on the true failure rate of the system λ, called state of nature, which belongs to a given set Λ and which is conditioning the environment of the decision-maker.

The overall loss function is then :

$$L(a,\lambda) = C(a,\lambda) + D(a,\lambda) \qquad (3)$$

With the assumption made concerning the variation of C and D as a function of a, there is (see Figure II) an optimum decision $\hat{a}(\lambda)$, minimizing the loss L. This optimum decision is a well defined function of the failure rate λ.

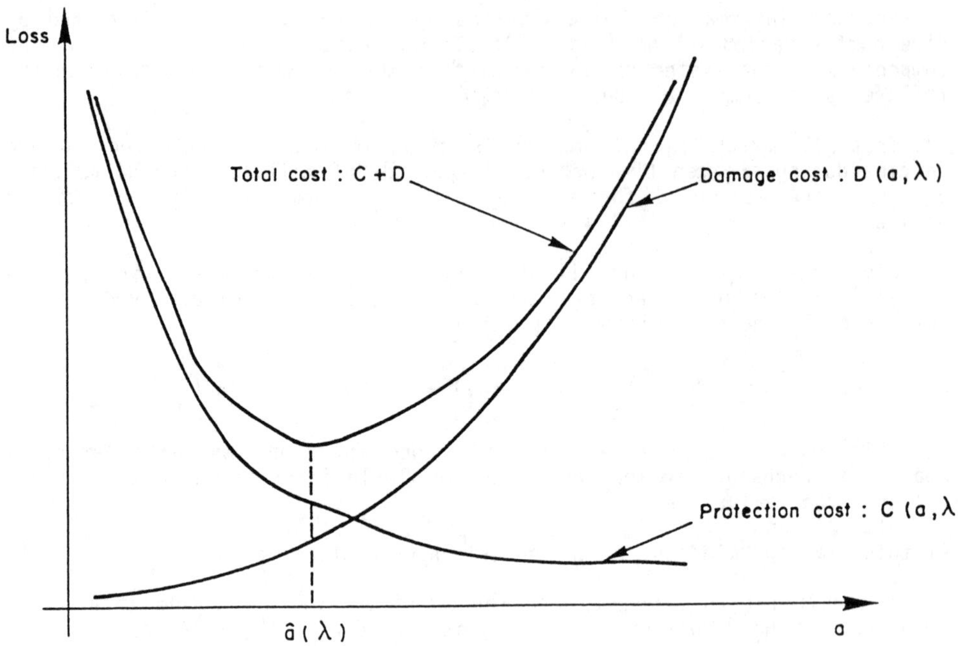

Figure II : Protection and Damage cost functions

The previous analysis assumes perfect information about λ , that is to say λ is assumed to be known exactly.
Due to uncertainties, the chosen decision a can be different from the optimum $a(\lambda)$ so that the decision maker is subject to an extra opportunity loss OL equal to :

$$OL(a,\lambda) = L(a,\lambda) - L\left[\hat{a}(\lambda),\lambda\right] \qquad (4)$$

This opportunity loss is an essential ingredient of the uncertainty analysis. An approximate formula is useful when the possible deviation from the optimum is small (for example, large quantity of information available or large value of T).

This approximation is obtained by expanding the opportunity loss around the optimum, so that :

$$OL(a,\lambda) \quad c\left[a - \hat{a}(\lambda) \right]^2 \qquad (5)$$

$$\text{with :} \quad c = - \frac{\partial^2 L(a,\lambda)}{\partial a^2} \quad a = \hat{a}(\lambda)$$

In a problem of this type, where there is a direct one-to-one relationship between the state of nature λ and the optimum decision $\hat{a}(\lambda)$, knowing λ is equivalent to choosing a particular a ; also the state of nature λ for which a is optimum can be calculated by the inverse of the function $\hat{a}(\lambda)$.

Hence the decision a is equivalent to estimating λ as the quantity ℓ and afterwards choosing a such that : $a = \hat{a}(\ell)$

Then the approximate relationship (5) defining the opportunity loss can be written as :

$$OL(a,\lambda) = OL(\ell,\lambda) = c\left[\frac{d\,\hat{a}(\lambda)}{d\lambda}^2 \right](\ell - \lambda)^2$$

so :

$$OL(\ell,\lambda) = c'(\ell-\lambda)^2 \qquad (6)$$

In this case, the estimation problem where λ is taken equal to ℓ and the decision problem of choosing a, given $\lambda = \ell$, can be solved separately using loss function (6) in the estimation problem.

The previous analysis is valid as long as one single-dimensional quantitative criterion is used for judging the protection decision. But in reliability problems, decisions often can be assessed using different figures of merit so that the decision involves multiple criteria decision making. Attempts have been made for developing statistical decision theory in line with multiple criteria (Davis et al. (2)) but the classical theory is usually presented with a one-dimensional loss function ; this will be the point of view adopted here.

Another type of loss function appears where the decision space is a dichotomy, such as whether to protect or not to protect against a specific failure mode. This case is common in the nuclear domain where decisions are taken with reference to a particular threshold λ_o for failure rate (for instance 10^{-6} or 10^{-7} per year). The concept of admissible or inadmissible failure mode can be embedded in a loss decision criterion by using losses as exhibited in Table I where c is the protection cost and D is the expected "damage cost" incurred by the decision maker when the inadmissible failure occurs. The use of a threshold criterion for admissible failure necessarily implies such a discontinuous consequence. The particular hypothesis of constant damage cost is adopted here for the sake of simplicity.

Table I : Example of discontinuous loss function

decision \ λ	$\lambda < \lambda_o$	$\lambda > \lambda_o$
a_1 : to protect	c	c
a_2 : not to protect	0	D

In this case the opportunity loss may also be given in a table form (Table II) :

Table II : Opportunity loss for admissible or inadmissible failure modes

decision \ λ	$\lambda < \lambda_o$	$\lambda > \lambda_o$
a_1 : to protect	C	0
a_2 : not to protect	0	D - C

In summary, the decisions set A, the states of nature set Λ and the loss function $L(a,\lambda)$ describe the decision problem with perfect information. But with uncertainties on λ, the decision must be made taking into account information N, the likelihood of which depends of the unknown state of nature λ.

IV. NORMAL FORM OF ANALYSIS OF UNCERTAINTIES

The basis of the normal form of analysis is the so-called decision rule describing the complete strategy of the decision maker and giving the decision a in A to be chosen for each outcome of the information gathering process (the experiment), i.e, in our example, for each value of N. Thus δ is a function of N :

$$a = \delta(N) \tag{7}$$

Henceforth, the set A of decisions will be taken as the set of estimated values ℓ of λ, which are equivalent to the protection decisions as noted before. So an example of the decision rule is the function :

$$\ell_1^* = \frac{N}{T} \tag{8}$$

For the purpose of comparing different decision rules, it is necessary to adopt a criterion, the so-called risk function :

$$R(\delta,\lambda) = E_\lambda\left[L(\delta(N),\lambda)\right] \tag{9}$$

or

$$R(\delta,\lambda) = E_\lambda\left[OL(\delta(N),\lambda)\right] \tag{10}$$

In these formulae, the mathematical expectation is with respect to the random value N, representing the outcome of the information gathering experiment.

Note that the risk function can be defined either with a loss or with an opportunity loss function giving two different criteria. Consider, for instance, the quadratic loss (6) with c' = 1.

In the case of the decision rule ℓ_1^* given by (8), the risk function associated with this quadratic opportunity loss is :

$$R(\ell 1^*, \lambda) = \frac{\lambda}{T}$$

Ranking of decision rules

Between two decision rules δ_1 and δ_2, we shall say that $\delta 1$ is R - better than 2, if $R(\delta_1, \lambda) < R(\delta_2, \lambda)$ for all states of nature λ, with strict inequality for some λ.
A decision rule δ is said to be admissible if there exists no R-better decision rule than δ.
Generally speaking, in any given problem, there are many admissible rules. In order to characterize a large class of admissible rules, the following procedure is useful :
Let us consider each state of nature λ as a value of a random variable. By using a so - called prior distribution with density $\pi(\lambda)$ on the set Λ of states of nature, we define the Bayes risk :

$$r(\delta, \pi) = \int_\Lambda R(\delta, \lambda)\ \pi(\lambda)\ d\lambda \tag{11}$$

For example given rule (8) and loss (6) with c' = 1, we have :

$$r(\delta, \pi) = \frac{1}{T} \int_\Lambda \lambda \pi(\lambda)\ d\lambda = \frac{E(\lambda)}{T}$$

where $E(\lambda)$ is the prior expectation of λ with respect to the prior distribution π.

We call Bayes decision rule with respect to a given prior distribution $\pi(\lambda)$, the rule δ_π such that $r(\delta_\pi, \pi)$ minimizes $r(\delta, \pi)$:

$$r(\delta_\pi, \pi) < r(\delta, \pi) \qquad \text{for any } \delta.$$

One of the main results of statistical decision theory states that any Bayes rule with $\pi(\lambda) > 0$ is admissible.

In the normal form of analysis, the prior probabilities have no operational interpretation except as pure mathematical devices allowing the generation of admissible rules. Note also that such a criterion as risk function based on mathematical expectation tends to measure how well the decision rules would perform in a long series of results of identical information experiments.

V. EXTENSIVE FORM OF ANALYSIS OF UNCERTAINTIES

Extensive form produces a constructive method for finding Bayes rules. Furthermore it is based on an operational use of a prior distribution $\pi(\lambda)$ able to represent the uncertainties about the state of nature in terms of likelihood degree given to each value of λ.

The main tool in this analysis is Bayes theorem giving the posterior probability density $\pi^N(\lambda)$, given the outcome N and prior $\pi(\lambda)$:

$$\pi^N(\lambda) = \frac{P(N, \lambda)\ \pi(\lambda)}{\int_\Lambda P(N, \lambda)\ \pi(\lambda)\ d\lambda} \tag{12}$$

The denominator of this formula : m (N) namely :

$$m(N) = \int_{\Lambda} P(N,\lambda)\ \pi(\lambda)\ d\lambda \qquad (13)$$

is the so called preposterior or predictive probability of the information N, unconditioned on the state of nature λ .

For each decision a (or ℓ) marked with a loss $L(a,\lambda)$ (or an opportunity loss function), the <u>posterior expected loss or opportunity loss</u> is calculated as :

$$E_a^N\ (L) = \int_{\Lambda} L(a,\lambda)\ \pi^N\ (\lambda)\ d\lambda \qquad (14)$$

or $$E_a^N\ (OL) = \int_{\Lambda} OL(a,\lambda)\,\pi^N(\lambda)\ d\lambda$$

Given the prior π , we can build the corresponding Bayes rule by choosing, for each outcome N, the decision a or ℓ minimizing the posterior expected loss or opportunity loss (the two minimizations give the same result), i.e :

$$\delta_{\pi}\,(N) : \quad N \rightarrow a \quad (\text{which minimizes } E_a^N\ (L) \text{ or } E_a^N\ (OL))$$

For instance, use of the opportunity loss given by (6) and estimates ℓ as decisions yields :

$$E_1^N(OL) = \int_{\Lambda} c'(\lambda)\left[\ell - E_{c'}^N(\lambda)\right]^2\ {}^N(\lambda)\ d\lambda + \int_{\Lambda} c'(\lambda)\left[\ell - E_{c'}^N(\lambda)\right]^2\ {}^N(\lambda)\ d\lambda$$

where $E_{c'}^N(\lambda)$ is the posterior weighted expectation :

$$E_{c'}^N(\lambda) = \frac{\int_{\Lambda} c'(\lambda)\ .\ \lambda\,\pi^N(\lambda)\ d\lambda}{\int_{\Lambda} c'(\lambda)\,\pi^N(\lambda)\ d\lambda}$$

Minimizing this expression shows that the Bayes rule is :

$$\delta_2^* = \ell_2^* = E_{c'}^N\ (\lambda) \qquad (15)$$

so that the minimum posterior expected loss is

$$\min\ E^N{}_{\ell}(OL) = \int_{\Lambda} c'(\lambda)\left[\ \lambda - E_{c'}^N\ (\lambda)\ \right]^2\ {}^N(\lambda)\ d\lambda \qquad (16)$$

that is to say the posterior weighted variance.

In our example, a possible prior distribution for the failure rate (as discussed later) is the gamma prior :

$$\pi(\lambda) = \frac{\alpha'^{\gamma'}}{\Gamma(\gamma')}\ e^{-\alpha'\lambda}\ \lambda^{\gamma'-1}$$

depending on 2 parameters α' and γ'.

With the likelihood given by formula (2), the posterior density has the same gamma distribution depending on two parameters α'' and γ'' such that :

$$\alpha'' = \alpha' + T$$
$$\gamma'' = \gamma' + N$$

Using the opportunity loss (6) with c' = 1, for simplification purpose, we have :

$$\delta_2^* = \ell_2^* = \frac{\gamma' + N}{\alpha' + T} \qquad (17)$$

and minimum posterior expected opportunity loss :

$$\min E_\ell^N (OL) = \frac{\gamma' + N}{(\alpha' + T)^2} \qquad (18)$$

Let us emphasize the following. Bayes rules such as the one given by formula (17), are justified in terms of normal form of analysis and risk function, the interpretation of which is a result averaged in a series of experiment outcomes, called the initial precision of decision rule by Berger (1). The extensive form of analysis can be itself in a Bayesian framework provided that the prior probabilities have an operational value. Then the accuracy of a decision (not a decision rule) must be judged in terms of final precision, after the data set or information N has been observed.

VI. PREPOSTERIOR ANALYSIS

The link between normal and extensive form of uncertainty analysis is through preposterior analysis, that is decisions and their performance are evaluated before the result or outcome of the information gathering experiment is known.

The first result along this line of analysis is the relationship between Bayes risk $r(\delta,\pi)$ and posterior expected loss of each decision $a = \delta(N)$ defined for each N through the given decision rule :

$$r(\delta,\pi) = \sum_N m(N) \; E^N_{(N)}\left[L(\delta(N),\lambda)\right] \qquad (19)$$

But the main result of preposterior analysis is the evaluation of worth of information. This evaluation is based on the interpretation of opportunity loss $OL(a,\lambda)$ as the benefit earned by the decision maker when, starting from the decision a, he learns about the "true state of nature" (the perfect information) and then he changes his decision into the optimum $\hat{a}(\lambda)$.

So the opportunity loss can be interpreted as the value of perfect information λ when the decision is a. Use of the prior and posterior distribution allows us to define :

- prior forecast of value of perfect information :

$$\underset{a}{\text{Min}} \int_\Lambda OL(a,\lambda)\,\pi(\lambda)\,d\lambda$$

- posterior forecast of value of perfect information

$$\underset{a}{\text{Min}} \int_\Lambda OL(a,\lambda)\,\pi^N(\lambda)\,d\lambda$$

In each case the use of the minimum is justified because in extensive analysis the decision maker must choose his decision by minimizing either the prior or the posterior expectation of his loss.

The difference between the two previous expressions is, by definition, the posterior expected value of information carried by N :

$$E^N(VI) = \underset{a}{Min} \int_{\Lambda} OL(a,\lambda)\pi(\lambda)\, d\lambda - \underset{a}{Min} \int_{\Lambda} OL(a,\lambda)\pi^N(\lambda)\, d\lambda \quad (20)$$

Next, for operational purposes in order to choose a better experiment or collection of data scheme, before the outcome is known, the preposterior value of information must be estimated :

$$E^T(VI) = \sum_N m(N)\, E^N(VI) \quad (21)$$

using the preposterior probabilities of N. The index T means that the expectation (21) is calculated with the assumption of a given type of experiment, indexed by the duration T in our example.

Using once more our example of a quadratic loss and formulae (17) and (18), we find, in the case a of gamma prior :

$$E^N(VI) = \frac{N\alpha'^2 + 2\alpha'\gamma' T + \gamma'^2 T^2}{\alpha'^2(\gamma' + T)^2}$$

It can be shown, using formula (13), that the preposterior distribution of N is the so-called negative binomial distribution :

$$m(N) = \frac{\Gamma(N+\gamma')}{\Gamma(\gamma')\, N!} \; \frac{\alpha'^{\gamma'}}{(T+\alpha')^{N+\gamma'}} \quad (22)$$

for which :

$$E(N) = \sum_N N m(N) = \frac{\gamma'}{\alpha'} . T$$

so that :

$$E^T(VI) = \gamma' \frac{(3\alpha' T + \gamma' T^2)}{\alpha'^2 (\alpha' + T)^2}$$

We see that the preposterior value of information is an increasing function of T but with diminishing marginal value and that $E^T(VI)$ has a limit equal to $\frac{\gamma'^2}{\alpha'^2}$ depending only on the prior parameters.

VII. INTERPRETATION AND ESTIMATION OF PRIOR PROBABILITIES

There is a coherence between the justification of risk function in normal analysis as averaged result of decision rules in a long series of experiments and the classical justification of probability as limit of frequency of results in a long series of similar experiments. Coming before an experiment is made, prior probability and extensive analysis cannot have such a justification from the point of view of philosophy of approach.

The foundations of prior probabilities rest upon a new definition of probability, an operational one, resulting from the conditions of a bet made by the decision maker (or by an expert) in such uncertainty situations. For this personal or subjective probability to have an operational meaning and to obey the classical rules of mathematical probability theory, the behaviour of the decision maker must follow a few axioms. As a matter of fact, these are the same axioms which justify the use of utility functions as loss functions for decision making (Berger (1), Raiffa (13)).

Let us consider the practical methods for estimating prior probabilities. There are three principal methods for determining prior densities described as follows, for the case of univariate state of nature λ (Berger (1)) :

- The histogram method consists of dividing the entire set Λ into several subintervals, estimating the personal (or subjective) probability for each subinterval and plotting a histogram from this set of estimates.

- The relative likelihood approach compares the subjective likelihood of various values of λ , and a prior density may be plotted.

- The functional form method which uses a family of density functions depending on parameters fitted so that the resulting density has the properties adequately representing the prior knowledge of the decision maker (correct moments or percentiles for instance). The gamma family used in our example is such a convenient prior family. It belongs to the so-called conjugate family obtained from the likelihood of the information process by exchanging parameters and variables. However, as it will be seen in the companion paper prior distributions other than conjugate ones can also be convenient.

Even in the case where there exits no prior information at all, Bayesian extensive analysis can be useful. Then so-called non-informative priors must be used. There are many such non-informative priors described in the technical literature. They have to be judged from the point of view of robustness and sensitivity to the smallest possible information (see Berger (1)). Among the most useful non-informative priors are those obtained by taking limiting values of parameters of conjugate priors. For example, these limiting values represent the no-information situation (for instance $\alpha' = \gamma' = 0$ in the gamma distribution example for λ). Such a limit is a so-called improper prior because the integration on the entire set Λ is infinite, but the posterior distribution does exist in this case and the corresponding rule minimizing the existing Bayes risk is called <u>generalized Bayes rules</u>. Notice that in the case $\alpha' = \gamma' = 0$, the generalized Bayes rule corresponding to this limit of gamma prior is the classical estimator $\delta x_1 = N/T$.

An important theorem of statistical decision theory states that any admissible rule is either a Bayes rule or a generalized Bayes rule.

VIII. ANALYSIS OF SAMPLING UNCERTAINTY ABOUT FAILURE RATES

Let us study further the problem of protection against failure of systems through the estimation of the rate λ. We have already seen two decision rules.

- the classical estimate : $\ell_1^* = \frac{N}{T}$ (23)
- the Bayes rule with respect to a gamma prior : $\ell_2^* = \frac{\gamma' + N}{\alpha' + T}$ (24)

The first rule is a limiting form of the case of non-informative prior. These rules are the best ones for a particular quadratic opportunity loss.

Third type of decision rule

Now assume that damage results from failure that occurred too soon with a critical limit t_0 such that the damage is a direct function of the reliability function $R(t_0)$:

$$F(t_0, \lambda) = 1 - e^{-\lambda t_0}$$

The resulting approximate opportunity loss can still be quadratic in terms of :

$$OL\left[F(t_0, \ell) ; F(t_0, \lambda)\right] \simeq C(\lambda)\ (\ell - \lambda)^2$$

with $$C(\lambda) = \left(\frac{\partial F}{\partial \lambda}\right)^2 = e^{-2\lambda t}$$

A weight C has appeared so that the Bayes rule, defined by the formula (15), takes on a different form. Some algebra shows that the best rule is :

$$\ell_3^* = \frac{N + \gamma'}{\alpha' + T + 2t_0} \qquad (25)$$

Fourth type of decision rule

Let us now come back to the problem with a binary decision space : to protect or not to protect which, in case of perfect information about , is simply to compare the failure rate with the threshold λ_0. Taking into account the sampling uncertainty, what is the estimator ℓ_4^0 to be compared with the threshold λ_0 ?

In this case the posterior expected opportunity losses are, from Table II :

- for a_1 (to protect) : $E^N_{a_1}(OL) = C\int_0^{\lambda_0} {}^N(\pi)\, d\lambda$
- for a_2 (not to protect) : $E^N_{a_2}(OL) = (D-C)\int_{\lambda_0}^{\infty} \pi^N(\lambda) d\lambda$

so that the Bayes rule is :

a_1 : if $E^N{}_{a_1}(OL) < E^N{}_{a_2}(OL)$, a_2 otherwise

The previous condition can be written as :

$$\int_{\lambda_0}^{\infty} \pi^N(\lambda)d\lambda \quad > \frac{C}{D}$$

Defining ℓ_4^x as the percentile of posterior distribution of λ, given that N the probability of larger value than ℓ_4^* is C/D, the Bayes rule is to compare λ_0 to this percentile.

$$\begin{aligned} &- \text{decide } a_1 \text{ if } \ell_4^* > \lambda_0 \\ &- \text{decide } a_2 \text{ if } \ell_4^* < \lambda_0 \end{aligned} \qquad (26)$$

In this case the correct estimator to be used is the posterior percentile ℓ_4^*.

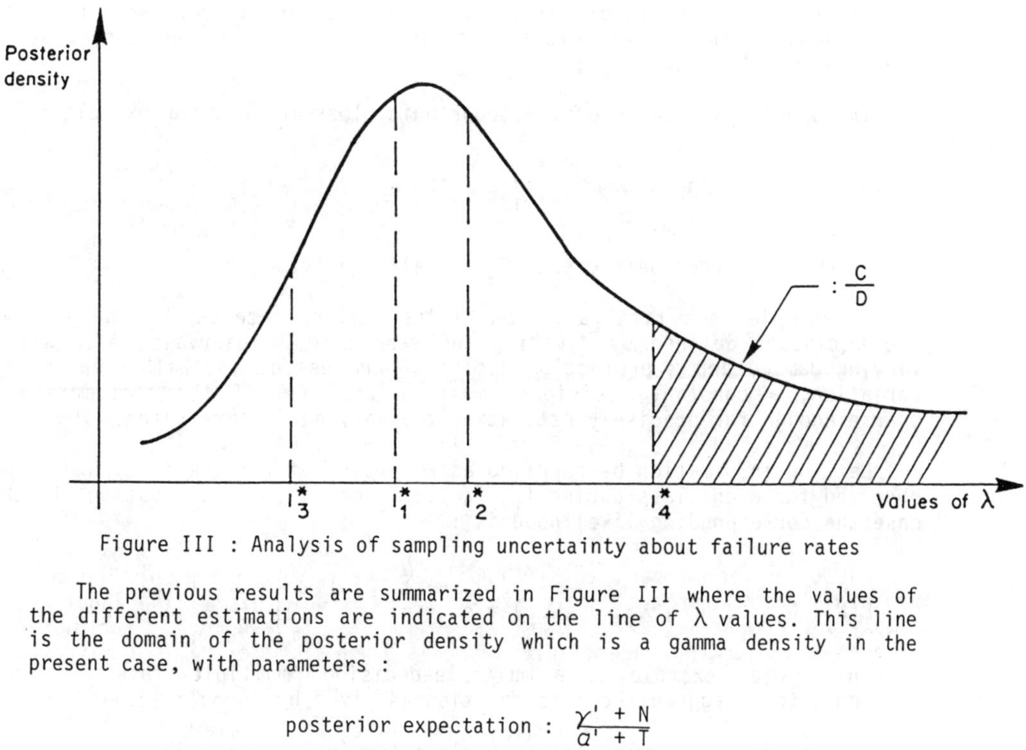

Figure III : Analysis of sampling uncertainty about failure rates

The previous results are summarized in Figure III where the values of the different estimations are indicated on the line of λ values. This line is the domain of the posterior density which is a gamma density in the present case, with parameters :

posterior expectation : $\dfrac{\gamma' + N}{\alpha' + T}$

posterior variance : $\dfrac{\gamma' + N}{(\alpha' + T)^2}$

Notice that, according to the different consequences or losses, there are different estimators, different ways to deal with sampling uncertainties. However the differences decrease when :

i) the amount of information (here measured by T) increases

ii) the randomness in the problem decreases.
In these Poisson-type problems, the randomness can be measured by the level of the failure rate itself because the relative variability (the coefficient of variation), is equal to $1/\lambda T$ so that the randomness can be estimated by the level of the inverse of N.

IX. ANALYSIS OF MODEL UNCERTAINTIES ABOUT FAILURE RATES

As noted before the uncertainties about a model can be seen as the failure of the decision-maker to identify correctly the factors that make the reliability parameters non constant. This can have consequences in the case where, if k classes of values of are necessary to represent correctly the variations of the parameters, decisions must be represented by a set of k estimates : $\ell_1, \ell_2 \ldots \ell_k$;

The approximate value of the opportunity loss may be, for example :

$$OL = \sum_{i=1}^{k} C_i(\ell_i - \lambda_i)^2 \qquad (27)$$

Assume that the coefficients C_i are all equal to 1.

An example is a time variation of the failure rate during the year and the k classes defined by dividing the year into k intervals, each with a varying damage and k protection decisions are needed. Notwithstanding this variation, a constant decision ℓ must be adequate if the information is sparse enough for precisely estimating the varying failure rates.

Let the information be supplied as k numbers of failure : $N_1, N_2 \ldots N_k$, observed for each class during $T_1, T_2 \ldots T_k$ periods of observation. In this case the corresponding likelihood is :

$$P(N_1, N_2 \ldots N_k ; \lambda_1, \ldots \lambda_k) = e^{-\sum_{i=1}^{k} i\, Ti} \prod_{i=1}^{k} \left(\frac{\lambda_i\, T_i{}^{Ni}}{N_i!}\right) \qquad (28)$$

This is an example of a multiple-decision, multiple-state of nature problem with an aggregated loss function as given by formula (27).

Several decision rules are possible ; for instance :

$$\ell_1^* = \frac{\sum_{i=1}^{k} N_i}{\sum_i T_i} \quad \text{for any } i. \qquad (29)$$

which is subject to model error, in this case the non-recognition of the variation of failure rate.

It is also possible to select a multiple valued decision rule such as, for each index i :

$$\ell_{2_i}^{x} = \frac{N_i}{T_i} \quad (i = 1,2 \ldots k) \tag{30}$$

Consider, for each i, Bayes rule knowing the assumed gamma prior distribution is constant (α' and γ' independent of index i) this assumption means that the different λ_i are values of a random variable having constant parameters α' and γ'. Bayes rule is written as :

$$\ell_{3_i}^{x} = \frac{\gamma' + N_i}{\alpha' + T_i} \quad (i = 1,2 \ldots k) \tag{31}$$

In this multiple type of problem, the estimation of prior parameters α' and γ' can be made either by means of prior considerations or by means of posterior analysis. As a matter of fact, let us make the simplifying assumption that the T_i are constants and equal to T. It can be shown that the preposterior distribution of the set of statistics N_i is such as these values are independent, each of them following the negative binomial distribution (22) with the same parameters α' and γ'. These parameters can thus be estimated by the method of moments applied to this negative binomial :

$$\alpha' = \frac{\bar{N}}{T(S^2 - \bar{N})} \quad , \quad \gamma' = \frac{\bar{N}^2}{S^2 - \bar{N}}$$

with :

$$\bar{N} = \frac{\sum N_i}{k} \quad , \quad S^2 = \frac{\sum (N_i - \bar{N})^2}{k}$$

Substituting the values of α' and γ' by these estimates into (31), gives :

$$\ell_{3_i}^{*} = \frac{1}{T}\left[Ni - \frac{\bar{N}}{S^2} (Ni - \bar{N})\right] \quad (i = 1,2 \ldots k) \tag{32}$$

This is a so-called <u>empirical Bayes rule</u>.

As a result, this decision rule is a mixture of rule ℓ_1^* and rule ℓ_2^* with weights a function of the observations N_i.

By calculating the posterior expected opportunity loss by (27), the gamma posterior and Bayes risk by the formula (15) knowing that the prior was gamma, one finds :

$$r(\ell_1^{x}, \pi) = \frac{\gamma' T}{\alpha'} \left[1 + \frac{k-1}{\alpha'}\right] \tag{33}$$

$$r(\ell_2^{x}, \pi) = \frac{k\gamma' T}{\alpha'} \tag{34}$$

$$r(\ell_3^{x}, \pi) = \frac{k\gamma' T}{\alpha'(1+\alpha')} \tag{35}$$

This last result for ℓ_3^* depends on an approximation using the true preposterior moments instead of $\bar{N}$ and S^2 in formula (32). (It is an asymptotic approximation.)

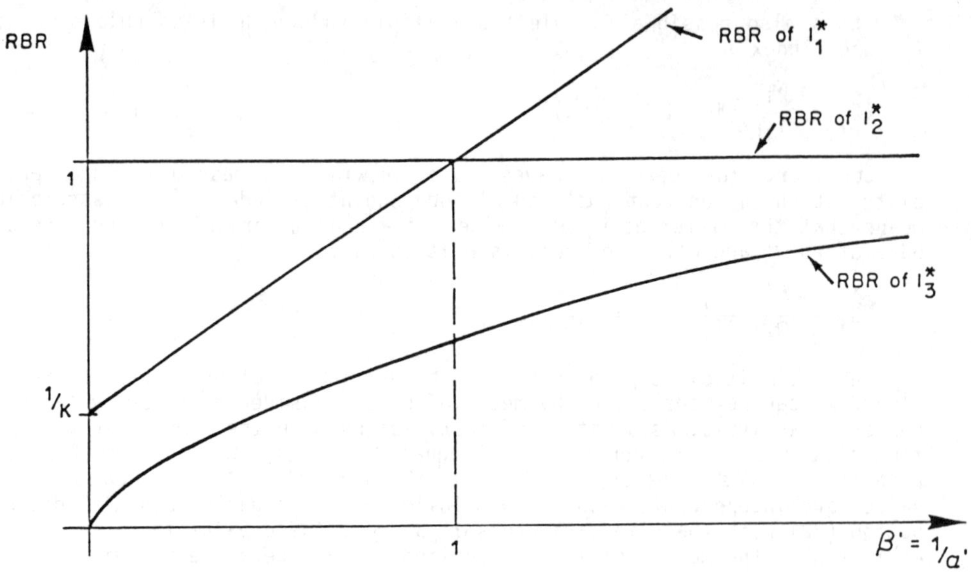

Figure IV : Analysis of model uncertainty about failure rate

RBR : relative Bayes risk = $\nearrow \dfrac{r(\ell_1^*,\pi)}{r(\ell_2^*,\pi)}$ $\searrow \dfrac{r(\ell_3^*,\pi)}{r(\ell_2^*,\pi)}$

β' : prior estimate of $\sum_{i=1}^{k} \dfrac{(\lambda i - \bar{\lambda})^2}{k}$ for gamma prior

Note that $\dfrac{r(l_1^*,\pi)}{r(l_1^*,\pi)} > 1$ for any β'

Figure IV shows the variation of the three Bayes risks with prior parameter $\beta' = \dfrac{1}{\alpha'}$, given a constant value of γ'. According to the interpretation of α' in terms of moments, we have :

$$\beta' = \frac{1}{\alpha'} \simeq \sum_{i=1}^{k} \frac{(\lambda i - \bar{\lambda})^2}{K\, \bar{\lambda}} \qquad \text{with } \bar{\lambda} = \frac{\Sigma\lambda_i}{k}$$

β' is a prior estimate of an index of dispersion of the k parameters λ_i (ratio of variance over mean value which is equal to 1 for a homogeneous Poisson process).

Figure IV shows that the constant rule $\ell^x{}_1$ can be better than the model error-free rule $\ell^x{}_2$ for a value of dispersion index ' smaller than 1. But we also see that the empirical Bayes rule $\ell^x{}_3$ is the best of the three rules under any circumstance. These properties of Bayes rules could be also interpreted in terms of risk function and admissibility. They are similar, for Poisson processes, to the properties emphasized by Stein and James (Efron and Morris, 1973) or (Judge and Bock, 1978) in the case of multiple estimation problems with gaussian distributions. The rule given by (32) is an example of Stein-type rule used in multiple decision, multiple states of nature problems which are frequent in water resources studies.

X. CONCLUSIONS

It is important to distinguish between the two types of uncertainties involved in analysis of water resources systems : natural uncertainty on one hand, technological uncertainties on the other hand. Statistical decision theory and Bayes methods deal with these technological uncertainties and allow the relationship between information, hydrological models and consequences of their uses to be examined.

SELECTED BIBLIOGRAPHY

1 Berger J.O (1980) Statistical Decision Theory - Springer-Verlag - New-York

2 Davis D, F. Szidarowsky and L. Duckstein (1983) A Distance - based Measure of the Effect of Uncertainty on Multiobjective Decision Problems. 11th IFIP Conference on System Modelling and Optimization - Copenhagen-Denmark -

3 Davis D, C.C. Kisiel and L. Duckstein (1972) Bayesian Decision Theory Applied to Design in Hydrology - Water Resources Research - Vol 8 (1)

4 Duckstein L, D. Davis and E. Plate (1983) Use of Climatological Information in River Basin Planning under Risk - Inter. Meet. on Statist. Climat. Lisboa - Portugal

5 Duckstein L, E. Plate and M. Benedini (1984) Incidents and Failure in Reservoir and Dams ; A Mathematical Systems Framework - American geophy. Fall Meeting-San Francisco -

6 Efron B. and C. Morris (1973) Stein Estimation Rules and its Competitors : An Empirical Bayes Approach - J. Amer. Statist. Assoc. Vol 68-

7 Gondran M. and A. Pagès (1980) Fiabilité des systèmes - Eyrolles - Paris

8 Judge G. G and M.E. Bock (1978) The Statistical Implications of Pre-Test and Stein Rule Estimators in Econometrics - North - Holland

9 Mac Cormick N. (1981) Reliability and Risk Analysis - Methods and Nuclear Power Applications - Academic Press

10 Miquel J. (1984) Guide pratique d'estimation des probabilités de crue - Eyrolles - Paris

11 North M. (1980) La simulation des processus hydrologiques intermittents par des modèles alternés inhomogènes - Hydro. Sciences Bulletin (AISH) n° 25 - 1 - 3

12 North M. (1980) Time Dependent Stochastic Model of Floods - Journal of ASCE - HY5

13 Raiffa H. (1968) Decision Analysis - Introducting Lectures on Choices under Uncertainty. Addison - Wesley ; Reading

14 Raiffa H. and R. Schlaifer (1961) Applied Statistical Decision Theory - Harvard Uni. Press

15 Todorovic P. and E. Zelenhasic (1970) A stochastic model for flood analysis. Wat. Res. Research Vol.n° 6

RELIABILITY ESTIMATION OF UNDERGROUND WATER CONTROL SYSTEMS UNDER NATURAL AND SAMPLE UNCERTAINTY

ISTVÁN BOGÁRDI
Civil Engineering Department
University of Nebraska - Lincoln
Lincoln, NE 68588 USA

On leave from
Tiszadata Consulting Engineers
Mikó u. 1, 1012 Budapest, Hungary

LUCIEN DUCKSTEIN
Department of Systems and Industrial Engineering
University of Arizona
Tucson, AZ 85721, USA

FERENC SZIDAROVSZKY
Department of Computer Sciences
University of Agriculture
Villanyi ut 29-35
1113 Budapest, Hungary

ABSTRACT

Reliability of a protection system against karstic water hazard is estimated. Such a system may consist of several interrelated elements (tunnels, pipelines, sediment settler and removal equipment, pumps). The failure of the system can be caused by a complex set of natural and engineering factors. Both the loading of the system - the karstic water inflow - and the resistances (capacities) of system elements are modeled as random variables. First a classical reliability analysis based on prior statistics of loading and resistances is performed using failure-tree analysis. However, the estimated reliability is very sensitive to the above statistics. To account for this fact, a Bayes reliability model is developed for the operation period of the system. The methodology is illustrated by a real-life example of an underground mine.

1. INTRODUCTION

Reliability of an underground water control system is estimated using a Bayesian approach. System reliability of engineering structures is very sensitive to the accuracy of loading and resistance statistics because the numerical value of reliability is generally found in the upper tail of the distribution, where the probability estimates are most uncertain. At the same time, the loading and resistance statistics are themselves uncertain because they are based on small samples, or worse on regional data or experience. This motivates the use of Bayesian distributions which account for this parameter uncertainty.

The construction of underground spaces such as subways or tunnels and the operation of mines are often subject to water hazard. If an underground space is being planned or already operating below groundwater level, an inflow control system should be provided. The main elements of this system are drainage equipment, grouting facilities, water conveyance, sediment settling and removal equipment, and pumping stations. In general, either system load or its resistance, or both, are random variables.

In the next section, the problem is formulated using the example of mining under water hazard. The classical, non-Bayesian reliability model is described in the section, "Non-Bayesian Reliability Model," and the Bayesian model is given in the subsequent section, "Bayes Reliability Model." The failure-tree method is used for constructing the model and a simulation algorithm is applied to provide the solution. A real-life example is provided in the Application section. In the last section, results of the analysis are discussed and conclusions are drawn.

2. PROBLEM FORMULATION

Reliability theory has been developed in various areas of engineering such as structural design (Shooman, 1968) or airplane and rocket design (Bazowsky, 1961). Reliability analysis has also been applied to hydrologic problems (Szidarovszky et al, 1975; Duckstein and Plate, 1985) and to mining engineering where several studies, such as the analysis of the haulage, rescue, and ventilation systems, can be found (Lucas and Adler, 1973).

Kesserü (1978) has pointed out the necessity of using reliability theory to design mine water control systems whenever mining operations take place under water hazard. Mining specifications are often given in terms of an "economic reliability" for property protection and a "maximal reliability" (or a safety level) for life protection (Esztó et al, 1975). Reliability analyses have resulted in the specification of elements such as emergency storage spaces, rescue routes, or warning systems (Schmieder et al, 1975).

A non-Bayesian reliability model which has been developed and applied to mining conditions in Duckstein et al (1981) needs further development for the following reasons:

(a) statistical parameters characterizing loading and resistance conditions are uncertain because of the lack of direct observations during the design stage;

(b) as the construction or operation starts, observations on loading and resistance become available; this information should be utilized for an updating of the reliabilities estimated during the design stage.

It will be shown that features (a) and (b) can be accounted for by the Bayes reliability model. For illustration purposes, a protection system frequently used against karstic water hazard is now described.

The six elements of the underground water control system considered are (Fig. 1): (1) protection of faces; (2) water cut of faces; (3) block water cut; (4) mine water cut; (5) sediment settler and removal equipment; (6) central pumping station. These six elements are only the most important ones; a more detailed analysis would require further breakdown. The protection of faces (Element 1) against inrushes can be effected in several ways, such as preliminary or posterior grouting, or the "Instantan" method (Kapolyi, 1977) to be explained later. Water collected or drained from the faces is conveyed through water cuts (Element 2) to the block water cut (Element 3). In other systems, there may be special sediment settlers and booster pumps in every block, but gravitational conveyance of water and sediment is predominant in recently designed facilities. Water and sediment from block water cuts travel to the mine water cut (Element 4), which leads to the central sediment settler (Element 5) and the central pumping station (Element 6). Sediment is removed from the settler by special equipment and pumped to the surface by hydraulic means. The number of faces and blocks increases as the exploitation of the mine proceeds; the amount of inrushes and sediment also increases as new underground spaces are opened.

The failure of an underground flood control system can be caused by a complex set of natural and engineering factors. It is necessary to single out those failure events which are critical as far as the design and operation are concerned. More precisely, two types of events or incidents are distinguished: "disturbance of operation" and "flooding." The "disturbance of operation" corresponds to an incident which disrupts or decreases mining production but does not stop it. On the other hand, "flooding" is defined as a failure which stops production altogether. Depending on the location of the incident or failure, the following top events are defined:

(a) disturbance of operation and flooding in faces;

(b) disturbance of operation and flooding in blocks;

(c) disturbance of operation in the mine with simultaneous disturbances in several blocks;

(d) flooding of the mine.

In the next section, the classical reliability model is summarized.

3. NON-BAYESIAN RELIABILITY MODEL

System loading corresponds to inrush events which can be characterized by the following three quantities:

I. q = magnitude of an inrush event;

II. q_{max} = maximum inrush event yield into a volume of given vertical dimensions and rectangular area A of unit width;

III. $Q(A)$ = total yield of inrush events over area A.

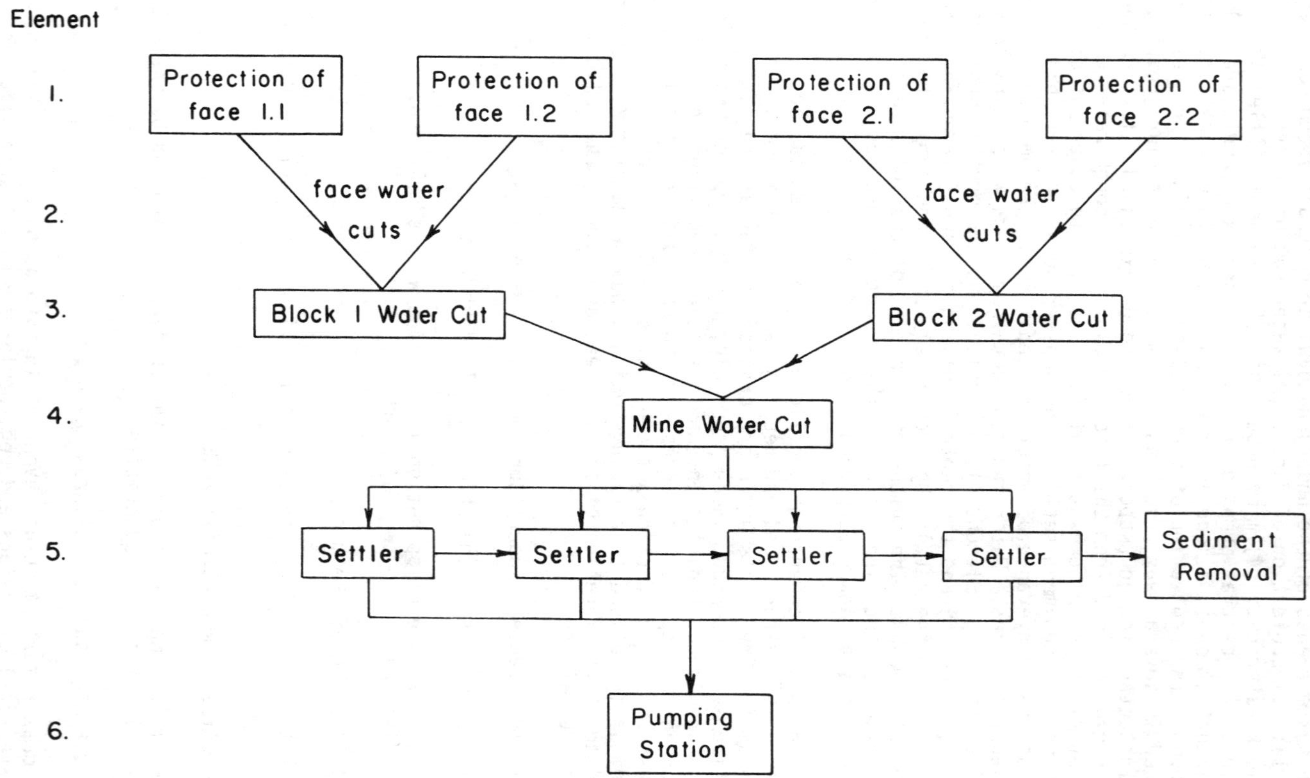

Figure 1. Elements of protection system.

The probability density function (pdf) of these quantities can be estimated as follows (Bogárdi et al 1980):

3.1 Yield of Inrush Events

A reasonable hypothesis based on physical reasoning and strengthened by observation data is that q follows a lognormal distribution.

3.2 Maximum Inrush Event Yield Over Area A

A second hypothesis based on phenomenological reasoning and reinforced by observation data is that N(A), the number of inrush events occurring over an area A, follows a Poisson distribution with mean A. Then the asymptotic distribution of qmax is derived from the distributions of N(A) and q as follows:

$$P(qmax \leq x) = F_{qmax}(x) = \exp\left[-\lambda A\,(1-F_q(x))\right] \tag{1}$$

It is assumed that qmax follows this asymptotic distribution.

3.3 Total Yield of Inrush Events Q(A)

The total yield for area A is calculated as the sum of a Poisson number N(A) of lognormal inrushes q(i).

$$Q(A) = \sum_i q(i), \quad i = 1,2,\ldots,N(A) \tag{2}$$

The distribution function (DF) of (Q) must be determined from the DF of q and N, since direct observation data on Q are rarely available. For this purpose, the simulation approach described in Szidarovszky et al (1976) appears to be appropriate. In particular, this method makes it possible to account for the spatial dependence between stochastic inrush events.

Note that the use of a statistical distribution of inrush characteristics is necessary only in the case where these inrushes come from a discrete number of natural fractures or geological faults. In ordinary fractured rocks, there is no specific need to identify particular characteristics of inrushes; it could be sufficient to consider the general aspects of the phenomenon. In this case, assuming that all the inrushes have the same average value would certainly simplify the problem without losing--presumably--much information.

The yield of inrush events can be controlled by various methods such as grouting and freezing (Schmieder et al, 1975). The impact of the control method on inrush yield should be considered in the reliability analysis. A decision rule or impact function expressing the effect of control strategy on an inrush may be defined as:

$$q'_{ij} = f_{ij}(q_{ij}, a_{1ij}, a_{2ij}, a_{3ij}) \tag{3}$$

where

q'_{ij} = controlled yield of inrush in face (i,j),

$f_{ij}(\cdot)$ = impact function for face (i,j),

q_{ij} = natural yield of event and

$a_{1ij}, a_{2ij}, a_{3ij}$ = parameters of the impact function.

As an example, consider the control method of grouting coupled with the following decision rule:

(a) if $q_{ij} < a_{1ij}$, do not grout; then $q'_{ij} = q_{ij}$
(b) if $a_{1ij} < q_{ij} < a_{3ij}$, grout a portion $(1-a_{2ij})$ of the inrush yield; $q'_{ij} = a_{2ij} \cdot q_{ij}$
(c) if $q > a_{3ij}$, grout completely the inrush; $q'_{ij} = 0$.

To estimate the protection system reliability, events CC_K (disturbance of operation of the mine) and MF (flooding of mine) are considered.

Event CC_K means that there is simultaneous disturbance of operation in at least K blocks:

$$CC_K: \bigcup_{k=K}^{m} \left[\bigcap_{1\le i_1 < i_2 < \dots i_K \le m} \left(C_{i_1} \cap \dots \cap C_{j_k} \right) \quad \underset{1\le t<k}{j \ne i_t} \quad \bar{C}_j \right] \tag{4}$$

where m is the total number of blocks and C_i is the disturbance of operation in block i. Event C_i occurs when disturbance of operation is present in every face of the block. Figure 2 shows the failure tree of event C in any of the blocks, with n being the number of faces in a block and

$$R_{ij}: (q\max > q_{ij}\ (0)) \tag{5}$$

where q_{ij} (0) is the threshold yield for face (i,j).

Event MF (flooding of mine) occurs when any one of six events E,F,FR,G,H,L occurs (Fig. 3). These events are defined as:

Event E: there is flooding in every block of the mine:

$$E: D_1 \cap D_2 \cap \dots \cap D_m \tag{6}$$

where $D_i = (Q_i > CV_i)$ = the event of flooding in block i; Q_i is the block mine water yield and CV_i is the capacity of block water cut i (element 3).

Event F: the sediment removal capacity CH of the mine is smaller than the maximum sediment inrush yield h. Experience shows that the following linear statistical relationship may be assumed between sediment volume h

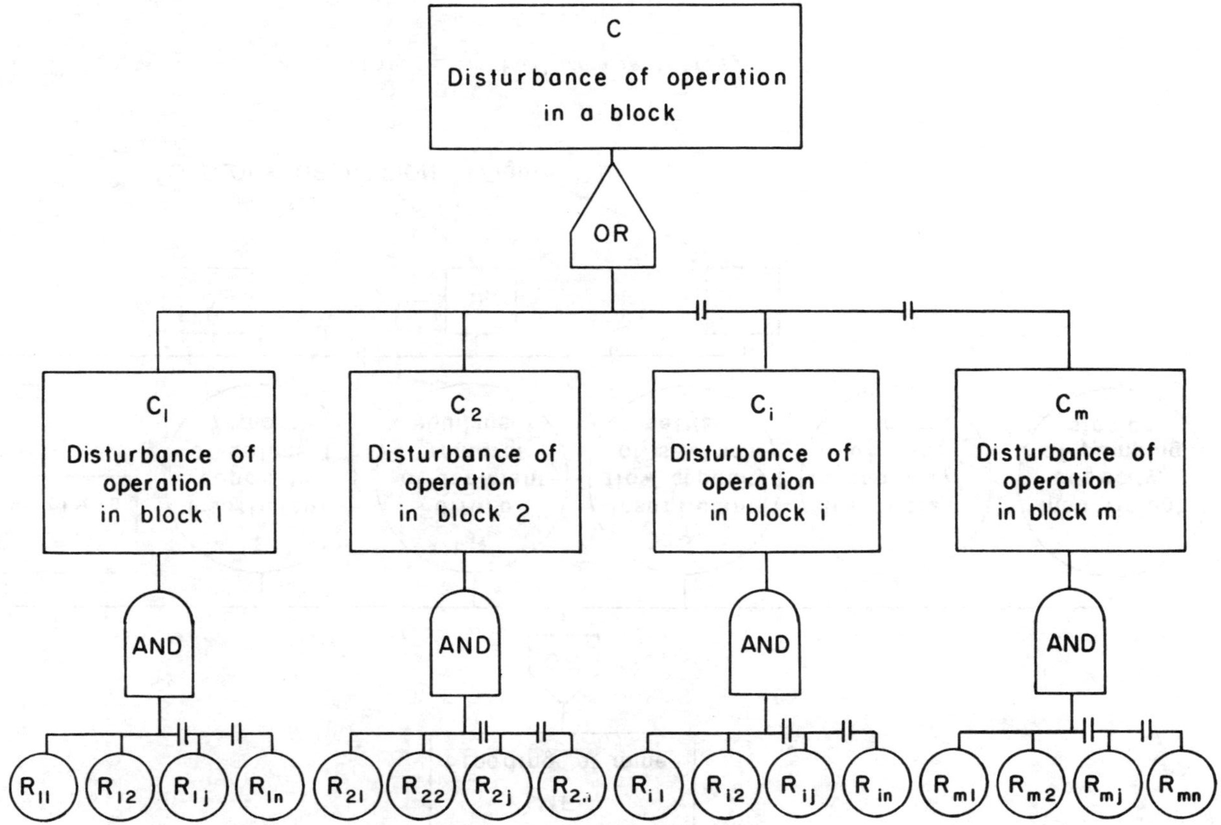

Figure 2. Sketch of disturbance events.

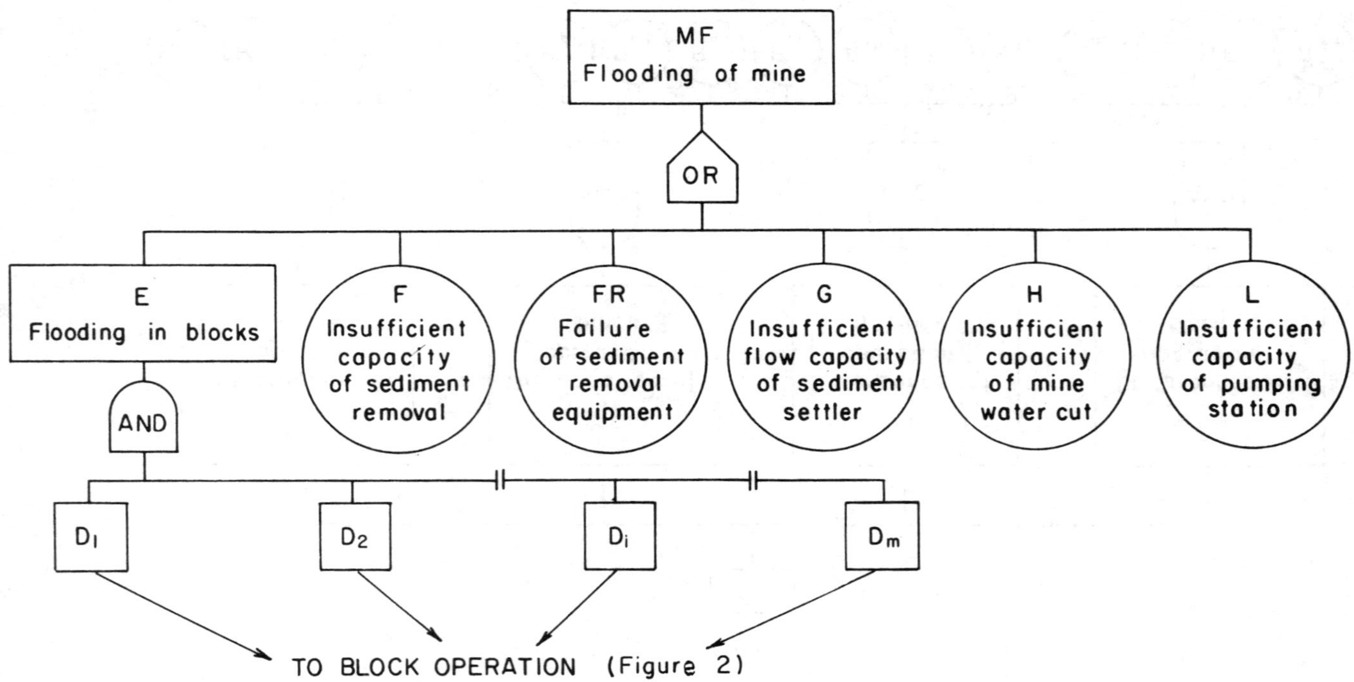

Figure 3. Failure event tree.

and water yield qmax:

$$h = \max_{ij} (k_0 \cdot qmax_{ij} + \varepsilon) \tag{7}$$

where k_0 is a specific sediment yield (tons of sediment per m^3 of water), and ε is an error term, assumed to be distributed normally with parameters (μ_1, σ_1). Thus, event F is:

$$F: (h > CH) \tag{8}$$

Event FR: a failure of the mine sediment removal equipment occurs:

$$FR: (\beta < t) \tag{9}$$

where β is the first failure time of the mine sediment removal equipment over time horizon t. The variate β is taken as exponential with mean λ_2 (the expected number of failures per unit time).

Event G: the total yield of mine water, Q^* is greater than the capacity CQ of the central sediment settler (element 5):

$$G: (Q^* > CQ) \tag{10}$$

Event H: the total mine water yield Q^* is larger than the capacity CV of the mine water cut (element 4):

$$H: (Q^* > CV) \tag{11}$$

Event L: the actual capacity α of the central pumping station (element 6) is smaller than the total yield of mine water, Q^*:

$$L: (\alpha < Q^*) \tag{12}$$

This event may be caused by an excessive flooding of the mine, or failure of some of the pumps; in either case, the real capacity α of the pumping station is smaller than Q^*. The pump failure events are assumed to be exponentially distributed with parameter λ_3, which is the average failure rate of one pump. The number of pumps remaining in operation is a binomially distributed variate, and α is the product of this binomial variate and the nominal capacity of one pump.

Since all possible failure events have been defined, the event MF of mine flooding can be written as:

$$MF: E \cup F \cup FR \cup G \cup H \cup L \tag{13}$$

Using a Monte-Carlo simulation method, failure probabilities can be estimated for several periods from a single computer run.

4. BAYES RELIABILITY MODELS

The statistical parameters in the reliability model are: the mean μ and variance σ^2 of inrush event yields q_1, the specific number λ, of inrushes, the variance $\sigma_1^2(\varepsilon)$ of sediment yield (Eq. 7), the specific number λ_3 of failures of a pump, and the specific number λ_2 of failures of the sediment removal equipment.

If the values of these parameters were known, the reliability estimates would fully account for randomness or natural uncertainty. However, in the design stage, only indirect information is available, leading to uncertainty in parameter estimation. Thus, reliability estimates are subject to this parameter uncertainty which decreases as operation starts and more and more observation data become available.

In the following, a Bayesian approach is used to account for both natural and parameter uncertainty (Raiffa and Schlaifer, 1961: Benjamin and Cornell, 1970; Musy and Duckstein, 1976; Martz and Waller, 1982; Ang and Tang, 1984). Such a Bayesian approach has been applied to reliability engineering, among others in Mann et al (1974), Chang and Thompson (1976), Metler (1980), and Tillman et al (1982). However, its application to the reliability analysis of complex systems, such as the one considered herein, appears to be quite infrequent.

The present study accounts for parameter uncertainties in loading statistics. Uncertainties in λ_2 and λ_3 can be taken into consideration in a similar way. First, the design stage is considered, and second, the operation stage.

In the design stage, parameter uncertainty is present, and parameters $\underline{\theta}$ are taken as random variables; a Bayesian distribution $\overline{f}(.)$, which accounts for both natural and parameter uncertainty, can be estimated as:

$$\overline{f}_X(x) = \int f_X(x|\underline{\theta})\ f_{\underline{\theta}}(\underline{\theta})\ d\underline{\theta} \tag{14}$$

where $f_X(x|\underline{\theta})$ is the model distribution given parameter vector $\underline{\theta}$, and $f_{\underline{\theta}}(\underline{\theta})$ is the distribution of the parameter. The procedure described in detail by Bogardi et al (1982) generates Bayesian values of the three loading components q, qmax, and Q by Monte-Carlo simulation.

4.1 Magnitude of Inrush Events q

Since log q follows a normal distribution with unknown parameters (μ,σ), the joint conjugate distribution of (μ,σ) is a normal-gamma distribution (Raiffa and Schlaifer, 1961; Benjamin and Cornell, 1970). This conjugate distribution has four parameters: the mean and variance of μ, and the mean and variance of σ. However, only three of these parameters are independent. Furthermore, any one of the parameters may be replaced by n, the number of data points (or sample size).

Prior information on the above parameters may be determined by regional estimation, geological analogy, hydraulic calculations and/or literature data. As shown in Bogárdi et al (1980), the sample size n', the mean m' of m and the variance $s'^2(m')$ can be estimated from prior information.

The Bayesian simulation algorithm is as follows:

1. Generate a pair (μ, σ) using the normal-gamma distribution with prior m', s' and n'.

2. Generate a realization of a normal variate η_j using parameters (μ, σ) generated in step 1.

3. A prior Bayesian random value of q is $q_j = \exp \eta_j$.

4.2 Maximum Inrush Event Yield over Area A = q max.

The additional uncertainty caused by parameter λ(Eq. 1) is entered into the estimation of the DF of qmax.

The conjugate distribution of λ is a gamma-2 distribution with two prior parameters α' and β' such that:

$$m'(\lambda) = \frac{\alpha'}{\beta'} \text{ and } s'^2(\lambda) = \frac{\alpha'}{\beta'^2} \tag{15}$$

The parameters $m'(\lambda)$ and $s'^2(\lambda)$ are estimated from available prior information (Bogardi et al, 1980). The same type of gamma-2 distribution defines the conjugate distribution of λ_2 and λ_3.

A Bayesian distributed value N_j can be simulated as follows:

1. Generate a value of λ using a gamma-2 distribution with prior parameters $\alpha'$and β' (Eq. 15).

2. Generate a (prior) Bayesian Poisson variate N_j using parameter λ found in step 1.

The simulation procedure for estimating a Bayesian distribution of qmax is:

1. Generate a random value of N.

2. Generate a number N of prior (or posterior) realizations of q: $(q_1, q_2, \ldots, q_n)$.

3. Find the value $q_{max} = \max_{1 \leq i \leq N} \{q_i\}$; iterate and obtain an empirical distribution of qmax.

4.3 Total Yield Q of Inrush Events Over Area A.

Using Equation (2), a Bayesian distribution of Q can be simulated in the following way:

1. Generate a Bayesian distributed value of N.

2. Generate a number N of Bayesian distributed values of q: $q_1, q_2, \ldots q_N$.

3. A Bayesian sample value of Q is: $Q = \sum_{i=1}^{N} q_i$.

Next, in the operation stage, assume that a number n of inrush events has been observed over an area A_d up to period $t < T$. Let m be the mean of the logarithm of observed inrush yields, and s^2 the variance of the same quantity. Bayesian distributions are used to obtain more accurate reliability estimates for the subsequent periods ($t > T$). Since prior distributions belong to a conjugate family, the posterior (updated) parameter distribution is of the same type as the prior, and the parameters can readily be calculated as before. Updated Bayesian variates can then be generated by the algorithm given for the design stage.

5. APPLICATION

The reliability model has been applied to a mine located at a depth of 300 meters below karstic water level. Three alternative mine water control strategies are considered: two alternatives refer to the application of "Instantan" protection combined with grouting, while the third alternative consists of a passive control method. The "Instantan" method provokes inrushes at selected locations and evacuates the water via specially designed conveyance cuts. Passive control means that inrushes are left to occur into facilities in which a drainage system has been provided, but added costs emerge as a result of production slowdown.

Parameters of the impact function (Eq. 3) are given in Table 1. The same decision rule is used for every face of the mine in every period. Some of the input data for the reliability estimation are given in Table 2. Note that such data are routinely available in mines or tunnels subject to water hazard.

TABLE 1 -- Parameters of the Impact Function

Parameters m^3/min	Alternatives Passive Control	 Moderate Grouting	 Intensive Grouting
	A	B	C
a_1	30	0.35	0.2
a_2	1	0.5	0.3
a_3	100	9	1.3

TABLE 2 -- Data for the Estimation of System Reliability

Time Period		Number of blocks	Number of faces in a block	Area of a face	Specific number of inrushes	Data on inrush events		
						Average yield	Statistics of the lognormal distribution	
Number	Length years			A	λ			
				km^2	$1/km^2$	m^3/min	m	s
1	2	1	1	0.10	40	1.21	-0.3	0.48
2	1	2	2	0.15	40	1.2	-0.3	0.48
3	3	4	2	0.25	40	1.2	-0.3	0.48
4	10	5	2	0.6	60	2	0	0.48
5	10	5	2	0.6	60	2	0	0.48
6	4	5	1	0.2	40	2	0	0.48

The non-Bayesian reliability model utilizes the loading statistics given in Table 2. Table 3 gives prior information on loading and resistance statistics for the Bayes model. Tables 4 and 5 illustrate results of applications of the non-Bayesian and the Bayesian models, respectively. Prior information on λ_2 and λ_3 is inferred from performance data of submerged pumps and sediment removal equipment, respectively.

The effect of different grouting strategies is reflected in the numerical values of failure probabilities for both non-Bayesian and Bayesian models. On the other hand, the Bayesian reliability estimates are smaller than the non-Bayesian ones, because, as can be expected, parameter uncertainty increases the estimates of failure probabilities.

TABLE 3 -- Prior Information on Loading and Resistance

	Loading statistics				
Time periods	m'	s'	n'	m'(Aλ)	s'(Aλ)
1	-0.3	0.4	9	4	2
2	-0.3	0.4	9	6	3
3	-0.3	0.4	9	10	5
4	0	0.4	9	36	18
5	0	0.4	9	36	18
6	0	0.4	9	8	4

Resistance statistics:
For all six time periods

$m'(\lambda_2) = 0.11$, $s'(\lambda_2) = 0.05$,

$m'(\lambda_3) = 0.02$, $s'(\lambda_3) = 0.007$

TABLE 4 -- Non-Bayesian Failure Probabilities for the Whole Mine

Probability of Disturbance of operation	Time periods 1	2	3	4	5	6
A. Passive control	0.0	0.0	0.0	0.0	0.0	0.0
B. Moderate grouting	0.0	0.0	0.0	0.0	0.0	0.0
C. Intensive	0.0	0.0	0.0	0.0	0.0	0.0
Probability of Flooding						
A.	0.065	0.137	0.466	1.0	1.0	1.0
B.	0.060	0.136	0.299	0.981	1.0	1.0
C.	0.060	0.136	0.299	0.569	0.981	1.0

TABLE 5 -- Bayesian Failure Probabilities for the Whole Mine

Probability of Disturbance of operation	Time periods 1	2	3	4	5	6
A.	0.006	0.006	0.006	0.006	0.006	0.059
B.	0.0	0.0	0.0	0.0	0.0	0.0
C.	0.0	0.0	0.0	0.0	0.0	0.0
Probability of Flooding						
A.	0.086	0.167	0.683	1.0	1.0	1.0
B.	0.080	0.153	0.462	0.990	1.0	1.0
C.	0.072	0.145	0.362	0.662	0.995	1.0

TABLE 6 -- Results of the Updated Bayesian Failure Probabilities

Probability of disturbance of operation	Time periods 4	5	6
A	0.0	0.0	0.0
B	0.0	0.0	0.0
C	0.0	0.0	0.0
Probability of flooding			
A	0.380	1.0	1.0
B	0.360	1.0	1.0
C	0.360	0.594	0.824

In the next step, a mine that has been operating for 6 years is considered, with input data given in Table 2 and prior loading and resistance given in Table 3. During this time, n = 78 inrush events have been observed over the total area of faces, $A_d = 2.6\ km^2$. The mean and standard deviation of the logarithm of observed inrush yields are: m = 0.4, s = 0.5. Table 6 shows the updated Bayesian reliabilities estimated for the subsequent time periods (4, 5, 6), without any further change in the original input data. It can be seen that expected failure probabilities have become smaller, because of the smaller number and magnitude of observed inrushes than predicted in prior information (Table 3). At the same time, the failure probabilities are sensitive to the different grouting strategies.

6. DISCUSSION AND CONCLUSIONS

A Bayesian reliability system model has been applied to a real-life underground water control system. In contrast with a classical reliability model, this Bayesian approach accounts for the uncertainty in statistical information on loading and/or capacity. The fact that Tables 5 and 6 exhibit fairly high values of estimated system failure probabilities prompts the following two remarks:

a) One of the critical events leading to mine flooding is the failure of the sediment removal equipment. As numerical results show, for a specific failure time λ_2 = 0.02/year, it is quite probable that such a failure event will occur during the lifetime of the mine. Higher quality equipment or redundancy would be necessary in order to decrease this failure probability.

b) The other weak point in the system is the failure event of submerged pumps. In the numerical example no standby was considered, while in reality, at least two standby pumps are available for the given mining capacity. Using standard reliability theory (Shooman, 1968), including a given number of standbys into the reliability model is straightforward.

The results presented in this paper point to the following conclusions:

1. The reliability estimation of an underground engineering system, such as a mine water control system, may account for both natural and sample uncertainty in both loadings and resistances.

2. The estimation procedure can be represented by a failure tree.

3. A spatial event-based stochastic model can be used to characterize water inflow quantities, that is, the loading of the mine water control system.

4. System reliabilities for both the non-Bayesian and Bayesian models can be estimated by the simulation techniques developed herein.

5. The effect of loading control in the case of grouting can be accounted for by an impact function.

6. Bayesian reliability estimation is to be used in cases where (a) uncertain loading and/or capacity statistics are available in the design stage, and (b) earlier operational experience is available and more accurate reliability estimates are sought for subsequent operation periods.

ACKNOWLEDGEMENTS

Research leading to this paper was supported in part by funds from the U.S. National Science Foundation under Grant INT-78-12184, "Decision-Making in Natural Resources Management," and from the Ministry of Industry, Budapest, Hungary.

REFERENCES

Ang, A.H.S. and W.H. Tang (1984) Probability Concepts in Engineering Planning and Design - Vol II - Decision, Risk and Reliability, J. Wiley & Sons, NY, 562 pp.

Bazowsky, I. (1961) Reliability Theory and Practice, Prentice-Hall, Inc., 420 pp.

Benjamin, J.R. and C.A. Cornell (1970), Probability, Statistics and Decision for Civil Engineers, McGraw-Hill, New York, 684 pp.

Bogárdi, I., L. Duckstein, A. Schmieder and F. Szidarovszky (1980) Stochastic forecasting of mine water inrushes, Advances in Water Resources, Vol. 3, March, pp. 3-8.

Bogárdi, I., L. Duckstein and F. Szidarovszky (1982) Bayesian analysis of underground flooding, Water Resour. Res., Vol. 18, pp. 1100-1116

Chang, E.Y. and W.E. Thompson (1976) Bayes analysis of reliability for complex systems, Operations Research, Vol. 24. Jan-Feb.

Duckstein, L.I. Bogárdi and F. Szidarovszky (1981) Reliability of an underground flood control system, Journal of the Hydraulics Division, ASCE, Vol. 107, HY5.

Duckstein, L. and E.J. Plate (1985) A system framework for hydrologic reliability with application to the design of structures, Int. Assoc. for Hydr. Sciences Journal, Vol. 147.

Esztó, M., Z. Kesserü and L. Tárkány (1975) New Mining Regulations for the Protection against Water Hazards, in Hungarian, 12 Bányászati Lapok, Vol. 12, Budapest.

Feller, W. (1968) An Introduction to Probability Theory and its Applications, John Wiley and Sons, New York, Vol. 1, p. 264.

Kapolyi, L. (1976) New trends in mine water protection, Proceedings, VII Int. Conf. on Mine Water Protection, Vol. 1, Budapest, Hungary.

Kesserü, Z. (1978) Remarks on new developments in decision analysis in mining, in Hungarian, Bányászati Lapok, Budapest, 3.

Lucas, J.R. and L. Adler (1973) Groundwater and groundwater control, J.R. Lucas and L. Adler, Editors, Section 26 in SME Mining Engineering Handbook, New York, pp. 26.01-26.55.

Mann, N.R., R.E. Schafer and N.D. Singpurwalle (1974) Methods for Statistical Analysis of Reliability and Life Data, John Wiley and Sons, New York, Chapter 8, pp. 379-418.

Martz, M.F. and R.A. Waller (1982) Bayesian Reliability Analysis, John Wiley & Sons, New York, p. 745.

Metler, W.A. Estimation of Large-Scale Complex Systems Reliability, unpublished Ph.D. dissertation, Department of Systems and Industrial Engineering, University of Arizona, Tucson, Arizona 85721.

Musy, A. and L. Duckstein (1976) A Bayesian approach to tile drain design, Journal of the Irrigation and Drainage Div., ASCE, Vol. 102 IR3, Proc. Paper 12390, pp. 317-334, September.

Raiffa, H. and R. Schlaifer (1961) Applied Statistical Decision Theory, Harvard University, Boston, Massachusetts, 356 pp.

Schmieder, A., Z. Kesserü, J. Juhász, T. Willems and F. Martos (1975) Water hazard and water management in mining in Hungarian, Müszaki Könyvkiadó, Budapest.

Shooman, M.L. (1968) Probabilistic Reliability: An Engineering Approach, McGraw-Hill, New York.

Szidarovszky, F., L. Duckstein and I. Bogárdi (1975) Levee system reliability along a confluence reach, ASCE, Journal of the Engineering Mechanics Div., Vol. 101, EM5, pp. 609-622.

Szidarovszky, F., S. Yakowitz and R. Krzysztofowicz (1976) A Bayes approach for simulating sediment yield, J. Hydrological Sciences, pp. 33-45.

Tillman, F.A., W. Kao, C.L. Hwang and D.L. Grosh (1982) Bayesian reliability and availability: a review, IEEE Trans. on Reliability, Vol. R-31(4): pp. 362-372, October.

TARGET-RELATED RELIABILITY IN SURFACE WATER SYSTEM OPERATION

Roman Krzysztofowicz

Department of Systems Engineering
University of Virginia
Charlottesville, Virginia 22901, USA

ABSTRACT

We present fundamentals of planning seasonal water supply volume (target) under conditions of risk. A simple static decision model is formulated. The target is set based on the criterion of maximization of the expected utility of outcomes. Modeling and estimation of the utility function are discussed. Relationships between the optimal target rules and reliability of the actual supply are studied. The decision-theoretic approach based on a utility function is contrasted with a direct approach based on tradeoff judgments in the reliability-target space. The developments are illustrated by an analytic solution of the model with a two-piece linear utility function and a normal distribution of the seasonal runoff volume.

1. INTRODUCTION

1.1 Modeling Principles

This lecture is devoted to fundamentals of planning seasonal water supply under conditions of risk. These fundamentals are rooted in Statistical (Bayesian) Decision Theory. The theory admits the fact of life that optimality is not a universal but rather relative property of a decision - relative to one's preferences over possible outcomes and information about uncertain states of the world. These determinants of optimality enter into normative decision models via two principles. (i) Principle of decomposition: preferences of a decision maker are quantified in the form of a utility function, while uncertainties are quantified in terms of a probability distribution function; these quantifications should be independent. (ii) Principle of rationality: given the utility and distribution functions, that decision is optimal which maximizes the expected utility of outcomes.

1.2 Target Setting Problem

The volume a of water planned for delivery from a hydrosystem during a specified time period is often called the target. It is a decision variable. The volume r of water that is actually supplied (output) is the sum of the volume already in storage y which is known with certainty, plus the future inflow to the reservoir ω which is uncertain. The consequences of system operation typically depend upon both a and $r = y + \omega$. Since at the time of decision making r is a random variable, the problem of setting the target is a classical example of decision making under risk. In this lecture, we study (i) target rules which are optimal with respect to the criterion of maximization of the expected utility of outcomes, and (ii) relationships between the optimal target rules and reliability of the actual supply.

The fundamentals presented herein hold for a broad class of

target problems. However, our exposition is confined to a static model that arises in the context of seasonal water supply planning by an irrigation company. The distribution of the supply throughout the season and its consequences are not accounted for. The adequacy of this assumption depends upon the hydrosystem. The assumption is likely to be adequate for systems with large reservoirs that can ensure the desirable distribution of outflow throughout the season, no matter the distribution of inflow.

1.3 Decision Context

At the beginning of each year, an irrigation company must plan the volume of water $a \in A$ (a target) to be supplied to farmers during the irrigation season. On the basis of a the farmers plan their crop production (acreage to be irrigated, types of crops, etc.). At the time these plans are made, the actual supply volume $r \in R$ is a random variable due to uncertainty in the seasonal (e.g., April-July) snowmelt runoff. It is assumed that uncertainty in the runoff prior to the snowmelt is negligible in comparison to uncertainty in the snowmelt runoff. These conditions are typical of the Western United States.

The benefit derived by the farmers from irrigation is a function of both the actual supply r and the planned supply a. The relationship is generally as follows: for a fixed a, the benefit increases with r; for a fixed r, the benefit attains the maximum at $a = r$ and decreases in both directions. The irrigation company, as a benevolent decision maker, wishes to set the planned supply volume (target) based on a rational criterion. It is apparent that the criterion function must be defined on the product space $R \times A$.

2. EXPECTED UTILITY CRITERION

2.1 Definition of Utility

The total consequences of the actual supply $r \in R$, when the planned supply has been $a \in A$, are evaluated in terms of the utility $u(r,a)$. For a fixed $a \in A$, the utility function $u(\cdot,a)$ represents the decision maker's preferences on R conditional upon decision a. Specifically, $u(\cdot,a)$ encodes the value of water and risk attitude of the decision maker.

A three-dimensional illustration of a utility function u is shown in Figure 1. The function u is defined on a bounded domain $R \times A = [r^o,r^s] \times [a^o,a^s]$, with $a^o = r^o$ and $a^s = r^s$. The preference order between the corner points is always: $u(r^s,a^s) > u(r^s,a^o) \geq u(r^o,a^o) > u(r^o,a^s)$. For a given $a \in A$, $u(\cdot,a)$ is increasing on R; for a given $r \in R$, $u(r,\cdot)$ is increasing on $[a^o,r]$ and decreasing on $[r,a^s]$.

2.2 Components of Utility

The function $u(\cdot,\cdot)$, defined along the main diagonal $\{(r,a) : r = a\}$, represents utilities of "perfect decisions": the actual supply r is equal to the planned supply a. (Such decisions would be possible if perfect forecasts of the seasonal snowmelt runoff were available.) The quantity

$$v(r) = u(r,r) \tag{1}$$

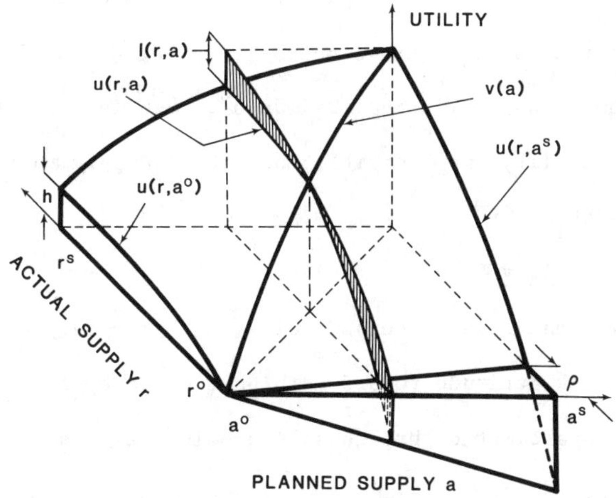

Figure 1. Spatial illustration of the utility function u, its components, and the opportunity loss function ℓ. This is a scaled plot of u for the Owyhee Irrigation Project. Component functions v, $u(\cdot,a^o)$, and $u(\cdot,a^s)$ are also shown in Figure 2A.

may be interpreted as the relative value of the supply volume r. The function v is termed a value function, and its shape (concave, convex, linear) implies the monotonicity of the marginal value function (decreasing, increasing, constant).

The function $u(\cdot,a^o)$ defined along the line $\{(r,a^o) : r \in R\}$, represents utilities of the actual supplies r given the "extremely pessimistic decision": the planned supply is at the lowest value a^o. The constant $h = u(r^s,a^o)$ represents the utility of the maximum volume r^s when it is supplied after the decision a^o to supply the minimum has been made.

The function $u(\cdot,a^s)$, defined along the line $\{(r,a^s) : r \in R\}$, represents utilities of the actual supplies r given the "extremely optimistic decision": the planned supply is at the highest value a^s. The function $u(\cdot,a^s)$ has a nonpositive segment on $[r^o,\rho]$; i.e., when the planned supply is the maximum a^s, then in order to realize a positive utility (or benefit), the actual supply must be greater than ρ, with $\rho > r^o$.

2.3 Estimation of Utility

If the decision maker is concerned solely with monetary consequences and his risk attitude toward gambles over money is neutral, then u is estimated through an economic analysis of irrigation benefits.

If economic data for estimating benefits are not available, the decision maker is concerned with nonmonetary consequences (in addition to monetary consequences), or his attitude toward gambles over money is not risk neutral, then u should be estimated via assessment of the decision maker's strength of preference for various outcomes and risk attitude toward gambles over these outcomes. The assessment can be structured so that all relevant dimensions of consequences (sometimes called objectives) are taken into consideration. The resultant u then provides an aggregate representation of preferences.

Figures 2A and 2B show functions v, $u(\cdot,a^o)$, and $u(\cdot,a^s)$ obtained from economic benefit functions for two irrigation projects: the Owyhee Project in Idaho and the Salt River Project in Arizona. They can be compared with Figures 2C and 2D which show functions obtained for a case study in India via assessment

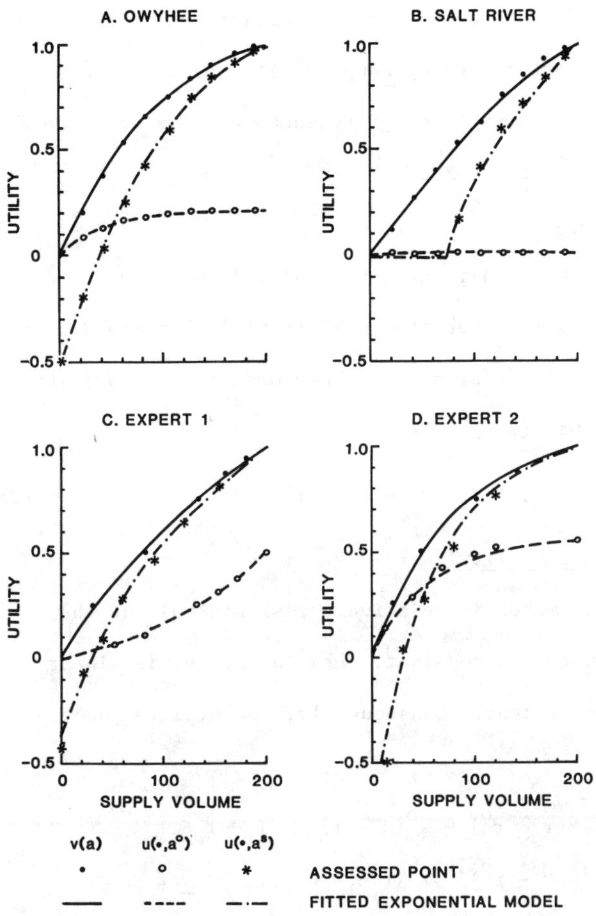

Figure 2. Component functions v, $u(\cdot,a^o)$, and $u(\cdot,a^s)$ of the utility function u. (A) and (B): u obtained from an economic benefit function. (C) and (D): u obtained through assessment of decision maker's preferences.

of preferences of two water resource experts. Not surprisingly, the utility function u is distinct in each case. However, the structural similarity between them is apparent.

Conclusion. Adequate quantitative decision criteria can be obtained through expert reasoning coupled with analysis, modeling, and assessment of the strength of preference and risk attitude of an expert decision maker.

2.4 Opportunity Loss

Given the outcome (r,a), the opportunity loss $\ell(r,a)$ is defined as the difference between the utility v(r) that would be realized had a perfect decision a = r been made and the utility of the actual outcome u(r,a). Hence

$$u(r,a) = v(r) - \ell(r,a) . \tag{2}$$

2.5 Two-piece Linear Utility

If the decision maker is relatively risk neutral and the marginal value of water is constant, then in (2) v is linear and ℓ is two-piece linear. Consequently, the utility function takes the form:

$$u(r,a) = \begin{cases} k(r - r^o) - k^o(a - r) & \text{if} \quad r \leq a , \\ k(r - r^o) - k^u(r - a) & \text{if} \quad r \geq a , \end{cases} \tag{3}$$

where r^o is a scaling constant, $k > 0$ is the marginal utility (benefit) of "perfect target" (r = a), $k^o > 0$ is the marginal opportunity loss caused by overestimation of the actual supply by the target (r < a), and $k^u > 0$ is the marginal opportunity loss caused by underestimation of the actual supply by the target (r > a). Figure 3 provides an illustration.

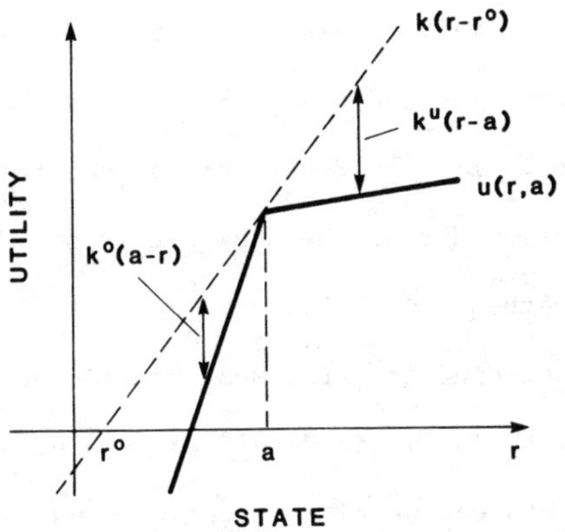

Figure 3. Two-piece linear utility function $u(\cdot,a)$ for a fixed a.

When u is scaled by $u(r^o,a^o) = 0$ and $u(r^s,a^s) = 1$, then

$$k = \frac{1}{r^s - r^o} , \tag{4a}$$

$$k^o = \frac{\rho - r^o}{(r^s - r^o)(r^s - \rho)} , \tag{4b}$$

$$k^u = \frac{1 - h}{r^s - r^o} , \tag{4c}$$

where h and ρ are defined in Figure 1. Clearly, the two-piece linear utility function is completely defined by either (h,ρ) or (k^o,k^u); k is a proportionality constant.

3. GENERAL DECISION MODEL

3.1 Optimal Target Problem

Let y denote the volume of water available with probability one for supply at the beginning of the snowmelt season (sure supply). The variable y may represent a firm groundwater supply, water already in storage, water expected in storage before the snowmelt season, or a combination thereof. Let ω denote the actual runoff volume measured at a fixed station during the snowmelt season. The actual supply volume is then $r = y + \omega$.

The commitment concerning target a must be made before ω is observed. The uncertainty in ω is described by a probability density function g. This can be either a prior density estimated from a climatological record of runoffs, or a posterior density conditional upon a snowmelt runoff forecast. The corresponding distribution function is denoted by G.

The target is set based on the criterion of maximization of the expected utility of outcomes:

$$T(y) = \max_a \int u(y + \omega, a) g(\omega) d\omega \ . \tag{5}$$

When solved for every y, (5) yields the target rule α^* which prescribes the optimal planned supply:

$$a^* = \alpha^*(y) \ . \tag{6}$$

The target rule, in turn, implies the probability P of the surplus (reliability of the supply):

$$\begin{aligned} \pi^*(y) &= P(y + \omega > \alpha^*(y)) = 1 - P(\omega \leq \alpha^*(y) - y) \\ &= 1 - G(\alpha^*(y) - y) \ . \end{aligned} \tag{7}$$

The reliability $\pi^*(y)$ is optimal in the sense that it is consistent with the decision maker's risk attitude encoded in the utility function u.

Conclusion. Formulation of the target problem in terms of a utility function u and a distribution function G implies, via the optimal target rule α^*, the reliability of the supply π^* which is consistent with the risk attitude of the decision maker.

3.2 Inverse Problem

Suppose that instead of a utility function u, the decision maker specifies the least acceptable reliability $\hat{\pi}(y)$ for every y. Function $\hat{\pi}$ implies, via (7), a target rule

$$\hat{\alpha}(y) = y + G^{-1}(1 - \hat{\pi}(y)) \ . \tag{8}$$

Principles of rationality require that if both u and $\hat{\pi}$ are assessed, then for every y we must have

$$\alpha^*(y) = \hat{\alpha}(y) \ . \tag{9}$$

Conclusions. (i) Reliability function $\hat{\pi}$ provides information sufficient for solving the target problem. (ii) The solution $\hat{\alpha}$ is indistinguishable from the solution α^* that would be obtained by maximizing the expectation of a suitably chosen utility function u.

3.3 Preferences in the Reliability-Target Space

Before the target $\hat{a} = \hat{\alpha}(y)$ is determined, $\hat{p} = \hat{\pi}(y)$ represents the probability of the event $(r > \hat{a})$, where r is random and $\hat{a}$ is fixed but unknown. Such an abstract event may have little meaning to the decision maker. Hence the feasibility of assessing $\hat{\pi}$ directly and consistently with the decision maker's preferences over the set $R \times A$ of the outcome-target pairs (r,a) is doubtful, from the cognitive standpoint. To formulate a more sensible assessment method, we should consider preferences over the set $[0,1] \times A$ of the reliability-target pairs (p,a).

Formally, we could assess a utility function μ on $[0,1] \times A$. This function would have the minimum at $(0,a^o)$, the maximum at $(1,a^s)$, and would be increasing in both p and a so that the isopreference curves would look like those shown in Figure 4. Then we could formulate the following optimization problem:

$$\max_a \mu(p,a) \tag{10}$$

subject to

$$p = 1 - G(a - y) . \tag{11}$$

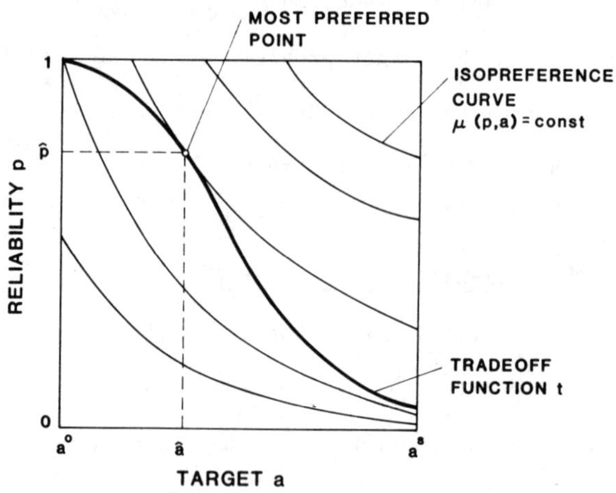

Figure 4. Isopreference curves implied by a utility function μ in the reliability-target space, and the admissible tradeoff function t for a fixed sure supply y.

Solving (10)-(11) for every y would yield the optimal target rule $\hat{\alpha}$ and the implied reliability function $\hat{\pi}$.

This formulation suggests still another approach. It is based on direct tradeoff judgments in the reliability-target space $[0,1] \times A$, and does not necessitate the assessment of μ.

3.4 Direct Tradeoff Judgments

(i) Fix y, and for all $a \in A$, compute

$$p = t(a) = 1 - G(a - y) \ . \tag{12}$$

Function t, illustrated in Figure 4, represents the locus of admissible tradeoffs between the magnitude of target a and its reliability p.

(ii) Ask the decision maker to indicate the most preferred point $(\hat{p},\hat{a})$ on the tradeoff function t.

(iii) Repeat tasks (i) and (ii) for several values of y. Then interpolate the assessed points to obtain the target rule $\hat{\alpha}$ and the associated reliability function $\hat{\pi}$. The target problem is thereby solved.

3.5 Should We Assess Utility or Tradeoff Judgments?

I am not aware of any experimental test of the utility approach against the direct judgment approach in the context of the reliability-target problem. However, the general laws of cognitive psychology strongly suggest that we should expect inconsistencies: $\alpha^* \neq \hat{\alpha}$. Furthermore, it is reasonable to hypothesize that the direct assessment of $\hat{\alpha}$ is more susceptible to biases and random errors of perception than the assessment of u.

Advice. If the decision maker is concerned with both the frequency and magnitude of the deviations of the actual supply r from the target a, and the consequences (monetary or otherwise) of these deviations, then assess u. If the decision maker is concerned solely with the frequency of the deficit $(a - r) > 0$, relative to target a, then the direct tradeoff judgment approach may be suitable.

4. LINEAR DECISION MODEL

4.1 Optimal Target Problem

When u is two-piece linear, as specified by (3), then the maximum expected utility is

$$T(y) = k(y + E(\omega) - r^o) - k^u E(\omega) + (k^u + k^o) \int_{-\infty}^{\alpha^*(y)-y} \omega g(\omega) d\omega , \quad (13)$$

where E denotes the expectation, and α^* is the optimal target rule:

$$\alpha^*(y) = y + G^{-1}\left(\frac{k^u}{k^u + k^o}\right) . \quad (14)$$

The implied optimal reliability is

$$\pi^* = 1 - \frac{k^u}{k^u + k^o} . \quad (15)$$

This relationship is illustrated in Figure 5.

Conclusions. (i) T and α^* are increasing linear functions of the sure supply y. (ii) α^* decreases with k^o and increases

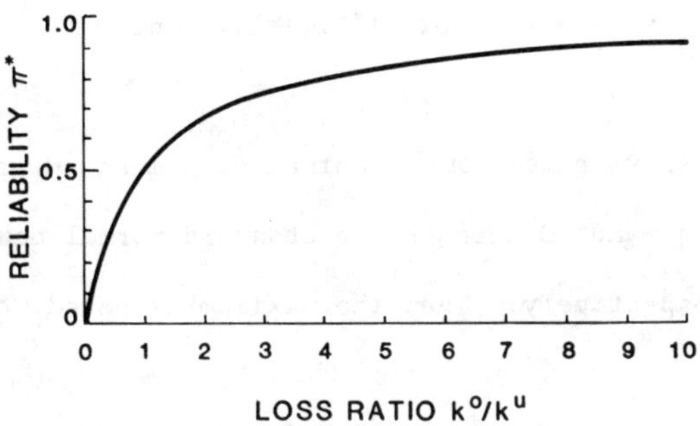

Figure 5. Relationship between the optimal reliability, π^*, and the ratio of the marginal opportunity losses caused by overestimation, k^o, and underestimation, k^u, of the actual supply by the target.

with k^u. (iii) α^* and π^* are uniquely determined by the ratio of k^o to k^u. (iv) π^* depends neither on y nor on the distribution G. In other words, the optimal target rule α^* ensures a constant reliability regardless of the sure supply y and the uncertainty in ω. (v) When $k^o = k^u$, the optimal target $\alpha^*(y)$ is the median of ω augmented by y, and the implied reliability is $\pi^* = 1/2$.

4.2 Inverse Problem

When reliability $\hat{\pi}$ is specified, then from (14) and (15)

$$\hat{\alpha}(y) = y + G^{-1}(1 - \hat{\pi}) \ . \tag{16}$$

The method described in Section 3.4 can be applied to assess the most preferred point $(\hat{p},\hat{a})$ for some fixed y. By assessing several points $(\hat{p}_i,\hat{a}_i)$, each for a different y_i, $(i=1,\ldots,n)$, one can verify whether the reliability is independent of the sure supply. The answer is affirmative whenever $\hat{p}_i = \hat{\pi}$ for all $i=1,\ldots,n$, and consequently $\hat{a}_i - \hat{a}_j = y_i - y_j$ for all $i,j=1,\ldots,n$.

4.3 Normal Case

Suppose the density g of ω is normal with mean M and variance S^2. Let q and Q denote the standard normal density and distribution, respectively. Then the maximum expected utility is

$$T(y) = k(y + M - r^o) - (k^u + k^o)q\left(Q^{-1}\left(\frac{k^u}{k^u + k^o}\right)\right)S , \tag{17}$$

the optimal target rule is

$$\alpha^*(y) = y + M + Q^{-1}\left(\frac{k^u}{k^u + k^o}\right)S , \tag{18}$$

and the implied optimal reliability π^* is given by (15).

Conclusions. (i) T decreases with uncertainty, as measured by S. (ii) α^* increases with uncertainty S if $\pi^* < 1/2$, is independent of uncertainty S if $\pi^* = 1/2$, and decreases with uncertainty S if $\pi^* > 1/2$.

4.4 Real-World Reliabilities

It is instructive to examine the values of reliability π^* implied by the marginal opportunity losses k^o and k^u found in real-world applications or hypothetical case studies intended to approximate real situations. Table 1 contains estimates retrieved directly or computed based on data from the literature, under the assumption that the two-piece linear utility model (3) holds. Interestingly, the estimates obtained through formal

Table 1. Estimates of the reliability π^* implied by the marginal opportunity losses k^o and k^u for a two-piece linear utility model.

Data Source	π^*	$Q^{-1}(1-\pi^*)$
A. FORMAL ESTIMATES		
Benefit function for the Owyhee Project (SCS, 1977)	0.237	0.716
Benefit function for the Salt River Project (SCS, 1977)	0.359	0.361
Utility function of Expert 1 (Krzysztofowicz, 1982)	0.788	-0.800
Utility function of Expert 2 (Krzysztofowicz, 1982)	0.809	-0.873
Schramm et al. (1974) reporting, after Nelson, an economic analysis by Shelton	0.640	-0.358
B. INTUITIVE OR ARBITRARY ESTIMATES		
Davis and Patten (1975)	0.857	-1.078
Davis and Patten (1975)	0.917	-1.385
Vicens et al. (1975)	0.909	-1.335
Wood (1978)	0.984	-2.145

methods are on the average much lower than those resulting from intuitive or arbitrary assumptions. Behavioral research points out that direct intuitive estimates are often severely biased; our sample is too small to justify further inference. However, the very fact that k^o and k^u are marginal losses implies that they are difficult to estimate through unaided judgment.

4.5 Asymmetry of Opportunity Losses

When sure supply y = 0, then from (15) and (18) the optimal target is $a^* = M + Q^{-1}(1-\pi^*)S$. For a symmetric opportunity loss function, $k^o = k^u$, $\pi^* = 1/2$, and thus the optimal target is equal to the mean runoff volume: $a^* = M$. In every case displayed in Table 1, the opportunity loss function is asymmetric. Consequently, the optimal target a^* differs from the mean M. In seven out of nine cases, the optimal target a^* is less than the mean runoff volume M; the largest difference equals 2.145 standard deviations S.

Conclusions. (i) Real-world opportunity loss functions are often quite asymmetric. (ii) Proper estimation of the asymmetry in the opportunity loss function is critical to optimal setting of the target.

5. CONCLUSION

Principles of normative rationality of decision theory imply the existence of a well defined relationship between the optimal planned supply (target) a, and the reliability p of the actual supply (outcome) r.

The target setting problem can be solved by assessing the decision maker's preference structure either over the set of the output-target pairs (r,a) or over the set of the reliability-target pairs (p,a). Decision models that require elements of both preference structures (e.g., chance constrained programming) or treat both target a and reliability p as decision variables (e.g., reliability programming) are either overspecified (if the preference structures are consistent) or sanction irrationality (if the preference structures are inconsistent).

ACKNOWLEDGMENTS

I am grateful to Vit Klemeš for the many penetrating and delightful discussions on the target-related problems. This material is based upon work supported by the National Science Foundation under Grant ECE-8352536, "Knowledge-Based, Computer-Aided Engineering Decision Support Systems." A part of this material is reproduced from the article "Expected Utility, Benefit, and Loss Criteria for Seasonal Water Supply Planning," Water Resources Research, March 1986.

SELECTED BIBLIOGRAPHY

Anderson, R.L. and Maass, A., A Simulation of Irrigation Systems, Technical Bulletin No. 1431, Economic Research Service, U.S. Department of Agriculture, Washington, D.C., 1971.

Castruccio, P.A., Loats, H.L., Jr., Lloyd, D. and Newman, P.A.B., Cost/Benefit Analysis for the Operational Applications of Satellite Snowcover Observations (OASSO). In Operational Applications of Satellite Snowcover Observations, NASA Conference Publication 2116, Proceedings of a Final Workshop held at Sparks, Nevada, 239-254, April 1979.

Davis, D.R. and Patten, R., Bayesian Decision Analysis Applied To River Basin Studies, Proceedings, Seminar on Colorado River Basin Modeling Studies, Utah State University, Logan, Utah, 549-567, July 1975.

Hashimoto, T., Stedinger, J.R. and Loucks, D.P., Reliability, Resiliency, and Vulnerability Criteria for Water Resource System Performance Evaluation, Water Resources Research, 18(1), 14-20, 1982.

Hufschmidt, M.M. and Fiering, M.B., Simulation Techniques for Design of Water-Resource Systems, Harvard University Press, Cambridge, Massachusetts, 1966.

Klemeš, V., Value of Information in Reservoir Optimization, Water Resources Research, 13(5), 837-850, 1977.

Klemeš, V., Storage Mass-Curve Analysis in a Systems-Analytic Perspective, Water Resources Research, 15(2), 359-370, 1979.

Krzysztofowicz, R., Utility Criterion for Water Supply: Quantifying Value of Water and Risk Attitude. In Unny, T.E. and McBean, E.A. (Eds.), Decision Making for Hydrosystems: Forecasting and Operation, Water Resources Publications, Littleton, Colorado, 43-62, 1982.

Krzysztofowicz, R., Strength of Preference and Risk Attitude in Utility Measurement, Organizational Behavior and Human Performance, 31, 88-113, 1983.

Krzysztofowicz, R. and Jagannathan, E.V., Stochastic Reservoir Control with Multiattribute Utility Criterion. In Unny, T.E. and McBean, E.A. (Eds.), Decision Making for Hydrosystems: Forecasting and Operation, Water Resources Publications, Littleton, Colorado, 127-144, 1982.

Loucks, D.P., Stedinger, J.R. and Haith, D.A., Water Resource Systems Planning and Analysis, Prentice-Hall, Englewood Cliffs, New Jersey, 1981.

Maass, A., Hufschmidt, M.M., Dorfman, R., Thomas, H.A., Jr., Marglin, S.A. and Fair, G.M., Design of Water-Resource Systems, Harvard University Press, Cambridge, Massachusetts, 1962.

McBean, E., Lenton, R. and Schaake, J.C., Jr., Simulation Analyses: the Basic Model. In Major, D.C. and Lenton, R.L. (Eds.), Applied Water Resource Systems Planning, Prentice-Hall, Englewood Cliffs, New Jersey, 82-94, 1979.

Moore, J.L. and Armstrong, J.M., The Use of Linear Programming Techniques for Estimating the Benefits From Increased Accuracy of Water Supply Forecasts, Water Resources Research, 12(4), 629-639, 1976.

Schramm, G., Fenton, R.W., Moore, J.L., Hughart, D. and Moore, G.R., Decision Making Under Uncertainty: Economic Evaluation of Streamflow Forecasts, Final Report to the Office of Water Resources Research, U.S. Department of the Interior, University of Michigan, School of Natural Resources, Ann Arbor, Michigan, July 1974.

SCS (Soil Conservation Service), U.S. Department of Agriculture, An Evaluation of the Snow Survey and Water Supply Forecasting Program, Washington, D.C., February 1977.

Vicens, G.J., Rodriguez-Iturbe, I. and Schaake, J.C., Jr., Bayesian Generation of Synthetic Streamflows, Water Resources Research, 11(6), 827-838, 1975.

Wood, E.F., Analyzing Hydrologic Uncertainty and its Impact Upon Decision Making in Water Resources, Advances in Water Resources, 1(5), 299-305, 1978.

Young, R.A. and Bredehoeft, J.D., Digital Computer Simulation for Solving Management Problems of Conjunctive Groundwater and Surface Water Systems, Water Resources Research, 8(3), 533-556, 1972.

BAYESIAN ANALYSIS : FURTHER ADVANCES AND APPLICATIONS

Jacques Bernier

Laboratoire National d'Hydraulique - 6 quai WATIER - 78400 Chatou - FRANCE

ABSTRACT

Bayesian methods which have been presented in a previous paper are applied to the study of risk situations induced by geophysical extreme events. The relationships between decisions and consequences are shown. Examples are given concerning the introduction of various types of information, such as historical data. It is also shown how parameters varying with time are taken in account.

I. INTRODUCTION

In a previous lecture, Bayesian methods have been presented and illustrated with simple examples where analytical calculations were possible. As a matter of fact, these methods also can be useful in more complex situations where posterior probabilities and posterior parameters are estimated by means of approximate or Monte Carlo methods. In particular, the latter have been extensively used in the Bayesian uncertainty analysis of complex nuclear systems (Pages and Gondran 1980). But one of the main fields of application of Bayesian Analysis is the study of risk situations induced by geophysical extreme events such as earthquakes, waves and storm surges near seacoasts (Miquel and Bernier 1979), precipitation and river floods (Bernier and Lencioni, 1978 ; Bernier and Miquel, 1979 ; Miquel - 1984). We shall present the flood problem but it must be understood that the models can be applied to other risk situations. Thus let us consider the following classical flood protection problem : a given hydraulic structure (levee, spillway of a dam, industrial platform) is threatened with river floods. Large floods can induce damages when some characteristics such as maximum discharge, volume or flood duration exceeds a given threshold. If, as usual in such problems, we choose maximum flood discharge as a performance index, damage index in fact, the state of nature set, say Θ, includes not only the rate of occurrence λ of damaging floods but also the parameter ρ describing the distribution function $F(q\ ;\rho)$ of maximum discharge q of each damaging flood.

So Θ, the state of nature set is multiple valued as (λ,ρ) ; the resulting process, of which Figure I shows a sample, is called risk process. For describing further the decision problem, we have to define the decision set A as the set of critical values of a maximum discharge under which the protection of the structure is insured and above which damages are supposed to occur. As usual, a is interpreted as the maximum discharge of the design flood, or design discharge in short.

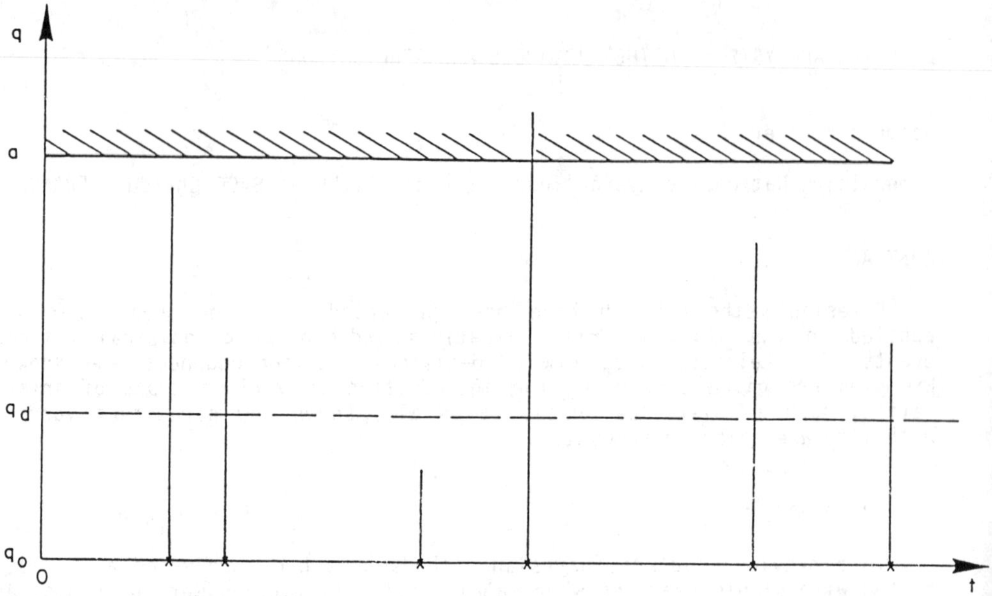

Figure I : Risk Models

II. DESCRIPTION OF THE LOSS FUNCTION AND UNCERTAINTY PROBLEM

Let the consequences of decision a be :

- an investment cost C(a), increasing with a, supposed to be incurred at the time o when the protection structure is built.
- residual damages cost D(q, a, t) expressed as a decreasing function of decision a, with value o if the flood with discharge q occuring at time t is lower than the design discharge a. The generality of the problem is not significantly reduced if the damage is a function of the discharge excess q-a, say D(q-a, t). Depending on q, damages are random and distributed along future time from the origin o where the structure is built. Taking in account these facts, it is necessary to calculate the present value (at time o) of expected future costs D as :

$$\bar{D}(a\ ;\lambda,\rho) = \int_0^\infty \lambda(t)e^{-rt}\,dt \int_a^\infty D(q-a,\ t)\,dF(q\ ;\rho) \qquad (1)$$

where :

- $F(q\ ;\rho) = \text{Prob}\left[Q < q\right]$ is the distribution function of discharge Q depending on parameter ρ.

- r is the so-called continuous discounting rate, allowing the calculation of the present value of damages distributed in time.

- $\lambda(t)$ is the time varying rate of occurence of damaging floods.

In practical applications, the rate r and damages are often taken as constant during the year, in which case the first integral is replaced by a sum. However, formula (1) allows us to accommodate the case where damages and the economic value of the floodplain, depend on time or on season, such as agricultural damages. Thus the loss function, as it is called in Bayesian terminology, is written as :

$$L(a\ ;\lambda,\rho) = C(a) + \overline{D}(a\ ;\lambda,\rho) \qquad (2)$$

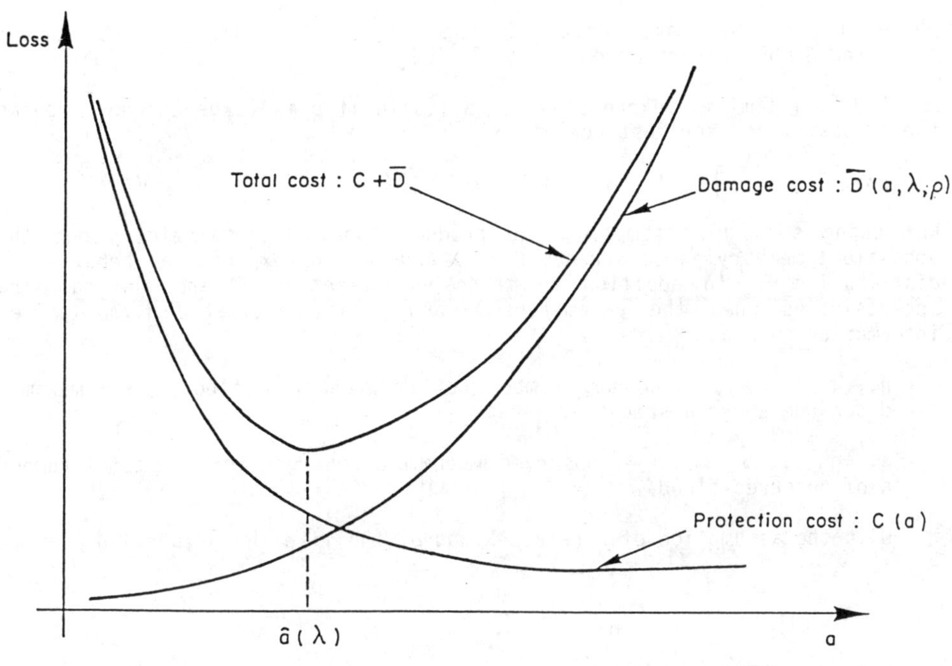

Figure II : Protection and Damage cost functions

Formulae (1) and (2) assume that an economical interpretation of the consequences of action a may be made, namely that all of them may be expressed in monetary units. In river flood problems, not all consequences of decision a can be expressed in such monetary units. What about human losses for instance ? It may be better to use multiple criteria. But here we shall only deal with the single criterion (2) assuming the damages expressed in economic terms are representative enough for making adequate decisions.

In order to calculate the criterion function (2) it is necessary to make assumptions about the flood model. A classical one is the following (Bernier and Miquel, 1979 ; Miquel, 1984 ; Todorovic, 1978) :

1 : the occurence rate function is assumed constant along the year
2 : the distribution function of Q is exponential :

$$F(q\ ;\rho) = 1 - e^{-\rho(q-qd)} \tag{3}$$

where qd is the damage threshold for the maximum discharge.

With the further assumption of independence of successive damaging discharges above the threshold, it can be shown (Bernier and Miquel, 1979) that the annual maximum discharge has the following distribution function :

$$\text{Prob.}\left[Q_{Max} < q\right] = G(q\ ;\lambda,\rho) = \exp\left[-\lambda T\left[1 - F(q\ ;\rho)\right]\right] \tag{4}$$

so that with exponential function (3) one finds :

$$G(q\ ;\lambda,\rho) = \exp\left[-\lambda T\, e^{-\rho(q-qd)}\right] \tag{5}$$

where T is the time unit of one year. This is the so called truncated Gumbel distribution.

A useful limiting expression for G (valid if q is large enough, such as the 10 year flood for instance) is :

$$\text{Prob.}\left[Q_{Max} > q\right] = 1 - G \simeq \lambda T\, e^{-\rho(q-qd)} \tag{6}$$

But using such an assumption can produce model uncertainties about the possible time varying occurrence rate λ and the non-exponential behavior of distribution F. In addition to these uncertainties, there are sampling uncertainties when the parameters λ and ρ are estimated from limited information such as :

- $n_1, n_2 \ldots n_\ell$: annual numbers of occurence of floods with maximum discharge above a specified limit.

- $q_1, q_2 \ldots q_j \ldots q_n$: observed maximum discharges for the total number n of observed floods.

With the assumption of constant λ and ρ, the likelihood function is :

$$L = \frac{e^{-\ell\Pi T}\,(\lambda T)^{n}}{\prod_{i=1}^{\ell} n_i\,!_{t}} \cdot \rho^{n}\, e^{-\rho \sum_{j=1}^{n}(qj - qo)} \tag{7}$$

This formula supposes that the observed q 's, above qo (usually smaller than the critical threshold qd) are also exponentially distributed with the same parameter ρ as the damaging q.

III. PERFECT INFORMATION SOLUTION

In the case where the exact values of λ and ρ are known, there is a single optimum â, solution of the equation :

$$\frac{\partial L}{\partial a} = c'(\hat{a}) - \lambda\rho \int_0^{\infty} e^{-rt} \, dt \int_{\hat{a}}^{\infty} \frac{\partial D}{\partial a} (q-\hat{a},\ t)\ e^{-\rho(q-qd)} \, dq = 0$$

provided the marginal damage cost $\frac{\partial D(u,t)}{\partial u}$, which is the derivative of the function D(u,t), exists ; otherwise another expression for formula (8) given below to be used. Then :

$$c'(\hat{a}) - \lambda e^{-\rho(\hat{a}-qd)} \int_0^{\infty} e^{-rt} \, dt \int_0^{\infty} \frac{\partial D}{\partial u} (u,t) \,.\, \rho e^{-\rho u} \, du = 0$$

The optimum value â is solution of :

$$T e^{-(\hat{a}-qd)} = p = \frac{c'(\hat{a})\ T}{\overline{D}'\ (\rho)} \tag{8}$$

$$\text{with } \overline{D'}(\rho) = \int_0^{\infty} e^{-rt} \, dt \int_0^{\infty} \frac{\partial D}{\partial u} (u,t) \,.\, \rho e^{-\rho u} \, du \tag{9}$$

which is the present value of the expected marginal damage cost $\frac{\partial D}{\partial u}$, expectation defined with respect to the exponential distribution depending on ρ .

Comparison of formula (8) with (6) shows that the optimum design discharge is defined as the value of annual maximum discharge, the probability p of which is equal to $\frac{c'T}{\overline{D}'(\rho)}$.

But this is not the usual criterion where p is chosen as a conventional value because even if the marginal investment c' is constant, the present expected marginal damage cost D' is not constant and depends on the parameter ρ which is varying with each case. Solving equation (8) gives :

$$\hat{a} = qd - \frac{1}{\rho} \text{Log} \frac{c'(\hat{a})}{\overline{D}'(\rho)} \tag{10}$$

noted as â(λ , ρ) function of λ and ρ in the sequel.

A useful formula is the opportunity loss

$$OL(a\ ;\lambda,\rho) = L(a\ ;\lambda,\rho) - L\left[\hat{a}(\lambda,\rho)\ ;\lambda,\rho\right]$$

expressed in approximate terms by an expansion around the optimum value :

$$OL(a\ ;\lambda,\rho) \simeq c'(a) \,.\, \left[a - \hat{a}(\lambda,\rho)\right]^2 \tag{11}$$

provided the marginal investment cost c' stays approximately constant near the optimum.

IV. BAYESIAN ANALYSIS

As required by Bayesian machinery for studying uncertainties, let us consider that the states of nature λ and ρ are random variables such that their probability distributions are :

- at the prior level, the conjuguate joint density with respect to the likelihood (7) :

$$\Pi(\lambda,\rho) = \frac{a'^{g'}}{\Gamma(g')} e^{-a'\lambda} \lambda^{g'-1} \cdot \frac{b'^{\gamma'}}{\Gamma(\gamma')} e^{-b'\rho} \rho^{\gamma'-1} \qquad (12)$$

that is to say, independent gamma priors for λ and ρ with parameters (a',g') and (b',γ') respectively.

Let us recall that these forms are convenient enough in practice for representing any type of prior knowledge about states of nature λ and ρ. For instance the case of a non-informative prior can be dealt with by selecting the prior values :

$$a' = g' = b' = \gamma' = 0 \qquad (13)$$

- at the posterior level, using Bayes theorem, the resulting probability densities are such that the parameters are posteriorly independent and still distributed as gamma laws G(a", g") and G(b", γ ") with respective parameters :

$$\begin{aligned} a'' &= a' + T \\ g'' &= g' + n \\ b'' &= b' + \sum_{j=1}^{n} (q_j - q_o) \\ \gamma'' &= \gamma' + n \end{aligned} \qquad (14)$$

Useful posterior characteristics for the sequel are the expectations of λ and ρ, and the variance of ρ, that is :

$$\begin{aligned} E''(\lambda) &= \frac{g' + n}{a' + \ell T} \\ E''(\rho) &= \frac{\gamma' + n}{b' + \sum_{j=1}^{n} (q_j - q_o)} \\ Var''(\rho) &= \frac{\gamma' + n}{\left[b' + \sum_{j=1}^{n} (q_j - q_o)\right]^2} \end{aligned} \qquad (15)$$

In line with the Bayesian analysis, the next steps are to compute the posterior expected loss or opportunity loss and to determine the decision which minimizes this loss. In our case the posterior expected loss can be expressed as :

$$E''(L) = c(a) + E''(\lambda) \int_o^{\infty} e^{-rt} dt \int_o^{\infty} \frac{\partial D}{\partial u}(u, t)\, E''\left[e^{-\rho(u+a)}\right] du$$

resulting from equation (2) in the case of λ being constant in time with subjectively defined random parameters λ and ρ assumed to be independent. In this formula the posterior expectation with respect to, $E''\left[e^{-\rho(u+a)}\right]$ can be calculated by means of the so called Laplace transform $f(s) = \frac{(b'')\gamma''}{(b''+s)^{\gamma''}}$ with $s = u+a$, relative to the posterior gamma distribution of the state of nature ρ.

The Bayes decision $\hat{a}_B$ can be determined by maximizing the previous equation and is a solution of the following equation :

$$c'(\hat{a}) - \gamma''(b'')^{\gamma''} \cdot E''(\lambda) \int_0^{\infty} e^{-rt}\, dt \int_0^{\infty} \frac{\partial D(u,t)}{\partial u} \cdot \frac{du}{(b''+u+\hat{a})^{\gamma''-1}} = 0 \qquad (16)$$

This equation is fairly complicated but can be solved in practical cases with numerical integrations and iterative techniques.

But notice that, according to (16) the optimum decision depends only on the posterior expectation of the rate of occurence λ but decision $\hat{a}_B$ depends on the entire posterior distribution of ρ and especially of the posterior dispersion of ρ.

It is possible to investigate a little further in general terms the behavior of the Bayes decision rule here :

From the very form of likelihood (7) we see there are sufficient sample statistics, which are : n and $\sum_{j=1}^{n} (q_j - q_o) = n(\bar{q} - q_o)$ with $\bar{q}$ equal to the average value of the q_i. So the Bayes rule is defined as the value $\hat{a}_B$ solution of (16) for all values of the previous statistics which are arguments of the posterior parameters. We note that, if the information is large enough, that is to say if n is large enough, the posterior variance of ρ is small and then the parameter ρ is distributed near its posterior expectation ; consequently an approximation of $\hat{a}_B$ is possible based on an expansion near the optimum value (perfect information case) $\hat{a}\,[E''(\lambda), E''(\rho)]$ corresponding to the posterior expectation. A limited expansion (up to the second term) of the solution of (16), together with the simplified assumption of constant marginal investment cost c' yields :

$$a_B \simeq \hat{a}\left[E''(\lambda), E''(\rho)\right] + \Delta \qquad (17)$$

with Δ calculated as :

$$\Delta = \left[\frac{(a-qd)^2}{2} \frac{\int_0 \frac{\partial D(o,t)}{\partial u} e^{-rt}\, dt}{\bar{D}'\left[E''(\rho)\right]} + \right.$$

$$\left. \frac{1}{2c'} \int_0^{\infty} e^{-rt}\, dt \int_0^{\infty} \frac{\partial^2 D(u,t)}{\partial u^2} . u^2\, e^{-E''(\rho).u}\, du \right] \frac{Var''(\rho)}{E''(\rho)} \qquad (18)$$

In this approximate formula, $\hat{a}$ is given by (10) in which λ and ρ are replaced by their posterior expectations, $\bar{D}'[E''(\rho)]$ is given by (9) and the posterior expectations and variances, by (15).

For analyzing the result, take an illustrative example with the following assumptions :

- The information is carried by ℓ = 44 years of observations for a French river, la Sioule (Pont du Bonchet gauging station) with a fairly high threshold q_o equal to 120 m³/s (daily discharge) :

  $n = 25$

  $n(\bar{q} - q_o) = 1672$ m³/s $\qquad (\bar{q} = 186,3)$

so that with non-information priors :

  $E''(\lambda) = 0,57$ (number of event per year)

  $E''(\rho) = 0,015$ (1/m³/s)

  $Var''(\rho) = 894\ 10^{-8}$

- The annual discount rate is 0,04. We use the assumption of a stationary damage function :

  $D(q\text{-}a,\ t) = Do(q\text{-}a)^s$

Do and the constant marginal investment cost c' are such that a (equation 10) is equal to the 1000-year flood in the case s = 1 where the damage is marginally constant. We also examine the effect of varying marginal damage by taking s = 1,1 (See Figure III).

Note that the perfect information solutions with posterior expectation values of the parameters are respectively :

  s = 1,1 : $\hat{a} - q_d = 448$ m³/s

  s = 1 (1000 years flood) : $\hat{a} - q_d = 423$ m³/s

The slight increase of marginal damage has a small effect in this case.

Example (Figure III)

1 - $D(q\text{-}a,\ t) = D_o(q\text{-}a)^s$

2 - Optimum decision :

  s = 1,1 $\qquad \hat{a} - q_d = 448$ m³/s

  s = 1 (1000 years flood) $\qquad a - q_d = 443$ m³/s

With these assumptions and in order to take into account the over-estimation due to sampling uncertainty, we have to calculate the Bayes decision and the necessary increase in design flood by equation (17).

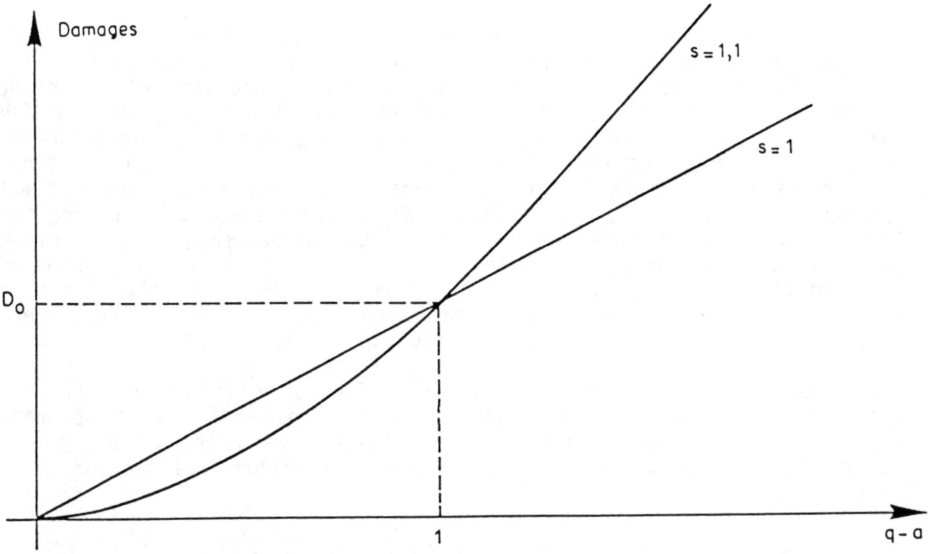

Figure III : Example of Damage function

Figure IV shows the variations of $\Delta / \hat{a}-q_d$ with respect to n, number of floods above the threshold for given $\overline{q}$ equal to 186.3 m³/s.

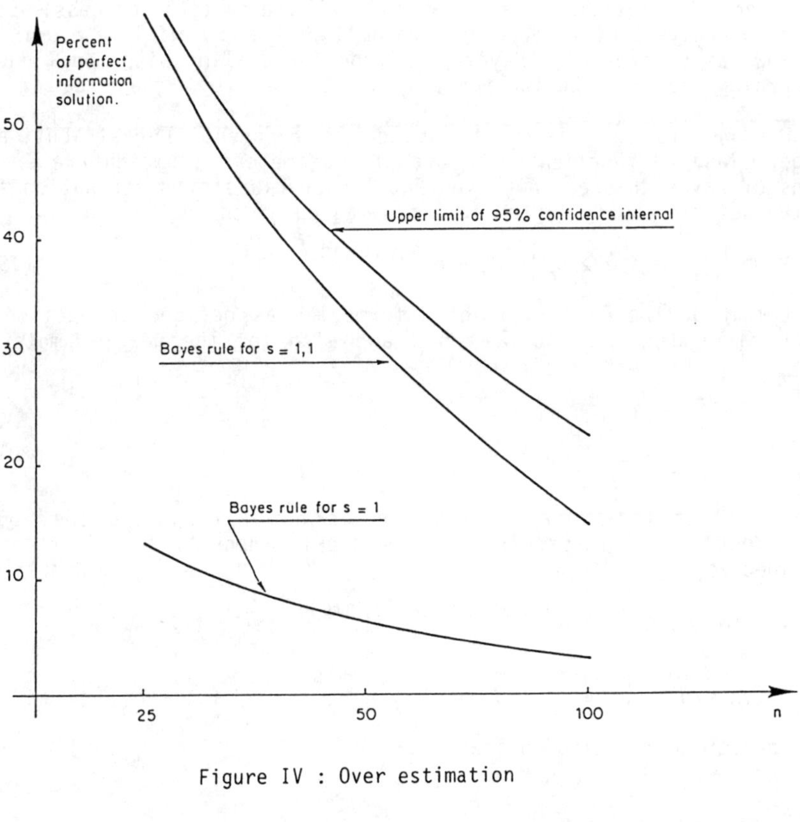

Figure IV : Over estimation

This graph also shows the limits of the traditional overestimation in such problems. The traditional method for dealing with sampling uncertainty is to calculate the upper limit of a confidence interval, with confidence probability α , say α = 95 %. Here we only give an approximate limit, taking into account the sample uncertainty of parameter ρ which as noted before is the most sensitive parameter for estimating the design discharge. In the case of constant marginal damage, the necessary overestimation appears to be generally lower than indicated by the classical confidence limit. Moreover, as shown by equation (18), Δ decreases with decreasing posterior variance of, and not with, the standard deviation of ρ upon which the confidence interval depends. But, above all, the main effect is that of marginal damage variation. These conclusions are felt to be fairly general in view of the approximate nature of equations (17) and (18).

In this problem it is possible to perform a preposterior analysis and calculate the predictive value of information either by exact opportunity loss function or approximate quadratic losses. We refer to Davis et al. (1972) for a discussion of these Bayesian concepts related to a flood problem.

V. REDUCING TECHNOLOGICAL UNCERTAINTIES

Owing to the fact that design discharge corresponds to a fairly small probability of exceedance for annual maximum discharge, it is very sensitive to model uncertainties. In many circumstances, the homogeneous Poisson process (constant λ) and exponential distribution for performance index q do not exactly represent the "reality" of the actual process. Later the Bayesian analysis of time varying parameters λ and ρ will be examined.

Now let us present a Bayesian method for dealing with model uncertainty concerning the distribution function of q.

One way to do this is to imbed the exponential distribution into a larger class of functions, the overall parameters of which are analyzed by means of Bayes theorem. A convenient larger class of distribution functions is the Weibull family fitted to the damaging discharge :

$$\text{Prob}\left[Q < q \,/\, Q > q_d\right] = 1 - e^{-\rho(q-qd)^{\rho}} \tag{19}$$

Combining the likelihood of information as defined in Section II with prior gamma densities for λ and ρ and also for the new parameter p, that is :

$$\Pi\,(p) = \frac{c'^{v'}}{\Gamma(v')}\, e^{-c'p}\, p^{v'-1} \tag{20}$$

Note that the gamma distribution is not a conjugate one, but it is convenient in this case. The joint posterior density for (λ, ρ, p) can be obtained as :

$$\Pi^{n,q}(\lambda, \rho, p) = k.\lambda^{n+a'-1}\rho^{n+b'-1}\,p^{n+c'-1}\prod_{j=1}^{n}(q_j - q_o)^{p-1} \;x\ldots$$

$$\ldots x \exp\; - \left[(\ell T + a')\lambda - \rho b' - pc' - \sum_{j=1}^{n}\rho\,(q_j - q_o)^{p}\right] \tag{21}$$

with proper normalizing constant k.

We refer to Miquel (1984) for presentation of properties of this posterior distribution. As a matter of fact, there exists an approximate method for studying it using a trivariate normal distribution centered at the posterior mode and maximizing the posterior density (20) with variances and covariances calculated through a limited expansion around this mode.

This general methodology is interesting because it allows the introduction of various types of prior information that are able to reduce sampling and model uncertainties. Such prior information can be :

- data about generating factors of floods, such as precipitation (Bernier, 1967 ; Bernier and Miquel, 1979).
- regional data about floods observed on neighboring drainage areas (Kuczera 1982, 1983). In the case of one neighboring basin with a more complete series of observations on a longer period, it is possible to use bivariate risk models (Bernier and Lencioni, 1978) taking into account the correlation between concomitant events. Such bivariate models have been used in the field of design wave estimation (Bernier and Graff, 1983).
- historical information concerning sparse large events observed during 50, 100 or 200 years before the recent period of systematic records.

Now some details are presented about the use of this historical information. These data are often neglected because they are considered as unreliable but it can be shown that, even with errors of measurement or recording, these data are of the uppermost interest for reducing both sample and model uncertainties.

VI. USE OF HISTORICAL INFORMATION

Assume that the M largest floods in N years before the systematic observation period have been recorded. Let these floods be ranked in decreasing order of their maximum discharges $Q'_1 \geqslant Q'_2 > \ldots \geqslant Q'_M$

A useful method for taking into account these floods is to use the posterior density $\Pi^{data}(\lambda, \rho, p)$ combining the likelihood of systematic observations as used in formula (20) together with the likelihood of the more complete information which is equal to :

$$L_2 = K'(\lambda . \rho . p)^M \prod_{j=1}^{M} (Q'_j - q_o)^{p-1} x\ldots$$

$$\ldots x \exp\left[- \prod_{j=1}^{M} (Q'_j - q_o)^p - N e^{-\rho (Q'_M - qo)} \right] \qquad (22)$$

So, with a gamma prior, the posterior density is :

$$\Pi^{n,N,q,Q'}(\lambda, \rho, p) = \Pi^{n,q}(\lambda, \rho, p) \times L_2$$

Using the same techniques as indicated in the previous section, we can obtain (see Miquel, 1984) the approximate characteristics of this posterior density. Figure V shows the result for the Moselle river at Hauconcourt, located in the eastern part of France. Here the estimates of parameters have been taken as the modal values of posterior densities and the indicated Bayesian confidence limits have been calculated with the normal assumption. The decreasing width of the confidence interval indicates the advantage of using historical information from the point of view of sample uncertainty. In this case, the exponential assumption $p = 1$ has been used. Figure VI shows another result for the Garonne river where the Weibull assumption has been used. (Miquel, 1984).

VII. MODEL UNCERTAINTIES AND TIME VARYING RISK PARAMETERS

Let us return to the risk model characterised by the occurrence rate of damaging floods and the distribution function $F(q, \rho)$ of successive independent maximum damaging discharges. The stationary model with constant values of parameters λ and ρ can be subject to uncertainties due to the possible variation of these parameters with time.

Now consider the so-called non homogeneous marked Poisson process of damaging floods (above the threshold q_d) with occurence rate $\lambda(t)$ and distribution function $F(q, \rho(t))$. As a result of classical properties of this Poisson process, the process of selected floods above the new threshold q still is Poisson with rate of occurence :

$$\lambda(t, q) = \lambda(t)\left[1 - F\left[q, \rho(t)\right]\right] \tag{23}$$

But the distribution function $G(q\ ; \lambda, \rho)$ of the annual maximum discharge is the probability of maximum discharge being lower than q for any flood during the time period T, that is to say, the probability of a no-flood event, during T. Given the distribution of the number of floods during T, which is Poisson with parameter $\lambda(t, q)$, it follows that the distribution of annual maximum discharge is :

$$G(q\ ; \lambda, \rho) = \exp\left[-\int_0^T \lambda(t)\left[1 - F\left[q, \rho(t)\right]\right] dt\right] \tag{24}$$

This model was applied by North (1980) to flood problems. Note that :

. $\int_0^T \lambda(t)dt = \lambda_0 T$ is the Poisson parameter for the annual number of selected floods above the threshold q_d.

$$\frac{\int_0^T \lambda(t)\, F\left[q, \rho(t)\right] dt}{\int_0^T \lambda(t)dt} = Fo(q)$$

is the mixed distribution function for any flood with occurs during the year so that equation (24) becomes :

$$G = e^{-\lambda oT\left[1 - Fo(q)\right]}$$

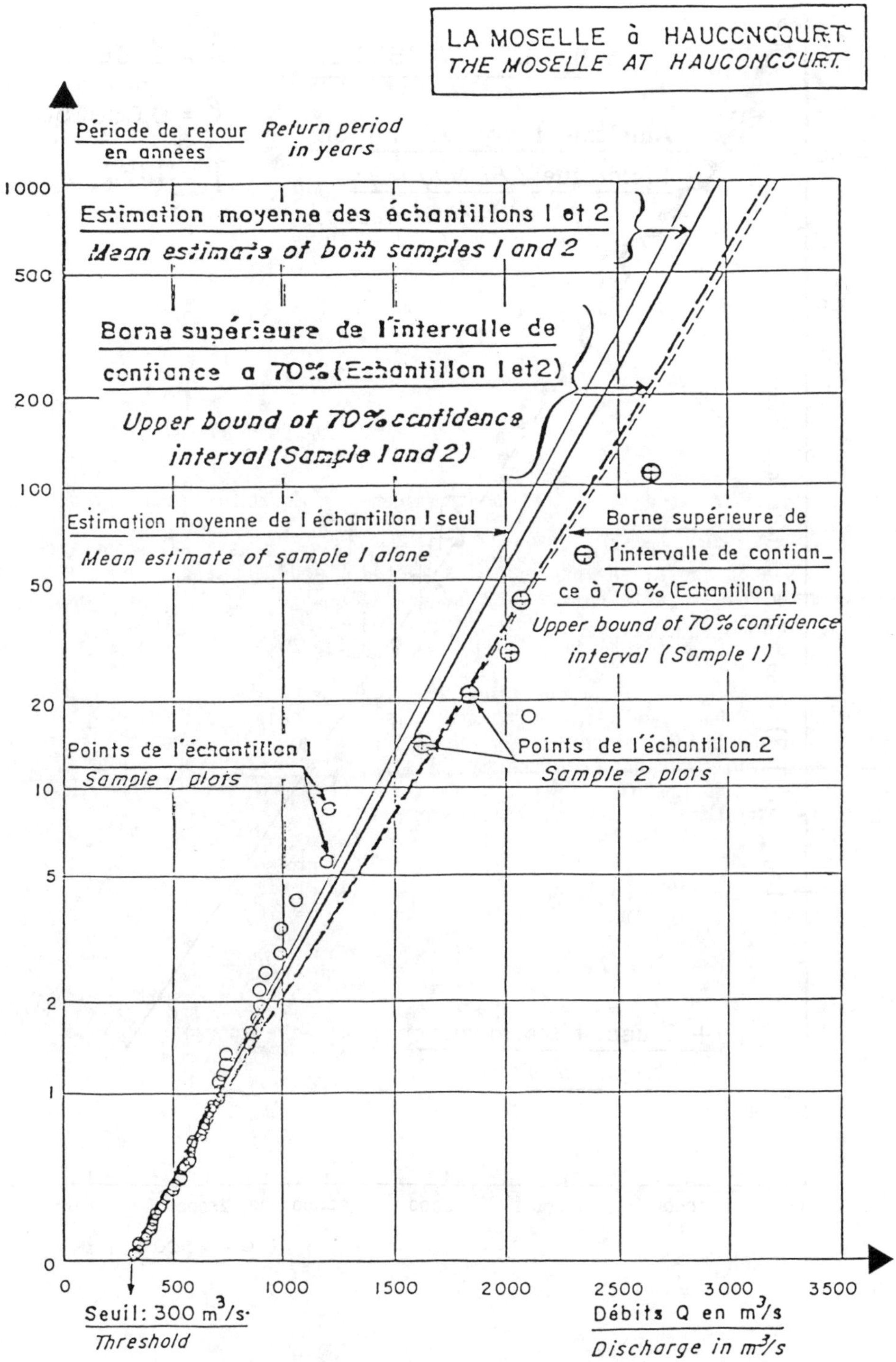

Fig V : The Moselle river at Hauconcourt

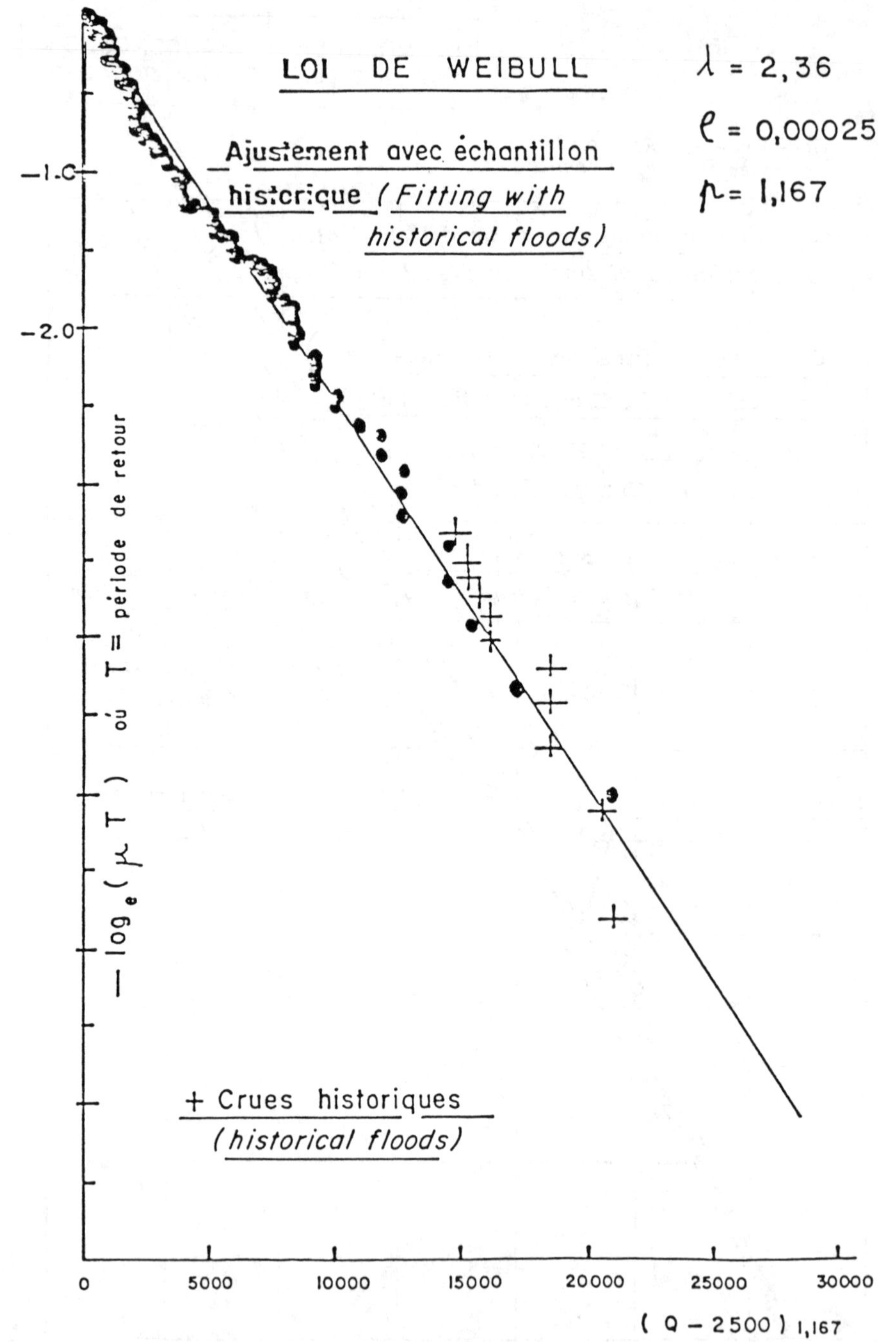

Figure VI : Risk model applied to the floods of Garonne river at Mas d'Agenais which is equivalent to equation (4), which gives the distribution function of annual maximum discharge in the stationary case.

So, provided the mixed distribution has an unbiased estimate, the stationary model can be used even if there are seasonal variations of λ and ρ. But this result is only valid when the annual maximum is a good damage index. This is not the case when the damages depend on the season. Then the entire function $\lambda(t)$ and $\rho(t)$ must be estimated.

Without making a complete Bayesian analysis, which however is possible here, it is interesting to show how Bayes methods can be used for having posterior probabilities and estimates of functions $\lambda(t)$ and $\rho(t)$ with the simple exponential assumption for function $F(q, \rho)$.

The principle of the method is to subdivide the year into k small intervals, the different sizes of the intervals being : $\Delta_1, \Delta_2 \ldots \Delta_i \ldots \Delta_k$

Let :

- n_i = total number of floods observed observed in ℓ years for each interval Δ_i.
- $z_i = \sum_{j=1}^{n_i} (q_{ji} - q_o)$, (sum of discharges above the threshold) observed in ℓ years for each interval.

To each interval there corresponds an average parameter $\overline{\lambda}_i$ and an average parameter $\overline{\rho}_i$. So we have a multiple parameter estimation problem for which we can apply the Bayes methods introduced in the first paper on Bayes analysis and the so-called Stein rule type of estimation.

For the $\overline{\lambda}_i$ and $\overline{\rho}_i$, the likelihoods are :

$$L(n, \overline{\lambda}_1, \overline{\lambda}_2, \ldots) = e^{-\sum_{i=1}^{\ell} \ell\Delta_i\overline{\lambda}_i} \prod_{i=1}^{\ell} \frac{(\ell\Delta_i\lambda_i)^{n_i}}{n_i\,!} \tag{25}$$

$$L(z, \overline{\rho}_1, \overline{\rho}_2, \ldots) = e^{-\sum_{i=1}^{\ell} \overline{\rho}_i z_i} \prod_{i=1}^{\ell} \frac{(\overline{\rho}_i z_i)^{n_i}}{n_i\,!} \cdot \frac{n_i}{z_i} \tag{26}$$

It must be noted that, except for functions of n_i and z which are constant in the Bayesian interpretation, these two likelihoods (25) and (26) have the same expression with respect to the parameters $\overline{\lambda}_i$ and $\overline{\rho}_i$; so we can deal with the estimation problem by using a general form of likelihood :

$$L = Ke^{-\sum_{i=1}^{k} Z_i\theta_i} \prod_{i=1}^{k} \frac{(Z_i\theta_i)^{n_i}}{n_i\,!} \tag{27}$$

with : $\theta_i = \overline{\lambda}_i$ (or $\overline{\rho}_i$) and $Z_i = \ell\Delta_i$ (or z_i)

Although we have a Poisson type multiple estimation problem, we shall use other prior densities than the conjugate ones. The reason is that generally the time variation of functions $\lambda(t)$ and $\rho(t)$ is smooth, a behavior which is incompatible with the independence of the θ_i, consequence of prior conjugate distributions in this case. Using the approach of Leonard (1973), the following procedure will be used :

. Define the parameter transformation $\varphi_i = \text{Log}\, \theta_i$

. Use a multivariate normal prior for the φ_i characterised by expectations $(\mu_1 = E(\varphi_1), \ldots \mu_k = E(\varphi_k))$ and by a prior matrix of variances and covariances V. A convenient assumption for this matrix V is :

$$V = w^2 \begin{vmatrix} 1 & \rho & \rho^2 & \ldots\ldots & \rho^{k-1} \\ \rho & 1 & & \ldots\ldots & \\ \rho_2 & & 1 & \ldots\ldots & \\ \rho_{k-1} & \ldots & \ldots & \ldots\ldots & \\ \rho & \ldots & \ldots & \ldots\ldots & \end{vmatrix}$$

equivalent to a Markov chain assumption for the successive φ_i.

. The prior expectation μ_i can incorporate any prior assumption for the variation of transformed parameters φ_i such as a constant or a sinusoidal form.

The prior parameters, μ_i, w^2 and ρ can be estimated by means of an empirical Bayes method (see the companion paper). The details are given in Vernos (1982) from which we present an application to the maximum daily discharges of floods observed in 59 years at the gauging station Vielle Brioude on the Allier River. Figure VII shows the risk process sample with 59 superimposed years. The seasonal effects are apparent. Figure VIII shows the estimation of function $\lambda(t)$ and $\rho(t)$ by means of the mode of the posterior distribution. The second graph gives the inverse value of $\rho(t)$, which is the instantaneous expected value of maximum discharge.

It must be noted that the seasonal effect on the rate of occurence λ, is quite strong, while this effect hardly is apparent with the function $\rho(t)$.

VIII - CONCLUSIONS

We have presented some examples of applications of Bayesian methods to floods problem which show the versatility of these techniques for dealing with different types of available information and for taking into account consequences of decisions. The efficient and more complete use of available data or even qualitative information like expert advice is necessary in water resources problems when the uncertainties are large. Bayes methods emphasize the interrelationship between data, model choice and decisional aspects. A result of this interplay is the bias of the optimum estimates of hydrological models. The Bayes estimates are biased. This is apparent in flood problems but the bias exists in any problem where the uncertainties are not reduced to zero, that is to say any hydrological problem (Bernier 1985). In the classical approach when the hydrological problem of model estimation and the decision problem are separated, it is possible to use unbiased estimates and interpret them in phenomenological terms but they can be inefficient in decision problems. If, by means of Bayes methods, we use the optimum estimates with respect to the decisional consequences, these estimates are biased and have no phenomenological interpretation at all.

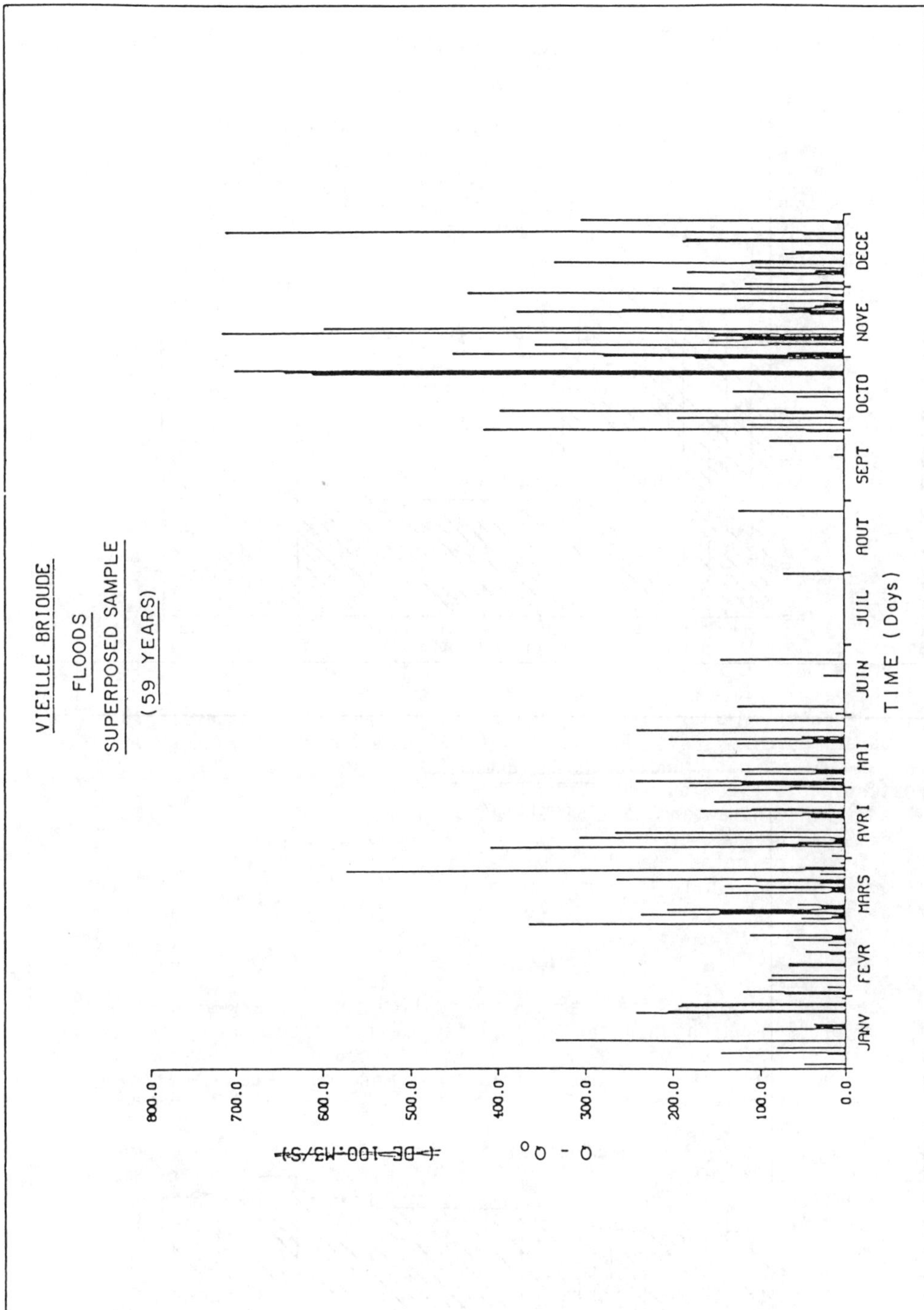
VIEILLE BRIOUDE
FLOODS
SUPERPOSED SAMPLE
(59 YEARS)
Q - Qo
(>DE=100.M3/S)
800.0
700.0
600.0
500.0
400.0
300.0
200.0
100.0
0.0
JANV
FEVR
MARS
AVRI
MAI
JUIN
JUIL
AOUT
SEPT
OCTO
NOVE
DECE
TIME (Days)

Figure VII

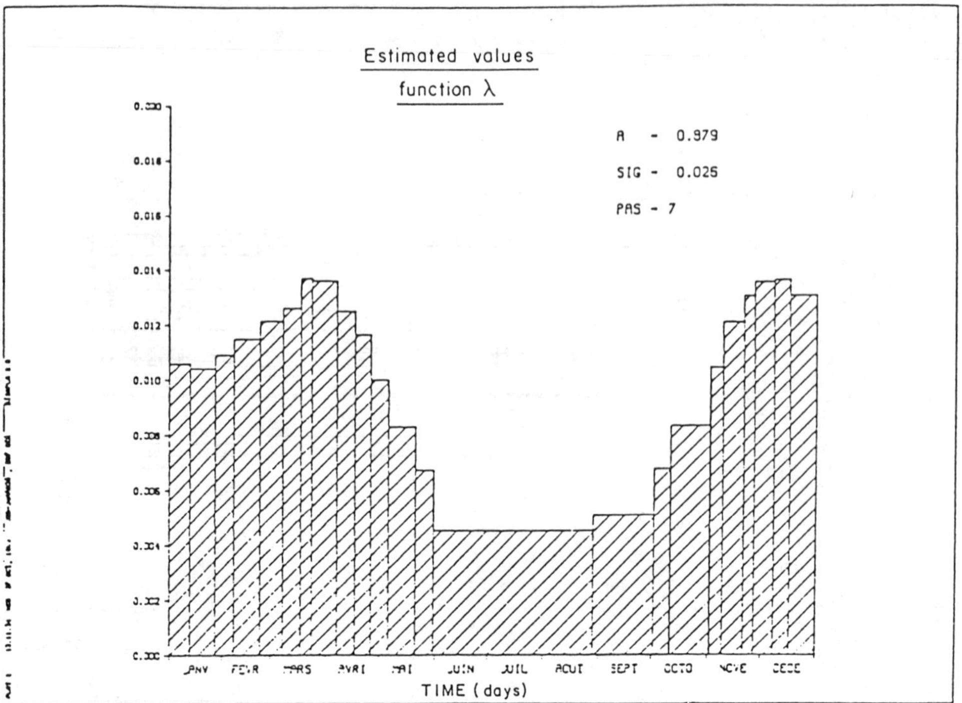
Estimated values
function λ
A - 0.979
SIG - 0.025
PAS - 7
TIME (days)

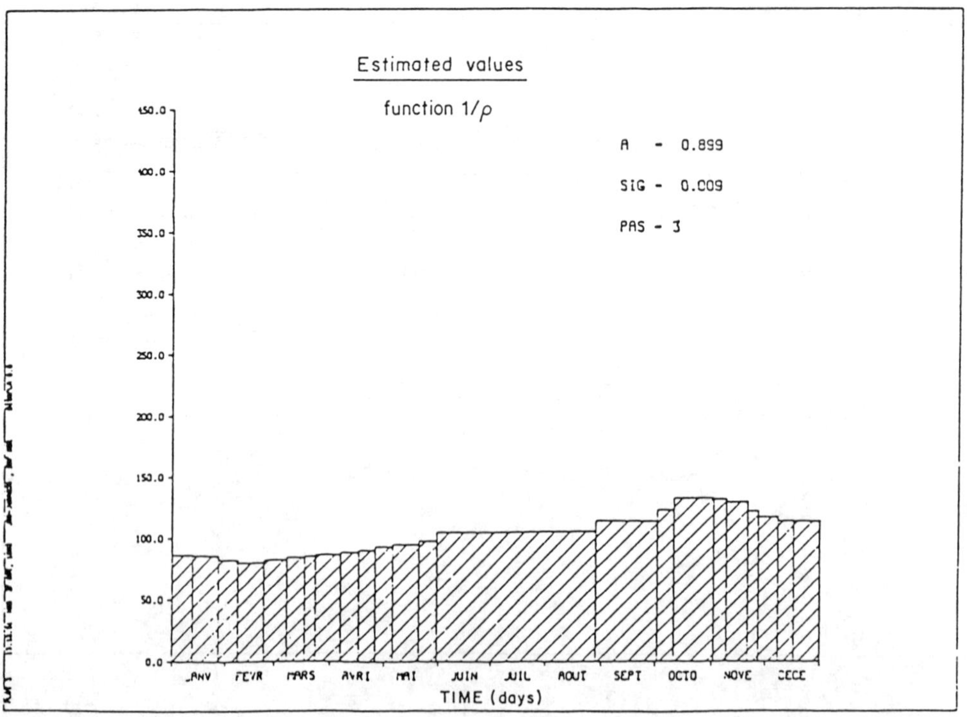
Estimated values
function 1/ρ
A - 0.899
SIG - 0.009
PAS - 3
JANV
FEVR
MARS
AVRI
MAI
JUIN
JUIL
AOUT
SEPT
OCTO
NOVE
TIME (days)

SELECTED BIBLIOGRAPHY

1 BERNIER J. (1967)
Les méthodes Bayesiennes en hydrologie statistique - Intern - Hydrol. Symp. - Fort Collins -

2 BERNIER J. (1985)
Prise en compte des incertitudes d'information et des objectifs dans l'estimation des modèles hydrologiques - to appear in : Revue Internationale de Science de l'Eau, Vol 1, n° 1/4 pp 3-10

3 BERNIER J. and M. GRAFF (1983)
Méthodes d'estimation de la houle de projet avec données incomplètes - internal report - EDF - LNH - HE42/83.34

4 BERNIER J. and P. LENCIONI (1978)
Un modèle probabiliste multidimensionnel applicable à la description des processus hydrologiques à l'échelle fine - La Houille Blanche - n° 6 - Paris

5 BERNIER J. and J. MIQUEL (1979)
Incertitudes de l'information et décisions en situation de risque - Cas de problèmes de protection contre les crues - La Houille Blanche - n° 4-5 - Paris

6 DAVIS D.R., C.C. KISIEL and L. DUCKSTEIN (1972)
Bayesian Decision Theory Applied to Design in Hydrology - Water Resources Research - vol 8

7 DAVIS D.R., F. SZIDAROVSZKY and L. DUCKSTEIN (1983)
A Distance-based Measure of the Effect of Uncertainty on Multiobjective Decision Problems - 11th IFIP Conference on System Modelling and Optimization-Copenhaguen - Denmark -

8 KUCZERA G. (1982)
Combining Site - Specific and Regional Information : An Empirical Bayes Approach - Water Resources Research - Vol 18 - n° 18 -

9 KUCZERA G. (1983)
A Bayesian Surrogate for Regional Skew in Flood Frequency Analysis - Water Resources Research - Vol 18 - n° 3 -

10 LEONARD T. (1973)
A Bayesian Method for histograms - Biometrika - n° 60 - 2 -

11 MIQUEL J. and J. BERNIER (1979)
Security of Coastal Power Stations in Relation with the State of the Sea - Symposium on Marine Forecasting - Liege-Belgium - 1979

12 MIQUEL J. (1984)
Guide pratique d'estimation des probabilités de crues - Eyrolles - Paris

13 NORTH M. (1980)
Time Dependent Stochastic Model of Flood - Jour. of ASCE-HY5

14 PAGES A. and M. GONDRAN (1980)
Fiabilité des systèmes - Eyrolles - Paris

15 TODOROVIC P. (1978)
Stochastic Models of Floods - Water Resources Research - Vol 14 - n° 2

16 VERNOS D. (1982)
Application des modèles stochastiques dynamiques linéaires à la description et à la simulation des processus des débits des rivières aux échelles saisonnières et journalières - thesis - Université P. et M. Curie - Paris -

RISK ASPECTS IN THE DETERMINATION OF OPTIMAL CROPPING PATTERNS

Harald Hiessl
Institut für Hydrologie und Wasserwirtschaft Universität Karlsruhe
D-7500 Karlsruhe, Kaiserstr. 12, FRG

ABSTRACT

In semi-arid regions water is the limiting factor for agricultural production. To reduce this disadvantageous effect of the natural water supply, efficient water resources management is required. One possible way to reach this goal in irrigated systems is to adapt the cropping pattern of the irrigation area to the system's water supply. Due to the stochastic character of the supply, this adaptation can be optimal only in a stochastic sense, which means there is a probability greater than zero that the water demand of the system cannot be satisfied at a given moment (i.e., a failure occurs). A highly flexible methodology (mixed simulation-optimization approach) is suggested to optimize cropping patterns of rather complex irrigation systems with respect to yield, failure, and nonproductive water losses. The optimization is achieved using a standard parameter optimization algorithm.

1. INTRODUCTION

The motivation for the research described in part in the present paper is the fact that in semi-arid regions water is the most important limiting factor for agricultural production. The limiting effect of the natural water supply has three main causes, which are:

a) only a small quantity of natural water supply,
b) an unfavourable distribution of water supply in time, and
c) a high stochastic variablility of a) and b).

To overcome this unfavourable situation, irrigation systems are built and operated with the main goal being to make efficient use of the natural water supply. The systems considered here consist of two parts as shown in Figure 1: the irrigation area and a reservoir supplying water to the irrigation area. Inputs into the reservoir are inflows and sediment load. It is assumed that all loads are sedimented in the reservoir, so that the effective reservoir capacity decreases with time. There are two kinds of outflows from the reservoir. The first consists of losses due to evaporation from the water surface and spilling. The second is the delivery of water from the reservoir to the irrigation system. The significant characteristics of the irrigation area vary in space and time because of coincident planting of different crops on different soils and of crop rotation in the fields during the year (cropping pattern).

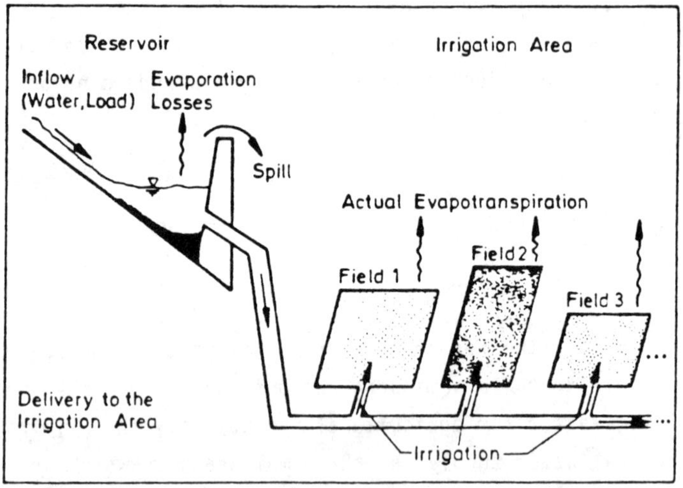

Figure 1. The Irrigation System

In order to make efficient use of the natural water supply, a balance between supply and demand has to be reached. This can be achieved, basically, in two ways.

a) Adjustment of the transient response of the irrigation system´s water requirement to that of the natural supply by selection of an appropriate cropping pattern. Cropping Pattern Problem (CP).
b) Adjustment of the transient response of the natural water supply to that of the system´s demand by determination of an appropriate control of the system´s decision variables. Distribution Problem (DP).

Obviously, CP addresses the lower frequency fluctuations of the natural water supply whereas DP adresses the higher ones.

It is the intention of the present paper to show a possible solution for the CP. Three main objectives for the explicit selection of an "optimal" cropping pattern where formulated. A cropping pattern has to be selected which:

a) maximizes the total area under cultivation with each crop,
b) maximizes the "reliability" of the water supply to the crops, and
c) minimizes the unproductive losses of water (spilling, evaporation).

These objectives are in conflict with one another so that the problem at hand is a multiple objective problem (MOP).

To illustrate the suggested methodology, a currently planned irrigation system in the South Tihama (Saudi Arabia) is used. Three main assumptions are made.

a) The irrigation area consists of 2 fields with 2 crops in sequence during a year on each field.
b) The types of crops to be cultivated are predetermined and the growing season of each crop is fixed.
c) The natural water supply (the inflows into the reservoir) is

modeled using the SCHMIDT-TREIBER simulation model for daily discharge in wadis in semi-arid regions (Schmidt and Treiber, 1980).

2. APPROACH

To solve both the CP and the DP a mixed simulation-optimization approach is used. The simulation part of the model is used for mathematical representation of the system. The application of the optimization part serves to find the optimal design of the cropping pattern according to the above objectives. The general approach (for both DP and CP) is sketched in Figure 2. The detailed approach for the solution of CP is shown in Figure 3 where Q(t) is the inflow into the reservoir, and x(t), z(t), and u(t) are the vectors of state variables, output variables, and control variables of the system, respectively.

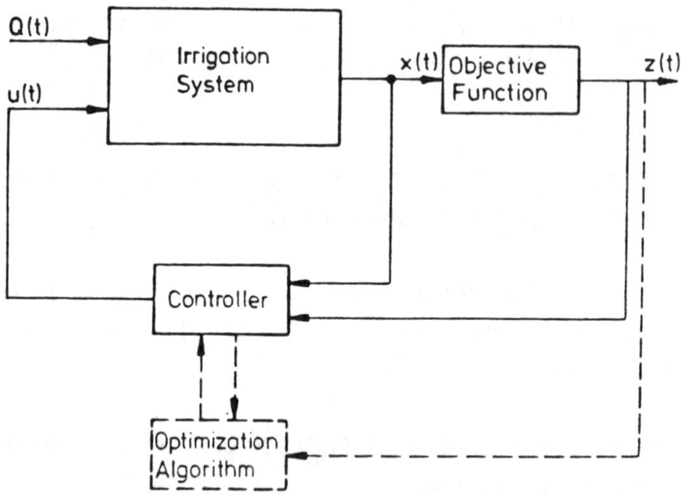

Figure 2. General Mixed Simulation-Optimization Approach

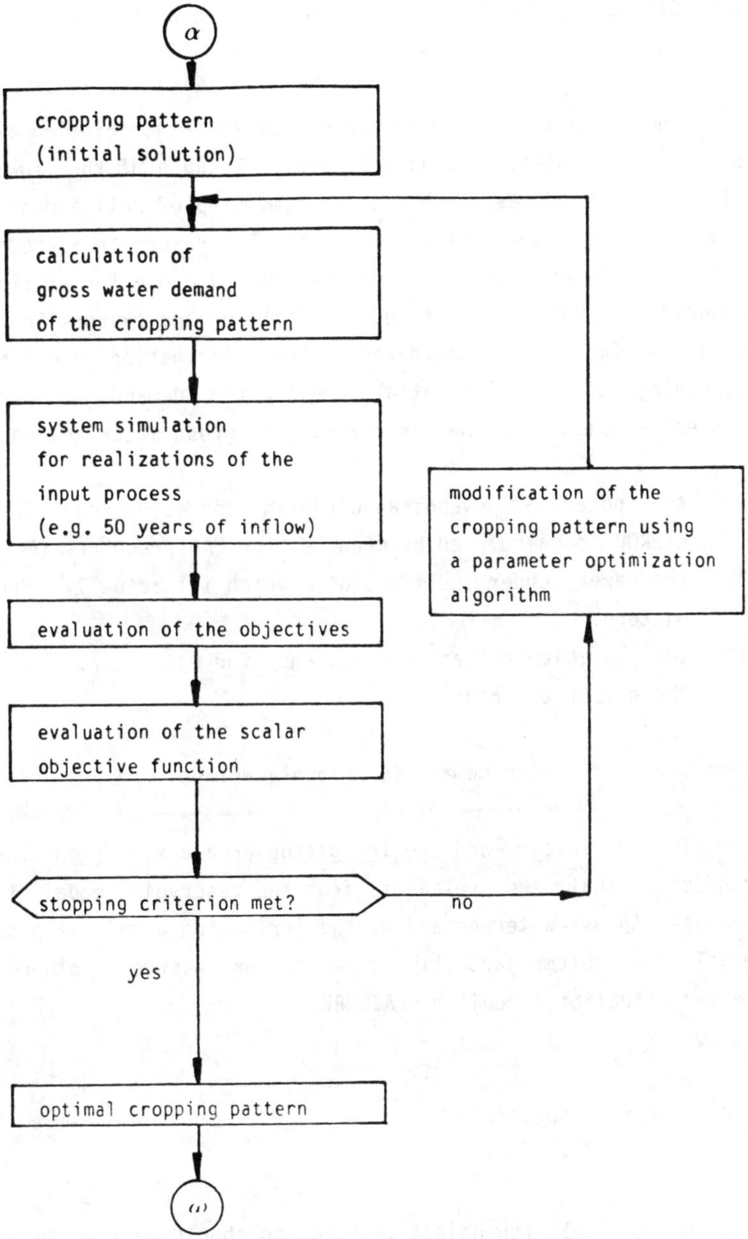

Figure 3. Simulation - Optimization Approach for the Cropping Pattern Problem

2.1. Simulation model for the cropping pattern problem

The simulation model is a mathematical description of the real system's essential characteristics. Because of the low frequency fluctuations addressed by CP, a time step of half a month for the simulation of the system is chosen. The reservoir is modeled using one mass balance equation for the water and one for silting of the reservoir. Water losses due to spilling and evaporation from the water surface are considered. The irrigation area is modeled according to its (potential) gross water demand which determines target releases from the reservoir. The gross water demand depends on

1. the potential evapotranspiration for each crop (ETCROP) (the ETCROP is calculated by means of the FAO procedure (FAO, 1977)).
2. the area under cultivation, which is actually the cropping pattern.
3. the possible irrigation efficiency for each crop.
4. the number of crops.

Thus the gross water demand is uniquely defined.

The reservoir and the irrigation area are coupled through the condition that the releases from the reservoir equal the target releases (gross water demand of the irrigation area), if possible, and equal the volume actually stored in the reservoir otherwise. The latter situation is called "FAILURE."

2.2. Objective function

In general, the objective function should measure the effects of failure situations on the productiveness of the whole system. If the quantity of water available for the crops is less than their ETCROP yield reductions occur. The yield reduction depends qualitatively on 1) frequency, 2) duration, and 3) intensity of the failure situation.

The smaller the area under cultivation, the less often failure occurs; also, the amount of water lost from the reservoir increases with decreasing delivery to the irrigation area. A general discussion of failures in relation to water resources systems is given by Hashimoto et al. (1982), Duckstein and Plate (1983), and Duckstein et al. (1985). To quantify these objectives 10 numerical indices are formulated:

C(1),...,C(4)	: area under cultivation for each single crop	MAX!
C(5)	: mean annual water supply deficit	MIN!
C(6)	: mean annual evaporation losses	MIN!
C(7)	: mean annual spilling losses	MIN!
C(8)	: frequency } of failure situation	MIN!
C(9)	: mean duration } of failure situation	MIN!
C(10)	: mean intensity } of failure situation	MIN!

Each of these indices is a function of the cropping pattern and can be calculated by simulation of the system over a given time period. Hence, the resulting vector valued objective function, C, for the MOP is

$$C = (C(1),\ldots, C(10)).$$

In the second step, the MOP is reduced to a scalar optimization problem (i.e., a problem with a scalar objective function) using the Compromise Programming approach (ZELENY, 1973). Therefore, it is necessary to

1) define the ideal rating $C^*(I)$: $C^* = (C^*(1),\ldots,C^*(10))$
2) define the worst case rating $C^W(I)$: $C^W = (C^W(1),\ldots,C^W(10))$
3) measure the distance Z_p between the rating of the actual solution C and the ideal rating C^* using a metric.

In Compromise Programming a generalized Minkowski metric is used:

$$Z_P = \left\{ \sum_I \left(W(I) * \frac{C^*(I) - C(I)}{C^*(I) - C^W(I)} \right)^P \right\}^{1/P} \quad (1)$$

with P as the metric-parameter and W the set of weights.

The final optimization problem is to minimize Z_P between the rating of the actual cropping pattern and the ideal rating. Any parameter optimization algorithm (capable of considering restrictions on independent variables) that finds the extremum of a function without using analytical derivatives may be used.

In applying the approach, two possibilities may appear:

1. The weighting set W characterizing the preference structure of the decision maker is known explicitly.
2. There is no knowledge about the weights at hand.

In the first case, the procedure is straightforward and minimization of (1) gives a nondominated solution for the MOP, i.e., an efficient cropping pattern. In the second case, the approach can be used in the sense of a generation method for the efficient (solution) set of the MOP (Cohon, 1978). The efficient set may then be used with an appropriate procedure to seek a solution that is actually preferred by the decision maker (Cohon, 1978; Goicoechea et al., 1982).

3. RESULTS

A cropping pattern optimized in the way described in Section 1 is input into the optimization model, the control of the system (the DP-model). A detailed description of this model is given elsewhere (Hiessl, 1985). This model computes an optimal controller (i.e., solution of the DP) on a daily basis. The simulation part of the model considers the soil water balances and actual evapotranspiration rates of each field and each crop in detail and uses a yield model (Jensen, 1968) to simulate the effect of water stress on plant yield.

A comparison between a cropping pattern PAT1 used as an initial solution (Figure 3) and a pattern PAT2 optimized with the approach described above from the initial solution can be made. The results for 6 arbitrarily selected pairs (PAT1, PAT2) are shown in Table 1. The so called "Mean Annual Net Yield" (MANY) is defined as a double-weigthed sum of the area of each crop grown in the pattern weigthed by the crop´s relative yield (actual yield relative to a crop specific maximum yield). Two different controllers (CONT1, CONT2) are actually implemented in the DP-model. Both are closed loop controllers. They differ in their complexity: CONT1 is more complex than CONT2 but both are easy to optimize and easy to use (i. e., without any computing facility) in real applications under conditions typical for developing countries.

From Table 1 it can be seen that it is possible to improve MANY by optimization of the cropping pattern with the procedure described above. Obviously the improvement depends on the quality of the initial solution PAT1 and one could argue that it would be sufficient to select a "good" initial solution in order to decrease the effect of the above optimization procedure. That may be true in irrigation systems with only a small number of crops (say 2) actually grown in the total area over one year. But in real systems with a large number of crops grown in the irrigation area, the possibility of selecting a "good" initial solution vanishes (becomes less probable) because of the complexity of the system.

To show that not only the mean annual net yield increases through the optimization of the cropping pattern but also the probability for high yield in an individual year, a comparison for the third pair of cropping patterns (initial and optimized pattern) from Table 1 is shown in Figures 4a,4b. This is done by comparing the distributions of relative yield of each of the four crops grown in the irrigation area over a time horizon of 50 years. It can be seen that the optimized cropping pattern (solid line) overall has more favourable yield distributions (i.e., the probability of exceedance of a certain yield is greater for the optimized pattern) than the initial pattern (dashed line).

initial solution PAT1 optimized solution PAT2 (both in ha)				Mean Annual Net Yield (MANY): CONT1	increased yield	CONT2	increased yield
1100	400	300	1400	1181.63		984.90	
700	320	370	900	1377.42	+16.6%	1244.90	+26.4%
400	700	800	300	1617.72		1544.12	
520	880	940	330	1690.32	+4.5%	1553.45	+0.6%
1800	300	400	1400	1081.30		1023.31	
340	510	50	1350	1346.04	+22.5%	1235.12	+20.7%
100	600	100	100	810.35		842.72	
170	1620	300	200	1576.15	+94.5%	1590.13	+88.7%
1000	100	100	1000	1113.26		1027.28	
210	300	100	740	1224.57	+10.8%	1240.23	+20.7%
1000	400	800	1400	1304.52		962.72	
460	70	1320	60	1429.86	+9.6%	1448.43	+50.5%
				average :	+26.4%	average :	+34.6%

Table 1. Comparison of Unoptimized and Optimized Cropping Pattern (With an Optimal Controller Type 1 and Type 2)

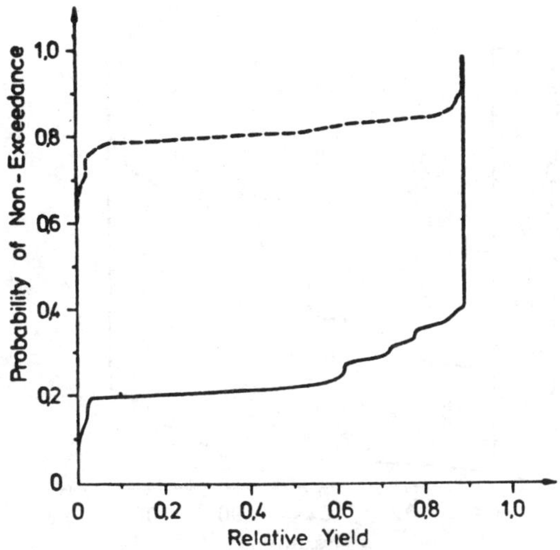

Field 1, Crop 1: IS = 1800 ha, OS = 340 ha

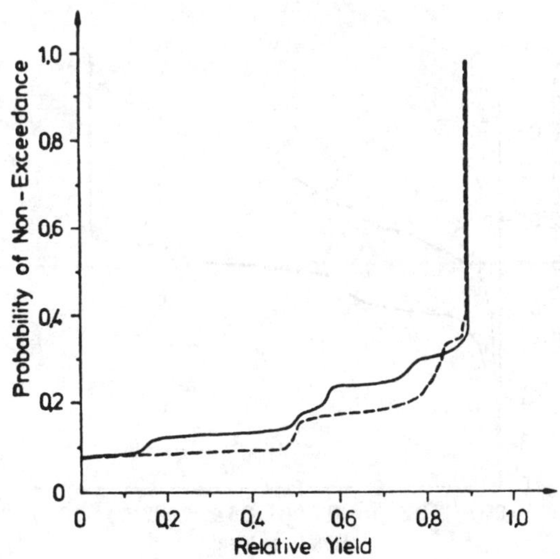

Field 1, Crop 2: IS = 300 ha, OS = 510 ha

Figure 4a. Comparison of the Distribution of Yield for Unoptimized and Optimized Cropping Pattern for Field 1.
Initial Solution (IS): — — —
Optimal Solution (OS): ————

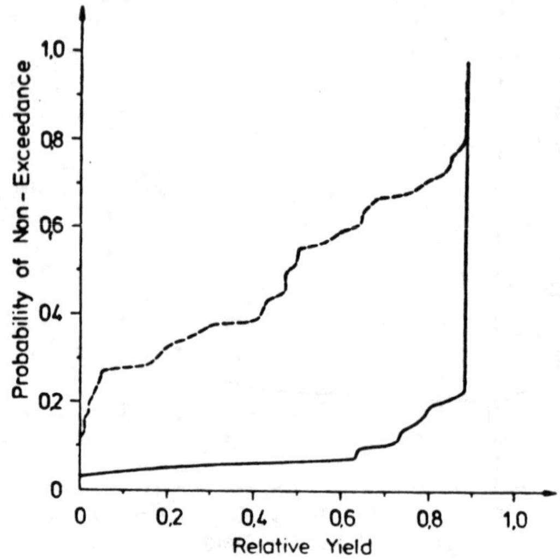

Field 2, Crop 1: IS = 400 ha, OS = 50 ha

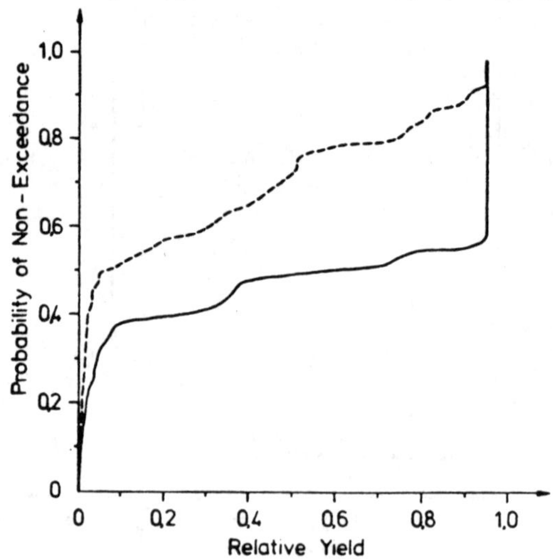

Field 2, Crop 2: IS = 1400 ha, OS = 1350 ha

Figure 4b. Comparison of the Distribution of Yield for Unoptimized and Optimized Cropping Pattern for Field 2.
Initial Solution (IS): — — —
Optimal Solution (OS): ———

4. CONCLUSIONS

With the approach suggested herein it is easy to optimally design a cropping pattern in a rather complex irrigation system (large number of fields and crop rotations on the fields within a year). The advantages of the approach are as follows.

1. Stochastic water supply can be considered (implicit stochastic programming).

2. The approach is easy to modify for both the description of the system (simulation model) and the design criteria of the system (objective function).

3. For the optimization, any preprogrammed parameter optimization algorithm can be used.

4. The computational expense of the simulation part increases approximately linearly with the number of crops. For the optimization, it depends on the optimization algorithm used.

5. The approach is applicable to the situation with a known preference structure of the decision maker (i.e., known weighting set) as well as to the situation with an unknown preference structure. In the latter case, the efficient solution set of the MOP is generated to which a further selection procedure may be applied.

5. REFERENCES

Cohon, J.L. (1978) : "Multiobjective Programming and Planning"; Mathematics in Science and Engineering, Vol. 140, Academic Press, New York - San Francisco - London.

Duckstein, L.; E.J.Plate (1983): "A System Framework for Hydrological Reliability with Application to Design Structures"; Proc. IAHS Symposium, Hamburg, Session 7, August 22-26,1983

Duckstein, L.; E.J.Plate; M.Benedini (1985): "Water Engineering Reliability and Risk: A System Framework"; NATO Advanced Study Institute on Engineering Reliability and Risk in Water Resources, Tucson, Arizona, USA, May 19 - June 2, 1985

FAO (1977): "FAO-Handbook No. 24: Crop Water Requirements"; Revised; FAO Rome, Italy

Goicoechea, A.; D.R.Hansen; L.Duckstein (1982) : "Multiobjective Decision Analysis with Engineering and Business Applications"; Wiley, New York.

Hashimoto, T; J.R.Stedinger; D.P.Loucks (1982): "Reliability, Resiliency, and Vulnerability Criteria for Water Resource System Performance Evaluation"; Water Resources Research (18)1, 14-20

Hiessl, H. (1985) : "Optimal Control of Irrigation Systems with Water Demand Varying in Space and Time" (in German); Mitteilungen Institut für Hydrologie und Wasserwirtschaft, Universität Karlsruhe, Karlsruhe, FRG. (to be published)

Jensen, M.E. (1968) : "Water Consumption by Agricultural Plants"; in: T.T.Koslowski (ed.) : "Water Deficit and Plant Growth"; Chapter I; Academic Press, New York.

Schmidt O.; B.Treiber (1980): "Ein Simulationsmodell für Tagesabflüsse in ariden Gebieten"; Wasserwirtschaft (70)1, pp 5-9.

Zeleny, M. (1973): "Compromise Programming"; in: J.L. Cochrane; M. Zeleny (eds.): "Multiple Criteria Decision Making"; Univ. of South Carolina Press, Columbia

RELIABILITY ASPECTS OF MULTICRITERION WATERSHED MANAGEMENT

Istvan Bogardi and Andras Bardossy

Department of Civil Engineering, University of Nebraska-Lincoln, Lincoln, Nebraska 68588-0531
Tiszadata Consulting Engineers, Miko u. 1., 1012 Budapest, Hungary

ABSTRACT

Risk and reliability aspects are incorporated into the multicriterion analysis of watershed management, which considers three conflicting objectives: I. to increase agricultural benefit by increasing effective precipitation, II. to decrease erosion and water pollution stemming from agricultural activities, III. to increase water yield in order to meet downstream water demand. To each of the objectives there corresponds one or more criteria such as, for objective II, sediment yield, dissolved phosphorus and other pollutants. Stochastic spatial parameters, such as soil properties and stochastic meteorological elements, influence the criterion or output functions. As a result, figures of merit (annual income, investment, P loading and yields of sediment and water) are also stochastic. The trade-off analysis considers risk of pollution versus reliability of water supplies and is performed by means of composite programming, an extension of compromise programming. The real-life example of an agricultural watershed illustrates the methodology.

1. INTRODUCTION

The purpose of this paper is to incorporate the effect of inherently stochastic characteristics (in space and time) of input into the trade-off analysis of conflicting objectives facing watershed management.

There is a growing competition between watershed agriculture, industry, environmental protection, and possibly other sectors for surface water runoff. This study considers the specific case of the management of agricultural watersheds which often serve conflicting objectives such as the maximization of agricultural revenue, minimization of erosion and water pollution, and maximization of surface water yield. The attainment of each of these objectives is generally measured by several criteria such as annual income, investment cost, sediment and nutrient yield.

In the next section, such a problem of multicriterion watershed management is described. Types of possible control actions include land-use, crop, mode of cultivation, erosion control, fertilizer use and engineering interventions such as silting basins. The stochastic input is composed of spatial parameters such as soil properties and meteorological elements such as rainfall and temperature. The systems model is developed in three parts. First, the six elements of the model are defined: 1. time scale (days) and horizon (one year), 2. state vector, 3. input set, 4. state transition function, 5. output set (figures of merit) 6. output function. Next, methodological tools such as geostatistics and event-based stochastic modeling are specified to consider the spatial and temporal stochasticity of input. In the third part the systems model is developed by using composite programming in a probabilistic framework. Composite programming, an extension of compromise programming, consists of a two-level trade-off analysis. At the first level, a compromise is sought within each objective, for example between sediment yield and nutrient yield both being within the environmental objective; then the objectives of economics, water pollution and water yield are traded off.

The use of the methodology is illustrated by the real-life example of a 27 km^2 agricultural watershed. Numerical results are calculated for the trade-off of, first, expected value output functions and, second, the probabilities of not satisfying target output values.

2. MULTICRITERION WATERSHED MANAGEMENT

2.1 Objectives and Criteria

The management of watersheds often faces conflicting interests for the utilization of existing water supply. The competition among water users such as industry, agriculture and recreation has been the subject of heavy bargaining processes and might have led to unsatisfactory allocation of available water resources. Consider the case of semi-arid agricultural watersheds where there are at least three conflicting interests:

a) watershed agriculture, requiring more and more water to increase economic benefit,

b) control of the contamination of groundwater and surface water,

c) downstream water users requiring an increasing amount of watershed runoff.

In general, neither objectives can be neglected; thus, trade-off solutions are to be sought instead of separate optimum solutions. Table 1 shows that the three objectives have several components or criteria.

TABLE 1: Objectives of watershed management

Objective		Criteria
Economic	:	annual agricultural revenue, investment cost, cost-benefit ratio, etc.
Environmental	:	sediment yield, dissolved phosphorus, sediment-bound phosphorus, etc.
Hydrologic	:	seasonal runoff

More generally, it may also be necessary to consider other components such as nitrogen and heavy metals. In fact, in a complete systems analysis, one can define the triplet objectives - specifications - criteria, as it is done for example in Salis and Duckstein (1983). Even in the case of economic criteria, which can certainly be expressed as a single monetary term, it is often appropriate to keep each criterion in a separate account; for example, agricultural revenue, investment and losses usually correspond to different economic agents.

There is a crucial conflict in watershed management between the interests of watershed agriculturists and downstream water users. The former endeavour to increase crop production by increasing the efficiency of rainfall utilization over the watershed. On the other hand, downstream water users such as the operators of downstream reservoirs would prefer to increase or at least to maintain runoff to meet demands of water supply, recreation or power generation.

Several applications of multicriterion decision making (MCDM) in watershed management can be found. One of the first multiobjective models of watershed management was proposed by Miller and Byers (1973). Onishi and Swanson (1974) considered nitrate and sediment as constraints on the economic optimization of crop production. Das and Haimes (1979) have presented a multiobjective model of water quality and land management. Goicoechea et al (1979) used a special MCDM methodology called PROTRADE to select the best reclamation scheme under uncertainty for an open-pit mining area. Bogardi et al (1983) have considered a joint lake-watershed system and found a tradeoff between minimizing P loading and the costs related to various control alternatives.

2.2 Control Actions

A number of structural and non-structural control actions are available for the management of watersheds. Table 2 indicates common types and examples for control actions.

TABLE 2: Types of Control Actions

Type	Example
Land-use	reforestation
Crop	protective crop rotation
Cultivation	contour farming
Slope length	terracing
Fertilizer	deep broadcasting
Structural	sediment retarding basin

Specifically, reforestation may lead to smaller agricultural revenue, but would drastically reduce all elements of environmental pollution, and water yield. Protective crop rotation (from an erosion aspect) may result in smaller revenue in the short term, but would decrease water pollution and runoff. Erosion control measures would contribute to higher crop yield, lower water pollution, and also smaller runoff. Considering, however, the various elements of environmental pollution, erosion control usually reduces sediment yield to a much greater extent that it reduces soil-bound P due to the enrichment effect (Meta Systems, 1979). Terracing would result in higher crop yield but requires considerable investment. Both water pollution and water yield would be reduced. Fertilizer management, such as deep broadcasting may not affect crop yield very much but results in saving fertilizer, as contrasted with the higher cost of broadcasting. Both dissolved and sediment bound P would be smaller. Water yield would not be influenced. Finally, structural interventions such as sediment retarding would not influence crop yield but require investment. However, all components of water pollution would take on more favorable values while water yield may be slightly smaller.

These examples show the complexity of the interaction between control actions. In addition, within each type of control action, several alternatives are available. Although some of the control actions may be represented as continuous decision variables, for example the area allocated to protective crop rotation or to contour farming, it has been experienced that watershed managers often prefer to examine discrete alternatives (Goicoechea et al, 1982; Bogardi et al, 1983). These alternatives would already account for a number of additional factors such as the constraints of local farm size, slopes, machinery and manpower.

2.3 Stochastic Input

There are two types of input stochasticity influencing the three objectives:

- spatial variation of properties, such as meteorological, soil, topographic and biological parameters,
- time variability of meteorological elements, such as rainfall, temperature and radiation.

As a result, most of the criterion elements should also be regarded as random variables, and the probability of achieving target levels such as a prescribed outflow from the watershed, or an allowable amount of P loading is to be estimated. Thus, reliability or risk is an important motivating aspect of watershed management.

3. SYSTEM MODEL

3.1 System Elements

Multicriterion watershed management can be modeled as a discrete-time dynamic system with the following six system elements:

1. T : the time scale ($j = 1,2,\ldots,t,\ldots J$) corresponds to days and the horizon J is a year.

2. S : the state vector describing the system at time t includes, among others, soil moisture content and N distribution in the root zone,

3. X = (U, W) : the input set composed of the set of controllable elements U and the set of non-controllable elements W. Set U consists of the possible control actions as specified in Table 2. The set W includes

- environmental elements such as natural factors (rainfall events, temperature, radiation), physical properties of the area (soil, topography, crop, etc.), ecological factors (pollution limits, transpiration demand, etc.),
- economic and social elements such as noncontrollable resources (capital, machines, manpower), target levels (water and crop yield),

4. F : the state transition function calculates the state at time (t + 1) as a function of the previous state S(t) and the previous input X(t),

$$S(t+1) = F(S(t), X(t)) \tag{1}$$

The state transition function includes the mass (water) and chemical (N) balance equations, such as the numerical solution of the differential equation describing water movement in the unsaturated zone,

5. Y : the output set consisting of three subsets:
- economic,
- environmental and
- hydrological objectives.

Figures of merit (FM) correspond to annual values of the output such as annual income and P loading, as well as the probabilities of not satisfying certain target output levels.

As a consequence of stochastic input (in time and in space) the state transition function and the output set are both stochastic. One has to decide on the type of the output values with due regard to the spatial and time input stochasticity. It is possible to use the expected values but also other statistics such as the variance, or given percentile values. These latter quantities could express the risk awareness of the farmers, or just the reliability of pollution control actions. Clearly, the selection of the proper form of the output function depends mostly on local watershed conditions and could be the subject of a further trade-off analysis even within a single criterion (Wrather and Yu, 1982; Bogardi et al, 1983).

In general, for stochastic output $y_1, \ldots y_u, \ldots y_n$ the following formulation can be used by introducing the operator: $G_u\ (y_u)$ which can be for example:

$$G(y) = \begin{cases} E(y) & \text{or} \\ P(y > \gamma) \end{cases}$$

where γ is the target level.

The first case corresponds to expected output, while the second one, to risk analysis or probabilistic decision making.

6. O : the output function provides the algorithm to calculate the output y(t) given the present state and input

$$y(t) = O(S(t), X(t)) \tag{2}$$

Output functions for multicriterion watershed management may be calculated with the help of a comprehensive physical model (Fig. 1), as described in detail by Bogardi et al (1984).

3.2 The Consideration of Spatial and Time Variability of Input

Several input elements of watershed management, such as properties of soil, crop and topography are stochastic in space. The problem is, given point observations on these properties to calculate output function values referring to the whole watershed. In other words, how can the spatial sto-

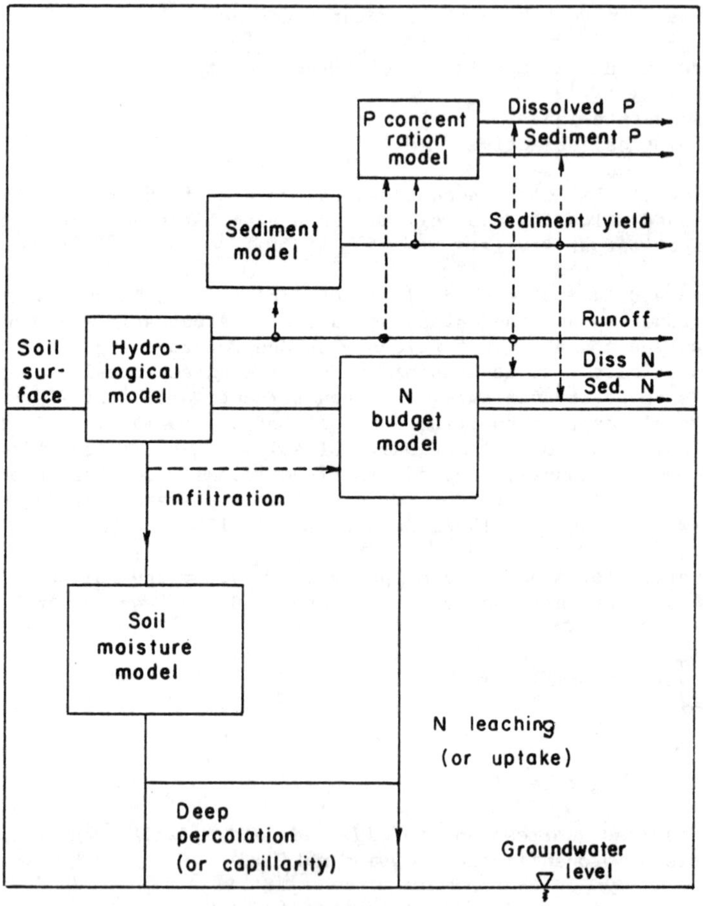

Figure 1: Elements of the physical model

chastic character of some input data be considered in calculating the output function.

It is possible to use geostatistics, that is, statistics of spatially correlated random variables (Matheron, 1971) to solve the above problem. Several attempts have been made to use geostatistics for describing soil parameters (Delhomme, 1979; Warrick and Nielsen, 1980; Russo and Bresler, 1981; Bogardi et al, 1983; de Marsily, 1984) or even to incorporate the spatial stochastic character of input into the investigated process (Dagan and Bresler, 1983; Warrick and Gardner, 1983). The geostatistical approach can be extended to outputs such as dissolved and sediment-bound P.

The stochastic time variability of input can be considered within an event-based framework (Todorovic and Zelenhasic, 1970; Woolhiser, 1973; Duckstein et al, 1979). However, some elements of the input may exhibit stochasticity both in time and space such as meteorological elements. As a consequence, it would be necessary to combine geostatistics and temporal event-based stochastic analysis (Bastin et al, 1984) in order to estimate outputs such as average spatial and temporal annual P loading.

In the present case the model is specified for consideration of time variability of input only. The annual sediment yield is estimated by routing hydrologic events through the modified Universal Soil Loss Equation (MUSLE) (Williams, 1975). This method uses rainfall events defined by storm depth, duration as input and provides fairly accurate results (Bardossy et al, 1984). Then the annual water yield is calculated by the SCS method (Soil Conservation Service, 1971).

The annual amount of dissolved P is estimated as the annual sum of a random number of random loading events characterized by the product of runoff volume per event and the dissolved P concentration. The latter concentration can be estimated from a physico-chemical model (Bogardi and Bardossy, 1984) which considers in particular: a) adsorption and desorption of dissolved and sediment P; b) sediment enrichment; c) fertilizer application; d) P leaching from plant residues; e) base-flow contribution; f) point-source loading. The annual amount of soil-bound available P is calculated by a method analogous to the one used for dissolved P loading. However, loading events are characterized by the product of sediment yield per event and labile P concentration per event. The latter quantity can be obtained in conjunction with the dissolved P concentration. Finally, the annual crop revenue is calculated as the product of crop yield and price.

3.3 Model Development

The watershed is to be managed so as to optimize simultaneously the three objectives: economics, pollution control and hydrology (Fig. 2). Moreover, in addition to the conflicting objectives, the various criteria within the objectives may also be in conflict (Table 1). Thus the following problem can be specified, with, here, $n = 6$:

$$\min G_u (y_u), \; u = 1,\ldots,n \tag{3}$$

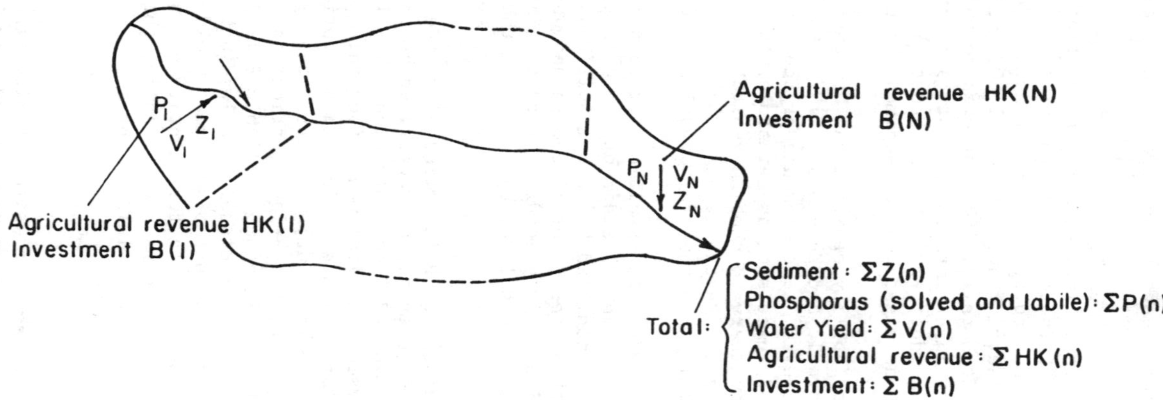

Figure 2. Multicriterion Watershed Managements

where:

y_1:	- net annual agricultural income	Economic
y_2:	annual investment	objective
y_3:	annual sediment yield	Environmental
y_4:	annual soluble P yield	objective
y_5:	annual sediment-bound P yield	
y_6:	- annual water yield	Hydrologic objective

Now let a (0-1) decision variable be introduced:

k_{ij} = 1 if alternative j is selected for subwatershed i, and 0 otherwise.

Only one alternative can be selected per subwatershed, thus:

$$\sum_{j=1}^{N_i} k_{ij} = 1, \; i = 1,\ldots,N \tag{4}$$

Here N is the number of subwatersheds and N_j is the number of alternatives for watershed i.

Now values of G_u (y_u) corresponding to a selected set of alternatives $\underline{\underline{k}} = \{k_{ij} : i=1,\ldots,N;\ j = 1,\ldots,\ N_i\}$ are calculated.

For the selection of the "best" control action, composite programming is used. Composite programming was first used in Bogardi and Bardossy (1983) for geologic exploration planning. Later, it was applied to network design (Bogardi et al, 1985); Bardossy (1984) showed that composite programming is a proper extension of compromise programming (Zeleny, 1974). The first step of composite programming is to select an ideal point, and then define a composite metric. In the example the optimal solution will be obtained as the solution minimizing the distance between the ideal point G*(Y) and the point G(Y) representing the solution. The ideal point is:

$$G^*(Y) = (G_1^*,\ G_2^*,\ G_3^*,\ G_4^*,\ G_5^*,\ G_6^*,) \tag{5}$$

where $G_u^* = \min_{\underline{\underline{k}}} G_u\ (y_u\ (\underline{\underline{k}})\ )$

The distance (G*, G) is defined stepwise as stated in the composite programming formulation. First, in each groups of objectives, a compromise solution and then the overall compromise are successively sought. For this purpose, in each group, one selects a value of the compromise programming parameter p and a set of weights $\{\alpha\}$. Then the overall objective function with parameter q and weight set $\{\beta\}$ is defined. This procedure yields the objective functions:

$$CF_1(\underline{\underline{k}}) = \left\{ \alpha_{11} \left[\frac{G_1(y_1(\underline{\underline{k}})) - G_1^*}{G_{1*} - G_1^*} \right]^{p_1} + \alpha_{12} \left[\frac{G_2(y_2(\underline{\underline{k}})) - G_2^*}{G_{2*} - G_2^*} \right]^{p_1} \right\}^{1/p_1}$$

$$CF_2(\underline{\underline{k}}) = \left\{ \alpha_{21} \left[\frac{G_3(y_3(\underline{\underline{k}})) - G_3^*}{G_{3*} - G_3^*} \right]^{p_2} + \alpha_{22} \left[\frac{G_4(y_4(\underline{\underline{k}})) - G_4^*}{G_{4*} - G_4^*} \right]^{p_2} + \alpha_{23} \left[\frac{G_5(y_5(\underline{\underline{k}})) - G_5^*}{G_{5*} - G_5^*} \right]^{p_2} \right\}^{1/p_2}$$

$$CF_3(\underline{\underline{k}}) = \frac{G_6(y_6(\underline{\underline{k}})) - G_6^*}{G_{6*} - G_6^*} \tag{6}$$

where $G_{u*} = \max G_u(y_u(\underline{\underline{k}}))$

The overall composite goal function is written as

$$CF(\underline{\underline{k}}) = \{\beta_1 . CF_1(\underline{\underline{k}})^q + \beta_2 . CF_2(\underline{\underline{k}})^q + \beta_3 . CF_3(\underline{\underline{k}})^q\}^{1/q} \tag{7}$$

The choice of the composite programming parameters p and q should reflect the nature of the criteria; thus, say $p_2 = 3$ to emphasize the strong influence of the "worst" pollution element. To optimize the composite goal function one may use, in continuous cases, standard methods such as the gradient method (Bogardi et al, 1985), or quadratic programming (Bardossy, 1984), in discrete cases, as here, a branch and bound algorithm or, (0-1) programming (Bogardi et al, 1984) can be used.

4. CASE STUDY EXAMPLE

4.1 The Watershed

Fig. 3 shows the 27 km^2 Vacszentlaszlo watershed in Hungary. This watershed is divided into six subwatersheds with basic characteristics as given in Table 3. Topographical, soil and agricultural conditions differ among subwatersheds but even within subwatersheds heterogeneity rather than homogeneity is the rule. Large-scale farming indicated by the magnitude of fields includes mostly wheat and corn. A storage reservoir of 640 x 10^3 m^3 serving irrigation water-supply and offering recreational possibilities is located at the outlet of the watershed.

The need to harmonize the conflicting interests of managing this watershed is illustrated as follows. Under existing conditions, sediment yield stemming mostly from surface erosion adversely affects agricultural

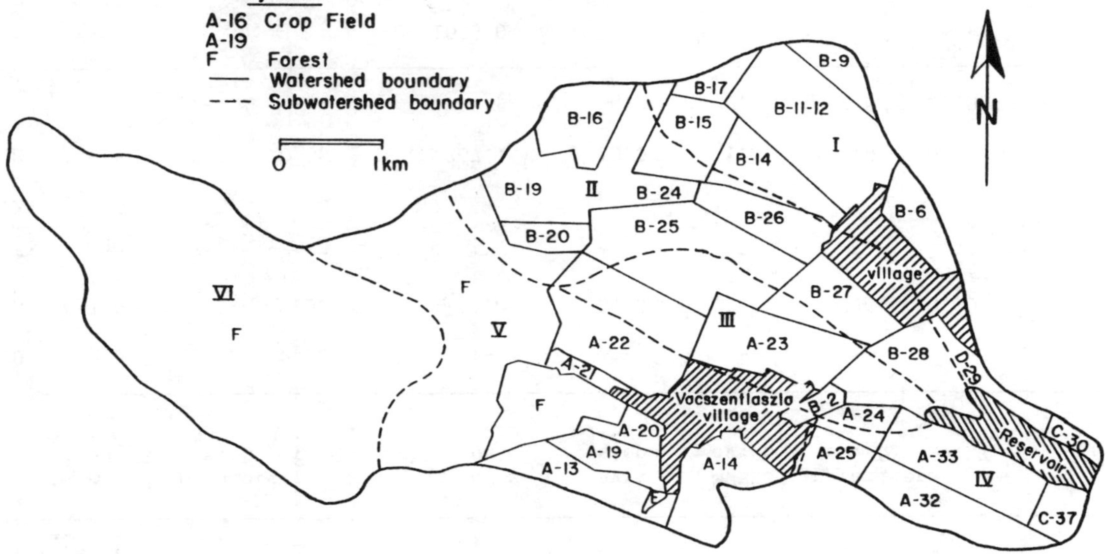

Figure 3: Vacszentlaszlo Watershed: existing land-use

TABLE 3: Basic Data for the Vacszentlaszlo Watershed

Subw.	Cropland					Forest				
	Area, Ha	Avr. slope %	MUSLE av. S	parameters av. L	av. K	Area, Ha	avr. slope %	MUSLE av. S	parameters av. L	av. K
I	176.7	16.0	2.3	3.3	0.36	-	-	-	-	-
II	330.6	12.5	1.6	3.5	0.32	41.0	14.0	1.9	3.6	0.30
III	202.5	8.5	0.8	3.1	0.34	-	-	-	-	-
IV	246.8	11.2	1.3	3.3	0.40	-	-	-	-	-
V	208.8	5.8	0.5	3.1	0.36	267.1	10.0	1.2	2.9	0.34
VI	-	-	-	-	-	767.9	8.0	0.8	2.8	0.30
Total	1165.4					1076.0				

production, gradually decreases the storage place of the reservoir, and substantially contributes to the average annual sediment-bound P loading. Both dissolved and sediment-bound P loadings into the reservoir cause increasing eutrophication, endangering its recreational functions. Intensive agriculture over the watershed has resulted in an average annual net agricultural revenue of 11,598 x 10^3 forints (1 dollar is worth about 50 forints). It is an important regional goal to maintain or possibly increase this revenue. On the other hand, farmers are not encouraged to borrow money for investment purposes, which would be necessary to implement some of the control actions. Concerning the third criterion (hydrology), water supply requirement from the reservoir has been growing. Thus one also seeks to increase the annual water yield from the watershed.

Target levels γ_u for the environmental and the hydrological objectives can be specified (Table 4). Note that no target levels have been given in the present example to the economic criteria.

TABLE 4: Target Levels

Criterion	Target level γ_u
Annual sediment yield	2×10^6 kg
Annual dissolved P	1×10^3 kg
Annual sediment P	2×10^3 kg
Annual water yield	90×10^3 m^3

4.2 Control Actions

A discrete number of alternative interventions over each subwatershed has been envisaged. Table 5 indicates the number of alternatives in each action category. The total number of alternatives is 46,268,820, precluding the use of total enumeration.

The encoding of management alternatives is specified as follows. Each variable k_{ij} is characterized by a six-digit code wherein:

- the first digit indicates the number of subwatersheds: 1,...,6

- the second digit shows land use: 1: existing; 2: total afforestation; 3: total meadow; 4: cropland, > 25% forest on slopes; 5: cropland < 17%, 17-25% meadow, forest > 25%; 6: cropland, > 25% meadow;

- the third digit refers to crop rotation: 0: existing; 1: soil protective

- the fourth digit indicates the mode of cultivation: 0: up-and-down slope culture; 1: countour farming, 2: contour farming and crop-residue

TABLE 5: Basic Alternatives for the Vacszentlaszlo Case

Types of alternatives	Number of alternatives for subwatersheds					
	I	II	III	IV	V	VI
Land use	6	6	5	5	4	1
Crop	2	2	2	2	2	1
Cultivation	3	3	3	3	3	1
Slope-length decrease	4	4	4	4	4	1
Amount of fertilizer (existing and 50% decrease)	2	2	2	2	2	1

- the fifth digit refers to erosion control by slope length control
  0: no control
  1: slope length: 400-1000 m
  2: slope length: 250- 700 m   depending on steepness
  3: slope length: 200- 500 m

- the sixth digit refers to fertilizer control
  1: existing amount of P, 2: 50% decrease.

4.3 Numerical Results

The MCDM analysis has been performed using, first expected value output functions (Case 1), then expected economic criteria and minimum risk environmental and hydrological criteria (Case 2). Thus

Case 1: $G_u = E\,(y_u)\ ;\ u = 1,\ldots,6$

Case 2: $G_u = E\,(y_u)\ ;\ u = 1,2$

$G_u = P\,(y_u > \gamma);\ u = 3,4,5,6,$

Composite programming parameters p and q are: $p_1 = 2$, for trade-off between agricultural revenue and investment, $p_2 = 3$, for trade-off among environmental elements, to emphasize the preponderance of the worst element, q = 2, a quadratic trade-off among the three objectives. Numerical results have been obtained for the four sets (A, B, C, D) of weights (α,β) given in Table 6.

TABLE 6: Sets of Weights for Composite Programming

	α_2	α_2	β
A	0.9, 0.1	0.3, 0.4, 0.3	0.4, 0.3, 0.3
B	0.9, 0.1	0.4, 0.3, 0.3	0.2, 0.6, 0.2
C	0.9, 0.1	0.3, 0.2, 0.5	0.1, 0.1, 0.8
D	0.9, 0.1	0.3, 0.1, 0.1	0.1, 0.8, 0.1

4.4 Expected Value: Case (1)

Table 7 indicates the existing, ideal and worst values of the criteria for the expected value case.

TABLE 7: Output Function Values for the Vacszentlaszlo Case (Expected Values)

Objective	Criterion	Actual	Ideal	Worst
	Expected annual agricultural revenue 10^3 forint	6390	17250	5611
Economic	Investment 10^3 forint	0	0	46616
	Expected annual sediment yield 10^6 kg	2.814	0.012	2.814
Environmental	Expected annual dissolved P 10^3 kg	1.39	0.02	1.38
	Expected annual sediment-bound P 10^3 kg	4.84	0.0016	4.84
Hydrologic	Expected annual runoff 10^3 m^3	120	120	57

TABLE 8: Results of Composite Programming (Expected Value Case)

Objective	Criterion	Set of Weights A	B	C	D
Economic	Revenue	15762	16494	12512	17219
	Investment	11508	13280	7662	18021
Environmental	Sediment	0.547	0.496	1.412	0.415
	Dissolved P	0.57	0.54	0.67	0.47
	Sediment P	0.58	0.52	1.42	0.42
Hydrologic	Runoff	105	102	112	96
Composite solution		161211	161221	110002	141221
		261211	261211	261211	241221
		341211	341211	310002	341221
		441211	441221	441211	441221
		541211	541221	541211	541221
		610001	610001	610001	610001

Numerical results for the expected value case are presented in Table 8. Trade-off solutions are found to have the following main features:

Since subwatershed 6 is total forest, it will be kept intact in every solution. Sets of weights A and B correspond to a balanced importance of objectives, as a result, the composite solutions are quite similar:

- same land use changes over the subwatersheds (1-5),
- use of protective crop rotation everywhere (1-5),
- contour farming and plant residues everywhere (0-5),
- slight (1) to medium (2) erosion control,
- existing amount of P-based fertilizer application.

Sets C and D give high preference to the hydrologic and environmental objectives, respectively. With highly preferred runoff maximization (set C), the composite solution changes considerably:

- maintenance of existing situation over subwatersheds 1, 3 and 6,
- decrease of fertilizer use over subwatersheds 1 and 3.

On the other hand, with a high preference for the environment (D) the solution corresponds to uniform land use (4) and erosion control (medium) over each subwatershed except No. 6.

4.5 Results of the Risk Analysis Approach (Case 2)

Table 9 contains the existing, ideal and worst values of the probabilities of not-satisfying target levels γ_3, γ_4, γ_5, and γ_6.

For illustration purposes the numerical calculation of the composite distance CF is shown with set of weights A:

TABLE 9: Output Function Values for the Risk Analysis

Objective	Criterion	Actual	Ideal (probabilities)	Worst
Environmental	Annual sediment yield	0.60	0.01	0.60
	Annual dissolved P	0.65	0.01	0.65
	Annual sediment P	0.71	0.01	0.71
Hydrologic	Annual water yield	0.10	0.10	0.73

$$CF_1 = \left[0.9 \left(1 - \frac{14735 - 5611}{17250 - 5610} \right)^2 + 0.1 \left(\frac{11852 - 0}{46616 - 0} \right)^2 \right]^{1/2}$$

$$CF_2 = \left[0.3 \left(\frac{0.1 - 0.01}{0.6 - 0.01} \right)^3 + 0.4 \left(\frac{0.07 - 0.01}{0.65 - 0.01} \right)^3 + \right.$$

$$\left. + 0.3 \left(\frac{0.09 - 0.01}{0.71 - 0.01} \right)^3 \right]^{1/3}$$

$$CF_3 = \frac{0.3 - 0.1}{0.73 - 0.1}$$

$$CF = (0.4\, CF_1^{\,2} + 0.3\, CF_2^{\,2} + 0.3\, CF_3^{\,2})^{1/2} = 0.235$$

Numerical results of case 2 calculations are given in Table 10.

The comparison of composite solutions for case 1 and 2 shows that the solutions are different for all set of weights. However, for set of weights A there are only slight differences:

- over subwatershed 1 land use alternative 4 (case 2) instead of 6 (case 1);
- over subwatershed 3, decreased amount of fertilizer P as compared to the existing amount.

Composite solutions are quite different for sets of weights A and B (Table 10) as compared to case 1 solutions (Table 8). In case 2, output function values are similar for sets of weights B and D, (both representing environment-preferred weights) but the composite solution for weights B corresponds to higher fertilizer P control and for weights D, to higher erosion control. Note that weights D emphasize the control of sediment yield (Table 6).

The number of simulation runs necessary to reach a given accuracy level of the selected statistics G can be estimated by the Kolmogorov-Smirnov test. Accordingly, in case 1 (expected values) 570 years were simulated in order to obtain an accuracy level of ± 2% on the 80% confidence level. In case 2 (risk analysis) the necessary number of simulation increases up to 10 000 sample elements for a maximum error of ± 1%.

TABLE 10: Results of Composite Programming (Risk Analysis)

Objective	Criterion	Set of Weights A	B	C	D
Economic	Revenue	14735	14210	13121	16815
	Investment	11852	9877	8319	18134
Environmental	Sediment	0.10	0.01	0.21	0.01
	Dissolved P	0.07	0.02	0.23	0.02
	Sediment P	0.09	0.01	0.17	0.01
Hydrologic	Runoff	0.03	0.33	0.15	0.37
Composite solution		141211	151212	161201	151221
		261211	251202	210001	251221
		341212	351202	341211	341222
		441211	451212	441211	441221
		541211	541212	541211	541221
		610001	610001	610001	610001

5. CONCLUSIONS

1. The management of agricultural watersheds involves conflicting objectives which, in the case considered, are agricultural economics, the environment and hydrologic runoff for water-supply.

2. To each objective, there corresponds one or more criteria to be evaluated for a discrete set of alternative plans under input uncertainty caused by stochastic rainfall and spatial variability in soil, topography and crop parameters.

3. Elements of a discrete dynamic system model can be used for defining the following figures of merit: annual values of revenue, investment, sediment, yield, dissolved and sediment-bound P, and water yield.

4. Due to the sto hastic input, figures of merit to be traded off are also stochastic.

5. Several possible statistics of the figures of merit can be considered such as, in the present case, expected values and the probabilities of not satisfying target levels.

6. Composite programming is shown to be applicable to either formulation of the figures of merit.

7. The present application refers to two cases: 1. all figures of merit are expected values, but environmental and hydrologic ones are the probabilities of not satisfying target levels.

8. The application of composite programming results in realistic alternatives for a 27 km^2 agricultural watershed facing conflicting economic, environmental and hydrologic objectives.

9. Composite solutions for the above two cases are different since they reflect the properties of expected value decision-making in all figures of merit versus partial risk-awareness in the environment and water supply objectives.

6. REFERENCES

Bardossy, A., I. Bogardi and I. Duckstein (1984) Accuracy of sediment yield calculation. Working paper 84-6, Systems Dept.Univ. of Arizona,Tucson, AZ.

Bardossy, A. (1984) Mathematics of composite programming. Working Paper 84-18, Systems Dept. Univ. of Arizona, Tucson, AZ.

Bastin, G., B. Lorent, C. Duque and M. Gevers (1984) Optimal estimation of the average areal rainfall and optimal selection of rain gauge locations. Water Resources Research 20 4, pp. 463-470.

Bogardi, I., L. David and L. Duckstein (1983) Trade-off between cost and effectiveness of control of nutrient loading into a waterbody. Research Report RR-83-19, IIASA, Laxenburg, Austria, July.

Bogardi, I., L. Duckstein, and A. Bardossy (1984) Trade-off between cost and efficiency of pollution control. 6th Int. Conf. on MCDM, Proceedings, Cleveland, Ohio, June.

Bogardi, I. and A. Bardossy (1984) Stochastic forecasting of N budget in the unsaturated zone. Proceedings, RIZA Symposium, Eds, Udluft, P., Merkel, B. and K. -H. Prosl, Technical Univ. of Munchen, FRD, pp. 743-756.

Bogardi, I. and A. Bardossy (1984) A concentration model of phosphorus stemming from agricultural watersheds. Working Paper 84-8, Tiszadata, Budapest.

Bogardi, I., A. Bardossy and L. Duckstein (1985) Multicriterion network design using geostatistics. Water Resources Research 21 2, pp. 199-208.

Dagan, G. and E. Bresler (1983) Unsaturated flow in spatially variable fields, 1., 2., 3. Water Resources Research 19 2 , pp. 413-420.

Das, P. and Y. Y. Haimes (1979) Multiobjective optimization in water quality and land management. Water Resources Research 15 6, pp. 1313-1322.

Delhomme, J. P. (1979) Spatial variability and uncertainty in groundwater flow parameters: a geostatistical approach. Water Resources Research 15 2, pp. 269-280.

Duckstein, L., M. Fogel and I. Gogardi (1979) Event-based models of precipitation for semiarid lands, in the hydrology of areas of low precipitation, IAHS-AISH Publ. 128, pp. 51-64.

Goicoechea, A., L. Duckstein and M. Fogel (1979) Multiple Objectives Under Uncertainty: An Illustrative Application of PROTRADE. Water Resources Research 2, pp. 203-210.

Goicoechea, A., D. Hansen and L. Duckstein (1982) Introduction to Multiobjective Analysis with Engineering and Business Applications. John Wiley and Sons, New York, N. Y., p. 511.

de Marsily (1984) Spatial variability of properties in porous media: stochastic approach, in "Fundamentals of Transport Phenomena in Porous Media," edited by J. Bear, The Hague.

Matheron, G., (1971) The Theory of Regionalized Variables and Applications. Ecole des Mines, Fontainebleau, France, p. 211.

Meta Systems, Inc. (1979) Costs and water quality impacts of reducing agricultual nonpoint source pollution. EPA-608/5-79-009. August.

Miller, W. L. and D. M. Byers (1973) Development and display of multiple objective projects impacts. Water Resources Research 9 4, pp. 11-20.

Onishi, H. and E. R. Swanson (1974) Effect of nitrate and sediment constraints on economically optimal crop production. Journal of Environmental Quality 3 3, pp. 234-238.

Russo, D. and E. Bresler (1981) Soil hydraulic properties as stochastic processes. I. An Analysis of field spatial variability. Soil Sci, Soc. Am. J. 45, pp. 682-687.

Salis, M. and L. Duckstein (1983) Mining under a limestone aquifer in southern Sardinia: a multiobjective approach. Int. J. of Mining Eng. I, pp. 357-374.

Soil Conservation Service (1971) National Engineering Handbook, Section 4, Hydrology USDA, Washington D.C.

Todorovic, P. and E. Zelenhasic. (1970) A stochastic model for flood analysis. Water Resources Research 6 6, pp. 1641-1648.

Warrick, A. W. and D. R. Nielsen (1980) Spatial variability of soil physical properties in the field, in Hillel, D. ed. "Applications of soil physics" Academic Press, N. Y., pp. 319-344.

Warrick, A. W. and W. R. Gardner (1983) Crop yield as affected by spatial variations of soil and irrigation. Water Resources Research 19 1, pp. 181-186.

Williams, J. R. (1975) Sediment yield prediction with universal equation using runoff energy factor. Agr. Res. Serv., ARS-S-40 USDA Washington, D.C. pp. 244-252.

Woolhiser, D. A. (1973) Hydrologic and watershed modeling - state of the art. Trans. Amer. Soc. Agric. Engrs Paper 71-747, pp. 553-559.

Wrather, C. and P. L. Yu. (1982) Probability dominance in random outcomes. Journal of Opt. Theory and Appl. 38 3, pp. 315-334.

Zeleny, M. (1974) A comcept of compromise solution and the method of the displaced ideal. Computer and Oper. Res. 1 4, pp.479-496.

A MIN-MAX OPERATING RULE FOR THE MANAGEMENT OF A MULTIPURPOSE RESERVOIR

G. Guariso°, S. Orlovski°°, S. Rinaldi° e R. Soncini-Sessa°

° Centro Teoria dei Sistemi, CNR, Via Ponzio 34, Milano, Italy
°° International Institute for Applied System Analysis, Laxenburg, Austria

ABSTRACT

A risk-averse approach which avoids extreme failures in the management of a multipurpose reservoir is presented and discussed in the paper. The method explicitly requires the real-time forecast of the future inflows to the reservoir. The main advantage of the approach is that it suggests a whole range of possible decisions which guarantee the efficient performance of the system. This allows the manager to choose that release of water which is more in keeping with the additional information he may have and to consider secondary objectives which were not included in the formulation of the problem. The method is applied to the case of Lake Como (Italy). A comparison of the results with historical events indicates that both flooding of the lake shores and water losses could have been significantly reduced.

1. INTRODUCTION

In this paper we present a risk-averse approach which is adequate for solving multiobjective optimal control problems of storage systems. In order to clearly point out the advantages and potentials of the method, we discuss in some detail a case study: the real-time operation of Lake Como (northern Italy). In this problem there are only two management objectives, namely, satisfaction of the water demand of the downstream users and attenuation of the floods on the lake shores. Since these goals are conflicting and incommensurable, the solution of the problem will be a set of efficient operating rules (Cohon and Marks, 1975).

Many different methods for determining such operating rules have been developed so far. They differ in the formal description of the objectives and constraints, in the way they deal with uncertainties, and in the solution algorithms (linear programming, dynamic programming, discrete dynamic programming, optimal control, etc.). One common feature of these methods is the explicit or implicit use of the notion of probability to describe future inflows and evaluate systems performance. For this reason, all these methods could be considered as different formal expressions of the stochastic approach to reservoir management.

Very frequently, managers react in an unfavourable way to the above optimization methods (Helweg et al., 1982). Reasons for this may be the inadequate description of the physical characteristics of the system, the complexity of the proposed algorithms, and the need for on-line optimization. Other reasons may be the lack of confidence in sophisticated stochastic techniques and the feeling that single-value operating rules are tools too rigid for solving complex, but soft, decision-making processes. Finally, detailed analyses made in different fields of management science have proved that decision makers very seldom consider long-term expected values of physical and/or economic indicators as representative measures of system performance. Indeed, very often, a manager focuses his attention and effort on avoiding dramatic failures when the system is under stress. In other words, in most cases, decision makers are risk-averse, even if this yields a poorer average performance of the system.

For all these reasons, in this paper we use a deterministic (min-max) approach to the management problem, which overcomes many of the above criticisms. The theoretical basis of the approach can be found in Orlovski et al. (1983), while the case of reservoir management is described in Orlovski et al. (1984) and Guariso et al. (1984a). In this approach, the performance of the system is evaluated with reference to a few specific inflow sequences suggested by the manager. These inflow sequences may be some real or synthetic inflow records or some hypothetical sequences of inflows which the manager considers as particularly well-suited for testing the reliability of any operating rule. The efficient solutions suggested by this approach will be shown to have some definite advantage with respect to those obtained by means of the stochastic methods. In fact, they do not require complex algorithms and on-line optimization; they make reasonable use of the real-time forecast of the inflows; and they suggest a whole interval of possible releases instead of a single value. The last such property is particularly important, since it introduces some flexibility into the decision-making process.

The simulation of the daily operation of Lake Como for a period of 15 years will show that both floods and water shortages would have been much smaller than shown in history if the method had been systematically applied.

2. THE ACTUAL OPERATION OF LAKE COMO

Lake Como stretches for almost fifty kilometers beneath the spectacular peaks of the central Alps near Italy's border with Switzerland (Figure 1). By June each year the lake is filled with the snowmelt from a catchment area in the Alps of over 4500 square kilometers. Over the hot dry summer the lake is drained by the River Adda, which flows 140 kilometers across the Padana plain southeast of Milan before merging with the River Po. In autumn the lake is refilled by the rains.

Since Roman times this typically alpine cycle has benefited farmers downstream of the lake, where the yields of wheat, corn, and forage from 144000 hectares of agricultural land can be greatly improved by using the waters of the River Adda for irrigation. The users remained subject to the caprices of nature, however, until the Second World War, when a dam was constructed at Olginate. The dam was built to regulate the lake's outflow for agricultural purposes over the summer and to increase the power production of the seven downstream hydroelectric plants, which have a total generating capacity of 92 megawatts.

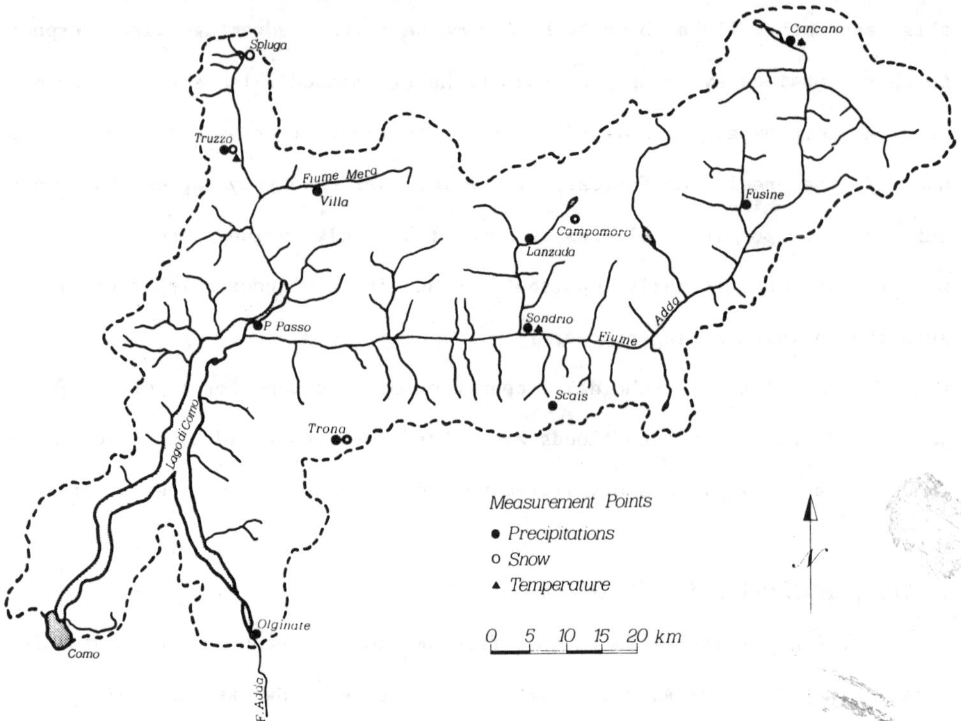

Figure 1 - Lake Como catchment and measurement network.

In the 1960s, the main square in the town of Como at the sourthern end of the lake began to subside, probably as a result of over-pumping from the underground aquifer. It is now about 60 centimeters below its 1946 level, and this has greatly increased the flooding of the square during the periods of high inflows due to snowmelt in the early summer and rainfall in the autumn. In the first twenty-five years of the dam's operation, only three floods lasted longer than five days; between 1972 and 1981 there were ten such floods. Since the main square is the commercial centre of Como, and the lake-side road links the eastern and western parts of the town, which has a population of about 100000, flood protection has become a matter of great concern. It has been the subject of much discussion between the Municipality of Como and the manager responsible for the day-to-day operation of the dam.

The dam at Oliginate is operated subject to the terms of a government licensing act which specifies that the manager decides the release each day provided that the lake level lies between two fixed limits. This control range defines the so-called "active storage," which is the volume of water the manager controls. The figure for Lake Como is about 250 million cubic meters.

Since the lake is filled twice during the year, first by snowmelt and later by rainwater, the manager is responsible for a total "budget" of 500 million cubic meters of water a year. When the level reaches the lower limit, the manager must release no more than the inflow into the lake, so that the level does not drop further to jeopardize navigation and cause sanitation problems. At the upper limit, the manager is obliged to open all the dam gates to alleviate flooding of the lake shore.

Retaining water in the reservoir to irrigate agricultural land in the summer and to generate hydroelectric power in the winter runs the risk of disruptive flooding in Como and around the lake shore when the inflows suddenly increase. Thus, managing Lake Como is a process of making tradeoffs between conflicting objectives: maximizing the water supply for irrigation (hydroelectric power generation is economically negligible) and minimizing the risk of flooding.

Maximizing agricultural production through irrigation corresponds to minimizing the total annual water deficit in the six agricultural districts supplied with water from Lake Como. The natural deficit figure, before the building of the dam, was 270 million cubic meters per year; this was reduced to an average of about 200 million cubic meters by the operation of the dam since 1946.

As far as environmental protection is concerned, two indicators best represent the economic and social effects of the flooding. The peak levels of the floods correspond more closely with property damage, but the duration of the flooding better characterizes the disruption of commercial life. In addition, a statistical analysis of all the floods since 1946 showed that the two indicators are highly correlated. This means that we can just consider the number of days of flooding per year without distorting the picture. This has averaged about ten days in the past, whereas the natural figure for the same period, had the dam not been built, would have been about twenty-three days. Estimates indicate that the cost of one day of flooding is not great in comparison with the economic gains in agriculture resulting from improved irrigation; however, the number of people directly involved in farming is rather small.

Some detailed statistical analysis of the historical data and discussions with the reservoir manager revealed that the manager is often most concerned with avoiding extreme failures of the system during severe hydrological episodes, even at the cost of a somewhat worse performance in terms of the average values of the indicators. This is why we followed an approach (see next section) which minimizes the maximum ill-effects rather than achieving the optimum mean values of the indicators. The same discussions and analysis also revealed that the manager uses all the information available in real-time on the reservoir catchment to make a decision, since this information can help in better evaluating the future availability of the resource. The way the information is processed and the decision is made can be interpreted in terms of the classical control scheme depicted in Figure 2, where s_t is the level of the lake at the beginning of the t-th day, and u_t and w_t are release and inflow during the same day. In such a scheme, in addition to a feedback channel, there is a feedforward channel bringing information (a vector z_t) on the state of the catchment to the manager. In the case considered in this paper such information is

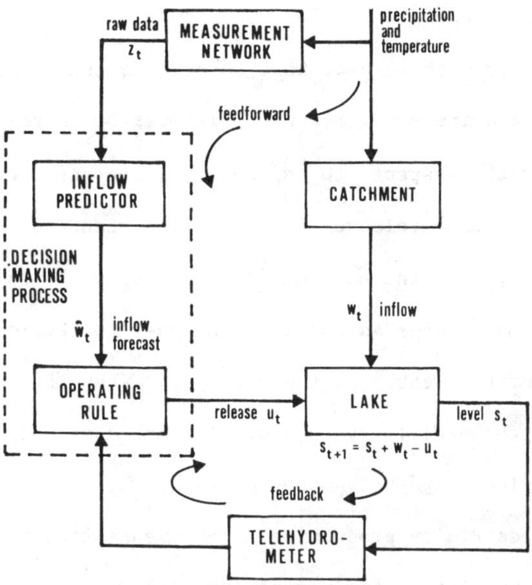

Figure 2 - Feedback and feedforward control scheme with inflow predictor.

precipitation and temperature at different points of the catchment (see Figure 1). By looking at the raw data z_t, $z_{t-1}, \ldots$, at the present and past measurements of the level s_t, $s_{t-1}, \ldots$, and the last values of the release u_{t-1}, $u_{t-2}, \ldots$, at the beginning of each day t the manager develops, on the basis of his experience, a forecast $\hat{w}_t$ of the inflow for the current day. Thus the overall decision-making process is decomposed into two parts. A first part is devoted to forecasting the inflows and a second part concentrates on the control aspects of the problem. Later, this control block will be formally described by an operating rule of the type

$$u_t = u(t, s_t, \hat{w}_t, F_t, D_t)$$

where F_t and D_t are the cumulative number of days of flood and water deficit from the beginning of the current year. (It is important to note that in the past the manager did not explicitly use an operating rule although his decisions are rather consistent in time and might be interpreted as the output of an implicit operating rule (Guariso et al., 1986).) The operating rule must be periodic with respect to t because of the yearly periodicity of inflows and water requirements. Moreover, it must be increasing (or at least non-decreasing) with respect to s_t and $\hat{w}_t$, because the manager will certainly increase the release from the lake when there is more water available. Finally, when the current flood indicator F_t is high the manager will release as much water as possible in order to avoid further floods during the rest of the year. On the contrary, when the current cumulative deficit D_t is high he will try to avoid further damage to the downstream users, thus releasing only that amount of water which is strictly needed for irrigation and electricity production. This means that the operating rule must be increasing with F_t and decreasing with D_t. All these structural properties of the decision-making process will be satisfied by the operating rules suggested by the min-max method described in the next section.

3. THE MIN-MAX APPROACH

Up to now, stochastic optimization methods have been proposed in the literature for reservoir management. On the contrary, actual operation of multipurpose reservoirs seems to prove that in most cases there is no great interest in optimizing the expected value of the objectives. Indeed, reservoir operation is very often aimed avoiding extreme and unacceptable failures of the objectives when the system is under stress. This is why reference is often made to very specific situations like "the driest (wettest) year of the century," "the highest recorded flood" and so on. In fact, it seems that managers, when selecting an operating rule, prefer to evaluate its performance by making reference to a particularly severe episode (or sequence of episodes) they have directly or indirectly experienced in the past. Consistently, in order to be safe, managers like to adopt that operating rule which best performs during that particular reference episode, even if this entails a reduction of the average performance of the system.

Lake Como in northern Italy is no exception. When the results of a detailed optimization study based on stochastic optimal control (Guariso et al., 1986) were presented to the manager, he recognized that he was not completely satisfied with the three objectives selected in that study (mean yearly agricultural deficit, average number of days of flood per year, and mean yearly hydroelectric production). Being risk-averse the manager showed a definite preference toward the possibility of avoiding failures of the system during severe and extreme hydrological episodes, like those he had experienced in the past.

In this section we briefly describe an approach which represents a first attempt to define operating rules which better account for this facet of the management attitude. As already stated, details on the theoretical aspects of the method can be found in Orlovski et al. (1983), Orlovski et al. (1984), and Guariso et al. (1984a).

3.1 Problem Formulation

The risk averse management problem is formulated as an optimal control problem, the solution of which guarantees a certain performance in terms of the objectives. The optimal operating rules u(.) will be selected by making explicit reference to particularly troublesome and specific situations formally defined by a set I of n one-year long daily inflow sequences, i.e., $I = \{ [w_t^i] \; ; \; t=0,\ldots,364; \; i=1,\ldots,n \}$. (In the following s_t^i will represent the level of the lake at the beginning of day t of year i, and w_t^i and u_t^i will be the inflow and the release on the same day). The reference set may contain recorded or synthetic sequences of inflows that the manager considers as particularly critical. For instance, in the case of the reservoir already being in operation, one might consider the sequences of the reference set to be those corresponding to the wettest and driest years experienced by the manager. In doing so, the proposed operating rules may also be compared with the performance the manager was able to achieve in practice.

Let us now indicate with D^i and F^i the water deficit in agriculture and the number of days of flood obtained by applying an operating rule u(.) during year i (i=1,...,n) of the reference set. Since the amount of water q_j required by the downstream users each day is a priori known, the value D^i is the sum over year i of the daily water deficits d_j^i given by $q_j-u_j^i$ if $u_j^i<q_j$, and zero otherwise. Consistently, F^i is the number of days of flood in year i (we consider day j as a flooding day if $s_j^i > s_c$, where s_c is the level at which there are the first detectable flood damages: in our case the level of the main square of Como).

Thus, the problem can be formulated as a two-objective optimal control problem in which the highest water deficit (max D^i) and the highest number of days of flood (max F^i) are minimized, that is

$$\min_{[S_o,u(.)]} \quad \max_{[1\leq i\leq n]} \{D^i\} \quad , \quad \max_{[1\leq i\leq n]} \{F^i\} \qquad (1)$$

where S_o is a set of initial storages with non-empty intersection with the control range. The constraints of the problem are:

$$s_{t+1}^i = s_t^i + w_t^i - u_t^i \qquad (2)$$

$$0 \leq u_t^i \leq N(s_t^i) \qquad (3)$$

$$s_t^i \geq \underline{s} \; ; \; u_t^i = N(s_t^i) \quad \text{if } s_t^i > \bar{s} \qquad (4)$$

$$s_{365}^i \in S_o \qquad (5)$$

with t=0,...,364; i=1,...,n.

Equation (2) is the continuity equation, and Equation (3) is the physical constraint expressing the fact that the release must be non-negative and smaller than or equal to the maximum amount of water which can be released in one day by keeping all the gates of the dam permanently open. This amount is approximately given by the so-called stage-discharge function $N(s^i_j)$. Equation (4) is the legal constraint which says that the level of the lake cannot be lower than a prescribed value $\underline{s}$ (-0.50m at the Fortilizio hydrometer) and that the manager must open all the gates of the dam to release as much water as possible when the level of the lake rises above a prescribed value $\bar{s}$ (1.20m at the Fortilizio hydrometer). Finally, Equation (5) is the terminal constraint which is imposed in order to avoid a good performance in one year being followed by a very poor performance in the next year.

3.2 Feasible Solutions

A feasible solution of problem (1-5) is a set S_o of initial storages and an operating rule u(.) statisfying constraints (2-5). In general, the operating rule u(.) will be a function of the information currently available, i.e.

$$u^i_j = u(j, s^i_j, w^i_j, D^i_j, F^i_j) \tag{6}$$

where D^i_j is the current water deficit in agriculture, and F^i_j is the current number of days of flood.

In order to solve problem (1-5) we will first analyse two simpler problems. The first one is called satisfaction of demand and consists of determining sets S^{D*}_o of initial storages, and operating rules of the form

$$u^i_j = u(j, s^i_j, w^i_j, D^i_j, D^*) \tag{7}$$

which are such that constraints (2-5) are satisfied and all yearly water deficits D^i are bounded by a given value D^*, i.e.

$$D^i < D^* \qquad i = 1,\ldots,n \tag{8}$$

The second problem is called flood protection and consists of finding sets S_o^{F*} of initial storages, and operating rules of the form

$$u_j^i = u(j, s_j^i, w_j^i, F_j^i, F^*) \tag{9}$$

which can guarantee that the number of days of flood at the end of all reference years will be at most equal to a given value F^*, i.e.

$$F^i < F^* \qquad i = 1,\ldots,n \tag{10}$$

The solutions of the two above problems will automatically point out operating rules of the more general form

$$u_j^i = u(j, s_j^i, w_j^i, D_j^i, F_j^i, D^*, F^*) \tag{11}$$

which can guarantee the satisfaction of both constraints (8) and (10) at the same time. In general, these operating rules are not unique. Thus, given the current value of information $(j, s_j^i, w_j^i, D_j^i, F_j^i)$, the solution algorithm suggests a whole range of possible releases u_j^i. This means that, under normal conditions, the manager still has a certain freedom in making the final decision. He might, for instance, take into account secondary objectives which were neglected in the formal description of the problem. However, we will see that when hydrological conditions become critical, i.e., when the reservoir is almost empty or full, this freedom might disappear.

3.2.1 Demand satisfaction. Let us now consider the problem of demand satisfaction, namely, the problem of determining a set of initial storages S_o^D and a set of operating rules of the form (7) which can guarantee the satisfaction of constraint (8). Obviously, solutions to this problem exist at least for sufficiently high values of D^*. One of these solutions is the so-called minimum release policy which corresponds to discharging during day j an amount of water which is as close as possible to the agricultural water demand q_j. Taking into account all physical and legal constraints, it is easy to check that such a policy is given by (see Figure 3)

$$u_{min}(j,s_j^i,w_j^i) = \begin{cases} \min\,[w_j^i, N(\underline{s})] & \text{if } s_j^i = \underline{s} \\ N(s_j^i) & \text{if } \underline{s} < s_j^i < S^{-1}(q_j) \\ q_j & \text{if } S^{-1}(q_j) \le s_j^i < \bar{s} \\ \min\,[\max(w_j^i,q_j), N(\bar{s})] & \text{if } s_j^i = \bar{s} \\ N(s_j^i) & \text{if } s_j^i > \bar{s} \end{cases} \qquad (12)$$

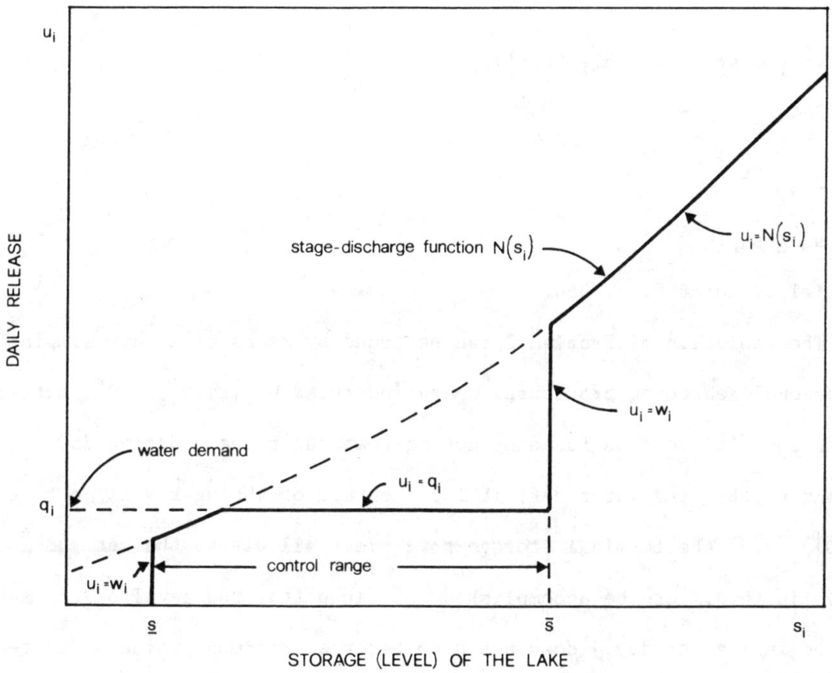

Figure 3 - The minimum release policy $u_{min}(j,s_j,w_j)$.

where $S^{-1}(.)$ denotes the inverse of the stage-discharge function. Notice that the minimum release policy does not fully exploit the information currently available since it does not depend upon D_j^i and D^*. The set $S_o^{D^*}$ corresponding to the operating rule (12) can easily be determined by recognizing that the yearly water deficit in agriculture obtained by applying a given operating rule is a non-increasing function of the initial storage s_o. Thus, the set $S_o^{D^*}$ will have the form

$$S_o^{D^*} = \{ s_o : s_o^D(D^*) \le s_o \} \qquad (13)$$

where the lower limit $s_o^D(D^*)$ is the solution of the following mathematical programming problem.

Problem 0

$$s_o^D(D^*) = \min s_o \tag{14}$$

subject to

$$\underline{s} \le s_o^i = s_o \le \bar{s} \tag{15}$$

$$s_{t+1}^i = s_t^i + w_t^i - u_{min}(t, s_t^i, w_t^i) \tag{16}$$

$$\sum_{t=0}^{364} d_t^i \le D^* \tag{17}$$

$$s_o \le s_{365}^i \tag{18}$$

with i=1,...,n; t=0,...,364.

The solution of Problem 0 can be found by means of a very simple one-dimensional searching procedure. Operating rules $u(j, s_j^i, w_j^i, D_j^i, D^*)$ satisfying constraint (8) can be found by noting that the release during day j must guarantee that the water deficit for the rest of the year will not exceed $(D^*-D_j^i)$ and the terminal storage s_{365}^i will fall within the set $S_o^{D^*}$. But this, in turn, can be accomplished provided that the level of the lake at the beginning of day j does not drop below a minimun value, denoted by $s_j^D(D^*, D_j^i)$, which can be computed by solving the following mathematical programming problem.

Problem j (j=1,...,364)

$$s_j^D(D^*, D_j^i) = \min s_j \tag{19}$$

subject to

$$s_j^i = s_j \ge \underline{s} \tag{20}$$

$$s_{t+1}^i = s_t^i + w_t^i - u_{min}(t, s_t^i, w_t^i) \tag{21}$$

$$\sum_{t=j}^{364} d_t^i \le (D^*-D_j^i) \tag{22}$$

$$s_o^D(D^*) \leq s_{365}^i \tag{23}$$

with i=1,...,n; t=j,...,364.

This problem can be solved by the same one-dimensional searching scheme used for Problem 0. Note, however, that the solution of Problem j requires the knowledge of the value $s_o^D(D^*)$ (see Equation (23)). Thus, Problem 0 must be solved first, while all other Problems j (j=1,...,364) are independent of each other.

Finally, one can notice that a volume of water greater than the current demand q_j can be released without any consequence on the management performance, provided the lake is sufficiently full and/or the inflow is sufficiently high. In fact, if

$$s_j^i + w_j^i - s_{j+1}^D(D^*,D_j^i) \geq q_j \tag{24}$$

any release u_j^i between q_j and $s_j^i+w_j^i-s_{j+1}^D(D^*,D_j^i)$ will leave the current value of the water deficit unchanged (in fact if $u_j^i \geq q_j$, then $D_{j+1}^i = D_j^i$). Moreover, if the release u_j^i is lower than or equal to

$$s_j^i + w_j^i - s_{j+1}^D(D^*,D_j^i) = s_j^i + w_j^i - s_{j+1}^D(D^*,D_{j+1}^i),$$

it will generate a storage

$$s_{j+1}^i = s_j^i + w_j^i - u_j^i \geq s_{j+1}^D(D^*,D_{j+1}^i)$$

which is indeed (by definition) the minimum value of s_{j+1}^i, which can guarantee the satisfaction of the demand (see Equations (19-23)). On the contrary, if the lake is so empty that a release $u_j^i \geq q_j$ is unfeasible, then an amount $u_{min}(j,s_j^i,w_j^i)$ of water is discharged and the current value of the deficit is updated. All this can be summarized (see also Figure 4) by saying that any release u_j^i such that

$$u_{min}(j,s_j^i,w_j^i) \leq u_j^i \leq \min \{ N(s_j^i), \max [s_j^i+w_j^i-s_{j+1}^D(D^*,D_j^i), u_{min}(j,s_j^i,w_j^i)] \} \tag{25}$$

will satisfy constraints (5) and (8). In other words, given the current information (j,s^i_j,w^i_j,D^i_j) and the required performance D^*, all operating rules which satisfy Equation (25) will guarantee that $s^i_{365} \in S^{D^*}_o$ and $D^i \leq D^*$ for all years i. For sufficiently high values of the storage s^i_j this implies the existence of a whole interval of feasible releases u^i_j. On the contrary, if the lake is too empty Equation (25) suggests a unique value for the release, namely $u_{min}(j,s^i_j,w^i_j)$.

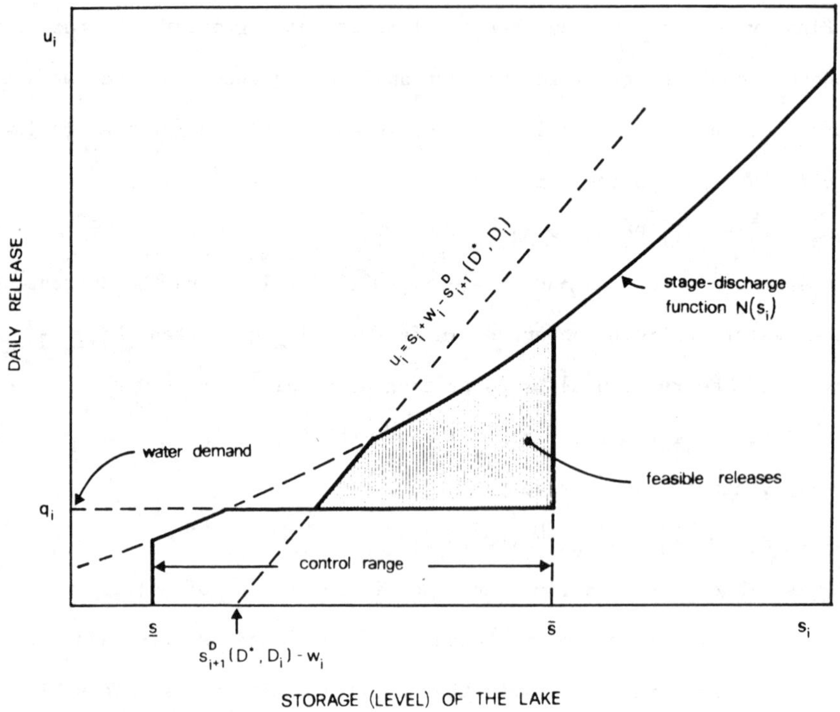

Figure 4 - The set of releases u_j that guarantee an agricultural deficit smaller than or equal to D^*.

3.2.2 Flood Protection. The problem of flood protection consists of finding a set $S^{F^*}_o$ of initial storages and a set of operating rules of the form (9) satisfying constraint (10) for a given value of F^*. In order to solve this problem we first introduce the maximum release policy

$$u_{max}(s_j^i,w_j^i) = \begin{cases} \min\ [w_j^i, N(\underline{s})] & \text{if } s_j^i = \underline{s} \\ \\ N(s_j^i) & \text{if } s_j^i > \underline{s} \end{cases} \tag{26}$$

which obviously minimizes the floods. Then, we notice that the number of days of flood F^i is a non-decreasing function of the initial storage s_o, so that the set $S_o^{F^*}$ is of the form

$$S_o^{F^*} = \{s_o : s_o \leq s_o^F(F^*)\} \tag{27}$$

where the upper limit $s_o^F(F^*)$ can be obtained by a mathematical programming problem similar to the above Problem 0. Then, the value $s_o^F(F^*)$ is used to formulate Problem j (j=1,...,364) which specifies the maximum storage at time j, called $s_j^F(F^*,F_j^i)$, for which operating rules exist of the form (9) which can guarantee no more than $(F^*-F_j^i)$ days of flood during the rest of the year and the satisfaction of the terminal constraint $s_{365}^i \in S_o^{F^*}$. These storages $S_j^F(F^*,F_j^i)$, j=1,...,364, allow a lower limit to the daily release from the reservoir to be defined. More precisely, one must note that the release u_j^i can be smaller than $u_{max}(s_j^i,w_j^i)$ provided the lake is sufficiently empty and/or the inflow is sufficiently low. In fact, if

$$s_j^i + w_j^i - s_{j+1}^F(F^*,F_j^i) \leq N(s_j^i) \tag{28}$$

and

$$s_{j+1}^F(F^*,F_j^i) < s_c \tag{29}$$

(recall that s_c is the threshold level defining the flood), then any release u_j^i between $s_j^i+w_j^i-s_{j+1}^F(F^*,F_j^i)$ and $N(s_j^i)$ will not give rise to a flood since $s_{j+1}^i \leq s_c$ (i.e., $F_{j+1}^i = F_j^i$). Moreover, the same release will generate a storage s^i_{j+1} smaller than or equal to $s^F_{j+1}(F^*,F^i_{j+1})$, which is indeed the maximum value of s^i_{j+1} that can guarantee the satisfaction of constraint (10).

Conversely, whenever inequality (28) is not satisfied, the release is set to the maximum feasible value $N(s^i_j)$ and the value of F^i_{j+1} is suitably updated. These observations can be summarized (see also Figure 5) by saying that any release u^i_j such that

$$\min \{ u_{max}(s^i_j,w^i_j),\ \max [s^i_j+w^i_j-s^F_{j+1}(F^*,F^i_j),0] \} \le u^i_j \le u_{max}(s^i_j,w^i_j) \qquad (30)$$

will satisfy constraints (5) and (10).

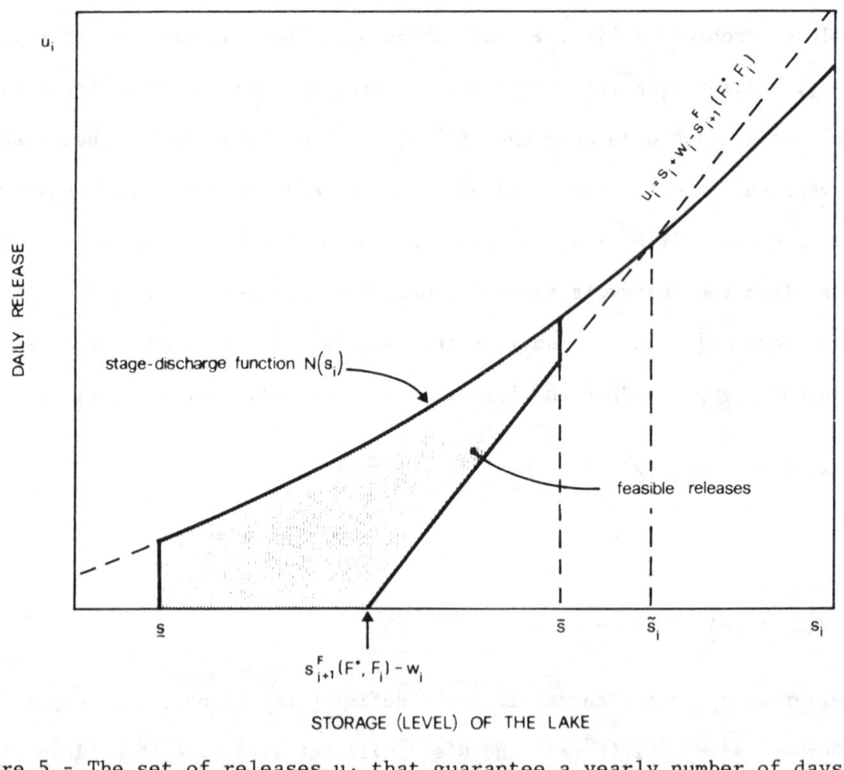

Figure 5 - The set of releases u_j that guarantee a yearly number of days of flood smaller than or equal to F^*.

For high values of the inflow w_j^i, the release $u_j^i = s_j^i+w_j^i-s_{j+1}^F(F^*,F_j^i)$ may equal the value of the stage-discharge function $N(s_j^i)$ at a level $\tilde{s}_j$ with $\tilde{s}_j < \bar{s}$. In such a case the manager would open all the gates of the dam even if he is not strictly obliged to do so by the license act (this has actually been done by the manager of Lake Como during the past few years).

3.2.3 Demand satisfaction and flood protection. Let us now consider the case in which the manager wants to guarantee specified values (D^*, F^*) of both the objectives (for example, D^* and F^* could be a percentage of the worst recorded values). If solutions to such problems exist, they will be constituted by the intersections of the sets of initial storages and operating rules which solve the problems of demand satisfaction and flood protection. Thus, the set of the initial storages is specified by (see Equations (13) and (27)):

$$s_o^D(D^*) \leq s_o \leq s_o^F(F^*) \tag{31}$$

while the release u_j^i is constrained by (see Equations (25) and (30)):

$$\begin{aligned} &\min \{ u_{max}(s_j^i,w_j^i), \max [s_j^i+w_j^i-s_{j+1}^F(F^*,F_j^i), 0] \} \\ &\qquad \leq u_j^i \leq \\ &\min \{ N(s_j^i), \max [s_j^i+w_j^i-s_{j+1}^D(D^*,D_j^i), u_{min}(j,s_j^i,w_j^i)] \} \end{aligned} \tag{32}$$

This means that very often there is the possibility of choosing the release in a relatively wide range (shaded area in Figure 6), but this freedom vanishes whenever the reservoir is too empty or too full, namely, when the achievement of one of the two targets (D^* or F^*) becomes critical.

3.3 Efficient Solutions

We can now point out a simple procedure for determining the solutions of the two-objective optimal control problem (1-5). For this we recall that the solution of a multiobjective problem (called efficient solution) is a feasible solution (S_o,u(.)) which is such that all other feasible solutions have at least one objective with a worse value. Thus, we must be able to determine those operating rules which can guarantee the minimum value of F^* (say F^o) for any given value of D^* (say D^o).

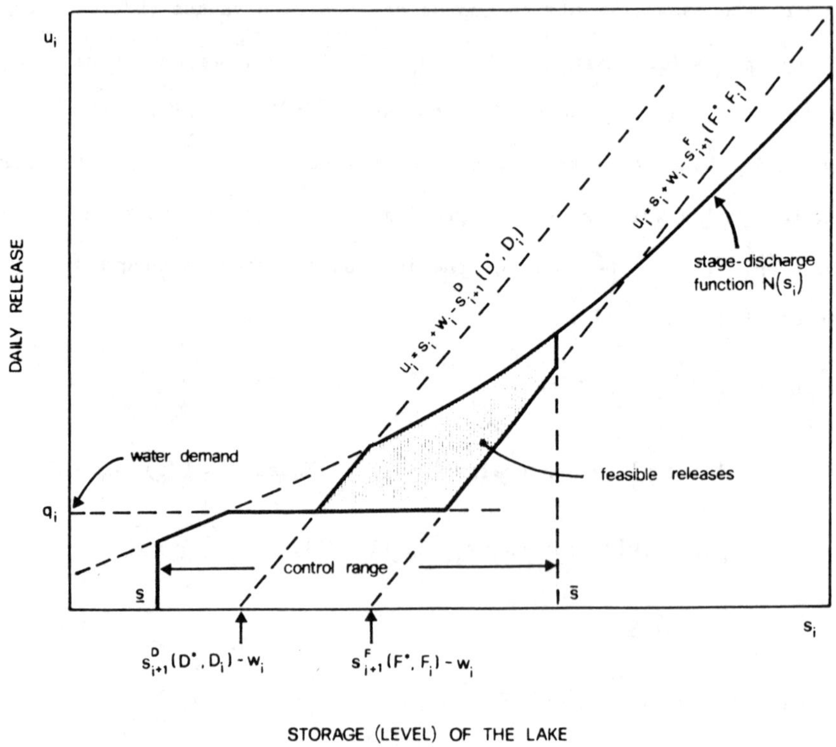

Figure 6 - Feasible releases proposed by the min-max approach for the two objective management problem.

For this, let us suppose that a value D^o of water deficit in agriculture is fixed. Thus, the set of initial conditions and the set of operating rules which can guarantee the satisfaction of the terminal constraint $s^i_{365} \in S^{D_o}_o$ and of the target $D^i \leq D^o$ are given by Equations (13) and (25) with $D^* = D^o$. The lowest storage within the set $S_o^{D^o}$ and the highest release satisfying Equation (25) obviously minimize the number of days of flood. Therefore, one can simulate the behaviour of the lake with initial storage $s_o^D(D^o)$ and operating rule

$$u_j^i = \min \{ N(s_j^i), \max [s_j^i+w_j^i-s_{j+1}^D(D^o,D_j^i), u_{min}(j,s_j^i,w_j^i)] \}$$

for all the inflow sequences of the reference set. Thus, a certain number of days of flood F^i is obtained for each sequence i and obviously $F^o = \max F^i$. The efficient pairs (D^o,F^o) could also be found by fixing the value F^o and searching for D^o, which simply implies simulation of the behaviour of the lake starting from $s_o^F(F^o)$ with the operating rule given by the lower bound of Equation (30) with $F^* = F^o$.

Once an efficient pair (D^o, F^o) of the objectives has been found, the efficient operating rule and the set of initial storages are simply determined by substituting D^o and F^o for D^* and F^* in Equations (31) and (32).

4. APPLICATION OF THE METHOD TO LAKE COMO

The method described in the preceding section has been applied to the management problem of Lake Como for which reliable data were available. The application consisted of three parts:

1. Solution of the mathematical programming problems (Problem 0 and Problem j) described above and determination of the efficient operating rules;
2. Development of suitable one-day ahead inflow predictors;
3. Simulation over a relatively long period of time of the performance of the system with a particular efficient operating rule and with the proposed predictor and comparison of the results of the simulation with the historical data.

These three parts of the study are now described in order.

4.1 Efficient operating rules

The seven most critical inflow sequences of the period (1946-1981) were chosen to define the reference set I. Among them we have 1976 and 1977, which are the driest and wettest years of the last decades. Historically, the highest number of days of flood was 42 in 1977, and the highest deficit was 850 million cubic meters in 1976. Problem 0 (see Eqs. (14-18)) was solved for different values of D^* by simulating the system behaviour for different values of the initial storage s_o selected by a one-dimensional search. It turned out that no solution existed for D^* smaller than 600 million cubic meters and that $s_o^D(600) = \underline{s}$ (obviously $s_o^D(D^*) = \underline{s}$ for all $D^* \geq 600$), which means that constraint (23) in Problem j is a priori satisfied. In other words, in the case of Lake Como, $s_j^D(D^*,D_j^i)$ is only a function of $(D^*-D_j^i)$. This peculiar characteristic allows determination of $s_j^D(D^*,D_j^i)$ in the following very simple way: For each initial storage s_j in the control range, simulate over the rest of the year the behaviour of the lake with the operating rule $u_{min}(j,s_j^i,w_j^i)$ for each inflow sequence i of the reference set, and store, as shown in the table of Figure 7, the maximum deficit obtained in this way. This value corresponds to the smallest deficit $(D^*-D_j^i)$ which can be guaranteed from that day on. By inverting the table of Figure 7, one can compute the function $s_j^D(D^*,D_j^i) = s_j^D(D^*-D_j^i)$, which in this case is shaped as in Figure 8.

In a very similar way one can determine the functions $s_j^F(F^*,F_j^i)$ needed to solve the flood protection problem (see Section 3.2.2). In fact Problem 0 showed that it is not possible to guarantee less than 18 days of flood per year and that $s_o^F(18) = \bar{s}$, which implies $s_o^F(F^*) \geq \bar{s}$ and $[\underline{s},\bar{s}] \in S_o^{F^*}$ for any $F^* \geq 18$. Since no flood has ever occured after November 15, and no flood has lasted more than ten days, the storage at the end of any year always falls within the control range and thus the terminal constraint is a priori satisfied also for the flood protection problem and $s_j^F(F^*,F_j^i) = s_j^F(F^*-F_j^i)$. One can thus simulate the behaviour of the system with initial level s_j and

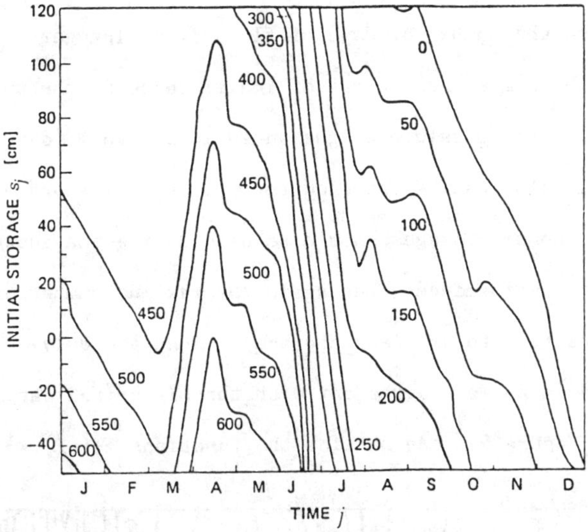

Figure 7 - The maximum agricultural deficit obtained by simulating the system behaviour in the period from j to the end of the year.

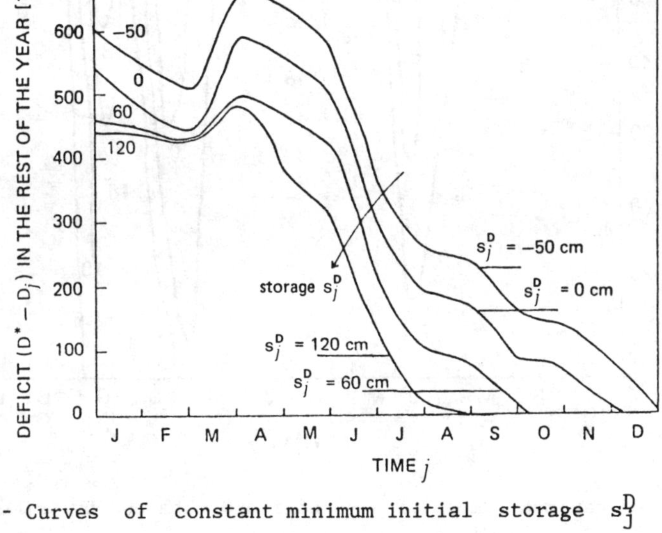

Figure 8 - Curves of constant minimum initial storage s_j^D needed to guarantee a prescribed deficit (D^*-D_j) from day j to the end of the year.

maximum release policy $u_{max}(s_j^i, w_j^i)$ for each inflow sequence of the reference set, and determine the maximum number of days of flood over the rest of the year as shown in the table of Figure 9. This table represents a step-wise function, since the number of days of flood is an integer. It appears from Figure 9 that there are large areas of indifference in the space (j, s_j). For example, it is not possible to guarantee less than 18 days of flood per year, but this value can be obtained for all initial storages s_j in January and February. Lower storages should be used during the snow-melt season to achieve the same performance, but again the maximum number of days of flood is rather insensitive to the lake storage in June and July. These seasonal variations are in perfect agreement with the historical data. By inverting the table of Figure 9, one obtains the functions $s_j^F(F^*.F_j^i) = s_j^F(F^*-F_j^i)$

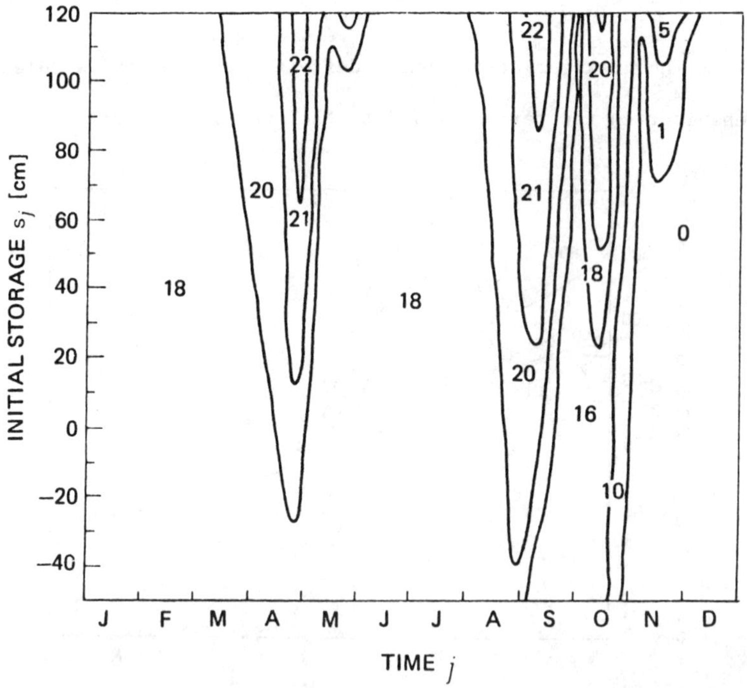

Figure 9 - The maximum number of days of flood obtained by simulating the system behaviour in the period from j to the end of the year.

which are shown in Figure 10.

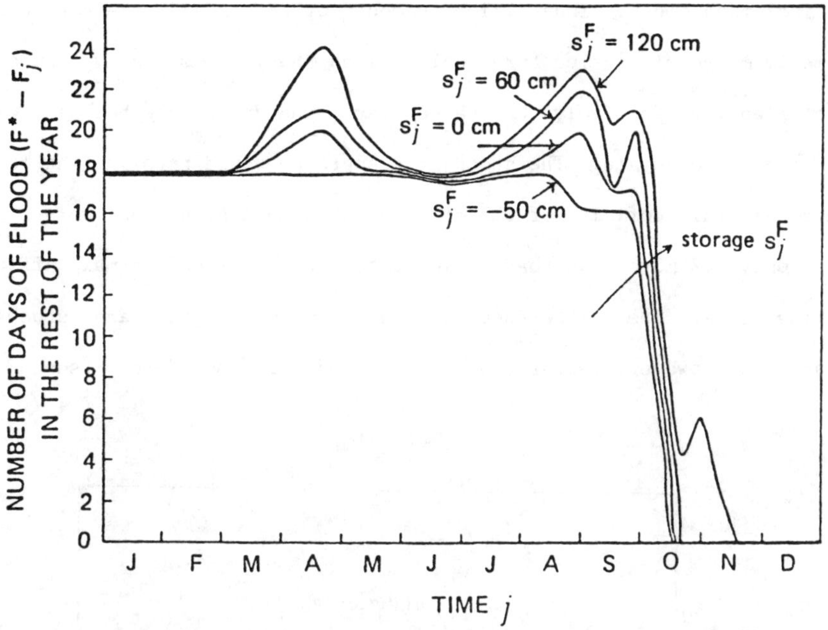

Figure 10 - Curves of constant minimum initial storage s_j^F needed to guarantee a prescribed deficit (F^*-F_j) from day j to the end of the year.

Figs. 8 and 10 contain all the information necessary to determine the efficient solutions (D^o, F^o) of the risk-averse management problem (1-5), provided the procedure outlined in Section 3.3 is used. The set of these efficient solutions is shown in Figure 11 in the space of the objectives. In the same figure one can find the performance of the historical management (point H, which corresponds to an agricultural deficit of 850 million cubic meters and to 42 days of flood) and the "utopia" point U, which represents the independent, and hence unfeasible, optimum of the two objectives (600 million cubic meters of deficit and 18 days of flood). Finally, point P represents the performance of the operating rule by means of the classical stochastic approach, which is discussed in Guariso et al. (1986). Among all efficient solutions, the one closest to segment HU (see point X) has been suggested to the manager. This solution is clearly superior to the historical management. Indeed, improvements of 27% and 52% are possible for the maximum yearly agricultural deficit and the maximum yearly number of days of flood, respectively. On the contrary, the difference between point X and P is only moderate. The stochastic approach would have in fact caused an agricultural deficit of 623 million cubic meters in the worst case, namely only 1% more than that achieved by the suggested min-max efficient operating rule. The difference between the deterministic and stochastic approach is however greater (20%) when floods are considered (see Figure 11).

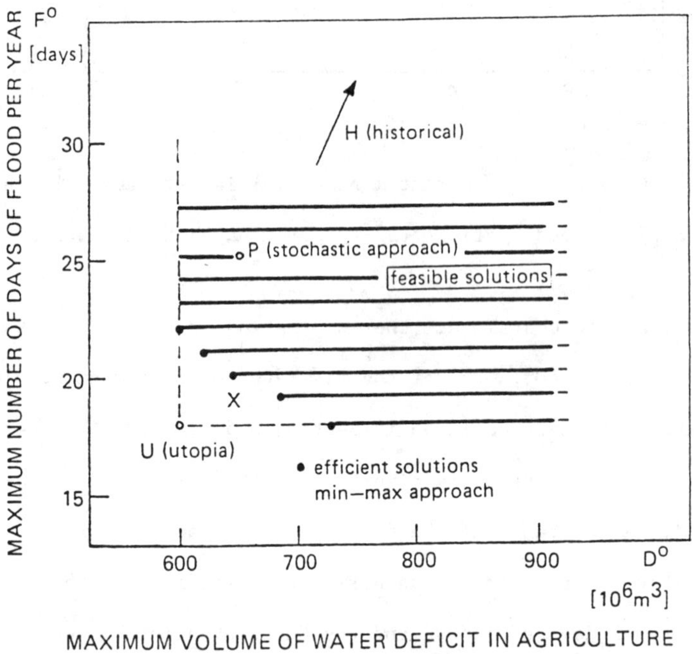

Figure 11 - Feasible solutions (horizontal lines) and efficient solutions of the min-max approach: point X is the proposed solution and point P is the solution of the stochastic approach.

Of course the results obtained can be guaranteed with reference to a very particular situation, namely the case in which the inflows are perfectly forecasted and are exactly equal to those of the reference set. For this reason it was mandatory to evaluate the system performance in different years and with real predictors.

4.2 Inflow Predictors

Several schemes have been developed for the one-day ahead forecast of the inflows of Lake Como. All of them are linear autoregressive schemes with exogenous inputs that relate the forecasted inflow $\hat{w}_t$ of day t to the computed values of the inflow on the preceding day ($w_{t-1} = s_t - s_{t-1} + u_{t-1}$) and to other hydrological variables z_{t-1}, $z_{t-2}, \ldots$ They differ from each other by the type of information they use and by the method employed to estimate the parameters. It turns out that a single predictor (i.e., a single autoregressive equation) cannot lead to satisfactory results in all hydrological conditions. Thus, the predictor that was finally used is constituted by a set of simple predictors which are used alternatively depending upon the hydrological conditions. Each day t, if the inflow-rate of the preceding day t-1 is smaller than the mean plus standard deviation of the inflow, the forecast is simply based on the recorded inflows of the past two days, i.e.,

$$\hat{w}_t = p_t^1 \, w_{t-1} + p_t^2 \, w_{t-2}$$

where the parameters p_t^1 and p_t^2 vary periodically during the year and are estimated by an off-line least square procedure. On the contrary, when the inflow of day t-1 is high, the forecast is also based on the total rainfall

y'_{t-1} on the catchment in the preceding day and on the water content y''_{t-1} of the upper layer of the ground. (These two terms were estimated on the basis of the measurement network shown in Figure 1). Thus, the structure of the predictor, is, in this case, the following:

$$\hat{w}_t = p_t^1 \, w_{t-1} + p_t^2 \, w_{t-2} + p_t^3 \, y'_{t-1} + p_t^4 \, y''_{t-1}$$

Also in this case the parameters p_t^i were estimated off-line and are given different values depending upon the trend of the inflow in the two hours before computing the forecast.

The performance of this predictor is quite satisfactory. The noise to signal ratio is 0.32 and the correlation between real and forecasted values is 0.95. If only the periods of fast inflow variations are considered (beginning of the floods), the above-mentioned ratio remains around 0.4 and the correlation is still 0.91. However, as is common to all types of autoregressive predictors, peak values are often underestimated. For instance, the highest recorded flood of October, 1976 is underestimated by about 35%.

4.3 Simulation

In order to evaluate the reliability of the proposed method, the system has been simulated for the 15-year period 1965-1979. This period is characterized by a great diversity of hydrological episodes (severe droughts in 1965 and 1976, and the biggest floods of the last decades in 1976 and 1977) and is therefore well-suited for testing the method. The simulation proceeds as follows: At the beginning of each day t a forecast w_t of the inflow for the current day is computed, and then used together with the measurement of the level s_t to determine the set $[\underline{u}_t^o, \bar{u}_t^o]$ of possible releases by means of the efficient operating rule (32). If this set is a single value, say u_t^o, then the level of the lake at the beginning of the next day is determined by the continuity equation in which the real inflow is used, i.e.

$$s_{t+1} = s_t + w_t - u_t^o.$$

If, on the contrary, the set of possible releases $[\underline{u}_t^o, \bar{u}_t^o]$ is a proper set, a selecting rule must be used first in order to specify the release u_t^o. The selecting rule can be given the form

$$u_t^o = (1-\lambda_t)\, \underline{u}_t^o + \lambda_t\, \bar{u}_t^o \qquad 0 \leq \lambda_t \leq 1$$

where the parameter λ_t is the relative weight given to the two objectives. If $\lambda_t = 1$, the release is the greatest possible release ($\bar{u}_t^o$): in other words, the manager uses all the freedom he has to alleviate future floods. On the contrary, if $\lambda_t = 0$, the manager will act in favour of agriculture. Curve (a) in Figure 12 shows the results obtained with time-invariant selecting rules (λ_t = constant over all the year). The indicators D and F on the axes are the mean values of the yearly deficit and of the yearly number of days of flood (4 days), while the point at $\lambda = 0$ has the minimum deficit (140 10^6 m^3). The historical values of the same indicators (see point H of Figure 12) are 10.2 days and 200 million cubic meters respectively. Thus, the improvements obtained with the min-max method are rather important.

Of course, one would expect, in general, better results when using more sophisticated inflow predictors. Nevertheless, the improvements one can obtain with better predictors are only marginal in this case. Indeed, if a perfect predictor ($\hat{w}_t = w_t$) were available, curve (b) in Figure 12 would be obtained instead of curve (a). Actually, these two curves intercept each other for $\lambda = 0.8$, and the results obtained with the real predictors are

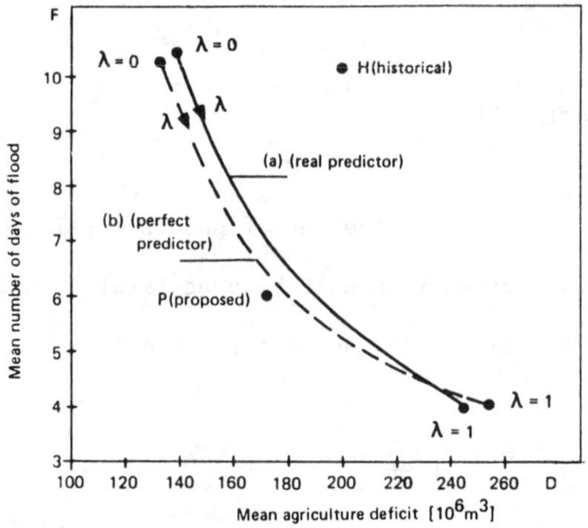

Figure 12 - Mean values of the objectives during the period 1965-1979 for different selecting rules.

slightly better than those obtained with the perfect predictor when the selecting rules are strongly in favour of flood protection. This rather surprising feature can be explained as follows. Predictors are low pass filters and, therefore, underestimate the sudden variations of the inflows during flood episodes. Thus, during these periods the release is smaller than the one that would be computed with a perfect predictor, the difference being a surplus of storage that can be used later to satisfy the demand of the downstream users. The fact that the performance of the system is rather insensitive to the reliability of the forecast is certainly a peculiar feature of Lake Como and not of the min-max method. In fact, the same qualitative conclusions were reached (Guariso et al., 1984b) when the possibility of improving, with an inflow predictor, the performance of another operating rule was investigated. The reason for this is that, in the case of Lake Como, the action of the manager is rather limited during the floods. In fact, as soon as the level of the lake is above the upper limit $\bar{s}$ of the control range, the manager has no option and must open the gates of the dam completely. Thus, in these periods, errors in the forecast do not influence the final decision, so that the overall performance of the system becomes rather insensitive to the forecasting techniques.

On the contrary, a detectable improvement can be obtained by using time-varying selecting rules λ_t which take into account the typical seasonal variations of the inflows. In Figure 12 point P represents the performance of one of these selecting rules used in conjunction with the real predictor. The relative weight λ_t is simply adapted (see Figure 13) to the seasonal variations of the mean plus standard deviation $(\mu_t + \sigma_t)$ of the inflows (Rinaldi, 1983). Thus, the manager releases as much as possible $(\lambda_t = 1)$ during the flooding seasons (May, late August and September) and is more concerned with the downstream users during the dry seasons (winter and July). The comparison between the historical data (point H in Figure 12) and

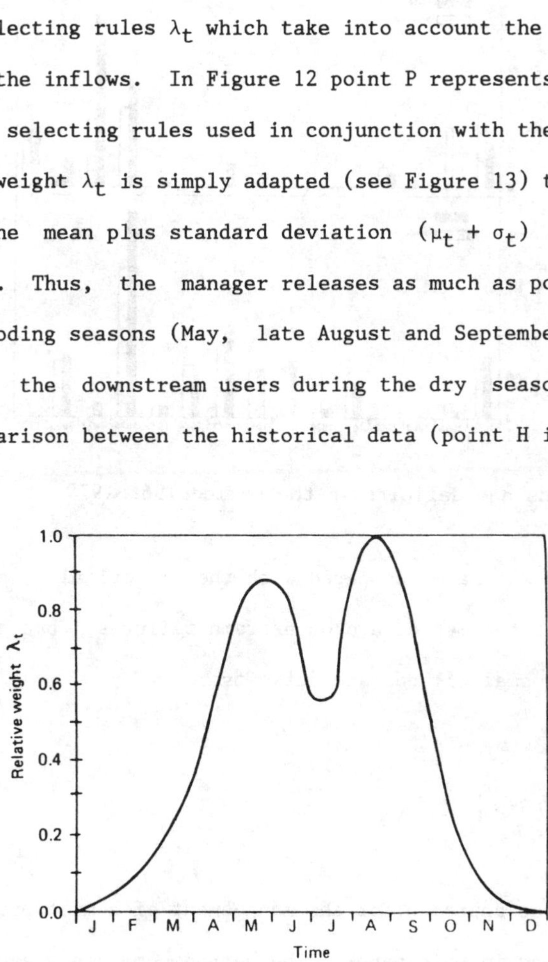

Figure 13 - The proposed time varying selecting rule λ_t (see point P in Figure 12).

the proposed operating rule (point P in Figure 12) is very satisfactory. On the average, floods are reduced by 40% and deficits are reduced by 14%. Even greater advantages are obtained when reference is made to the most critical years of the period (1977 for the floods and 1976 for the deficits) as shown in Figure 14, where floods and deficits obtained in every year with the

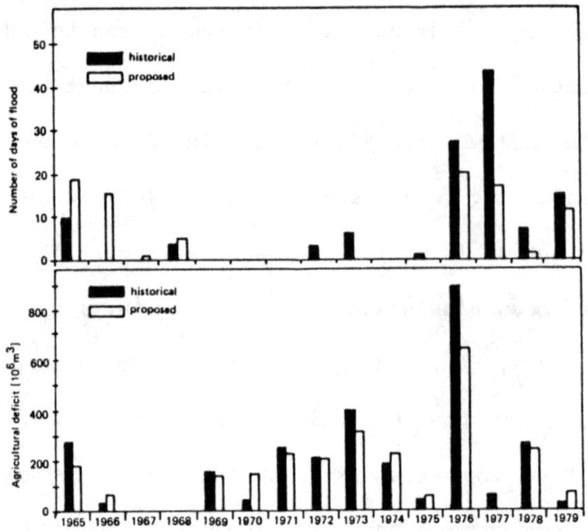

Figure 14 - Floods and deficits in the period 1965-1979.

risk-averse approach are compared with the historical ones. The figure clearly shows that the method avoids extreme failures, but does not pay so much attention to small floods and deficits.

5. CONCLUDING REMARKS

A risk-averse approach for the management of a multipurpose reservoir has been described in this paper. The method completely avoids statistical considerations on the inflows and minimizes the worst possible failures of the system when the inflows are equal to prespecified crucial values. The efficient operating rules can be obtained by solving off-line a series of simple mathematical programming problems. The use of the proposed method requires an inflow predictor to be developed which must be used in real-time in conjunction with the operating rule. An important feature of the method is that most of the time an interval of possible releases is suggested instead of a single value. The analysis of the Lake Como system (northern Italy) has shown that relevant advantages can be obtained with this method. Moreover, at least in this particular case, the overall performance of the proposed solution is not very sensitive to the reliability of the inflow predictor.

REFERENCES

Cohon, J. and D. Marks (1975) A review and evaluation of multiobjective programming techniques, Water Resour. Res., Vol. 11, pp.208-219.

Guariso, G., S. Orlovski, and S. Rinaldi (1984a) A risk-averse approach for reservoir management, Prep. IX IFAC World Congress, Budapest, Hungary, Vol. 4, pp. 183-188.

Guariso, G., S. Rinaldi, and P. Zielinski (1984b) The value of information in reservoir management, Appl. Math. and Comput., Vol. 15, pp. 165-184.

Guariso, G., S. Rinaldi, and R. Soncini-Sessa (1986) The management of Lake Como, Water Resour. Res., Vol. 22, pp. 109-120.

Helweg, O.J., R. W. Hinks, and D. T. Ford (1982) Reservoir systems optimization, J. Water Resour. Plan. Manag. Div. Am. Soc. Civ. Eng., Vol. 108, pp. 169-172.

Orlovski, S., S. Rinaldi, and R. Soncini-Sessa (1983) A min-max approach to storage control problems, Appl. Math. Comput., Vol. 12, pp. 237-254.

Orlovski, S., S. Rinaldi, and R. Soncini-Sessa (1984) A min-max approach to reservoir management, Water Resour. Res., Vol. 11, pp.1506-1514.

Rinaldi, S. (1983) Remarks on periodic stochastic linear reservoirs, Water Resour. Res., Vol. 3, pp. 855-859.

FORMAL INCORPORATION OF RISK INTO CONFLICT ANALYSIS

Keith W. Hipel
Department of Systems Design Engineering
University of Waterloo
Waterloo, Ontario, Canada N2L 3G1

Niall M. Fraser
Department of Management Sciences
University of Waterloo
Waterloo, Ontario, Canada N2L 3G1

ABSTRACT

The methodology of conflict analysis is presented in a manner which allows risk to be considered when ascertaining the politically and socially feasible solutions to a given large scale engineering project. In particular, conflict analysis can be employed to model a given conflict or game which may be concerned with risk and to predict the possible compromise solutions. Risk can be incorporated into the analysis since it may affect the preference structure of a specific participant for the possible feasible outcomes in the dispute. To demonstrate how the methodology works in practice, it is applied to the Garrison Diversion Unit conflict which is a serious environmental dispute involving American and Canadian interests.

1. INTRODUCTION

Most large-scale engineering projects involve some degree of risk. For example, within this paper, an enormous water resources undertaking is described where the risk of catastrophic environmental damage is very high. The Garrison Diversion Unit (GDU) is a mammoth irrigation scheme within the state of North Dakota in the United States (US); when the project comes into operation, the runoff from the irrigated fields could seriously pollute waters flowing into Canada. Because of this, the GDU has evolved into an international dispute between Canada and the US, and techniques are required for modelling the socio-political implications of this risky project. Consequently, the purpose of this paper is to show how the implications of risk can be considered when determining the politically and socially feasible solutions to a water resources project such as the GDU.

A situation where two or more groups are in dispute over some issue is called a conflict or game. Conflict analysis refers to a unique solution concept from non-cooperative game theory which is concerned with studying real world controversies (Fraser and Hipel, 1979a, 1984). By utilizing the GDU dispute, the technique of application of the conflict analysis method is clearly explained. Furthermore, it is shown how risk affects the preference structure of the participants or players among the possible feasible outcomes in the conflict. In this way, risk is indirectly incorporated into the socio-political considerations of a given conflict or game.

The concept of risk constitutes a contentious topic which has been addressed by many authors from a wide variety of disciplines. For example, to obtain an extensive survey of risk and hazard assessment one may wish to refer to the paper by Sage and White (1980) and the book edited by Lind (1982). Within this paper, an exact definition of risk is not required, because a particular participant or player involved in a dispute can define risk in a manner which is most relevant for himself. However, what is important is how a given player's interpretation of risk affects his ordinal preference structure among the feasible outcomes in the game. Generally, a risky outcome is one which could have adverse consequences upon one or more of the players in the conflict. Consequently, if a player considers a specific outcome to be risky for himself, this outcome will be usually less preferred than outcomes which involve less risk. Furthermore, outcomes which are risky for one player can be used for sanctioning purposes by the other players in order to stop the given player from improving his position.

In the next section, previous research in conflict analysis is briefly described and the advantages of employing conflict analysis are outlined. Following a description of the history of the GDU, conflict analysis is employed to model the GDU dispute and forecast the compromise resolutions. The manner in which risk can be formally considered in a conflict analysis study is thoroughly explained using the GDU application.

2. CONFLICT ANALYSIS

The systematic study of a conflict such as the GDU dispute involves two main steps. First, the conflict is modelled by identifying the players, the options or courses of action available to each player, and each player's preferences among the possible feasible outcomes. To model the conflict, it is necessary to possess a thorough understanding of the dispute by referring to the available published literature, and, if possible, contacting people who were directly or indirectly involved. Accordingly, the history of the GDU conflict is outlined in the paper, before the actual modelling of the conflict.

The second major stage in the procedure is to perform a stability analysis of the feasible outcomes in the game in order to predict the possible resolutions or equilibria. Thus, based upon the solution concept of conflict analysis, a stability analysis is executed to predict the possible resolutions to the GDU conflict. The output from a stability analysis, such as the one for the GDU, can be used as a basis for decision making by one or more of the players in the game.

The conflict analysis method developed by Fraser and Hipel (1979a, 1984) is a reformulation and extension of metagame analysis (Howard, 1971) which in turn possesses some linkages to classical game theory (Von Neumann and Morgenstern, 1953). Kilgour et al (1984) show the exact mathematical relationships among important game theory methods and clearly point out the many inherent theoretical advantages of conflict analysis. Although metagame analysis (Howard, 1971) may work reasonably for some applications (Hipel et al, 1974, 1976; Hipel and Fraser, 1980; Fraser and Hipel, 1980b), theoretical improvements to metagame analysis as well as developing flexible algorithms for applying the improved method in practice were required to create the comprehensive approach of conflict analysis (Fraser and Hipel, 1979a; 1984). To establish the fact that conflict analysis is a general systems science technique which is not dependent upon the particular area of application, the technique has been applied to a wide range of conflicts which include business (Stokes and Hipel, 1983; Stokes et al, 1985), labour-management (Fraser and Hipel, 1980d, 1980e, 1981), energy (Savich et al, 1983), water resource (Dufournaud, 1982; Hipel, 1981), environmental (Hipel and Fraser, 1982), military (Hipel and Fraser, 1986a; Fraser et al, 1983; Fraser and Hipel, 1983; Fang et al, 1986), political (Hipel and Fraser, 1986b; Meleskie et al, 1982), and peace treaty (Kuhn et al, 1983) disputes. To provide a more comprehensive framework for modelling conflicts and comparing and developing solution concepts, Kilgour et al (1986) have devised a flexible graph model for use in conflict studies. Fang et al (1986) have explained how conflict analysis, as well as other game theory methods, can be used as a decision support system.

In a conflict or game, two or more participants are in dispute over some resource or issue. The chosen participants in a conflict model are called players, and are considered to be entities that have actions that they can take which have significance to other players in the conflict. The possible actions of the players are called options. A set of options that can be taken by a player is called a strategy, and the stituation where each player selects a strategy is called an outcome. Each player has his own preference ordering of all the outcomes in the conflict. A game or conflict model consists of the players, their options and their preferences among the possible feasible outcomes. The analysis of the game is carried out by determining the stability of each outcome for every player. If an outcome is stable for a given player, it does not benefit the player to move unilaterally to any other outcome by changing his own strategy. An outcome that is stable for all players in the game model is an equilibrium and constitutes a possible resolution to the conflict. These definitions are clarified in the GDU application.

There are many distinct advantages for utilizing the approach of conflict analysis in practical applications (Fraser and Hipel, 1984). First, documented applications for both current and historical disputes confirm that conflict analysis produces realistic and useful results. For example, the international business conflict studied by Stokes and Hipel (1983) and Stokes et al (1985) was in fact analyzed while the conflict was in progress and the conflict analysis correctly predicted what eventually happened. Secondly, the method can easily be applied to any conflict with any finite number of players and options. A third attribute is that even though the technique is firmly founded upon mathematical concepts from set theory and logic, it is not necessary that a practitioner understand the underlying theory in order to apply the method to an actual dispute. Fourth, for each player, at most it is only necessary that ordinal preference information be available so that the feasible outcomes can be ranked from most to least preferred. If equally preferred outcomes are present or there are intransitive preferences where a player prefers outcome x to y, y to z but z to x, these kinds of preference information can also be handled. Fifth, to model the dynamics of a conflict as it evolves over time, the state transition model, which is based upon the static conflict analysis model, can be employed (Fraser and Hipel, 1983). A sixth advantage of conflict analysis is that in games where there are at least three players, the technique can be used for modelling coalitions where groups of two or more players may elect to join forces and thereby form coalitions (Kuhn et al, 1983). Seventh, although all calculations can be done easily by hand, a microcomputer can be utilized for conveniently performing extensive analyses within a decision support system framework (Fraser and Hipel, 1979b, 1980a, 1980c, 1980e, 1981; Fang et al, 1986).

An additional advantage of using conflict analysis is that it can be used to analyze hypergames. In a hypergame, one or more of the players may possess a mistaken view of the actual conflict where misunderstandings can be built upon others' misperceptions in order to form different levels of perception (Bennett, 1977, 1980; Takahashi et al, 1984). The players in a hypergame may have a false understanding of the preferences of the other players, have an incorrect comprehension of the options available to the other players, not be aware of all the players in the game, or have any combination of the foregoing faulty interpretations.

Because the conflict analysis algorithm of Fraser and Hipel (1979a, 1984) analyzes a conflict according to the manner in which each player perceives reality, the method can be used to analyze any type of hypergame with any finite number of players and options (Takahashi et al, 1984). The efficacy of conflict analysis for studying hypergames has been demonstrated by applications to water allocation (Okada et al, 1984, 1985), international trade (Stokes and Hipel, 1983), bargaining and negotiation (Fraser and Hipel, 1981) and military (Shupe et al, 1980; Wright et al; 1980, Takahashi et al, 1984; Fraser and Hipel, 1984) conflicts. Due to the fact that each player in a game involving risk may interpret risk in a different manner, hypergame analysis may be useful for studying certain games of risk.

Conflict analysis can be considered to be a comprehensive decision making methodology which falls within the realm of the systems science or operational research set of tools. In fact, conflict analysis can model decision making problems for which there are multiple decision makers and each decision maker can have multiple objectives. Because at most, only ordinal preference information is required, conflict analysis constitutes a nonquantitative procedure for modelling the social interactions among the decision makers and predicting the compromise resolutions. Hipel (1981) has explained how conflict analysis can be employed in conjunction with other operational research techniques for solving problems related to large scale engineering projects such as a system of water resources reservoirs and electrical generating plants.

Okada et al (1984) demonstrate how a hierarchical gaming approach can be employed for modelling decision making in water resources disputes. The efficacy of this hierarchical procedure is confirmed by applying it to the Lake Biwa dispute in Japan. This dispute is representative of a typical water allocation problem where the downstream users would like the upstream controllers of a lake to release more water for consumption downstream. To furnish an overall framework for hierarchical decision making, the strategic/tactical model can be utilized (Giesen and Radford, 1983). By employing appropriate operational research techniques, actual decision making situations can be realistically modelled at both the strategic and tactical levels. Because the most information one can often hope to obtain at the strategic level are the ordinal preferences or utilities for each player, conflict analysis (Fraser and Hipel, 1979a, 1984) is particularly useful for modelling the higher level decision making. For the case of the Lake Biwa dispute, the misunderstandings among the players are structured using a hypergame model and then conflict analysis is employed as a solution concept for predicting the compromise political solutions to the problem (Okada et al, 1984, 1985). At the tactical level where more precise and quantitative decisions are made, often for the purpose of implementing the overall political solution decided upon at the strategic level, appropriate quantitative operational research methods can be used. For example, nonlinear programming is used for modelling the tactical decision making in the Lake Biwa dispute. Okada et al (1984) maintain that a hierarchical gaming approach to decision making should prove to be useful not only in water allocation conflicts as well as other kinds of water resources disputes, but in virtually any field where complex socio-economic and political issues must be properly considered in the decision making process. Many of these conflicts may involve risk which can be appropriately considered at the strategic level using conflict analysis.

The conflict analysis of the GDU dispute models the decision making carried out at the strategic level in the presence of risk. After the history of the GDU is presented in the next section, the GDU dispute is analyzed using conflict analysis.

3. THE GARRISON DIVERSION UNIT

The GDU is a partially constructed, multi-purpose water resources project involving the transfer of water from the Missouri River basin to areas in central and eastern North Dakota that are mainly located within the Hudson Bay drainage basin. Figure 1 illustrates the major regions affected by the project. If the system becomes fully operational, water will be pumped from Lake Sakakawea on the Missouri River along the McClusky Canal to the Lonetree Reservoir which lies within the Hudson Bay drainage basin. From the Lonetree, water will proceed to the major irrigation areas via the Velva and New Rockford Canals. The reservoir will also augment the flow in the James River for downstream irrigation. Runoff from the irrigated fields will make its way to either the Souris or Red Rivers which flow across the international border into the Canadian province of Manitoba.

The GDU has a long history, dating back beyond the turn of the century, when the first written request to Congress for construction of a Missouri diversion was made. Numerous other plans similar to the one in Figure 1 were proposed and then rejected (Environmental Impact Assessment Project (EIAP), 1975). The Garrison Dam on the Missouri was completed by the Bureau of Reclamation of the US Department of the

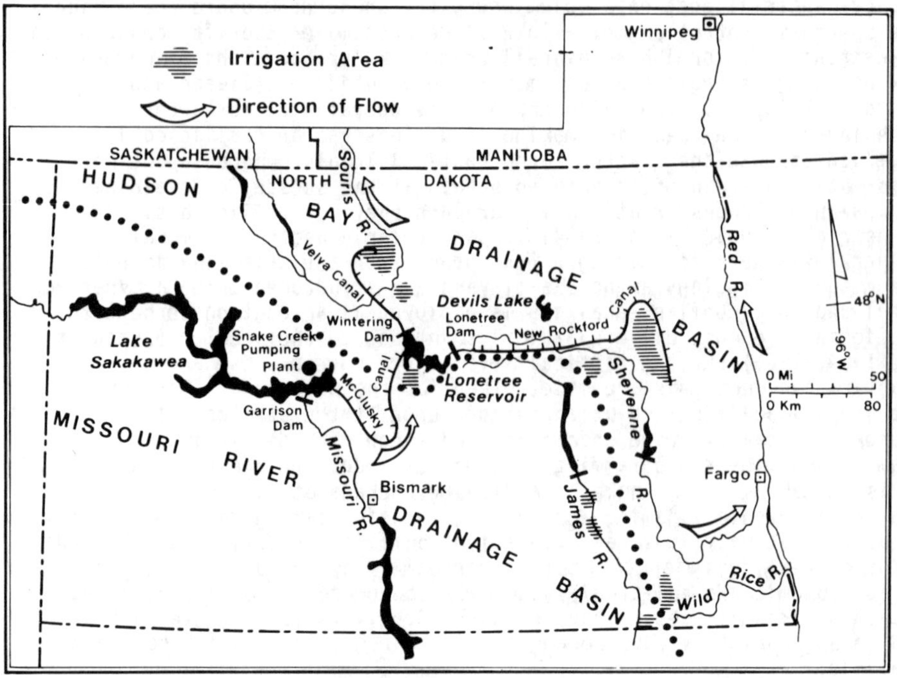

Figure 1. Map of the Garrison Diversion Unit

Interior in 1955. The resulting creation of Lake Sakakawea inundated some 120,000 hectares of fertile valley land as well as enormous areas of valuable wetlands and wildlife habitat (Mundry, in Leitch and Keleher, 1974), and apparently left a large unpaid political debt by the Bureau of Reclamation to the people of North Dakota. The Bureau of Reclamation seemed obliged to complete the entire GDU project so that the water from Lake Sakakawea could be used to restore by irrigation the agricultural land lost through the building of the Garrison dam.

Diversion plans with benefit-cost ratios of 0.53 to one and 0.76 to one were proposed in 1957 and 1959, respectively, but were rejected. Finally, a plan with a claimed benefit cost ratio of 2.51 to one received Congressional authorization in 1965 and construction was started by the Bureau of Reclamation in 1968 (EIAP, 1975). Kremers (1974) points out that this final plan converts 89,000 hectares of farmlands and wetlands into drains, ditches, service roads and other facilities to irrigate 100,000 hectares that already support relatively prosperous agriculture.

A major benefit of the water for the Americans is irrigation, but other benefits include the provision of municipal and industrial water supply, furnishing recreational and wildlife areas, and providing higher employment. The total economic value of the GDU for North Dakota is expected to reach $122 million per annum (Dorothy, 1973, p. 5).

The conflict over the GDU arises as a result of Canadian fears that the project may cause extensive environmental damage due to the runoff of the polluted irrigation waters which flow via the Souris and Red rivers into Canada (Boyd, 1975). In particular, the main reason why the Canadians consider the GDU to be a risky undertaking is that fish, fish diseases, fish parasites and other biota present in the Missouri River water could possibly destroy or have drastic consequences upon the natural aquatic life in the Hudson Bay drainage basin. Additionally, American environmentalists are concerned that the environmental damage caused by the project within the US itself has been grossly underestimated (EIAP, 1975, p.v.). This international conflict also involves the International Joint Commission (IJC), a nominally impartial political entity with special powers over political disputes between Canada and the US, as specified in the Boundary Treaty of 1909 between the two countries (IJC, 1965).

The first official action on the part of the Government of Canada was in April, 1969, when questions regarding Garrison were raised by the Canadian Embassy in Washington. This was followed in April, 1970, by a diplomatic note crystallizing Canadian concerns (IJC, 1977, p. 7). Diplomatic notes reiterating these concerns were submitted to the American Government in October 1971, October 1973, June 1975, October 1976, and April 1978. The October 1973 diplomatic note strongly requested a moratorium on construction of the GDU, and in a response in February 1974 the American government promised to halt all parts of the project affecting Canada. Evidently, the determination of the sections of the project which affect Canada is not clear because the October 1976 diplomatic note protests the continuation of the project, and, in particular, it objects to the building of the Lonetree Reservoir.

In February, 1977, the Carter administration stopped all funds for portions of the project affecting Canada, including the Lonetree reservoir (Associated Press, dateline Bismark, North Dakota, found in the Toronto Star, February 19, 1977a, p. 1), and a diplomatic note from the US State Department (1977) reiterated the commitment to undertake no construction of works affecting Canada. In spite of this, on June 22, 1977, the Senate approved "full funding" for the project (Associated Press, dateline Bismark, North Dakota, found in the Toronto Globe and Mail, June 23, 1977b, p. 2). As of 1980, no water has actually entered the Hudson Bay drainage basin but the Snake Creek Pumping Plant and the Wintering Dam are complete, and the McClusky Canal is virtually finished.

An international mediator exists between the US and Canada in the form of the IJC. The IJC is a commission composed of three members from Canada and three from the US which may make recommendations and judgements on boundary conflicts as detailed in the Boundary Treaty of 1909. The powers of the IJC in the Garrison conflict are outlined in Section 3.1 of the treaty. In 1975 the IJC was directed by the two governments to (IJC, 1977, p. 6):

1. report on existing conditions in the area to be affected by Garrison,

2. determine the impact of the GDU on these conditions,

3. make recommendations to ensure that Article IV of the Boundary Waters Treaty of 1909 is honored, and,

4. estimate the costs involved.

The report of the commission was completed in August 1977, following submission of the report by its International Garrison Diversion Study Board (IGDSB), in October, 1976.

Charges were brought against the builders of Garrison for violating US environmental legislation at least as early as 1972 (EIAP 1975, p. 5). In 1973 the Bureau of Reclamation was engaged in lawsuits by an environmental organization, an amalgamation of farmers, and a few professional people (Dorothy, 1973, p. 8). The controversial first Final Environmental Impact Statement (EIS) was released by the Bureau of Reclamation in January 1974 (Bureau of Reclamation, 1974). It was heavily criticized as being biased in support of a large diversion project (EIAP, 1975).

In May 1976 the National Audubon Society filed suit against the US Department of the Interior (of which the Bureau of Reclamation is part). An agreement was reached in May 1977 between the two parties which required a new EIS by January, 1978. The resulting EIS presented six alternative plans, one of which was recommended by the US Department of the Interior (US Department of the Interior, 1978a). A diplomatic note from the Canadian Department of External Affairs to the US State Department dated April 4, 1978, notes that the revised EIS "does not address substantive Canadian concerns." More detailed background information about the GDU and the conflict that surrounds it can be obtained by referring to the research of Hipel and Fraser (1979, 1980), or the IJC (1977).

4. MODELLING THE GARRISON CONFLICT

The IJC has said that the transboundary implications of the GDU constitute "one of the most difficult and intricate issues the Commission has ever been asked to consider" (IJC, 1978, p. 18). This high degree of complexity makes it difficult for a decision maker to fully comprehend the problem and eventually reach a reasonable decision. Consequently, the methodology of conflict analysis is needed to provide a structure for systematically modelling the problem and predicting the possible resolutions or equilibria to the dispute.

A conflict analysis of the GDU was previously performed by the authors using metagame analysis (Hipel and Fraser, 1979, 1980), and also employing conflict analysis (Hipel and Fraser 1982; Fraser and Hipel, 1984, Ch. 2). Background information for the following analysis was obtained from the large amount of published material concerning the GDU, although some preference information was developed directly from telephone conversations with individuals involved in the project. April 1976 is selected as the date for which the GDU project is analyzed.

As mentioned earlier, a conflict study is executed by following two major stages. First, the historical information is systematically organized according to a formal structure by ascertaining the players, options and preferences for each player among the possible feasible outcomes in the game. After this modelling process, the second stage is to perform a stability analysis in order to predict the possible resolutions or equilibria to the conflict. Each outcome is analyzed for stability for each player and those outcomes which possess some type of stability for all the players form the set of equilibria.

4.1 Players.

The more than forty explicitly named participants in the Garrison conflict can be grouped into four main players, with similar interests, shared powers, and common goals. These four players are:

1. US Support for Garrison

The greatest support for the GDU has historically been given by the Bureau of Reclamation of the US Department of the Interior. Friction within the Department of the Interior over this support resulted in the creation of a new entity, the aptly named Oversight and Management Group, to prepare the second EIS for the project (US Department of the Interior, 1978b, p. 14).

The Garrison Diversion Conservancy District is a well-funded and politically powerful organization which also supports the GDU. The Director of the Garrison Diversion Conservancy District firmly believes that the GDU will be completed.

Other parties have expressed support for Garrison include the State of North Dakota, the Garrison Diversion Irrigation Council, the National Park Service, and the North Dakota State Game and Fish Department.

2. US Opposition to Garrison

US opposition to Garrison consists mostly of environmental organizations and private individuals. Environmental organizations include the National Audubon Society, the Environmental Protection Agency, the Bureau of Sport Fisheries and Wildlife (now reorganized as the Fish and Wildlife Service and is also located within the Department of the Interior), and the President's Council on Environmental Quality (Kremers, 1974, p. 36). Relevant environmental legislation which could be used for altering or blocking construction of the GDU includes the National Environmental Policy Act, the Water Supply Act of 1985, the Federal Water Pollution Control Act, and the Migratory Bird Convention of 1916 (EIAP, 1975, pp. 80-89). Since 1976 there have been numerous other laws and Presidential Executive Orders protecting wetlands which strengthen the litigation grounds for the opposition.

Other organizations which have expressed opposition to Garrison on various grounds include the Committee to save North

Dakota, the North Dakota Farmers' Union, the National Farmers' Union, the North Dakota Students' Organization, the National Water Commission, the State Department, the Office of Management and Budget, and the General Accounting Office (Kremers, 1974, p. 36).

3. Canadian Opposition to Garrison

Canadian organizations that have expressed opposition to Garrison include the Canadian Federal Government; Environment Canada (the Canadian federal environmental ministry; the Manitoba Provincial Government; the Manitoba Ministry of Mines, Resources and Environmental Management; the Manitoba Environmental Council; the Manitoba Wildlife Federation; the Prairie Environmental Defense League; and a number of municipalities, industries and professionals. Due to the devastating environmental damage which could occur when the GDU comes into operation, the Canadians consider the project to be far too risky.

Legal tools available to oppose the Garrison project include the Boundary Treaty of 1909, the Doctrine of Equitable Utilization contained in the 1966 Helsinki rules, and the Migratory Bird Convention of 1916. The most powerful of these legal agreements is the Boundary Treaty of 1909, which is discussed further in the next section. Carter (1975, p. 180) affirms that the Boundary Treaty confers legal rights to Canadian citizens in US courts.

4. The International Joint Commission (IJC)

The IJC acts as an intermediary for water disputes between Canada and the US. The IJC was formed as part of Article VI of the "Treaty Between the United States and Great Britain Relating to Boundary Waters and Questions Arising Between the United States and Canada", commonly called the Boundary Waters Treaty of 1909. The IGDSB, which was responsible for assessing the transboundary implications of the GDU, was an investigative board commissioned by the IJC on October 23, 1975.

The detailed text of the Boundary Waters Treaty is published by the IJC (1965), while the key elements of the Treaty are described by Hipel and Fraser (1979, 1982). Because the legal basis for the IJC in the GDU conflict is Article IX of the Treaty, all recommendations by the IJC with respect to the GDU project are nonbinding upon the US and Canada. Nonetheless, given the special powers of the IJC and the fact that the IJC represents a unique point of view, it is informative to include the IJC as a separate player in the GDU conflict. Any recommendation by the IJC may affect the actions and preferences of one or more of the other players, especially the Canadian opposition.

4.2 Options

In April, 1976, the point in time for which the analysis will be done, the IJC recommendations had been called for by the governments of the US and Canada. The final IJC report would not be released for another nine months. The first EIS had been released by the Bureau of Reclamation more than two years previously, and numerous comments on this had been published. Officially, parts of the project affecting Canada had been halted for over two years, although construction had been continuing unabated. The lawsuit by the National Audubon Society had not yet been filed.

Based primarily on the IJC final report (1977), the options found in Table 1 may be selected. Note that the players are amalgamated into the four major parties defined in the last section.

Table 1. Players and Options for the Garrison Conflict.

Players	Options
U.S. support	1. Proceed to complete full GDU.
	2. Proceed to complete GDU modified to reduce Canadian impacts.
	3. Proceed to complete GDU modified to appease U.S. environmentalists.
U.S. opposition	1. Legal action based on environmental legislation.
Canadian opposition	1. Legal action based on the Boundary Treaty of 1909.
IJC (International Joint Commission)	1. Support completion of full GDU.
	2. Support completion of GDU modified to reduce Canadian impacts.
	3. Support suspension of the GDU except for the Lonetree Reservoir.
	4. Support complete suspension of the GDU.

In Table 1, the player "US Support" is considered to be the builders of the GDU project. The options available to this player are much more varied than the three presented, but these reflect the three major types of projects that would determine the responses of the other players. The "full GDU" option represents projects which are not significantly different from the one approved by Congress in 1965. The project modified to reduce Canadian impacts includes undertakings which have limited Canadian effects, similar to that suggested by the IGDSB in its report to the IJC (IGDSB, 1976). A GDU modified to appease environmentalists has not been clearly specified in the literature, but would be a project to which US environmentalists would not find fault.

The single option for "US Opposition" represents legal action based on any of the environmental legislation listed in the previous section under this player. Legal action based on the Boundary Treaty was selected as the only option for the "Canadian Opposition" because it is the most powerful legal tool at its disposal. Other options which might be brought to bear by the Canadians as potential sanctions do not appear in the literature, and so have not been included in this analysis.

The options of the "IJC" are somewhat precognitive, but the other parties would be aware of these possible recommendations of the IJC. The first and second options listed for the IJC refer to the same projects described in the first two options, respectively, for the US Support. The third IJC option is the support of a project incorporating irrigation areas only in the Missouri River basin. Athough the Lonetree Reservoir is located in the Hudson Bay drainage basin, it is required for irrigation in the Missouri basin via the James River (see Figure 1). By invoking its fourth option, the IJC would support suspending the entire project.

4.3 Outcome Removal

Outcomes are written as columns of ones and zeros listed opposite an ordering of players and options, where a 1 indicates that an option is taken by a player in the outcome, while a 0 indicates that the option is not taken. The outcomes are written horizontally in text, where the order of ones and zeros is understood to correspond to some meaningful ordering of options. A strategy is a selection by a player of the options under his control, and thus an outcome can be considered to be the situation where each of the players has selected a strategy.

For the Garrison conflict, the 9 options in Table 1 mean that $2^9 = 512$ outcomes are mathematically possible. In practice it is preferable to end up with not more than 25 to 50 outcomes after removing various kinds of outcomes from the total set of possible outcomes. This allows for exhaustive stability analyses to be conveniently done by hand or using a computer.

It is easy to develop sets of infeasible outcomes from an understanding of the conflict. For example, for the Garrison conflict the three options for the player US Support are all mutually exclusive. Therefore, no matter what the other players do, as long as the US Support selects more than one of its options the resulting outcomes are infeasible. All of these infeasible outcomes are contained in the three sets of infeasible outcomes represented by (11-,-,-,----), (1-1,-,-,----) and (-11,-,-,----). A dash (-) represents either a 1 or a 0 so that each of these sets stands for 2^7 or 128 outcomes to be removed. Table 2 lists all the outcomes which can be removed from the Garrison conflict model. The selection of outcomes to eliminate in a conflict can be challenging.

As explained by Hipel and Fraser (1982) and Fraser and Hipel (1984, Ch. 2), removable or infeasible outcomes can be categorized according to four different types. Using one of a number of simple algorithms which are available, the removable outcomes such as those in Table 2 can be removed from the game. This can be done by hand or else using a microcomputer. When all of the outcomes listed in Table 2 are eliminated from the GDU conflict, 23 outcomes remain in the game.

Table 2. Outcomes Removed from the Garrison Model

Removable outcomes	Reasons
(11-,-,-,----) (1-1,-,-,----) (-11,-,-,----)	Mutually exclusive options for U.S. support
(---,-,-,11--) (---,-,-,1-1-) (---,-,-,1--1) (---,-,-,-11-) (---,-,-,-1-1) (---,-,-,--11)	Mutually exclusive options for IJC
(000,-,-,----)	Some sort of diversion will be built
(---,-,-,0000)	The IJC will make a recommendation
(1--,0,-,----)	U.S. opposition will pursue legal action against full project
(--1,1,-,----)	U.S. opposition will not puruse legal action if appeased
(1--,-,0,0---) (-1-,-,0,--1-) (-1-,-,0,---1) (--1,-,0,--1-) (--1,-,0,---1)	Canadian opposition will pursue legal action if any project is built that is larger than that approved by the IJC

The 23 remaining outcomes for the GDU conflict are displayed in Table 3. The players and options in Table 3 refer to those presented in more detail in Table 1. The binary form of each outcome has been converted to its decimal representation in this table. Thus the outcome on the extreme left, written as (010, 0, 0, 1000) in binary notation, has a decimal equivalent of $0 \times 2^0 + 1 \times 2^1 + 0 \times 2^2 + 0 \times 2^3 + 0 \times 2^4 + 1 \times 2^5 + 0 \times 2^6 + 0 \times 2^7 + 0 \times 2^8 = 2 + 32 = 34$.

4.4 Preferences

The stability analysis tableau for the Garrison conflict is shown in Table 4. The decimalized outcomes are ordered from most preferred on the left to least preferred on the right to form the preference vector for each player. UI's (unilateral improvements) are determined as the outcomes that a player can improve to by changing the options under his control, given that the other players' strategies remain the same. The

Table 3. Feasible Outcomes for the Garrison Conflict.

U.S. support																							
Full	0	0	1	0	0	0	1	0	0	0	0	0	0	1	0	0	0	1	0	0	0	1	0
Reduced	1	0	0	1	1	0	0	1	1	0	1	1	0	0	1	1	0	0	1	1	0	0	1
Appease	0	1	0	0	0	1	0	0	0	1	0	0	1	0	0	0	1	0	0	0	1	0	0
U.S. oppos.																							
Legal	0	0	1	1	0	0	1	1	0	0	1	0	0	1	1	0	0	1	1	0	0	1	1
Can. oppos.																							
Treaty	0	0	0	0	1	1	1	1	0	0	0	1	1	1	1	1	1	1	1	1	1	1	1
IJC																							
Full	1	1	1	1	1	1	1	1	0	0	0	0	0	0	0	0	0	0	0	0	0	0	0
Reduced	0	0	0	0	0	0	0	0	1	1	1	1	1	1	1	0	0	0	0	0	0	0	0
Lonetree	0	0	0	0	0	0	0	0	0	0	0	0	0	0	0	1	1	1	1	0	0	0	0
Suspend	0	0	0	0	0	0	0	0	0	0	0	0	0	0	0	0	0	0	0	1	1	1	1
Decimal.	34	36	41	42	50	52	57	58	66	68	74	82	84	89	90	146	148	153	154	274	276	281	282

Table 4. Stability Analysis Tableau for the Garrison Conflict.

U.S. support																						
E	×	×	*E*[a]	×	×	×	×	×	×	*E*	×	×	×	×	*E*	×	*E*[a]	×	*E*	×	*E*[a]	×
r	*r*	*r*	*r*	*u*	*r*	*r*	*u*	*r*	*r*	*r*	*r*	*u*	*u*	*u*	*r*	*u*	*r*	*u*	*r*	*u*	*r*	*u*
41	57	34	36	42	50	52	58	66	82	74	90	68	84	89	148	146	153	154	276	274	281	282
				41			57					66	82	90		148		153		276		281
U.S. opposition																						
r	*r*	*r*	*r*	*r*	*r*	*r*	*r*	*r*	*r*	*r*	*r*	*r*	*r*	*r*	*r*	*r*	*u*	*u*	*u*	*u*	*u*	*u*
276	148	84	68	52	36	282	154	90	74	58	42	281	153	89	57	41	274	146	82	66	50	34
																	282	154	90	74	58	42
Canadian opposition																						
r	*r*	*r*	*r*	*r*	*r*	*r*	*r*	*r*	*u*	*u*	*u*	*r*	*r*	*r*	*s*	*s*	*u*	*r*	*r*	*r*	*r*	*u*
282	274	276	154	146	148	74	66	68	90	82	84	36	42	34	58	50	52	281	153	89	41	57
									74	66	68				42	34	36					41
IJC																						
r	*r*	*r*	*r*	*r*	*r*	*r*	*r*	*r*	*r*	*r*	*r*	*r*	*r*	*r*	*r*	*r*	*r*	*r*	*r*	*r*	*r*	*r*
34	36	41	42	50	52	57	58	66	68	74	82	84	89	90	146	148	153	154	274	276	281	282

[a]Outcomes 36, 153, and 281 are false equilibriums.

UI's are listed under the outcomes in the preference vector as a column with the most preferred UI from the outcome listed on top, and the least preferred UI written at the bottom. For example, as shown in Table 4, the US Support has a UI from outcome 42 to 41. To interpret the meanings of the decimalized outcomes in Table 4, it is convenient to refer to the binary translations given in Table 3. Notice that when outcomes 41 and 42 are compared in Table 3, only the American strategies are different in these two outcomes since the other players do not change their strategies.

A number of considerations determines the ordering of outcomes in the preference vectors in Table 4. First consider the preference structure of the player US Support. It is taken to be politically unacceptable for the US Support for Garrison not to build some sort of diversion project involving irrigating areas in the Hudson Bay drainage basin. This is supported by Robinson (1966, p. 465), who says of the early days of the current diversion plan:

> "Although [an economic analysis] raised serious doubts about irrigation in a region which received adequate rainfall in at least three years out of every four, it was given little attention. No one in North Dakota publicly questioned the benefits of diversion, any more than he would motherhood, virtue, or patriotism."

This view is maintained today, as evidenced through personal contact with individuals associated with the project. It is generally felt that the US Government has a binding contract to build the GDU and it is fully expected to do so.

The US Support would most prefer to build the full project, but there is no rationale available to distinguish preferences between a project modified to reduce Canadian impacts as described in the IGDSB report (IGDSB, 1976), and one modified to appease environmentalists. Because of this, they are considered to be equally preferred, except that when the IJC supports a project for irrigation in the Hudson Bay drainage basin, the US Support would prefer a project to match the IJC recommendation. Thus, the first eight outcomes in the preference vector for the US Support all involve the IJC supporting the full project. In the two most preferred outcomes, 41 and 57, the full project is built with the support of the IJC. Equally preferred outcomes are indicated by a bridge (i.e.⌜‾‾‾⌝) placed over the outcomes involved. An example of this is outcomes 34 and 36, which indicates the indifference felt by the US Support between a reduced project and one built to appease the environmentalists, given that the IJC is supporting a full project. When the IJC supports no project for irrigation in the Hudson Bay drainage basin, the US Support would prefer to ease US internal legal pressure by proceeding with a project which would appease environmentalists. Thus 148 is preferred to 146 and 276 is preferred to 274, but 34 and 36 are equally preferred as are outcomes 50 and 52.

The US Opposition to Garrison has numerous powerful legislative tools at their disposal, and is composed of many very motivated groups. It is reasonable to expect that one of these groups will press legal action as long as the project has any adverse and unmitigated effects on the environment. Thus, the US Opposition would invariably prefer to choose the option of legal action on environmental grounds if the full project were built, and would be likely to prefer to choose this option even if the reduced Canadian impacts project option is chosen by the US Support. If the US Support elects to modify the diversion project to appease environmentalists, the US Opposition would prefer not to press legal action. The first six outcomes of the US Opposition's preference vector in Table 4 are the outcomes where the US Support chooses to build a project that appeases environmentalists. These are all equally preferred for the US Opposition since this group is appeased and doesn't care what the IJC or the Canadians do. In the next two outcomes, 282 and 154, the US Opposition is also appeased, but is taking legal action, which it prefers not to do in this circumstance.

The Canadian Opposition prefers no diversion project to be built. However, it has no control in the matter other than to press legal action, via the Boundary Treaty of 1909. If any project were built, the Canadians would generally prefer to choose this option. On the other hand, if the IJC were to support a diversion project, the Canadian Opposition would likely be wasting its time and money pursuing legal action, because the IJC recommendation is relatively unassailable. Also, there is nothing to indicate that the Canadians are likely to prefer the project with supposedly reduced Canadian impacts over one to appease the American environmentalists. Thus, in the Canadian Opposition preference vector the sets of outcomes given by 274 and 276, 146 and 148, 66 and 68, 82 and 84, and 50 and 52 constitute groups of outcomes where the two outcomes in each set are equally preferred.

Notice in Table 4 that the least preferred outcome for the Canadian Opposition is outcome 57. From Table 3, it can be seen that this outcome is the situation where the IJC supports a full project which is constructed by the US Support while the Canadian and US Opposition go to court. Because outcome 57 could be environmmentally disastrous for the Canadians, this outcome involves a high degree of risk and is therefore least preferred by the Canadian Opposition. In fact, the five outcomes from Table 3 for which the Americans build the full GDU (i.e., outcomes 57, 41, 89, 153, and 281) are the five least preferred outcomes for Canada.

The IJC was created by a treaty which was to "...prevent disputes...between the United States and Canada..." (IJC, 1965, p. 13). It has an obligation to resolve conflicts between the two countries. According to the Reference of October 22, 1975, "Both the United States and Canada ascribe particular importance to the view of the Commission on this matter [the transboundary implications of Garrison]" (IJC, 1977, p. 132). This suggests that the recommendation of the IJC is expected to aid in the resolution of the conflict, a consideration borne out by an examination of the situation. Thus, it would definitely be preferred by the IJC to make a recommendation. However, since the IJC would be expected to retain impartiality, the best model for the IJC's preferences among its possible recommendations is to consider them all equally preferred. This is indicated in Table 4 by the bridge which connects all the outcomes in the IJC preference vector.

5. STABILITY ANALYSIS

All of the outcomes in Table 4 must be analyzed for stability for each of the four players. There are four types of stability possible for a game model of a conflict. Recalling that UI means "unilateral improvement", a particular outcome for a given player can be:

1. Rational (r). The player has no UI's from this outcome.
2. Sequentially Sanctioned (s). The player is deterred from taking a UI from this outcome because credible actions by the other players could result in a less preferred outcome for all of the player's possible UI's from the particular outcome.
3. Unstable (u). The player has at least one UI which is not deterred by credible actions on the part of the other players.
4. Simultaneously Sanctioned (/). This type of stability is calculated after the aforementioned three types of stability have been determined for all of the outcomes for each player. When an outcome is unstable for at least two players, simultaneous action by more than one player could cause a less preferred outcome to occur and thereby induce stability for an outcome which was previously thought to be unstable for a given player.

The algorithm of the stability analysis of a conflict with n players is shown in Figure 2. Note that a sanction is determined as any outcome formed by the other players consistently improving themselves from the outcomes under consideration. This means that not only do the UI's of the other players need to be examined, but all UI's leading from these UI's must also be assessed as possible sanctions. As an example of how the stability analysis proceeds, consider outcome 50 for each of the players. For the US Support and the IJC, 50 is rational because neither player has a UI available. To indicate this, an r is placed above 50 in the preference vectors of the two players.

The US Opposition has a UI from 50 to 58. From 58, the US Support has a UI to 57, which is also preferred by the US Opposition to outcome 50, and consequently no sanction has yet been found. From 57, the Canadian Opposition has a UI to 41, which is also preferred to 50 by the US Opposition, and is rational for all the players. Backtracking to 58, it can be seen that the Canadian Opposition has a UI from 58 to 42, which is also preferred by the US Opposition to 50, and from 42 the US Support can again go to 41. All possible outcomes brought about by consistent improvements by the other players have not resulted in an outcome less preferred to the original outcome by the US Opposition, and so 50 is unstable for this player. A u is placed above the 50 in the preference vector for the US Opposition. Note that outcome 50 for the US Opposition is the most complex search for sanctions in the Garrison analysis.

Finally, consider the stability of outcome 50 for the Canadian Opposition. The Canadians have a UI from 50 to 34. The US Opposition can improve from 34 to 42 and then the US Support has a UI from 42 to 41. Outcome 41 is less preferred to 50 by the Canadian Opposition, and

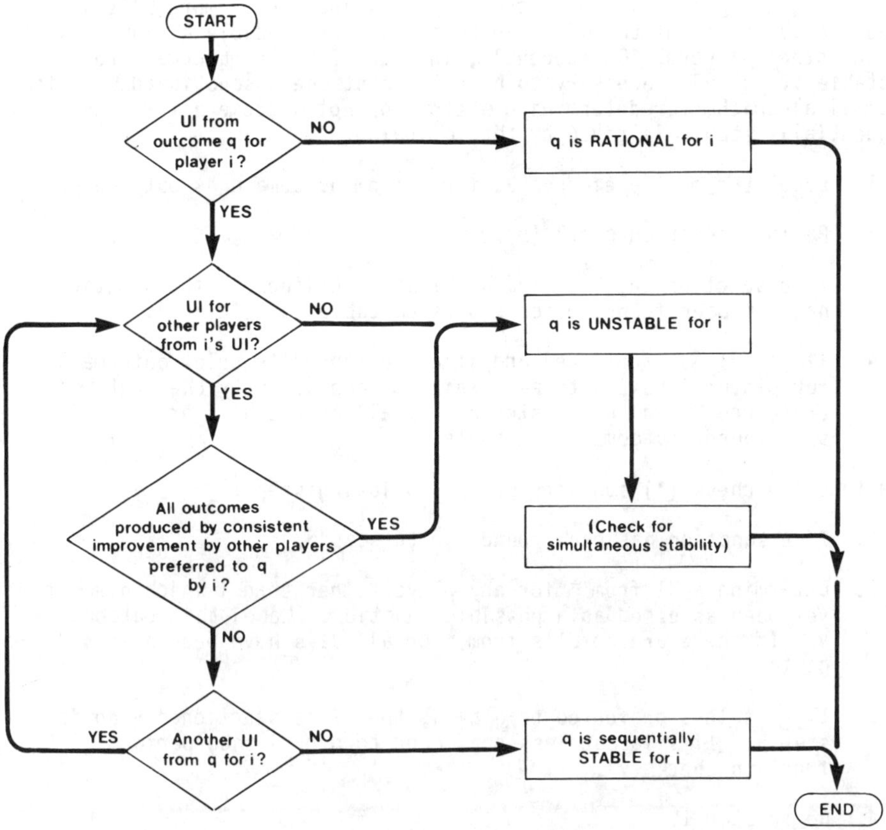

Figure 2. Stability of outcome q for player i in an n-player game

so 50 is sanctioned for the Canadians, as shown by the s written above it in Table 4. By referring to Table 3, it can be seen that outcome 41 is much more risky for the Canadians than outcome 50, since the full GDU project is constructed in outcome 41 while a project to reduce Canadian impacts is built in outcome 50. Consequently, outcome 41 is less preferred than 50 by Canada, and hence the threat of ending up at 41 if Canada moves from 50 to 34 constitutes a credible sanction. Notice, however, that the US Support does not consider outcome 41 to be environmentally risky for itself and this outcome is most preferred by the US Support. When checking for credible sanctions using the concept of consistent improvement, it is only necessary to find at least one credible sanction in order for stability to be induced. Note that this ensures that no equilibria are missed, although the dynamics of a conflict may indicate that a player may depart from an equilibrium in the hope that the expected sanction will not endure.

For a given player i, an outcome q possessing one or more UI's is stable only if all of the UI's can be blocked by credible actions levied by the other players. Consequently, in order for the outcome q to be unstable it is only necessary to have at least one unsanctioned UI. The overall algorithm for determining whether or not outcome q for player i is sequentially stable is given by the following:

1. Label the most preferred UI for i from outcome q as outcome p.

2. Perform function check (p).

3. If a sanction has been found, UI is sanctioned. If a sanction has not been found, outcome q is unstable for player i.

4. If the UI is sanctioned and if all of the UI's below outcome q for player i have not been examined, replace p by the next most preferred UI and go to step 2. If all of the UI's are sanctioned, outcome q is stable.

The function check (*) consists of the following steps:

1. If a sanction has been found, go to step 5.

2. Determine a UI from * for any player other than i which has not yet been assessed as a possible sanction. Label this outcome as v. If there are no UI's from * or all UI's have been assessed, go to 5.

3. If v is less preferred to q by i, the UI is sanctioned - go to step 5. If v is not less preferred to q by i then perform function check (v).

4. Go to step 1.

5. End.

In practice it is a straightforward procedure to apply the above algorithm, and, after it has been used a few times, the intricacies of the procedure become quite evident. Keep in mind that for a given UI from outcome q this algorithm checks all possible improvements by the other players from this outcome.

If an outcome is unstable for two or more players, one may check for stability by simultaneous sanctioning. When an outcome q is unstable for m out of n players in a n-player game, the simultaneous sanctioning calculations must be done for all possible combinations of two or more players for which q is unstable. Let A_i be the set of UI's for player i from outcome q which are not sequentially blocked and $a_i \in A_i$. Also, let M be the set of all players for which q is unstable and $|M| \geq 2$.

For every $\Gamma \subseteq M$, $|\Gamma| \geq 2$, calculate

$$p = \sum_{i \in \Gamma} a_i - (|\Gamma| - 1)q \qquad (1)$$

where p is determined for every $a_i \in A_i$ for each player in the set and all outcomes are decimalized. If anywhere during the calculations p is less preferred than q by $i \in \Gamma$, the UI given by $a_i \in i$ is blocked. If all $a_i \in A_i$ are blocked, q is simultaneously sanctioned for player i. This is indicated by putting a / through the u above the outcome q for player i. Stokes and Hipel (1986) provide an alternative definition regarding the concept of simultaneous sanctioning.

None of the outcomes in the Garrison conflict were found to be stable by simultaneous sanctioning, and none were unstable for more than 2 players. For example, in Table 4 consider outcome 84 which is unstable for the US Support and also the Canadian Opposition. The US Support has a UI from 84 to 82 while the Canadian Opposition has a UI from 84 to 68. The resulting outcome is given by (1) to be

$$p = (82 + 68) - ((2 - 1) \times 84) = 150 - 84 = 66$$

Since outcome 66 is preferred by both players over 84, outcome 84 is not simultaneously sanctioned for either player. Occasionally when checking for simultaneous sanctioning in models in which infeasible outcomes of a certain type have been removed, the outcome p will not be found in the preference vector. An examination of the meaning of such an outcome and its proper location in the preference vectors of the players will determine whether stability by simultaneous sanctioning is induced. A detailed discussion regarding this point is given by Fraser and Hipel (1984, Ch. 2).

The only special consideration required when performing an analysis in which equally preferred outcomes are included is in the assessment of sanctions. If a possible sanction is found to be equally preferred to the outcome for which the stability is being checked, then the outcome is indeed sanctioned, since there is no advantage for the player to move from one equally preferred outcome to another. In other words, outcomes which are considered equally preferred when ordering preferences are considered less preferred to each other when examining them for sanctions against UI's. In the analysis, if a possible sanction is under the same bridge as the outcome being examined, the outcome is sanctioned, even if the possible sanction is physically located to the left of the outcome in the player's preference vector.

In order for an outcome to be a possible resolution or equilbrium in the conflict, it must possess some type of stability for each player. For example, in Table 4 outcome 36 is rational for all four players and therefore an E is written above outcome 36 in the preference vector for the US Opposition in Table 4. On the other hand, outcome 50 is unstable for the US Opposition and therefore cannot constitute an equilibrium even though it is stable for the other players. To indicate this, an X is placed above outcome 50 in the preference vector for the US Support.

In Table 2, some outcomes which were removed are actually feasible but were eliminated in order to reduce the number of outcomes to examine for stability. However, removing realizable outcomes during outcome removal may result in the generation of false equilibria. The false equilibria are readily identifiable by examining the equilibria individually. For a false equilibrium to exist, an unsanctioned outcome to which a player can unilaterally improve from the false equilibrium must have been removed from the full set. For example, consider outcome 153 from Table 4. (Refer to Table 3 for a description of this outcome.) A consideration of the preferences of the four players reveals that there is an identifiable UI available for the player US Support from "build full GDU" to "appease environmentalists". The resulting outcome, (001,1,1,0010), was one of the group of outcomes removed on the basis of the judgement that the player US Opposition would always oppose any project except one designed to appease it. This improvement is clearly unsanctionable because any combination of feasible strategies by the players US Opposition and Canadian Opposition are still preferred to the original outcome, and the IJC has no UI's available.

After removing the three false equilibria (001,0,0,1000), (100,1,1,0010), and (100,1,1,0001), which are the decimalized outcomes 36, 153 and 281, four equilibria remain. These are outcomes 41, 74, 148, and 276. The equilibrium given by 41 suggests that if the IJC were to support the full GDU, the US Support would proceed to build the full project in spite of legal action by the US Opposition. The Canadian Opposition would not pursue legal action. From the equilibrium outcome 74 it is evident that if the IJC chose to support a project with reduced Canadian impacts, the US Support would proceed with such a plan, the US Opposition would pursue legal action and the Canadian Opposition would not.

In the situations where the IJC supports a project including the Lonetree Reservoir only, or supports total suspension of the GDU, the US Support would proceed to construct a project which appeases the US Opposition, so it does not pursue legal action. The Canadian Opposition would initiate legal action because of the IJC support. These situations are represented by outcome 148 where the IJC supports Lonetree only, and 276 where the IJC supports the complete suspension of the GDU.

As events turned out, this model predicts the conflict equilibrium well. The IJC recommended in its report of August, 1977, entitled, "Transboundary Implications of the Garrison Diversion Unit", that all portions of the project except the Lonetree Reservoir be discontinued (IJC, 1977, p. 121). This would suggest that the resolution to the GDU conflict, as assessed for April, 1976, should be the corresponding equilibrium outcome (001,0,1,0010), outcome 148. In other words, the conflict that existed in April, 1976, could be expected to result in the situation where the US Support builds a project designed to appease environmentalists, the US Opposition does not pursue legal action, and the Canadian Opposition does take legal action.

The US Support appears to have reached its resolution strategy. The US Department of the Interior released its Supplementary EIS to satisfy the May, 1977, agreement to the May, 1976, suit by the National Audubon

Society (US Department of the Interior 1978b) in January, 1978. However, this EIS "does not address substantive Canadian concerns" (Diplomatic Note, Canadian Department of External Affairs, Canada, April 4, 1978). The six alternative plans presented in this Statement all involve irrigation in the Hudson Bay drainage basin. The report on the plan recommended by the US Department of the Interior (US Department of the Interior, 1978a) makes only brief (one page) mention of the most significant Canadian concern, which is the environmental risk involved with the introduction of exotic fish into the Hudson Bay drainage basin. Minimizing environmental damage was an explicitly stated main objective of the recommended plan, while the "reduction" of interbasin biota transfer was accorded secondary consideration (US Department of the Interior, 1978a, p. 1). Hence, this appears to be a plan designed to appease American environmentalists.

If the US Support is successful in appeasing the US Opposition, the current state of no significant legal action on environmental grounds is likely to continue. Because the suit by the National Audubon Society was settled by the release of the Supplementary EIS, the US Opposition has now chosen its resolution strategy.

The Canadian Opposition has not firmly reached its final strategy. The equilibrium suggests that Canada will pursue legal action based on the Boundary Treaty. Action in the form of protests via diplomatic notes has been continued to the present, but no explicit court proceedings have yet been initiated. Court action may be expected as long as the game retains the format presented in this analysis.

6. CONCLUSIONS

As exemplified by the GDU application, risk can be formally incorporated into a conflict analysis study. Outcomes which involve a lot of risk for a given player will tend to be less preferred than other outcomes. Additionally, outcomes which are risky for a particular player can be utilized for sanctioning purposes by one or more of the other players in order to stop the given player from taking advantage of his possible UI's.

When designing a large scale engineering project such as the GDU, the undertaking must not only be physically, environmentally, financially and economically feasible, but also socially and politically viable. Although historically useful analytical techniques have been developed for the assessment of many types of feasibilities, only recently has the conflict analysis methodology become available for rigorously examining the social and political factors. Because of the increasing importance of social and political influences in decision making, the employment of conflict analysis should expand dramatically in the future. Furthermore, since conflict analysis methodology has been shown to work well in practice and it can readily handle risk considerations, the method constitutes a comprehensive and realistic approach for studying virtually any type of real world dispute.

REFERENCES

Associated Press (1977a) dateline Bismark, North Dakota, found in the Toronto Star, February 19, p. 1.

Associated Press (1977b) dateline Bismark, North Dakota, found in the Toronto Globe and Mail, June 23, p. 2.

Bennett, P.G. (1977) Toward a theory of hypergames, OMEGA, 5(6), pp. 749-751.

Bennett, P.G. (1980) Hypergames: the development of an approach to modelling conflicts, Futures 12(6) pp. 489-507.

Boyd, D.H. (1975) A scientific and policy review of the final environmental statement for the initial stage, Garrison Diversion, Unit, Volume II, The Impacts of the Garrison Diversion Unit on Canada, Manitoba Environmental Council, January.

Bureau of Reclamation (1974) Final environmental statement - Initial Stage Garrison Diversion Unit, Billings, Montana: U.S. Department of the Interior.

Canadian Department of External Affairs, Diplomatic Notes, October 23, 1973; June 23, 1975; October 12, 1976; April 3, 1978.

Carter, A.T. (1975) The boundary treaty of 1909: does it provide an environmental cause of action? South Dakota Law Review, 20, pp. 147-180.

Dorothy, R.E. (1973) Some engineering, environmental and social aspects of the Garrison Diversion Unit, North Dakota, Paper presented at the annual meeting, North Central Region American Society of Agricultural Engineers, Regina, Saskatchewan, Canada, October 12.

Dufournaud, C.M. (1982) On the mutually beneficial co-operative scheme: dynamic change in the payoff matrix of international river schemes, Water Resources Research, 18(4), pp. 764-772.

Environmental Impact Assessment Project of the Institute of Ecology (1975) A Scientific and Policy Review of the Final Environmental Statement for the Initial Stage, Garrison Diversion Unit, Volume I, United States Impacts, January.

Fang, L., K.W. Hipel, and D.M. Kilgour (1986) A comprehensive decision support system for two player conflicts, Large Scale Systems.

Fraser, N.M., and K.W. Hipel (1979a) Solving complex conflicts. IEEE Transactions on Systems, Man and Cybernetics, SMC-9(12), pp. 805-815.

Fraser, N.M., and K.W. Hipel (1979b) Computer assistance in conflict analysis, Proceedings of the 1979 International Conference on Cybernetics and Society, sponsored by the IEEE Systems, Man and Cybernetics Society, Conference held at Denver, Colorado, pp. 205-209, October 8-10.

Fraser, N.M., and K.W. Hipel (1980a) Computational techniques in conflict analysis, Advances in Engineering Software 2(4), pp. 181-185.

Fraser, N.M., and K.W. Hipel (1980b) Metagame analysis of the Poplar River conflict, Journal of the Operational Research Society, 31, pp. 377-385.

Fraser, N.M., and K.W. Hipel (1980c) The conflict analysis program, Technical Report No. 64-SFW-070580, Department of Systems Design Engineering, University of Waterloo, Waterloo, Ontario, Canada.

Fraser, N.M., and K.W. Hipel (1980d) Conflict analysis and bargaining. Proceedings of the 1980 International Conference on Cybernetics and Society, sponsored by the IEEE Systems, Man and Cybernetics Society. Conference held at Boston, Massachusetts, pp. 225-229, October 8-10.

Fraser, N.M., and K.W. Hipel (1980e) Computer analysis of a labour-management conflict. Technical Report No. 71-SM-270880, Department of Systems Design Engineering, University of Waterloo, Waterloo, Ontario, Canada.

Fraser, N.M., and K.W. Hipel (1981) Computer assistance in labour-management negotiations. Interfaces, 11(2), pp. 22-30.

Fraser, N.M., and K.W. Hipel (1983) Dynamic analysis of the Cuban missile crisis, Conflict Management and Peace Science, 6(2), pp. 1-18.

Fraser, N.M., and K.W. Hipel (1984) Conflict analysis: models and resolutions, North-Holland, New York, 377 pp.

Fraser, N.M., K.W. Hipel, and J.R. del Monte (1983) Approaches to conflict modeling: a study of a possible USA-USSR nuclear confrontation, Journal of Policy Modeling 5(3), pp. 397-417.

Giesen, M.O., and K.J. Radford (1983) Strategy and tactics: a new model for organizational decision making. INFOR 21(1), pp. 52-60.

Hipel, K.W. (1981) Operational research techniques in river basin management. Canadian Water Resources Journal 6(4), pp. 205-226.

Hipel, K.W., and N.M. Fraser (1979) The historical and political aspects of the Garrison Diversion Unit, Technical Report No. 60-SM-051179, Department of Systems Design Engineering, University of Waterloo, Waterloo, Ontario, Canada.

Hipel, K.W., and N.M. Fraser (1980) Metagame analysis of the Garrison conflict, Water Resources Research, 16(4), pp. 629-637.

Hipel, K.W., and N.M. Fraser (1982) Socio-Political implications of risk, In Technological Risk (Proceedings of the First University Symposium on the Evaluation of Risk in New Technologies), edited by N.C. Lind, pp. 41-74, University of Waterloo Press, Waterloo, Ontario, Canada.

Hipel, K.W., and N.M. Fraser (1986a) Conflict analysis in systems management, In Encyclopedia of Systems and Control, edited by M.G. Singh, Pergamon Press, Oxford.

Hipel, K.W., and N.M. Fraser (1986b) Modelling political uncertainty using the improved metagame analysis algorithm, Peace and Change, 12.

Hipel, K.W., R.K. Ragade, and T.E. Unny (1974) Metagame analysis of water resources conflicts, Journal of the Hydraulics Division, American Society of Civil Engineers, 100(HY10), pp. 1437-1455.

Hipel, K.W., R.K. Ragade, and T.E. Unny (1976) Metagame theory and its application to water resources, Water Resources Research, 12(3), pp. 331-339.

Howard, N. (1971) Paradoxes of Rationality, Theory of Metagames and Political Behaviour, M.I.T. Press, Cambridge, Massachusetts, 1971.

International Garrison Diversion Unit Study Board (1976) Report to the International Joint commission, October.

International Joint commission (1965) Rules of Procedure and Text of Treaty, Ottawa, Canada - Washington, D.C.

International Joint Commission (1977) Transboundary Implications of the Garrison Diversion Unit, Report to the Governments of Canada and the United States, Ottawa, Canada - Washington, D.C.

International Joint Commission (1978) Annual Report for 1977, Ottawa, Canada - Washington, D.C.

Kilgour, D.M., K.W. Hipel, and L. Fang (1986) The graph model for conflicts, Automatica.

Kilgour, D.M., K.W. Hipel, and N.M. Fraser (1984) Solution concepts in non-cooperative games, Large Scale Systems, 6, pp. 49-71.

Kremers, O. (1974) Garrison Diversion Unit, Manitoba Environmental Council Annual Report Two. Winnipeg: Manitoba Environmental Council.

Kuhn, J.R.D., K.W. Hipel, and N.M. Fraser (1983) A coalition analysis algorithm with application to the Zimbabwe conflict, IEEE Transactions on Systems, Man and Cybernetics, SMC-13(3), pp. 338-352.

Leitch, W.G., and J.J. Keleher (eds.) (1974) Garrison Diversion Project Presentations, Winnipeg: Manitoba Environmental Council.

Lind, N.C. (ed) (1982) Technological risk, Proceedings of the First University Symposium on the Evaluation of Risk in New Technologies, University of Waterloo Press, Waterloo, Ontario, Canada, 185 pp.

Meleskie, M.F., K.W. Hipel, and N.M. Fraser (1982) The Watergate tapes conflict: a metagame analysis, Political Methodology 8(4), 1-23.

Okada, N., K.W. Hipel, and Y. Oka (1984) A hierarchical gaming approach for the resolution of water resources conflicts, Technical Report No. 84013, Department of Applied Mathematics and Physics, Kyoto Univerity, Kyoto 606, Japan.

Okada, N., K.W. Hipel, and Y. Oka (1985) Hypergame analysis of the Lake Biwa conflict, Water Resources Research 21(7), pp. 917-926.

Robinson, E.B. (1966) History of North Dakota, Lincoln, Nebraska, University of Nebraska Press.

Sage, A.P., and E.B. White (1980) Methodologies for risk and hazard assessment: a survey and status report, IEEE Transactions on Systems, Man and Cybernetics, SMC-10(8), pp. 425-446.

Savich, P., K.W. Hipel, and N.M. Fraser (1983) The Alaskan gas pipeline conflict, Energy, The International Journal, 8(3), pp. 213-224.

Shupe, M.C., W.M. Wright, K.W. Hipel, and N.M. Fraser (1980) The nationalization of the Suez Canal: a hypergame analysis, Journal of Conflict Resolution, 24(3), pp. 477-493.

Stokes, N.W., and K.W. Hipel (1983) Conflict analysis of an export credit trade dispute, OMEGA 11(4) pp. 365-376.

Stokes, N.W. and K.W. Hipel (1986) Simultaneous sanctioning in non-cooperative games, Journal of the Operational Research Society 37(6).

Stokes, N.W., K.W. Hipel, and P.H. Roe (1985) The New York subway car dispute, INFOR,23(2), pp. 141-158.

Takahashi, M.A., N.M. Fraser, and K.W. Hipel (1984) A procedure for analyzing hypergames, European Journal of Operational Research, 18, pp. 111-122.

U.S. Department of the Interior (1978a) Report on recommended plan - Garrison Diversion Unit, February.

U.S. Department of the Interior (1978b) Garrison Diversion Unit - Draft Supplementary Environmental Statement, January.

U.S. State Department (1977) Diplomatic Note, February 18, 1977.

Von Neumann, J., and O. Morgenstern (1953) Theory of Games and Economic Behaviour, Princeton University Press, Princeton, New Jersey, U.S.A., Third Edition.

Wright, W., M. Shupe, N.M. Fraser, and K.W. Hipel (1980) A conflict analysis of the Suez Canal invasion of 1956. Conflict Management and Peace Science, 5(1), pp. 27-40.

INDEX